Advances in Robotics, Automation and Control

Edited by

Jesús Arámburo
and Antonio Ramírez Treviño

Advances in Robotics, Automation and Control
Edited by Jesús Arámburo and Antonio Ramírez Treviño

Republished by InTech
Janeza Trdine 9, 51000 Rijeka, Croatia

Technical Editor Teodora Smiljanic

Cover Designer Dino Smrekar

Additional hard copies can be obtained from orders@intechopen.com

Advances in Robotics, Automation and Control, Edited by Jesús Arámburo and Antonio Ramírez Treviño
p. cm.
ISBN 978-953-7619-16-9

For EU product safety concerns:
IN TECH d.o.o., Prolaz Marije Krucifikse Kozulić 3, 51000 Rijeka, Croatia,
info@intechopen.com or visit our website at intechopen.com.

Preface

Nowadays, production systems have become more complex since they require adapting to the market challenges, to introduce more flexible mechanisms and control strategies in their structure to ensure the efficient use of system resources and to allow system layout reconfiguration.

The book presents an excellent overview of the recent developments in the different areas of Robotics, Automation and Control. Through its 24 chapters, this book presents topics related to control and robot design; it also introduces new mathematical tools and techniques devoted to improve the system modeling and control. An important point is the use of rational agents and heuristic techniques to cope with the computational complexity required for controlling complex systems. Through this book, we also find navigation and vision algorithms, automatic handwritten comprehension and speech recognition systems that will be included in the next generation of productive systems developed by man.

We would like to thank all the authors for their excellent contributions in the different areas of the control, robotics and automation. It is their knowledge and enthusiastic collaboration that lead to the creation of this book, which we are sure that will be very valuable to the readers.

October 2008

Editors

Jesús Arámburo
Antonio Ramírez Treviño

Contents

1

Adaptive Control Optimization of Cutting Parameters for High Quality Machining Operations based on Neural Networks and Search Algorithms

J. V. Abellan, F. Romero, H. R. Siller, A. Estruch and C. Vila
Department of Industrial Systems Engineering and Design
Castellon, 12071, Spain

1. Introduction

In traditional Computer Numerical Control (CNC) systems, machining parameters are usually selected prior to machining according to handbooks or user's experience. These practices tend to select conservative parameters in order to avoid machining failure and assure product quality specifications. Less conservative practices try to find optimal machining parameters off-line to increase process productivity after conducting experimentation (Chien & Chou, 2001). However, variations during the machining process due to tool wear, temperature changes, vibrations and other disturbances make inefficient any off-line optimization methodology, especially in high quality machining operations where product quality specifications are very restrictive. Therefore, to assure the quality of machining products, reduce costs and increase machining efficiency, cutting parameters must be optimised in real-time according to the actual state of the process. This optimization process in real-time is conducted through an adaptive control of the machining process.

The adaptive control applied in machining systems is classified as (Liang et al., 2004; Ulsoy & Koren, 1989): Adaptive Control with Constraints (ACC), Geometric Adaptive Control (GAC), and Adaptive Control with Optimization (ACO). In the ACC systems, process parameters are manipulated in real time to maintain a specific process variable, such as force or power, at a constraint value. Typically, ACC systems are utilized in roughing operations where material removal rate is maximized by maintaining the cutting forces at the highest possible cutting force such that the tool is not in danger of breaking (Zuperl et al., 2005). In the GAC systems, the economic process optimization problem is dominated by the need to maintain product quality such as dimensional accuracy and/or surface finish (Coker & Shin, 1996). GAC systems are typically used in finishing operations with the objective of maintaining a specific part quality despite structural deflections and tool wear. Sensor feedback is often employed to measure surface roughness and dimensional quality between parts and adjustments, so tool offsets and feed overrides can be adjusted for the next part. In the ACO systems, machine settings are selected to optimize a performance index such as production time, unit cost, etc. Traditionally, ACO systems have dealt with adjusting cutting parameters (feed-rate, spindle speed and depth of cut) to maximise material removal rate subject to constraints such as surface roughness, power consumption, cutting forces, etc (Venu Gopal & Venkateswara Rao, 2003). Other ACO systems optimise a multi-objective function which are more practical in industrial applications (Zuperl & Cus, 2005). For example, it is quite often to search the optimal cutting parameters to minimize the cost of the operation, maximize the production rate and maximize the part quality. ACO systems are basically composed of several units which integrate the machine-tool system and the equipment required for acquiring real-time process measurements and adjusting the cutting parameters. Fig. (1) shows a simplified scheme of a basic ACO system presented in (Koren, 1983). Basically, the ACO system requires a sensor system which provides real-time data for tool wear diagnosis and part quality prediction. The real-time data are used by process models previously obtained from experimental data. Tool wear and part quality models are used in the multi-objective function together with cutting parameters. An optimizer unit is then applied for searching optimal cutting parameters, and the selected parameters are sent to the CNC system.

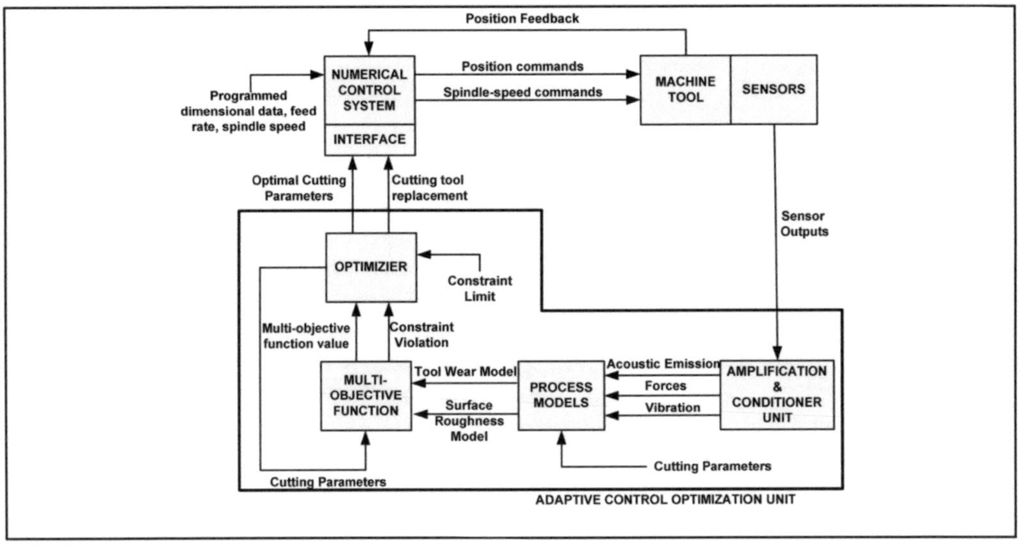

Fig. 1. Adaptive Control Optimization (ACO) scheme adapted from (Koren, 1983).

Interesting works related to ACO systems can be found in (Liu & Wang, 1999; Liu et al., 1999; Chiang et al., 1995). Liu (Liu & Wang, 1999) proposed an adaptive control system based on two neural network models, a Back-Propagation Neural Network (BP NN) and an Augmented Lagrange Multiplier Neural Network (ALM NN). The BP NN was used for modeling the state of the milling system, using as a single input the feed parameter and sensing the cutting forces on-line. The ALM NN was used for maximising the material removal rate which it was carried out adjusting the feed rate. Chiang (Chiang et al., 1995) presented a similar work for end-milling operations, but surface roughness was also considered as constraint. Both research works were based on theoretical formulas for training the neural networks, and both applied an ALM NN for optimization, which it is claimed to be an approach that can greatly reduce processing time in comparison to conventional optimal algorithms and make real-time control possible. Liu (Liu et al., 1999) also extended his previous work with a new optimization procedure based on a Genetic Algorithm (GA).

In spite of the potential application of ACO systems, their use in industry is limited due to the non-existence of reliable on-line monitoring systems for tool wear diagnosis and quality prediction (Azouzi & Guillot, 1997; Liang et al., 2004). Therefore, the optimal selection of cutting parameters is usually done off-line for the cutting-tool life-cycle (Ghani et al., 2004; Chien & Chou, 2001). The off-line parameters optimization is usually carried out through short cutting experiments which are later used to obtain an empirical model which could be optimized subjected to some constraints. Ghani (Ghani et al., 2004) optimized cutting parameters using a Taguchi's Design of Experiments in end milling operations. With a minimum number of trials compared with other approaches such as a full factorial design, the methodology presented reveals the most significant factors and interactions during cutting process which leads to choose optimal conditions. A similar methodology is described in (Zhang et al., 2007). However, both methodologies do not permit to evaluate quadratic or non-linear relations between factors, and the analysis is restricted to the levels analysed in each factor. A more generic approach although more costly in experiments is based on Response Surface Model (RSM) and Response Surface Model Optimization (RSMO). Suresh (Suresh et al., 2002) used RSM for modeling the surface roughness as a first and second-order mathematical model and the surface roughness optimization was carried out through GA. Cus (Cus & Balic, 2003) also applied GA for optimising a multi-objective function based on minimum time necessary for manufacturing, minimum unit cost and minimum surface roughness. All the process models applied in his research were empirical formulas from machining handbooks which were fitted through regressions. More complex models have also been applied for surface roughness and tool wear modeling to optimise off-line cutting parameters. Zuperl (Zuperl & Cus, 2003) also applied and compared feed-forward and radial basis neural networks for learning a multi-objective function similar to the one presented in (Cus & Balic, 2003). Choosing the radial basis networks due to their fast learning ability and reliability, he applied a large-scale optimization algorithm to obtain the optimal cutting parameters. Chien (Chien & Chou, 2001) applied neural networks for modeling surface roughness, cutting forces and cutting-tool life and applied a GA to find optimum cutting conditions for maximising the material removal rate under the constraints of the expected surface roughness and tool life.

These previous works are off-line optimization methodologies which can be efficient enough if tool wear effects have a minimal impact to surface roughness and/or a high surface roughness quality is not required. Otherwise, an on-line optimization methodology should be applied since optimal cutting conditions may vary during the cutting-tool life-cycle due to tool wear effects on surface roughness. In this chapter, an ACO system is presented for optimising a multi-objective function based on material removal rate, quality loss function related to surface roughness, and cutting-tool life subjected to surface roughness specifications constraint. The proposed system adjusts the cutting parameters during the cutting-tool life-cycle in order to maximise in real-time the multi-objective function. The core of the system is composed of three process models: a cutting-tool wear model for diagnosing the state of the cutting tool, a surface roughness deviation model for predicting the quality loss function and a cutting-tool life model. All models are developed using artificial neural networks to model the non-linear relationships in machining processes. Since the process models are black-box models, optimal cutting parameters are obtained applying genetic algorithms and mesh adaptive direct search algorithms. The proposed system is compared with 2 traditional methods for off-line cutting parameters selection: (1) selection based on suggested cutting parameters from handbooks, and (2) selection based on RSMO.

2. Experimental system

2.1 Machining process description

Machining hardened steels (hardness from 30 to 62 HRC) for moulds and dies with surface roughness specifications less than 0.3 microns are commonly applied in industry, and require costly and time-consuming traditional operations such as electro-discharge machining or grinding. Recently, some research studies have reported the use of high performance machining operations for these applications with important benefits as reducing lead times and costs (Siller et al., 2008). However, tool wear process impacts directly to surface roughness so optimal cutting parameters are difficult to obtain since they vary according to cutting-tool state. Therefore, although high performance machining can technically substitute grinding or electro-discharge machining, additional efforts should be conducted in order to tune cutting parameters for an optimal machining. For these applications, ACO techniques can improve the process significantly with respect to other non-adaptive optimization techniques.

The machining process studied in this paper is presented in Fig. 2, and it consists of a face-milling operation on workpieces of hardened AISI D3 steel (60 HRc) with dimensions 250x250 mm. The experiments were conducted on a CNC machining center suited for mould and die manufacturing, and the cutting tool used was a face milling tool with Cubic Boron Nitride (CBN) inserts. In order to generate a good surface finish and avoid run-out problems, a single insert was mounted on a tool body with an effective diameter of 6.35 mm.

Fig. 2. Machining process analysed

2.2 Monitoring system description

A monitoring system to estimate on-line tool wear and surface roughness is required to select the optimal cutting parameters according to the actual state of the machining process. In this chapter, the monitoring system implemented is a multi-component sensor system composed of a piezoelectric dynamometer, accelerometers and signal conditioners (Fig. 3). Two acquisition boards were used for data acquisition. The first board, an Iotech DaqBook 112, was used for acquiring cutting forces from the dynamometer and it was configured for a sample frequency of 3 kHz. A second board, an Iotech DaqBoard 3000, was used for vibration signal acquisition from accelerometers and it was configured for a sample frequency of 100 kHz. Cutting forces were amplified and filtered by a Kistler 5405 amplifier configured with a low-pass filter of 300 Hz cut-off frequency. Vibration signals were amplified by a PCB 482A22 amplifier. Root-mean-square of forces and vibrations were calculated for each cutting pass at the cutting-location x = 175 mm for a 2 seconds data acquisition. Surface roughness (Ra) was measured by a Mitutoyo Surftest 301 profilometer at the cutting-tool locations x = 40 mm, x = 110 mm, x = 175 mm every cutting pass (sampling length $\lambda = c/l = 0.8$ mm and number of spans n = 5). Cutting tool wear (Vb) was measured by a stereo-microscope Nikon MZ12 after each face-milling pass every 250 mm length of cut. Fig. 4 describes the machining process with the Ra and Vb sampling procedure.

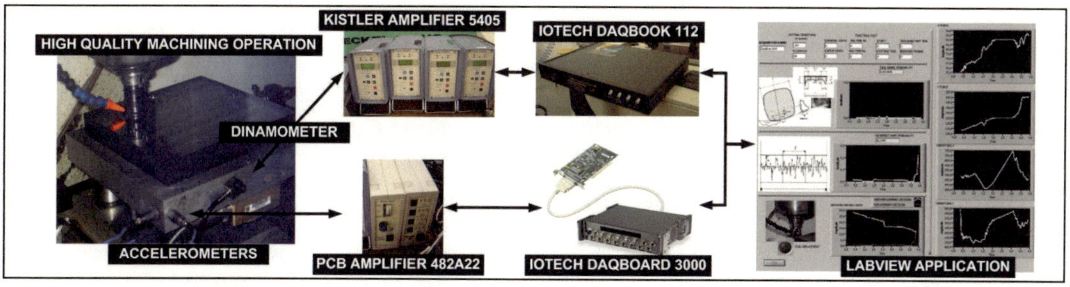

Fig. 3. Multi-component sensor system

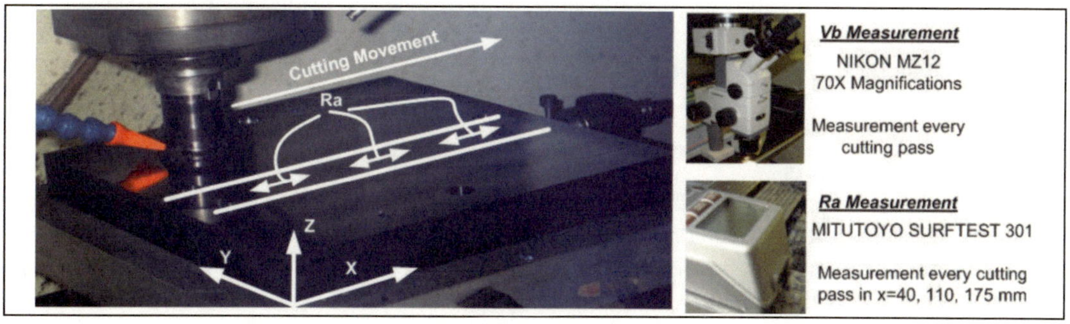

Fig. 4. Machining process and surface roughness and tool-wear sampling.

3. Design of experiments

In order to compare cutting parameters optimization by RSMO and AI approaches, it is necessary to carry out a Design of Experiments (DoE) to be useful for both. RSMO requires classical designs of experiments such as Box-Wilson Central Composites Designs (CCD) or Box-Behnken designs (Nist, 2006), in case that it is only considered linear and quadratic effects. On the other hand, AI approaches require enough data for training and testing, varying the factors in all its domain, but it does not require any specific DoE design.

The factors considered in the experimentation were the feed per tooth (f_z) and the cutting speed (V_c). The radial depth of cut (a_e) was considered constant, with a value of 31.25 mm to maximize the material removal rate. The axial depth of cut (a_p) was defined as a constant (0.4 mm) since the machining operation studied was a finishing operation. The minimal experimentation required to apply RSMO with two factors is a face centered CCD with one center point which is equivalent to a 2^3 full factorial design. For each experiment, the face-milling operation was carried out until the cutting tool edge was worn (Vb higher than 0.3 mm, usual value for finishing operations (ISO 8688-1, 1989)) or the surface roughness was outside specifications. Fig 5 shows the cutting conditions analysed and the order of the cutting experiments.

Experimental Run	Vc (code)	Vc (m/min)	fz (code)	fz (mm)
1	1	200	0	0.08
2	0	150	1	0.12
3	1	200	1	0.12
4	-1	100	0	0.08
5	-1	100	-1	0.04
6	0	150	0	0.08
7	1	200	-1	0.04
8	-1	100	1	0.12
9	0	150	-1	0.04

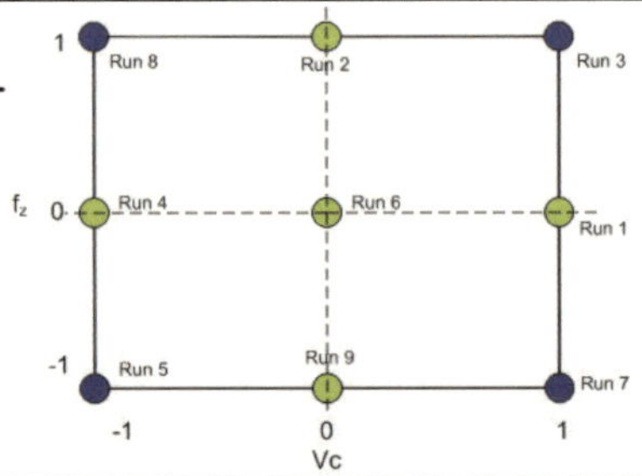

Fig. 5. Design of Experiments and run order. Face Centered Central Composite Design.

4. Definition of the optimization problem

The machining economics problem consists in determining the optimal cutting parameters in order to maximize/minimize an objective function. Typical objective functions to optimize cutting parameters are "minimize unit production cost", "maximize production rate", "maximize profit rate", etc. On the other hand, several cutting constraints have to be considered in machining economics, such as tool-life constraint, cutting force constraint, power, stable cutting region constraint, chip-tool interface temperature constraint and surface finish constraint (Cus & Balic, 2003).

4.1 Objective functions

Typically, three objective functions are considered in a cutting parameters optimization problem: (1) Material Removal Rate (MRR), (2) surface roughness and (3) cutting-tool life. MRR is a measurement of productivity, and it can be expressed by analytical derivation as the product of the width of cut (w), the feed velocity of the milling cutter (F) and the depth of cut (a_p) (Eq. (1)). Surface roughness is the most important criterion for the assessment of the surface quality, and it is usually calculated empirically through experiments. Some research works directly use the empirical relationship presented in Eq. (2), where Vc and f are the cutting speed and feed rate respectively and k, x_1, x_2, x_3 are empirical coefficients. Cutting-tool life is the other important criterion for cutting parameters selection, since several costs such as cutting-tool replacement cost and cutting-tool cost are directly related with tool life. The relation between the tool life and the parameters is usually expressed by the well-known Taylor's formula presented in Eq. (3), where K_T, α_1, α_2, α_3 are empirical coefficients.

$$MRR = w\, a_p\, F \tag{1}$$

$$Ra = k\ V_c^{x_1}\, f^{x_2}\, a_p^{\ x_3} \tag{2}$$

$$T = \frac{K_T}{V_c^{\alpha_1}\, f^{\alpha_2}\, a_p^{\ \alpha_3}} \tag{3}$$

However, for high quality machining operations using CBN cutting tools, both traditional surface roughness and tool life equations may not provide a good estimation. Machining a very low feed speeds produce that additional mechanisms influence the surface roughness generation such as vibrations, engagement of the cutting tool, built up edge, etc. (Siller et al., 2008). On the other hand, CBN tools have a different wear process than traditional cutting-tools such as high speed steels, so Taylor's formula may not be directly applied (Trent & Wright, 2000). For both reasons, other empirical models based on experimental data must be applied instead of Eqs. (2,3).
For the case study presented in this chapter which is a high quality face milling operation based on CBN tools, two alternative objective functions were applied. Instead of Ra model, it is applied the quality loss function described by Eq. (4). Considering a desired Ra value, the quality loss function is usually applied to estimate the cost of manufacturing with a quality variation. The loss function is defined as:

$$W = A_{rework}\frac{V^2}{\Delta^2} \quad (4)$$

where $\Delta = Ra_{max} - Ra_{target}$ with Ra_{max} the maximum Ra defined by specifications and Ra_{target} the Ra desired; V^2 is the mean squared deviation as $V^2 = ((Ra_{target} - y_1)^2 + \ldots + (Ra_{target} - y_n)^2)/n$, with n the number of samples; and A_{rework} is the part cost if the part is outside specifications. On the other hand, instead of the traditional Taylor's formula, it is applied an empirical model learnt from the experimentation which is defined by the Eq. (5), where f is the function learnt.

$$T = f(V_c, f, a_p) \quad (5)$$

4.2 Multi-objective function

The optimization problem for the case study is defined as the optimization of a multi-objective function which is composed of the objective functions defined by Eqs (1,4,5). Since these objective functions are conflicting and incomparable, the multi-objective function is defined using the desirability function approach. This function is based on the idea that the optimal performance of a process that has multiple performance characteristics is reached when the process operates under the most desirable performance values (Nist, 2006). For each objective function $Y_i(x)$, a desirability function $d_i(Y_i)$ assigns numbers between 0 and 1 to the possible values of Y_i, with $d_i(Y_i) = 0$ representing a completely undesirable value of Y_i and $d_i(Y_i) = 1$ representing a completely desirable or ideal objective value. Depending on whether a particular objective function Y_i is to be maximized or minimized, different desirability functions $d_i(Y_i)$ can be used. A useful class of desirability functions was proposed by (Derringer & Suich, 1980). Let L_i and U_i be the lower and upper values of the objective function respectively, with $L_i < U_i$, and let T_i be the desired value for the objective function. Then, if an objective function $Y_i(x)$ is to be maximized, the individual desirability function is defined as

$$d_i(Y_i) = \left\{ \begin{array}{cc} 0 & If\ Y_i(x) < L_i \\ \left(\frac{Y_i - L_i}{T_i - L_i}\right)^w & If\ L_i \le Y_i(x) \le T_i \\ 1 & If\ Y_i(x) > T_i \end{array} \right\} \quad (6)$$

with the exponent w is a weighting factor which determines how important it is to hit the target value. For w = 1, the desirability function increases linearly towards T_i; for w < 1, the function is convex and there is less emphasis on the target; and for w > 1, the function is concave and there is more emphasis on the target. If one wants to minimize an objective function instead, the individual desirability function is defined as

$$d_i(Y_i) = \begin{Bmatrix} 1 & If\ Y_i(x) < T_i \\ \left(\frac{Y_i - U_i}{T_i - U_i}\right)^w & If\ T_i \le Y_i(x) \le U_i \\ 0 & If\ Y_i(x) > U_i \end{Bmatrix} \tag{7}$$

Fig. (6) shows the individual desirability functions according to different w values. The individual desirability functions are combined to define the multi-objective function, called the overall desirability of the multi-objective function. This measure of composite desirability is the weighted geometric mean of the individual desirability for the objective functions. The optimal solution (optimal operating conditions) can then be determined by maximizing the composite desirability. The individual desirability is weighted by importance factors I_i. Therefore, the multi-objective function or the overall desirability function to optimize is defined as:

$$D = (d_1(Y_1)^{I1} d_2(Y_2)^{I2} \dots d_k(Y_k)^{Ik})^{\frac{1}{(I1+I2+\dots+Ik)}} \tag{8}$$

with k denoting the number of objective functions and I_i is the importance for the objective function I, where i = 1,2,…,k.

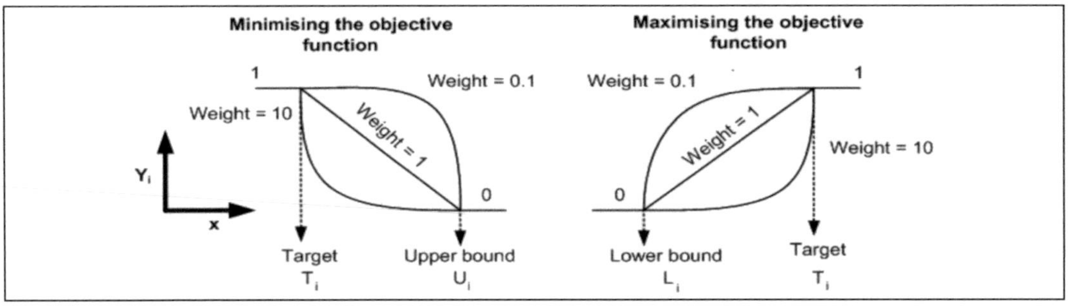

Fig. 6. Desirability functions according to the type of objective function

4.3 Constraints

Due to the limitations on the cutting process, manufacturers limit the range of the cutting parameters to avoid premature cutting-tool failures. Therefore, selected cutting parameters according to manufacturer specifications are constrained to:

$$V_{min} \le V_c \le V_{max} \tag{9}$$

$$f_{min} \le f_z \le f_{max} \tag{10}$$

$$a_p \le a_{max} \tag{11}$$

Surface roughness specification is also considered a constraint that can be expressed as

$$R_a \leq R_{spec} \rightarrow V^2 \leq (Ra_{target} - Ra_{spec})^2 \tag{12}$$

In addition, cutting power and force limitations are usual constraints, but they are commonly applied only for roughing operations.

4.4 Summary of optimization problem and numerical coefficients

The weights and the individual desirability coefficients for each objective function were chosen according to each objective function in the machining process. First the weights were defined considering how the objective function increases/decreases as the ideal value is not matched. Secondly, a comparison among individual desirability coefficients was done to define how much more important is each objective function than the other one. For the case study presented, the objective functions were considered linear (w=1) and the coefficient of importance were chosen to prevail productivity and surface roughness quality than cutting-tool cost and cutting-tool cost replacements. Therefore, importance factors I_1 and I_2 which are related to material removal rate and surface quality loss function were chosen as 1, whereas importance factor I_3 which is related to cutting-tool life was chosen as 0.5.
Considering the maximum and minimum values of each objective function obtained analytically, the desirability functions were defined as follows.

- MRR desirability function

$$d_1(MRR) = \frac{MRR - 398}{2387 - 398} \tag{13}$$

MRR_{target} = 2387 mm^3/min. $MRR_{minimum}$ = 398 mm^3/min. Importance factor I_1=1.

- Desirability function of Ra deviation objective function

$$d_2(W) = d_2(V^2) = \frac{V^2 - 0.012}{0.0001 - 0.012} \tag{14}$$

V^2_{target} = 0.0001 μm^2. $V^2_{maximum}$ = 0.012 μm^2. Importance factor I_2=1. Note that the desirability function of quality loss W for surface roughness can be defined by the surface roughness deviation V^2 since Eq. (4) relates W with V^2 by a constant coefficient of A_{rework}/Δ^2.

- Cutting-tool life desirability function

$$d_3(T) = \frac{T - 7.43}{46.7 - 7.43} \tag{15}$$

T_{target} = 46.7 min. $T_{minimum}$ = 7.43 min. Importance factor I_3=0.5.
The multi-objective function or the overall desirability function to be optimized is:

$$D = (d_1(MRR)^1 d_2(V^2)^1 d_3(T)^{0.5})^{\frac{1}{(1+1+0.5)}} \tag{16}$$

constrained to:

$$100 \text{ m/min} \leq V_c \leq 200 \text{ m/min} \tag{17}$$

$$0.04 \text{ mm/rev} \leq f_z \leq 0.12 \text{ mm/rev} \tag{18}$$

$$a_p = 0.4 \text{ mm} \tag{19}$$

$$V^2 \leq (0.1 - 0.2)^2 \ \mu m^2 \tag{20}$$

5. Parameter optimization based on handbooks

5.1 Description

Cutting tool parameters are traditionally chosen according to handbooks and cutting-tool data catalogs. For a given cutting-tool and workpiece material, a range of possible cutting-parameters are provided. The machinist chooses the parameters within the ranges using some well-known practices in shop-floor. Some of these practices are:

- Higher cutting speeds increase surface roughness quality but decrease cutting tool life.
- Higher cutting speeds decrease cutting tool life.
- Higher feed rates increase productivity as material removal rate is increased.
- Higher feed rates decrease surface roughness quality.
- Higher feed rates decrease cutting-tool life.
- Higher axial depth of cut increases productivity.
- Higher axial depth of cut decreases cutting-tool life.
- Very low axial depth of cut burns the workpiece surface and generates a low surface roughness quality and decreases cutting-tool life.

According to the final goal of the machining process, the machinist selects the best cutting-tool parameters combination. For example, if the only important constraint is a high cutting tool life, the machinist will select a low cutting speed, low feed rate and low-medium axial depth. Fig. 7a describes the typical optimization process based on handbooks.

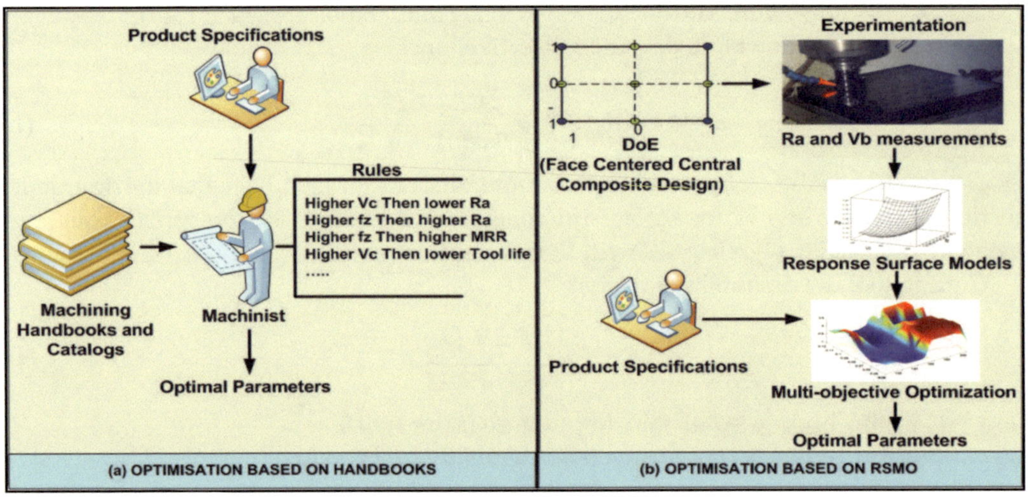

Fig. 7. Off-line cutting parameters optimization.

5.2 Optimization results

As it was explained above (section 4.4), the MRR is the most important objective function together with product quality whereas tool life is less important ($I_1=1$; $I_2=1$; $I_3=0.5$). A high feeding value increases MRR but decreases surface roughness, so it seems reasonably to fix the feeding value to an intermediate value. On the other hand, as V_c increases, both MRR and part quality increase, but cutting tool life decreases considerably. As both MRR and part quality are much important than cutting tool life, it seems reasonably to fix V_c at its maximum value. Therefore, the optimal cutting parameters based on catalogs and handbooks can be defined as V_c=200 m/min and fz=0.08 mm. Experimentation was conducted in order to check the overall desirability function at these cutting conditions. The experimental results showed a cutting tool life T=10.8 min, an average surface roughness deviation of $V^2=1.6 \cdot 10^{-3}$ μm^2, and a MRR=1592 mm^3/min. The overall desirability was 0.472, and the evolution of the overall desirability function due to surface roughness variability along cutting tool life-cycle is shown in Fig (8).

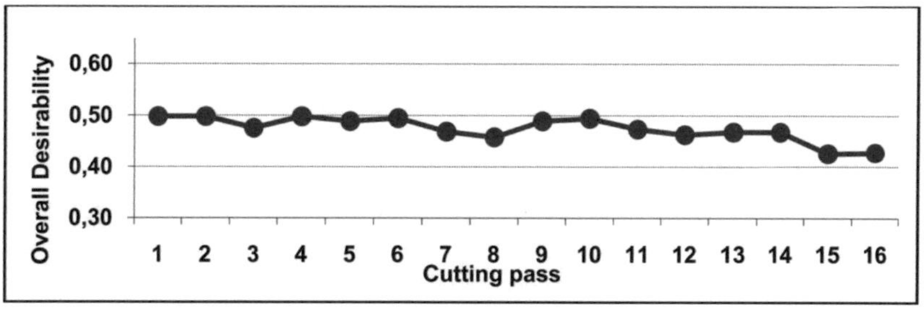

Fig. 8. Overall desirability function along cutting tool life-cycle. Parameter optimization based on handbooks. Vc=200m/min; fz=0.08mm; ap=0.4mm

6. Parameter optimization based on RSMO

6.1 Description

A less conservative method to optimize cutting parameters could be carried out through the Response Surface Model Optimisation methodology (RSMO). RSMO is a collection of mathematical and statistical techniques that are useful for the modeling and analysis of problems in which a response of interest is influenced by several variables and the objective is to optimize this response (Montgomery, 2001). In most RSM problems, the form of the relationship between the variable to be optimized (the response) and the independent variables is unknown. Thus, the first step in RSM is to find a suitable approximation for the true functional relationship between the response and the set of independent variables. First and second-order models are commonly applied to approximate the function through least squares. Considering a response surface y defined by $y=f(x_1,x_2 \ldots x_k)$ where $x_1, x_2, \ldots, x_k$ are independent variables, the optimal value of the response surface will be the set of $x_1, x_2, \ldots, x_k$ for which the partial derivatives $\frac{\partial \hat{y}}{\partial x_1} = \frac{\partial \hat{y}}{\partial x_2} = \cdots \frac{\partial \hat{y}}{\partial x_k} = 0$. This point is called the stationary point and it could represent a point of maximum response, a point of minimum response, or a saddle point. A general mathematical solution for the location of the stationary point can be obtained as follows. Given a second-order model in matrix notation as:

$$\hat{y} = \hat{\beta}_o + x'b + x'Bx \tag{21}$$

where

$$x = \begin{bmatrix} x_1 \\ \vdots \\ x_k \end{bmatrix} \quad x = \begin{bmatrix} \hat{\beta}_1 \\ \vdots \\ \hat{\beta}_k \end{bmatrix} \quad and \;\; B = \begin{bmatrix} \hat{\beta}_{11} & \frac{\hat{\beta}_{12}}{2} & \cdots & \frac{\hat{\beta}_{1k}}{2} \\ & & \ddots & \frac{\hat{\beta}_{2k}}{2} \\ sym & & & \hat{\beta}_{kk} \end{bmatrix} \tag{22}$$

That is, b is a (k x 1) vector of the first-order regression coefficients and B is a (k x k) symmetric matrix whose main diagonal elements are the pure quadratic coefficients ($\hat{\beta}_{ii}$) and whose off-diagonal elements are one-half the mixed quadratic coefficients ($\hat{\beta}_{ij}$, $i \neq j$).
The derivative of $\hat{y}$ with respect to the elements of the vector **x** equated to **0** is

$$\frac{\partial \hat{\mathbf{y}}}{\partial \mathbf{x}} = \mathbf{b} + 2\mathbf{B}\mathbf{x} = \mathbf{0} \quad (23)$$

The stationary point is the solution to Eq. (23), or

$$\mathbf{x}_s = -\frac{1}{2}\mathbf{B}^{-1}\mathbf{b} \quad (24)$$

To define whether the stationary point is a point of maximum or minimum response or saddle point, it is usually examined a contour plot of the fitted model. Refer to (Montgomery, 2001) for RSMO concepts.
In this section the RSMO methodology is applied to the high quality machining operation studied. The experimentation conducted through the DoE is used to obtained first and second order process models. The multi-objective function which is based on these models is optimized and the optimal cutting parameters are defined. Fig. 7b describes the optimization process based on RSMO in the machining process studied.

6.2 Machining process models based on first and second-order functions

RSM technique lets model the responses of interest W, T, and MRR as a first and second-order functions. After conducting the DoE shown in Fig. (5), a RSM was fitted for each response. The first and second order functions with its coefficient of determination were:

- Ra deviation (V^2) response:

$$V^2 = 0.0629 - 5 \cdot 10^{-4}V_c - 0.507f_z + 11.9 \cdot 10^{-6}V_c^{\,2} + 1.49f_z^{\,2} + 1.6 \cdot 10^{-3}V_c\ f_z \quad (25)$$

R^2_{adj}=89.1%

- T response:

$$T = 117.3 - 0.321V_c - 1372.3f_z + 5333.3f_z^{\,2} + 2.037V_cf_z \quad (26)$$

R^2_{adj}=80.4%

- MRR response

$$MRR = 99.47\ V_cf_z \quad (27)$$

R^2_{adj}=100%. Note that this is the exact analytical equation for MRR.
Fig. 9 shows the response surface models for each response analysed.

6.3 Optimization results

The optimum cutting parameters were obtained maximizing the overall desirability function, which depends on the previous first and second order models and the individual desirability and weights coefficients defined in section 4.4. The result of the optimization procedure showed the optimal cutting parameters V=165m/min and fz=0.12mm, with an overall desirability value of 0.61. To validate the RSMO result, experimentation was conducted with the expected optimal parameters. The experimental results showed a cutting tool life T=13.7 min, a surface roughness deviation function of $V^2=5.5 \cdot 10^{-3}$ μm^2, and a MRR=1967 mm^3/min. The overall desirability of the experiment was 0.495, and the evolution of the overall desirability function due to surface roughness variability along cutting tool life-cycle is shown in Fig (10). Note that the expected overall desirability was 0.610 whereas the real value was 0.495. This error is due to the inaccuracy of the process models developed through this methodology.

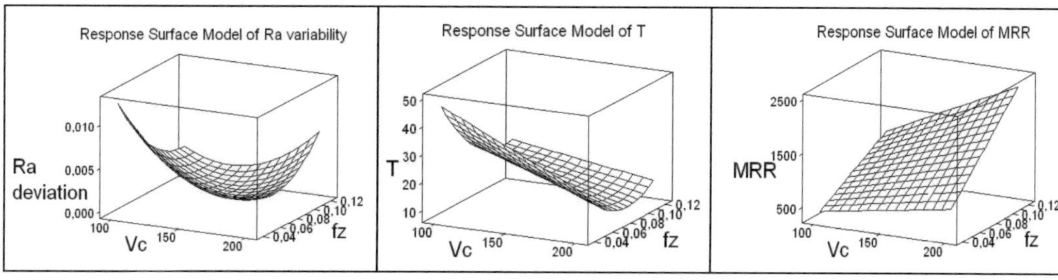

Fig. 9. Response Surface Models for deviation of Ra, T and MRR. Units: Vc (m/min); fz (mm); Ra deviation (μm^2); T (min); MRR (mm^3/min).

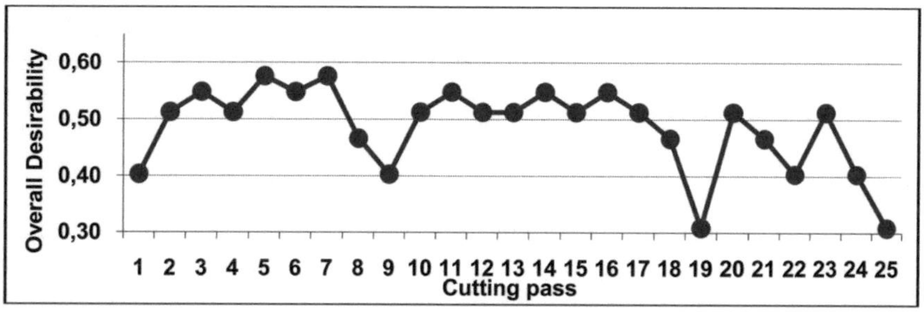

Fig. 10. Overall desirability function along cutting tool life-cycle. Parameter optimization based on RMSO. Vc=164m/min; fz=0.12mm; ap=0.4mm.

7. Parameter optimization based on AI

7.1 Description

Due to non-linearity, first and second order models from response surface methodology cannot be enough accurate to model tool-wear and surface roughness in machining processes. To overcome this limitation, AI techniques such as Artificial Neural Networks or Fuzzy Logic can be applied in order to deal with non-linearity. However, these AI models are called black-box models, and they cannot be optimized with conventional optimization methods. Due to this limitation, a cutting parameter optimization methodology based on AI models requires an advanced search methods for global optimization, such as Genetic Algorithms (GA) and Mesh Adaptive Direct Search (MADS) algorithms.
GAs are search algorithms based on the mechanics of natural selection and natural genetics, invented by (Holland, 1975), which can find the global optimal solution in complex multidimensional search spaces. A population of strings, representing solutions to a specified problem, is maintained by the GA. The GA then iteratively creates new populations from the old by ranking the strings and interbreeding the fittest to create new strings. So in each generation, the GA creates a set of strings from the previous ones, occasionally adding random new data to keep the population from stagnating. The end result is a search strategy that is tailored for vast, complex, multimodal search spaces. GAs

are a form of randomized search, in the way in which strings are chosen and combined is a stochastic process. This is a radically different approach to the problem solving methods used by more traditional algorithms, which tend to be more deterministic in nature, such as the gradient methods used to find minima in graph theory. However, although GA is an effective optimization algorithm, it usually takes a long time to find an optimal solution due to its slow convergence speed (Cus & Balic, 2003).
On the other hand, MADS algorithms are iterative search algorithms where the optimization is conducted through an adaptive mesh of points where the objective function is evaluated. At the first iteration, the mesh is built according to an initial point of the objective function. The algorithm computes the objective function at the mesh points until it finds one whose value is smaller than the objective function evaluated on the initial point. If a mesh point has a smaller value, the algorithm sets the next point in the sequence equal to this one and multiplies the current mesh size by a mesh expansion factor. The mesh is then expanded and the algorithm conducts a new iteration. In case none of the mesh points has a smaller objective function value than the value at the best current solution, the algorithm does not change the current point at the next iteration and the mesh size is contracted by a mesh contraction factor. After the re-size, a new iteration is conducted. The iterations are conducted until a stop condition is reached, typically when the mesh size reaches a minimum value. Refer to (Audet & Dennis, 2004) for concepts related to MADS.
In this section it is proposed an optimization methodology based on Artificial Neural Networks (ANN) models for modeling the machining process and GA-MADS algorithms to optimize the multi-objective function defined by Eq. (16) and the ANN models. The combination of GA and MADS algorithms lets reduce the computing time required for the optimization. Basically, GA is firstly applied in order to find the region where the multi-objective function is minimum. Then, the GA algorithm is interrupted and the MADS algorithm refines the search using the GA solution as the initial point of the mesh. The optimal cutting parameters are calculated when the MADS algorithm reaches the minimum mesh size. Fig. 11 describes the procedure of the optimization methodology proposed, based on process models using AI techniques and an initial Design of Experiments, and a GA-MADS optimization. Table (1,2) defines the main characteristics of the ANN models applied and the GA-MADS algorithms.

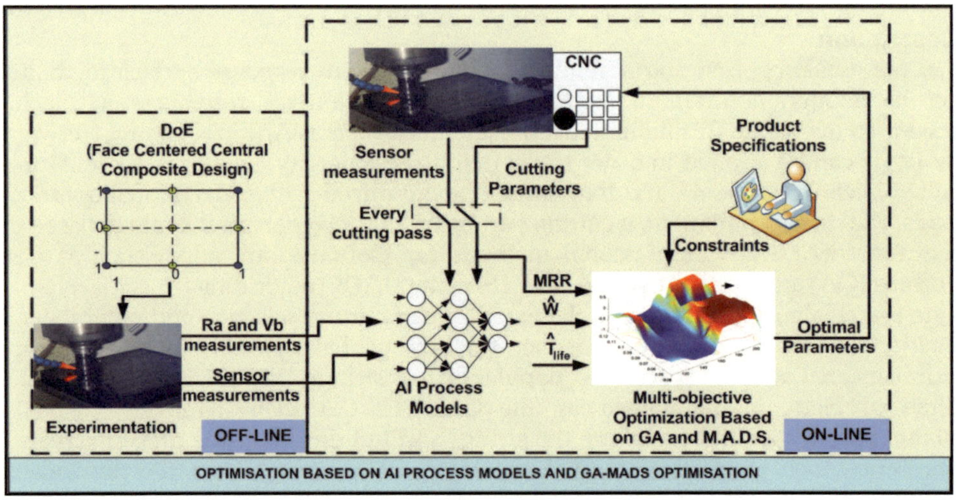

Fig. 11. On-line cutting parameters optimization based on AI

7.2 Machining process models based on AI

After conducting the DoE, an experimental database composed of 188 samples of 9 different cutting conditions was generated. Each sample was defined by: surface roughness deviation (V^2), cutting tool wear state (% of use), root-mean-square of cutting forces in X, Y direction (RMS Fx, RMS Fy), root-mean-square of cutting forces in XY plane (RMS Fxy), root-mean-square of vibrations in X direction (RMS Ax), root-mean-square of vibrations in Y direction (RMS Ay), cutting time (Tc), cutting speed (Vc) and feed per tooth (fz). The experimental database was used to learn three ANN process models: (1) Ra deviation model, (2) Cutting-tool life model, (3) Cutting-tool wear state model. Before training the ANN, a statistical study was conducted for each model in order to discard those inputs variables that were not significant. The final inputs variables applied for each model and the main characteristics of the ANN models are shown in Table 1. Note that vibrations signals were discarded for all models since they were close related to the cutting-tool position, and they cannot provide any objective information about the state of the process. Fig 12 shows the response of each ANN model according to V_c and f_z values.

Ra deviation model		**Cutting-tool life model**		**Cutting-tool wear state model**	
Type	Backpropagation	Type	Backpropagation	Type	Backpropagation
Inputs	Vc, fz, State	Inputs	Vc, fz	Inputs	Vc, fz, RMS Fxy
Output	Ra_deviation	Output	T_life	Output	State
Hidden Layers	1	Hidden Layers	1	Hidden Layers	1
Neurons	3	Neurons	3	Neurons	2
Mapping functions	Logsig	Mapping functions	Logsig	Mapping functions	Logsig
Training Method	Lev-Marq.	Training Method	Lev-Marq.	Training Method	Lev-Marq.
Epochs	300	Epochs	300	Epochs	300
Learning Rate	0.05	Learning Rate	0.05	Learning Rate	0.05

Table 1. Characteristics of ANN models

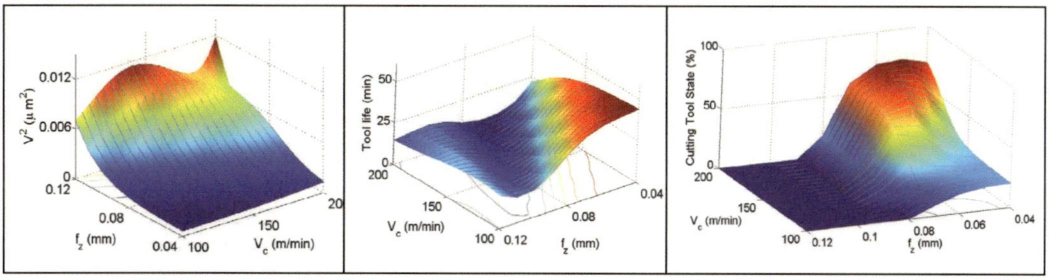

Fig. 12. ANN models of : (1) surface roughness deviation -V^2- for a cutting tool wear of 50%; (2) cutting tool life –T-; (3) tool wear state for a RMS Fxy of 60N.

7.2 Optimization results

The optimum cutting parameters were obtained maximizing the overall desirability function every cutting pass of 250 mm length of cut. During the cutting pass, the cutting forces were acquired and the cutting-tool wear state was predicted by the cutting-tool wear state model. With the cutting-tool state prediction, the GA-MADS optimization procedure was conducted to maximize the overall desirability function, which is based on the MRR model, and the ANN models of Ra deviation and cutting-tool life. The optimal solution was obtained after an average processing time of 6 seconds, and the new cutting parameters were sent to the CNC controller for the next cutting pass. The process was repeated until the cutting tool was worn or the surface roughness was outside specifications. The result of the optimization procedure showed that the optimal cutting parameters vary considerably, from 200m/min to 130m/min and from 0.12mm to 0.07mm for cutting speed and feed rate respectively, with an expected overall desirability value of 0.545. The real overall desirability value after measure surface roughness deviation and tool-life was 0.520. The evolution of the overall desirability function and the variation of the optimal cutting conditions along the cutting tool life-cycle are shown in Fig (13).

Genetic Algorithm		**Mesh Adaptive Direct Search Algorithm**	
Variables to optimise	Vc, fz	Variables to optimise	Vc, fz
Population Size	10	Initial Mesh Size	0.05
Generations	15	Max. Mesh Size	Inf
Crossover Frac.	0.8	Max. Func. Eval	Inf
Elite Count	2	Expansion	2
Mutation function	Gaussian	Contraction	0.5
Selection function	Stochastic	Poll Method	Positive Basis 2N
Initial ranges	Vc=[100,200] Fz=[0.04,0.12]	Polling order	Consecutive
Stop criterium	Stall Time: 6s Stall Generations: 7	Stop criterium	Tolerance Mesh: $5 \cdot 10^{-4}$

Table 2. Characteristics of search algorithms

8. Results and discussion

The results reported in the optimization methodology based on handbooks and catalogs show the most conservative cutting parameters as it was expected, where the overall desirability function has a value of 0.472. A simple RSMO through a 9 experimental runs provide enough information to improve cutting parameter selection and increase the overall desirability function to 0.495 which is an improvement of 5% in the desirability. The surface response models show that high cutting speeds increase surface roughness variability which were not taken into account by the machinist in the first optimization methodology. Although RSMO increases the overall desirability, the RSMO prediction is quite inaccurate as it is shown in the experimental validation. The predicted overall desirability by RSMO was 0.610 whereas the overall desirability function after the experimental validation was 0.495. The inaccuracy of the prediction is due to the inaccuracy of the process models, where surface roughness deviation and cutting-tool life models have been fitted by least squares with a low coefficient of determination, 89% and 80.4% respectively.

The methodology proposed in this chapter, based on AI techniques for modeling the machining process and the use of search algorithms to optimize the overall desirability function on-line, shows an overall desirability function of 0.520. This methodology compared with handbook optimization and RSMO increases the overall desirability in 10% and 5% respectively. The main benefits reported by this methodology are due to two factors. A first factor, the ANN process models let deal with non-linearity so this models are more accurate than response surface models for modelling high quality machining operations. Unlike RSMO where the error between the predicted overall desirability value and the experimental one was 19% (0.61 versus 0.495), the proposed methodology presents an error of 5% (0.545 versus 0.520). As a second factor, the on-line nature of the methodology lets adapt the cutting parameters every cutting pass so the system is more flexible to adapt any change in the objective function during the cutting-tool life. However, the ANN model for cutting tool state prediction based on cutting forces seemed to be low accurate due to the high variability of the cutting forces during machining. This variation produces that the cutting tool parameters selected each cutting tool pass were quite irregular. This effect is reflected in Fig. 13b, where cutting speed often varies from 195 to 150-130 m/min. The replacement of the direct measurement of cutting forces by indirect methods which are not so sensitive to cutting mechanisms such as current or power sensors might increase the overall desirability function and select optimal parameters with a more regular variation.

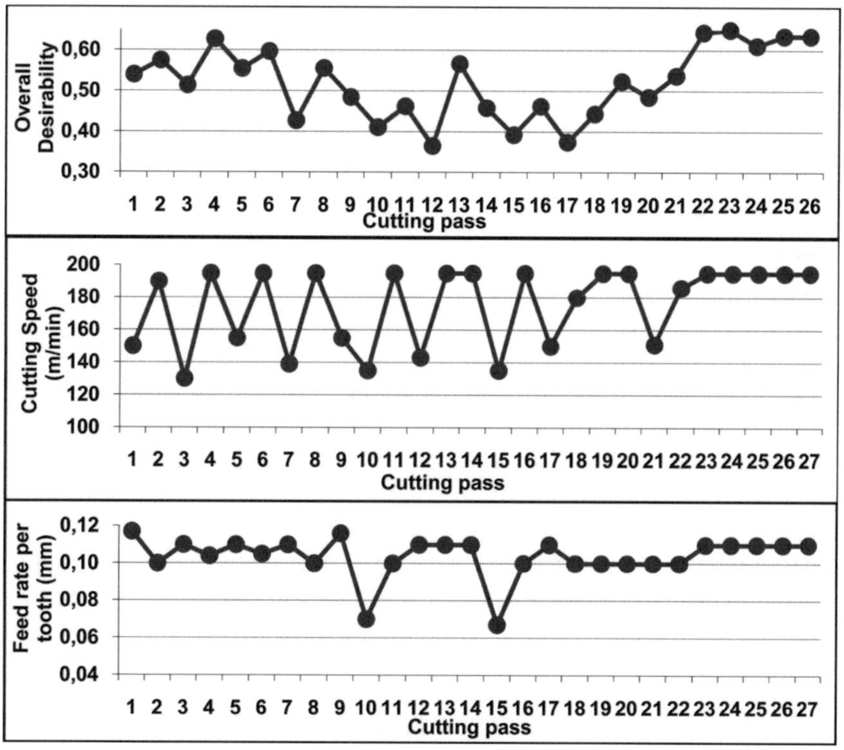

Fig. 13. (a) Overall desirability function along cutting tool life-cycle. Parameter optimization based on AI. (b) Adaptive cutting speed. (c) Adaptive feed per tooth

9. Conclusions

In this chapter, an Adaptive Control Optimization (ACO) system was presented for optimising a multi-objective function based on material removal rate, quality loss function related to surface roughness, and cutting-tool life subjected to surface roughness specification constraints. Unlike traditional optimization techniques, this methodology lets adapt the cutting parameters during the cutting-tool life-cycle in order to maximise in real-time the multi-objective function according to cutting-tool state. The core of the system is composed of three process models: a cutting-tool wear model for diagnosing the state of the cutting tool, a surface roughness deviation model for predicting the quality loss function and a cutting-tool life model. All models were developed using Artificial Neural Networks (ANN) to model the non-linear relationships in machining processes. The cutting parameter optimization is obtained applying genetic algorithms and mesh adaptive direct search algorithms. The system proposed was compared with 2 traditional methods for off-line cutting parameter selection: (1) selection based on suggested cutting parameters from handbooks, and (2) selection based on RSMO.

The results showed how conservative are the cutting parameters selected by off-line methodologies. A cutting parameter optimization based on handbooks provided an overall desirability function of 0.472, whereas cutting optimization through RSMO gave an overall desirability value of 0.495. The main inconvenient of RSMO is that this methodology is based on mathematical first and second order models which are not enough accurate for high quality machining operations. The new methodology proposed based on AI techniques increases the overall desirability function up to 0.520. The improvement is due to two effects: (1) the ANN process models deal with non-linearity so these models are more accurate than response surface models for modelling high quality machining operations; (2) the on-line nature of the methodology lets adapt the cutting parameters every cutting pass so the system is more flexible to adapt any change in the objective function during the cutting-tool life-cycle.

10. References

Audet, C. and Dennis, J. E. (2004). Mesh adaptive direct search algorithms for constrainted optimization. *Technical Report G-2004-04, Les Cahiers du GERAD*, Montréal.

Azouzi, R. and Guillot, M. (1997). On-line prediction of surface finish and dimensional deviation in turning using neural network based sensor fusion. *International Journal of Machine Tools and Manufacture*, 37, 9, 1201–1217, (1997), ISSN 0890-6955.

Coker, S. A. and Shin, Y. C. (1996). In-process control of surface roughness due to tool wear using a new ultrasonic system. *International Journal of Machine Tools and Manufacture*, 36, 3, 411–422, (1996), ISSN 0890-6955.

Cus, F. and Balic, J. (2003). Optimization of cutting process by ga approach. *Robotics and Computer-Integrated Manufacturing*, 19, 1-2, 113–121, (2003), ISSN 0736-5845.

Chiang, S. T. and Liu, D. I. and Lee, A. C. and Chieng, W. H. (1995). Adaptive-control optimization in end milling using neural networks. *International Journal of Machine Tools and Manufacture*, 35, 4, 637–660, (1995), ISSN 0890-6955.

Chien, W. T. and Chou, C. Y. (2001). The predictive model for machinability of 304 stainless steel. *Journal of Materials Processing Tech.*, 118, 1-3, (2001), 442–447, ISSN 0924-0136.

Derringer, G. and Suich, R. (1980). Simultaneous-optimization of several response variables. *Journal of Quality Technology*, 12, 4, 214–219, (1980). ISSN 0022-4065.

Ghani, A. K. and Choudhury, I. A. and Husni. (2002). Study of tool life, surface roughness and vibration in machining nodular cast iron with ceramic tool. *Journal of Materials Processing Technology*, 127, 1, 17–22, (2002), ISSN 0924-0136.

Ghani, J. A. and Choudhury, I. A. and Hassan, H. H. (2004). Application of Taguchi method in the optimization of end milling parameters. *Journal of Materials Processing Technology*, 145, 1, 84–92, (2004), ISSN 0924-0136.

Holland, J. (1975). *Adaptation in natural and artificial systems*. The University of Michigan Press, Ann Arbor.

ISO -Int. Standard Organization-, (1989). *Tool Life Test in Milling, Part 1: Face Milling. 1st Ed.*

Jain, L. C. and Martin, N. M. (1998). *Fusion of Neural Networks, Fuzzy Systems and Genetic Algorithms: Industrial Applications*. Crc Press edition, ISBN 0849398045.

Koren, Y. (1983). *Comp. Control of Manufact. Systems*. Mac.-Hill, ISBN 0070353417, New York.

Liang, S. Y. and Hecker, R. L. and Landers, R. G. (2004). Machining process monitoring and control: The state-of-the-art. *Journal of Manufacturing Science and Engineering-Transactions of the Asme*, 126, 2, (2004), 297–310, ISSN 1087-1357.

Liu, Y. M. and Wang, C. J. (1999). Neural network based adaptive control and optimization in the milling process. *International Journal of Advanced Manufacturing Technology*, 15, 11, 791–795,(1999), ISSN 0268-3768

Liu, Y. M. and Zuo, L. and Wang, C. J. (1999). Intelligent adaptive control in milling processes. *International Journal of Computer Integrated Manufacturing*, 12, 5, 453–460, (1999), ISSN 0951-192X.

Montgomery, D. C. (2001). *Design and Analysis of Experiments*. John Wiley & Sons, Inc. ISBN 0471316497.

Nist. (2006). Nist/sematech e-handbook of statistical methods. Electronic Citation, 2006.

Siller, H. R. and Vila, C. and Rodriguez, C. A. and Abellan, J. V. (2008). Study of face milling of hardened AISI D3 steel with a special design of carbide tool. *International Journal of Advanced Manufacturing Technology*, 2008.

Suresh, P. V. S. and Rao, P. V. and Deshmukh, S. G. (2002). A genetic algorithmic approach for optimization of surface roughness prediction model. *International Journal of Machine Tools and Manufacture*, 42, 6, 675–680, (2002), ISSN 0890-6955.

Trent, E. M. and Wright, P. K. (2000). *Metal Cutting*. Butterworth Heinemann, ISBN 075067069X.

Ulsoy, A. G. and Koren, Y. (1989). Applications of adaptive control to machine tool process control. *Control Systems Magazine, IEEE.*, 9, (1989), 33–37, ISSN 0272-1708.

Venu Gopal, A. and Venkateswara Rao, P. (2003). Selection of optimum conditions for maximum material removal rate with surface finish and damage as constraints in SiC grinding. *International Journal of Machine Tools and Manufacture*, 43, 13, 1327–1336, (2003), ISSN 0890-6955.

Wang, J. and Kusiak, A. (2000). *Computational Intelligence in Manufacturing Handbook*. CRC Press edition. ISBN 0849305926.

Zhang, J. Z. and Chen, J. C. and Kirby, E. D. (2007). Surface roughness optimization in an end-milling operation using the Taguchi design method. *Journal of Materials Processing Technology*, 184, 1-3, 233–239, (2007), ISSN 0924-0136.

Zuperl, U. and Cus, F. and Milfelner, A. (2005). Fuzzy control strategy for an adaptive force control in end-milling. *Journal of Materials Processing Technology,* 164, 1472–1478, (2005), ISSN 0924-0136 .

Zuperl, U. and Cus, F. (2003). Optimization of cutting conditions during cutting by using neural networks. *Robotics and Computer-Integrated Manufacturing,* 19, 1-2, 189–199, (2003), ISSN 0736-5845.

2

JPEG for Arabic Handwritten Character Recognition: Add a Dimension of Application

Abdurazzag Ali Aburas and Salem Ali Rehiel
International Islamic University Malaysia
Electrical and Computer Engineering
Malaysia

1. Introduction

The ultimate objective of any Optical Character Recognition (OCR) system is to simulate the human reading capabilities. That is why OCR systems are considered a branch of artificial intelligence and a branch of computer vision as well [Sunji Mori (1999)] Character recognition has received a lot of attention and success for Latin and Chinese based languages, but this is not the case for Arabic and Arabic-like languages such as Urdu, Persian, Jawi, Pishtu and others [Abdelmalek Z (2004)]. Researchers classify OCR problem into two domains. One deals with the image of the character after it is input to the system by, for instant, scanning in which is called Off-line recognition. The other has different input way, where the writer writes directly to the system using, for example, light pen as a tool of input. This is called On-line recognition.

The online problem is usually easier than the offline problem since more information is available [Liana M & Venu G (2006)]. These two domains (offline & online) can be further divided into two areas according to the character itself that is either handwritten or printed character. Roughly, the OCR system based on three main stages: preprocessing, feature extraction, and discrimination (called also, classifier, or recognition engine) Figure 1.1 depicts the block diagram of the typical OCR system. Traditional OCR systems are suffering from two main problems, one comes from features extraction stage and the other comes from classifier (recognition stage). Feature extraction stage is responsible for extracting features from the image and passing them as global or local information to the next stage in order to help the later taking decision and recognizing the character. Two challenges are faced; if feature extractor extracts many features in order to offer enough information for classifier, this means many computations as well as more complex algorithms are needed. Thus, long processor time will be consumed. On the other hand, if few features are extracted in order to speed up the process, insufficient information may be passed to classifier. The second main problem that classifier is responsible for, is that most of classifiers are based on Artificial Neural Networks (ANNs). However, to improve the intelligence of these ANNs, huge iterations, complex computations, and learning algorithms are needed, which also lead to consume the processor time. Therefore, if the recognition accuracy is improved, the consumed time will increase and vice versa.

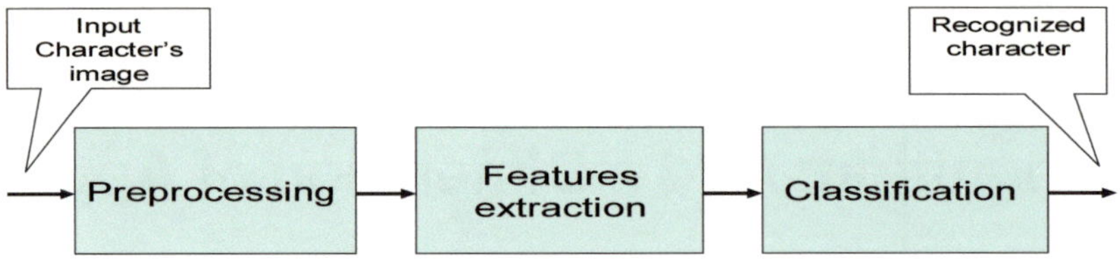

Figure 1.1 The typical OCR block diagram

To tackle these problems, a new OCR construction is proposed in this chapter, where neither features extractor nor ANN is needed. The proposed construction relies on the image compression technique (JPEG). Taking advantages of the compressor, that it compresses the image by encoding only the main details and quantizes or truncates the remaining details (redundancy) to zero. Then generates a unique vector (code) corresponding to the entire image. This vector can be effectively used to recognize the character since it caries the main details of the character's image. The importance of the main details is that they are common amongst the same character which is written by different writers.
The principle is illustrated and the algorithm is designed and tested in the related sections. In the next section a brief review of the related work is described, Arabic character characteristics are illustrated in section 3. In section 4, an overview of JPEG compression is shown. In section 5, the concept and the proposed algorithm are presented. The experimental results are shown and discussed in section 6. The conclusion of the chapter is drawn in section 7.

2. Review of the related work

Beside the main goal of any OCR system which is simulating human's reading capability, the accuracy and time consuming are very important issues in this aspect. Based on the latest survey which is published in May 2006 by Liana M. and Venu G. [Liana(2006)] all covered papers have presented their proposals seeking high accuracy and less time. Each one has treated the issue from different angle of view. Their work can be classified into three main categories: preprocessing problems, features extraction problems, and recognition (discrimination) problems. For preprocessing stage, where the image is often converted to a more concise representation prior to recognition the most common methods are: A skeleton which is a one-pixel thick representation showing the centrelines of the character. Skeletonization is also called "thinning" facilitates shape classification and feature detection. Many researchers have used skeletonization in their proposed preprocessing stages like for instant [S. Mozaffari et al., (2005)], [S. Alma'adeed et al., (2002)], [S. Alma'adeed et al., (2004)]. Another method which is known as "contour" is the Freeman chain code of the character's border. In this method chain code stores the absolute position of the first pixel and the relative positions of successive pixels along the character's border. This methods has been practiced by many like [R. Safabakhsh & P. Adibi (2005)], [L. Souici-Meslati & M. Sellami (2004)], [N. Farah et al., (2004)], however, skeletonization suffers from mislocalization of feature and ambiguities particular to each thinning algorithm whereas the contour approach avoids these problems since no shape information is lost. For the features extraction stage, where the main role is played and mostly the accuracy of recognition depends on the information passed from this stage to the classifier (recognizer). These information can be structural features such as loops, branch-points, endpoints, and dots or statistical which includes but is not limited to, pixel densities, histograms of chain code directions, moments, and Fourier descriptors. Because of the importance of this stage many approaches and techniques have been proposed. For instant, Clocksin in [W.F. Clocksin & P.P.J. Fernando (2003)] applied moment functions to image and polar transform image. Many like [A. Amin (2003)], [G. Olivier et al., (1996)], [I.S.I. Abuhaiba et al., (1998)], used loops, dots, curves, relative locations, height, sizes of parts of characters, loop positions and types, line positions and directions, and turning points. Where others used statistics from moments of horizontal and vertical projection like[H. Al-Yousefi & S.S. Udpa (1992)]. Histogram of slopes along contour is used by [M. Dehghan et al., (2001)]. Fourier descriptors have also been used in this stage by [R. Safabakhsh & P. Adibi (2005)]. Few used techniques such as wavelet or fractal like [Amir M. et al., (2002)], [Saeed M. et al., (2004 a)] and [Saeed M. et al., (2004 b)]. Artificial Neural Networks (ANNs) are the common seed of most if not all classifiers "recognition or discrimination stage". Many variations have been used in order to overcome the main disadvantage of ANNs which is time consuming. Sherif and Mostafa in [Sherif K. & Mostafa M (1996)], for example, presented a parallel design for backpropagation Neural Networks approach in order to accelerate the computation process. If we want to mention who did use ANNs in OCR, we may list all of them at least within our concern field in this chapter "off-line Arabic handwriting character recognition".

3. Arabic character characteristics

The Arabic alphabet contains basically 28 letters are written from write to left. Each letter can take from two to five different shapes, thus, roughly the alphabet set can expand to 84 different shapes according to the position of the letter (beginning, middle, end or isolated) as well as according to the style of writing (Nasekh, Roqa'a, Farisi and few others). This is one reason makes Arabic recognition complex. The second reason is the similarities among the different letters and the differences among the same latter. For example, letters Baa, Taa, and Thaa (number 2,3 & 4 respectively in Arabic Alphabet)are three different letters, but they have similar body shape, they only differ in number and position of dots (one, two or three dots below or above the body of the character). Also Jeem, Hhaa & Khaa (number 5, 6, & 7 respectively in Arabic Alphabet) differ only in one dot. On the other hand the differences among the same letter, for example, letter Haa (number 26 in the Arabic Alphabet) has three completely different shapes through its positions. Officially, there is a set of rules to write Arabic letters, but few follow. These rules may help OCR to extract features or segment text, such as so called base-line rule. This rule states that there are three lines. Each letter or group of letters has their lines where they should be lie.

4. JPEG compression background

The most important current standard for image compression is JPEG [W.B Pennebaker & J.L. Mitchell (1993)]. In the JPEG baseline coding system, which is based on the discrete cosine transform (DCT) and is adequate for most compression applications, the input and output images are limited to 8 bits, while the quantized DCT coefficient values are restricted to 11 bits. The human vision system has some specific limitations, which JPEG takes advantage of to achieve high rates of compression.

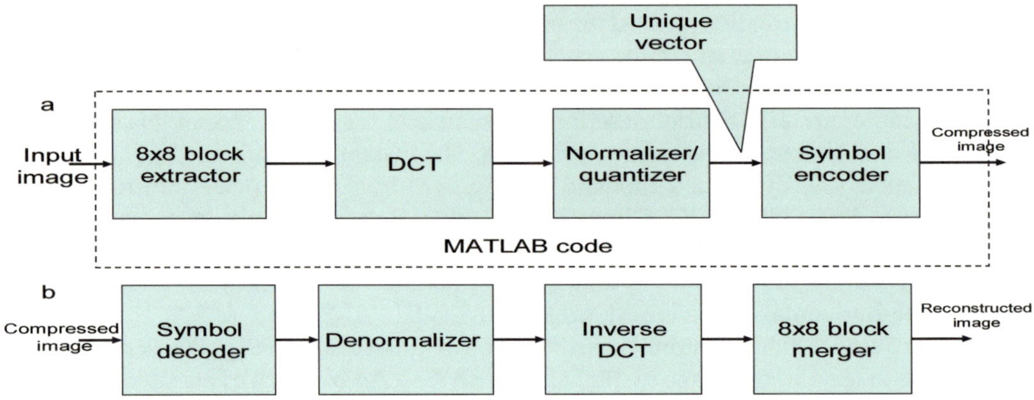

Figure 4.1 JPEG block diagram. (a) encoder and (b) decoder

As can be seen in the simplified block diagram of Figure 4.1, the compression itself is performed in four sequential steps: 8x8 sub-image extraction, DCT computation, quantization, and variable-length code assignment. Since we do not concern in this work about the reconstruction part, the only part of compression is used (dashed box) and the vector will be tapped immediately after quantization stage.

5. Proposed algorithm

The main concept of the proposed algorithm is based on the property that the JPEG compressed image is a vector which can uniquely represent the input image to be correctly reconstructed later at decompression stage. This property can be effectively used to recognize the character's image. A JPEG approximation MATLAB code is used to generate a vector of coefficients that represent the important information (or main details for the character's image). In the next sections the experimental work is illustrated

5.1 Experiment's tools and platform environment

The data consists of groups for the twenty eight Arabic alphabets, each group was written by 48 writers from different ages and educational backgrounds. Figure 5.1 shows one sample of one group "letter Sheen".

An 40x40 8-bit pixel colour image was used as input image. For experimental reason, the MATLAB version 7.2.0.232 (R2006a) has been used but in order to improve the speed of the proposed algorithm a prototype of C++ code might be used. The Intel Pentium 4 CPU, 3.00GHz clock and 960MB RAM of the computer, has been used

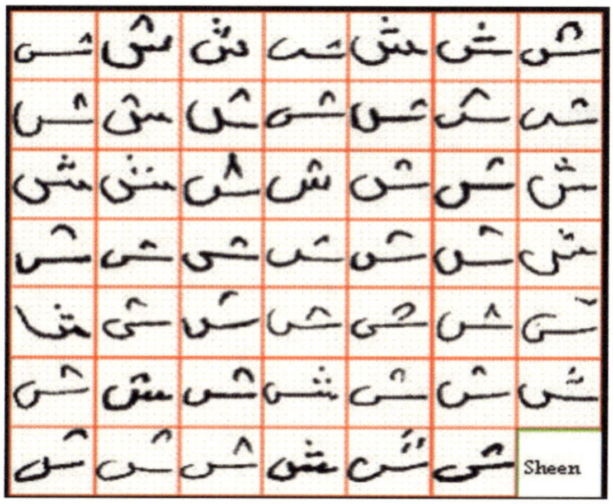

Figure 5.1 A group sample of letter Sheen

5.2 Proposed algorithm top-down flowchart

Figure 5.2 illustrates the sequence of the proposed algorithm's steps. After the character's image is scanned in the system the JPEG approximation MATLAB code will produce a vector. This vector is assumed to uniquely represent input image since it carries the important details of that image. Figure 5.3 shows a sample for letter Sheen and letter Ain. Then Euclidean distance between this vector and each vector in codebook will be measured. Finally, the minimum distance points to the corresponding character, and then the character is recognized.

5.3 Proposed system components

The two main components are the code of the compression stage (i.e., JPEG compressor in the flowchart shown in Figure 5.2) and the codebook

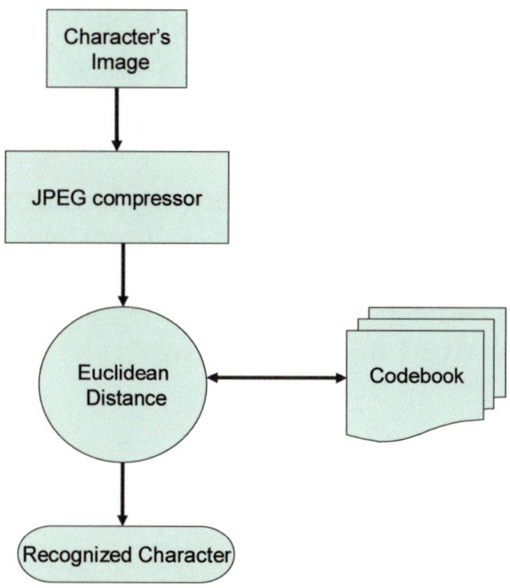

Figure 5.2 Flowchart diagram of the proposed algorithm

(A) Graph of a sample JPEG-approx. vector for letter Sheen ش

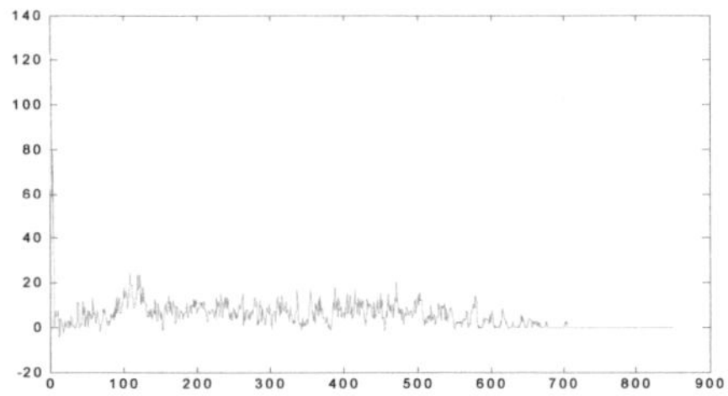

(B) Graph of a sample JPEG-approx. vector for letter Ain ع

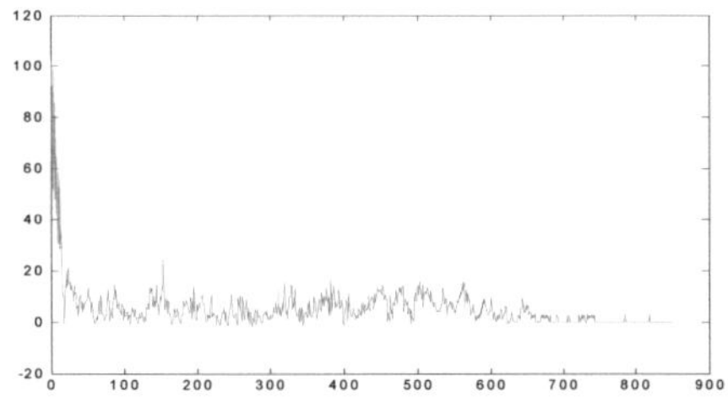

Figure 5.3 Sample from the produced JPEG-approximation vector; (A) for letter Sheen, (B) for letter Ain

5.3.1 Compressor MATLAB code

The code listed in List 5.1 is to compress an image (x) using a JPEG approximation based on 8x8 DCT transforms, coefficient quantization, and Huffman symbol coding. Input "Quality" determines the amount of information that is lost and compression achieved. Y is an encoding structure containing fields:

[y.size] is the size of the image (x)
[y.numblocks] is the number of the 8x8 encoded blocks
[y.quality] is the quality factor (as percent)
[y.huffman] is the Huffman encoding structure, as returned by mat2huff (see List 8.1)

```
% This function appeared in [R.C. Gonzalez, R. E. Woods, & S.L.Eddins (2003)]
function y=im2jepg(x,quality)
global vctr % vctr is the vector obtained after the image is
        % JPEG compreseed and before Huffman takes place
error(nargchk(1,2,nargin));   % check input argument
if ndims(x) ~=2 | ~isreal(x) | ~isnumeric(x) | ~isa(x,'uint8')
   error('the input MUST BE a UINT8 image');
end
if nargin<2
   quality=1;            % Default value for quality
end
m=[ 16 11 10 16 24 40 51 61
   12 12 14 19 26 58 60 55
   14 13 16 24 40 57 69 56
   14 17 22 29 51 87 80 62
```

```
    18 22 37 56 68 109 103 77
    24 35 55 64 81 104 113 92
    49 64 78 87 103 121 120 101
    72 92 95 98 112 100 103 99] * quality;
order=[ 1 9 2 3 10 17 25 18 11 4 5 12 19 26 33 ...
     41 34 27 20 13 6 7 14 21 28 35 42 49 57 50 ...
     43 36 29 22 15 8 16 23 30 37 44 51 58 59 52 ...
     45 38 31 24 32 39 46 53 60 61 54 47 40 48 55 ...
     62 63 56 64];
  [xm,xn]=size(x)
  x=double(x)-128;
  t=dctmtx(8);
%Compute DCTs of 8x8 blocks and quantize the coefficients
  y=blkproc(x,[8 8], 'P1*x*P2',t,t');
  y=blkproc(y,[8 8],'round(x./P1)',m);
  y=im 2col(y,[8 8], 'distinct');  %Break 8x8 blocks into columns
  xb=size(y,2);
  y=y(order,:);
  eob=max(x(:))+1            %Create end-of-block symbol
  r=zeros(numel(y)+size(y,2),1);
  count=0;
  for j=1:xb
     i=max(find(y(:,j)));
     if isempty(i)
        i=0;
     end
     p=count+1;
     q=p+i;
     r(p:q)=[y(1:i,j);eob];
     count=count+i+1;
  end
  count
  r((count+1):end)=[];  % Delete unused portion of r
  y.size=uint16([xm xn]);
  y.numblocks= uint16(xb);
  y.quality = uint16(quality *100);
  y.huffman =mat2huff(r);
  vctr=r;  % This is the produced vector for the proposed system
```

List 5.1 MATLAB code for the compressor stage

In accordance to the block diagram of Figure 4.1 (A) "dashed box", function **im2jpeg** processes distinct 8x8 sections or blocks of input image x one block at a time (rather than the entire image at once). Two specialized block processing functions - **blkproc** and **im2col** - were used to simplify the computations (more information available @ MATLAB Mathworks® documentary). Function **blkproc** is used to facilitate both DCT computation and coefficient denormalization and quantization, while **im2col** is used to simplify the quantized coefficient reordering and zero run detection. The function **im2jpeg** listed in List 5.1 uses an alternate matrix formulation

$$T = H\,F\,H \tag{1}$$

Where **F** is an 8x8 block of image f(x,y), **H** is an 8x8 DCT transformation matrix generated by dctmtx(8) (more details for MATLAB functions can be found in MATLAB® Documentary), and T is the resulting DCT of **F**. note that the **H** is used to denote the transpose operation.
The statement **y = blkproc(x,[8 8], 'P1*x*P2',h,h')** computes the DCTs of image x in 8x8 blocks, using DCT transform matrix **h** and transpose **h′** as parameters **P1** and **P2** of the DCT matrix multiplication, P1*x*P2

5.3.2 Codebook building

The codebook can be built as following:

1. Apply the MATLAB code listed in List 5.1 to all available database (our database contains 1968 written characters) to get 1968 vectors
2. Group the 1968 vectors (vctr) according to their represented character. For instance, the group of letter Sheen has (in our database) 48 different Sheen that were written by 48 different writers, so it will have 48 vectors (vctr)
3. Average each group (MATLAB function "**mean**" was used in our experiments). By now you should get one vector (vctr) for each group. These are the codes (vctrs) located in the codebook.

6. Results and discussion

Once the codebook is installed the algorithm is ready for testing. Test should cover all available data (1968 tests) to ensure that the code book can handle wide range of expected character images and recognize them with minor errors. Two criteria may be used to examine the system, accuracy and rate of recognition. The accuracy can be measured as the percent of the correctly recognized characters to the tested characters. The rate of recognition (or the speed) is the time system takes from the image inputs until it is recognized. Since the process is a character's shape-independent, the taken times will be equal for all images. Adding a timer code can return the time spent by the system. In addition to JPEG approximation MATLAB code the system needs to calculate Euclidean distance and the minimum value (the shortest distance). The Euclidean distance (*d*) between two vectors X and Y can be defined as :

$$d = \sum \sqrt{(x-y)^2} \tag{2}$$

Whose MATLAB function is

$$\text{D=dist(x, y)} \tag{3}$$

Table 6.1 shows the accuracy recognitions' percentage for each character obtained by the proposed algorithm. The overall average is also shown. The system was able to recognize the characters during short time comparing with any existing system using ANN because it saves time taken by features extractor as well as it uses codebook (lookup table) instead of ANN. Misrecognition occurred to some letters is because of the nature of the handwritten character itself neither the algorithm. Thus, there is no rejection state, the character must be recognized in all situations. It is either correct recognized or incorrect recognized this confusing in recognition may occur even to humans. To verify this conclusion the algorithm was examined with printed characters and obtained better results.

	character		Accuracy %		character		Accuracy %
1	ا	Alif	90.8333	15	ض	Dthad	51.2500
2	ب	Baa	32.5000	16	ط	Ttaa	45.0000
3	ت	Taa	20.0000	17	ظ	Dthaa	38.7500
4	ث	Thaa	80.4167	18	ع	Ain	28.3333
5	ج	Jeem	86.6667	19	غ	Ghen	32.5000
6	ح	Hhaa	100.0000	20	ف	Faa	30.4167
7	خ	Khaa	70.0000	21	ق	Qaf	22.0833
8	د	Dal	36.6667	22	ك	Kaf	45.0000
9	ذ	Theal	28.3333	23	ل	Lam	55.4167
10	ر	Raa	23.1250	24	م	Meem	71.0417
11	ز	Zaay	28.3333	25	ن	Noon	42.9167
12	س	Seen	73.1250	26	هـ	Haa	78.3333
13	ش	Sheen	90.8333	27	و	Waw	73.4583
14	ص	Sad	62.7083	28	ى	Yaa	20.6875
Average Accuracy							52.0975

Table 6.1 The recognition accuracy

Moreover, to test algorithm's intelligence some experiments were done such as taking off the dot from letter Khaa خ or Jeem ج the algorithm recognized them as Hhaa ح, and from letter Dhad ض the algorithm recognized it as Sad ص. Also from Theal ذ and Zaay ز they were recognized as Dal د and Raa ر respectively. Furthermore, when a character was turned upside-down, the farthest Euclidian distance; instead of the shortest Euclidian distance, pointed to that character

7. Conclusion

In this chapter we proposed a new structure of off line OCR system which is not based on ANN, to avoid the time consuming problems. Moreover, it benefited from the JPEG image compression property that is high compression ratio which produces minimum compressed image size and every compressed image has a unique vector which helps to identify each character. By using this unique vector, the proposed system has recognized the input character after measuring the Euclidean distance between the vector and the vectors in the codebook, then the shortest distance pointed to the corresponding letter. In addition to the advantage of speed using codebook, it can be universal by means of character's nature (language, writing mode) as well as character's image size. We used 40x40 8-pexil color image as input image. The result was considerably high in terms of accuracy and recognition rate. We used JPEG approximation MATLAB code . Future work could be done using in advanced on line OCR system, also could be implemented by using fast programming languages such as C++ and it is highly recommended to be used for different Biometrics applications.

8. Appendix: MATLAB code

```
% This function appeared in [R.C. Gonzalez, R. E. Woods, & S.L.Eddins (2003)]
function y=mat2huff(x)
```

```
if ndims(x)~= 2|~isreal(x) | (~isnumeric(x) & ~islogical(x))
  error('X must be a 2-d real numeric or logical matrix');
end
% store the size of input x
y.size = uint32(size(x));
% find the range of x values and store its minimum value biased
% by +32768 as a UINT16
x=round(double(x));
xmin=min(x(:));
xmax=max(x(:));
pmin=double(int16(xmin));
pmin= uint16(pmin+32768); y.min=pmin;
%compute the input histogram between xmin and xmax with uint
%width bins, scale to UINT16, and store
x=x(:)';
h=histc(x,xmin:xmax);
if max(h) > 65535
  h=65535*h/max(h);
end
h=uint16(h); y.hist=h;
%code the input matrix and store the result.
map = huffman(double(h));    % Make a Huffman code map
hx=map(x(:)-xmin+1);         % Map image
hx=char(hx)';                % Convert to char array
hx=hx(:)';
hx(hx==' ')=[];              % Remove the blanks
ysize=ceil(length(hx)/16);   % Compute encoded size
hx16=repmat('0',1,ysize*16); % Pre-allocate modulo-16 vector
hx16(1:length(hx))=hx;       % Make hx modulo-16 in length
hx16=reshape(hx16,16,ysize); % Reshape to 16-character words
hx16=hx16'-'0';              % Convert binary string to decimal
twos=pow2(15:-1:0);
y.code=uint16(sum(hx16.*twos(ones(ysize,1),:),2))';
```

List 8.1 MATLAB code (mat2huff)

9. Acknowledgements

All images of handwritten Arabic letters were provided by Prof. Zaki Kheder, University of Jordan. Authors would like also to thank Dr Rafael C. Gonzales for giving permission to use his MATLAB codes published in [R. C. Gonzalez et al (2003)]

10. References

A. Amin. (2003). Recognition of Hand-Printed Characters Based on Structural Description and Inductive Logic Programming" *Pattern Recognition Letters*, vol. 24, pp. 3187-3196.

Abdelmalek Z. (2004). ORAN: A Basis for Arabic OCR system, *Proceeding of 2004 International Symposium on Intelligent Multimedia, Video and Speech Processing,* Hong Kong, pp. 703-706.

Amir M., Karim F. & Abolfazl T. (2002). Feature Extraction with Wavelet Transform for Recognition of Isolated Handwritten Farsi/Arabic Characters and Numerals. *IEEE,* DSP, pp. 923- 926.

G. Olivier, H. Miled, K. Romeo, and Y. Lecourtier. (1996). Segmentation and Coding of Arabic Handwritten Words," *Proc. 13th Int'l Conf.Pattern Recognition,* vol. 3, pp. 264-268, 1996.

H. Al-Yousefi and S.S. Udpa. (1992). Recognition of Arabic Characters. *IEEE Trans. Pattern Analysis and Machine Intelligence,* vol.14,pp.853-857.

I.S.I. Abuhaiba, M.J.J. Holt, and S. Datta. (1998). Recognition of Off-Line Cursive Handwriting. *Computer Vision and Image Understanding,* vol. 71, pp. 19-38.

L. Souici-Meslati and M. Sellami. (2004). A Hybrid Approach for Arabic Literal Amounts Recognition. *The Arabian J. Science and Eng.,*vol. 29, pp. 177-194, 2004.

Liana M & Venu G. (2006). Offline Arabic Handwriting Recognition: A Survey. IEEE, *Transactions On Pattern Analysis and Machine Intelligence,* vol. 28, No. 5, pp. 712-724.

M. Dehghan, K. Faez, M. Ahmadi, and M.Shridhar. (2001). Handwritten Farsi (Arabic) Word Recognition: A Holistic Approach Using Discrete HMM. *Pattern Recognition,* vol. 34, pp. 1057- 1065.

N. Farah, L. Souici, L. Farah, and M. Sellami. (2004). Arabic Words Recognition With Classifiers Combination: An Application to Literal Amounts. *Proc. Artificial Intelligence: Methodology, Systems and Applications,* pp. 420-429.

R. C. Gonzalez, R. E. Woods, S. L. Eddins. (2003). *Digital Image Processing Using MATLAB.* Prentice Hall; 1st edition, September 5, 2003

R. Safabakhsh and P. Adibi. (2005). Nastaaligh Handwritten Word Recognition Using a Continuous-Density variable-Duration HMM. *The Arabian J. Science and Eng.,* vol. 30, pp. 95-118.

S. Alma'adeed, C. Higgens, and D. Elliman. (2002). Recognition of Off-Line Handwritten Arabic Words Using Hidden Markov Model Approach. *Proc. 16th Int'l Conf. Pattern Recognition,* vol. 3, pp. 481-484.

S. Alma'adeed, C. Higgens, and D. Elliman. (2004). Off-Line Recognition of Handwritten Arabic Words Using Multiple Hidden Markov Models. *Knowledge-Based Systems,* vol. 17, pp. 75-79.

S. Mozaffari, K. Faez, and M. Ziaratban. (2005). Structural Decomposition and Statistical Description of Farsi/Arabic Handwritten Numeric Characters. *Proc. Int'lConf. Document Analysis and Recognition,*pp. 237- 241.

Saeed M., Karim F, & Hamidreza R. (2004 a). Feature Comparison between Fractal Codes and Wavelet Transform in Handwritten Alphanumeric Recognition Using SVM Classifier. IEEE, *Proceedings of the 17th International Conference on Pattern Recognition* (ICPR'04).

Saeed M., Karim F., and Hamidreza R. (2004 b). Recognition of Isolated Handwritten Farsi/Arabic Alphanumeric Using Fractal Codes. *IEEE, 0-7803-83387- 7/04,* pp. 104-108.

Sherif K. and Mostafa M. (1996). A Parallel Design and Implementation for Backpropagation Neural Network Using MIMD Architecture. *IEEE,* pp. 1361-1366.

Sunji Mori, H. Nishida & H. Yamada. (1999). *Optical Character Recognition.* John Wiley & Sons

W.B Pennebaker and J.L. Mitchell. (1993). *The JPEG Still Image Data Compression Standard.* New York: Van Nostrand Reinhold.

W.F. Clocksin and P.P.J. Fernando. (2003). Towards Automatic Transcription of Syriac Handwriting. *Proc.Int'l Conf. Image Analysis and Processing,* pp. 664- 669.

3

Predicting Surface Roughness in Grinding using Neural Networks

Paulo R. Aguiar, Carlos E. D. Cruz, Wallace C. F. Paula
and Eduardo C. Bianchi
São Paulo State University – Unesp – Bauru Campus
Brazil

1. Introduction

High rates of manufactured items have been machined by grinding at some stage of their production process, or have been processed by machines whose precision is a direct result of abrasive operations. However, even being the grinding process the most used in industry for obtaining high level of surface quality, it remains as one of the most difficult and least understood processes (Wang et al., 2005). That maybe has origin in the mistaken faith the process is extremely complex to be understood due to the large number of cutting edges and irregular geometry, high cutting speed, and very small depth of cut which varies from grain to grain. In addition, according to (Haussi & Diniz, 2003), grinding is the process indicated when the workpiece demands good surface, dimensional and geometrical quality. Thus, the grinding process is usually one of the last steps in the machining operations chain. When the workpiece reaches this point, it has high aggregated value, which makes a possible rejection very expensive.

Monitoring of machining processes is mandatory for their optimization and control. Acoustic emission (AE) has become an increasingly popular monitoring technique. The sensors are inexpensive, easy to mount, and analog signal processing is comparatively simple, but good techniques for extracting reliable process information from the signals are still lacking (Hundt et al., 1997; Aguiar et al., 2002).

Electrical power signals have also been largely used in grinding researches. The signal can be monitored either by the electric current of the electric motor or by the product between voltage and current signals, which gives the electrical power consumed by the electric motor. Thus, an estimate of the cutting force can be easily obtained if a model of the electric motor is available (Aguiar et al., 2002).

Some researchers have shown the acoustic emission and the cutting power signals combined can provide significant results for monitoring the grinding process phenomena (Aguiar et al., 2002; Dotto et al., 2006; Kwak & Ha, 2004; Aguiar et al., 2006).

Neural network has attracted a special interest in grinding research owing to its functions of learning, interpolation, pattern recognition, and pattern classification. Various examples of applications into the production engineering field have been reported (Wang et al., 2005; Dotto et al., 2006; Kwak & Ha, 2004; Aguiar et al., 2006; Wang et al., 2001).

According to (Wang et al., 2005), surface roughness is one of the most important factors in assessing and determining the quality of a part. In practical, predicting and controlling the

roughness is difficult due to the fact that many variables are affecting the process and many of these variables are non-linear, interdependent, or difficult to quantify with crisp numeric precision. Still, taking into account that some grinding processes can only be represented by experimental data or linguistic descriptions, the usage of intelligent systems into the optimization of grinding processes appears to be inevitable.
The objective of this work was to use neural networks to predict the surface roughness of ground workpieces based on the analysis of several process output variables, such as acoustic emission, cutting power, and other statistics generated from these signals.

2. Literature review

Machining with grinding wheels is a very complex process affected by so many factors that a reproducible result is rarely obtained. The most important one is that the cutting ability of the grinding wheel changes considerably during the grinding time. In practice, the grinding process is carried out with cutting parameters which are safe but not optimal. The workpiece quality depends to a great extent on the experience of the operator (Lezanski & Rafalowicz, 1993).
The result of a grinding process can be subdivided into characteristics concerning the geometry and surface integrity of a ground component. The geometrical quantities are dimension, shape and waviness, as essential macro-geometric quantities; whereas the roughness condition is the main micro-geometric quantity. The surface integrity state can be described by residual stresses, hardness and structure of the material (Brinksmeier et al., 1998).
According to (Bhushan, 2001), solid surfaces, irrespective of their method of formation, contain irregularities or deviations from the prescribed geometrical form. The surfaces contain irregularities of various orders ranging from shape deviations to irregularities of the order of interatomic distances. No machining method, however precise, can produce a molecularly flat surface on conventional materials. Even the smoothest surfaces, such as those obtained by cleavage of some crystals, contain irregularities, the heights of which exceed the interatomic distances. For technological applications, both macro and micro-nanotopography of the surfaces (surface texture) are important.
A very general topology of a solid surface is seen in Figure 1. Surface textures that are deterministic may be studied by relatively simple analytical and empirical methods; their detailed characterization is straightforward. However, the textures of most engineering surfaces are random, either isotropic or anisotropic, and either Gaussian or non-Gaussian. Whether the surface height distribution is isotropic or anisotropic and Gaussian or non-Gaussian depends upon the nature of the processing method. Surfaces that are formed by cumulative processes (such as peening, electropolishing, and lapping), in which the final shape of each region is the cumulative result of a large number of random discrete local events and irrespective of the distribution governing each individual event, will produce a cumulative effect that is governed by the Gaussian form. It is a direct consequence of the central limit theorem of statistical theory. Single-point processes (such as turning and shaping) and extreme-value processes (such as grinding and milling) generally lead to anisotropic and non-Gaussian surfaces. The Gaussian (normal) distribution has become one of the mainstays of surface classification (Bhushan, 2001).
A typical parameter that has been used to quantify the quality of a surface topography is the surface roughness, which is represented by the arithmetic mean value, Ra, the root-mean-

square-average, Rq, and the maximum roughness height, Rt. In general, the longitudinal surface roughness has a lower value than the traverse surface roughness; therefore the latter is more frequently used in industry (Hecker & Liang, 2003). Still, according to (Kwak et al., 2006) the surface roughness obtained in grinding depends in a complex way upon the roughness of the wheel surface, the grinding parameters, and tribological interactions between the abrasive cutting points and the workpiece.

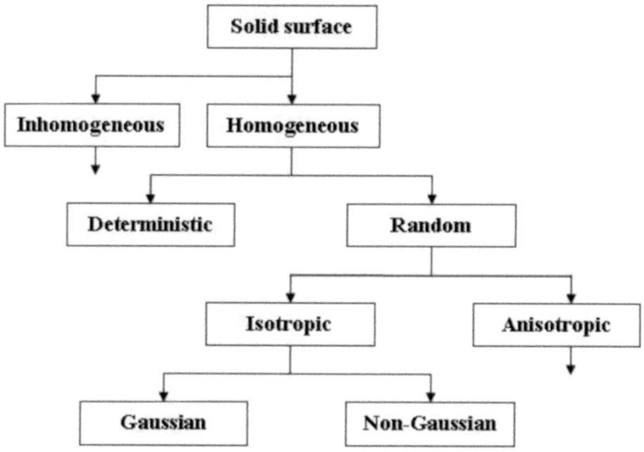

Fig. 1. General typology of surfaces (Bhushan, 2001)

Several sensing schemes for monitoring tool condition and cutting status in normal machining have been proposed and evaluated in the last two decades. However, the technique which involves monitoring the acoustic emission generated during the machining operation has been found to be very sensitive to elements of metal removal, such as sliding contact, plastic deformation, phase transformation, micro-cracking, fracture, impacts and so on (Dornfeld, 1985). The acoustic emission sensing technique uses the information contained in the transient elastic stress wave, which is generated by rapid release of energy within a material, to provide knowledge about the state of the process.

The application of signals from the cutting process became a more interesting tool when neural networks are used for their processing and interpretation. This tool has attracted interest of several researchers in the surface roughness prediction (Wang, 2005; Aguiar et al., 2006; Kwak et al., 2006; Fredj et al., 2002).

Artificial neural networks have been studied for many years in the hope of achieving the human-like performance in the field of the speech, image recognition and the pattern classification. These neural networks are composed of many non-linear computational elements operating in parallel. Neural Networks, because of their massive nature, can perform computations at a higher rate. Because of their adaptative nature using the learning process, neural networks can adapt to changes in the data and learn the characteristics of the input signals (Kwak & Ha, 2004). Still, according to the author, the ability to learn is a fundamental trait of the neural network. Although a precise definition of learning is difficult to formulate, the learning in a neural network means the finding an appropriate set of the weights that are connection strengths from the elements to the other layer elements.

What makes this work distinguished from others is the use of grinding parameters as input to the neural network, which have not been tested yet in surface roughness prediction by neural networks. Besides, a high sampling rate data acquisition system was employed to acquire the acoustic emission and cutting power.

3. Methodology

The materials, methods and equipments used for the development of the grinding tests, surface roughness measurements, training and validation of the neural networks will be presented in the next section.

3.1 Experimental set-up and grinding parameters

The workpieces for the grinding tests consisted of laminated bars of steel SAE 1020 ground in the shape of a prism with 150mm length, 12.7mm width and 43mm height. A surface grinding machine from Sulmecânica manufacturer, Brazil, model RAPH-1055 was used in the grinding tests. The grinder was equipped with an aluminium oxide grinding wheel, from Norton Manufacturer, Model ART-FE-38A80PVH.

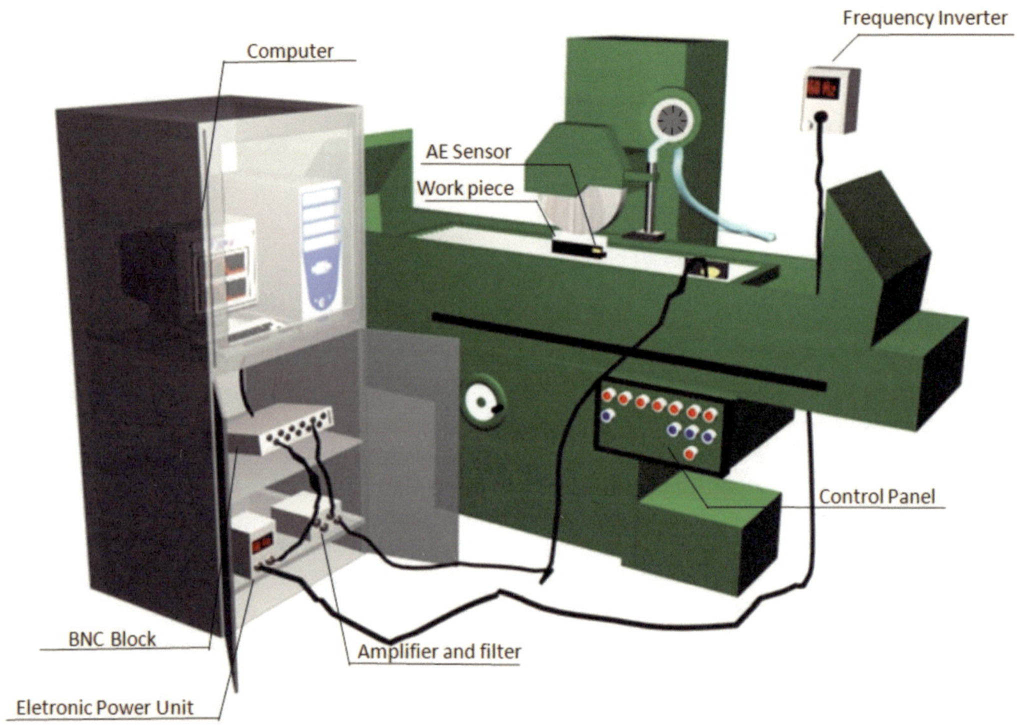

Figure 2. Schematic diagram of the grinding machine and instrumentation

A fixed acoustic emission sensor from Sensis manufacturer, model DM-42, placed near the workpiece and an electrical power transducer for measuring the electrical power consumed from the three-phase induction motor that drives the wheel were employed. The power transducer consists of a Hall sensor to measure the electric current and a Hall voltage sensor to measure the voltage at the electric motor terminals. Both signals are processed internally in the power transducer module by an integrated circuit, which delivers a voltage signal proportional to the electrical power consumed by the electric motor. The acoustic emission as well as the power signal are further sent to the data acquisition board from National Instrument, model PCI-6011, which is installed into a personal computer.

The LabVIEW software was utilized for acquiring the signals and storing them into binary files for further processing and analysis. The acoustic emission sensor used has a broad-band sensitivity of 1.0 MHz. Its amplifier also filtered the signal outside the range of 50 kHz to 1.0 MHz. Figure 2 shows the schematic diagram of the grinding machine and instrumentation used.

The tests were carried out for 15 different grinding conditions, using 5μm as the lowest depth of cut and 50 μm as the largest one. Dressing parameters, lubrication and peripheral wheel speed were adequately controlled in order to ensure the same grinding condition for the three repetitions of each test. The workpiece speed was set up at 0.043 m/s and the wheel speed at 30 m/s. The latter was maintained constant by adjusting the frequency of the induction motor on the frequency inverter, as the grinding wheel had its diameter decreased along the tests.

The dressing overlap ratio (U_d), which determines the wheel sharp level (Malkin, 1989), was set to 1 for all the tests. To control this parameter, the width of the dressing diamond tip was measured before each test. The measurements have been done using a profile projector which allows micrometer precision. Figure 3 shows the scheme used to measure the dressing diamond tip.

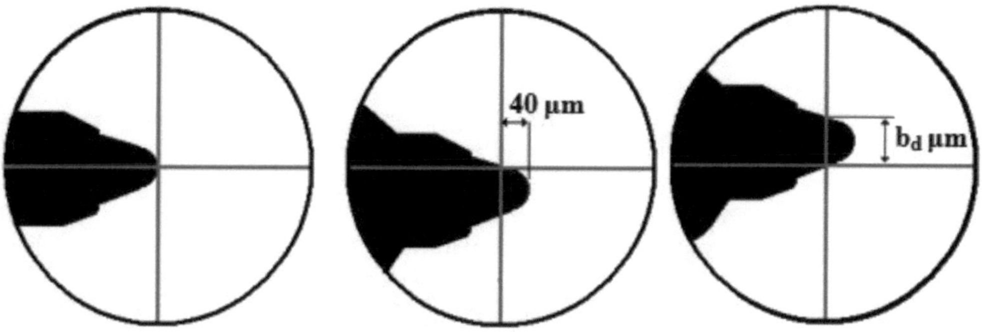

Figure 3. Method used to measure dressing tip width (b_d) on a profile projector.

Using equation 1 is possible to determine the dressing speed

$$V_d = \frac{n.b_d}{U_d.60} \tag{1}$$

Where: V_d is the dressing speed; L_r is the wheel width (31,75 mm); n the wheel rotation (1800rpm). For a better understanding, figure 4 was built to show some details on the dressing operation used in this work. It can be observed the single-point diamond moving across the wheel surface with direction given by the speed vector v_d.

A water-based fluid was used with 4% concentration. Each run consisted of a single grinding pass of the grinding wheel along the workpiece at a given grinding condition to be analyzed. The acoustic emission and cutting power signals were measured in real time at 2.5 millions of samples per second rate, and then stored onto binary data files for further processing. It is important to mention that the raw acoustic emission signal was acquired instead of the root mean square generally used.

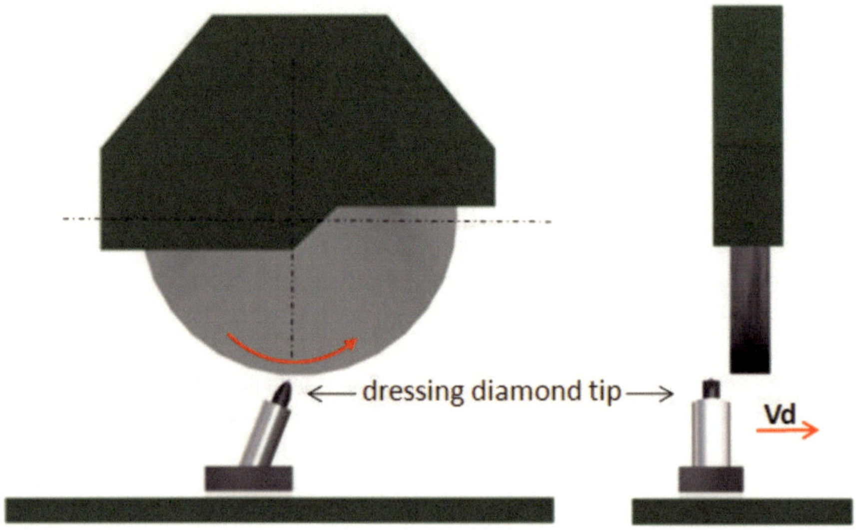

Figure 4. Dressing operation with single-point diamond.

3.2 Surface roughness measurements

Once all the grinding tests were done, surface roughness measurements were taken using a Surtronic 3+ tester and Taylor Hobson's TalyProfile software, version Lite 3.1.4, adjusted to a length of 8 mm sampling rate. Figure 5 shows the regions along the workpiece where the surface roughnesses were taken which accounted for 15 measurements. The surface roughness Ra was defined as the measurement parameter.

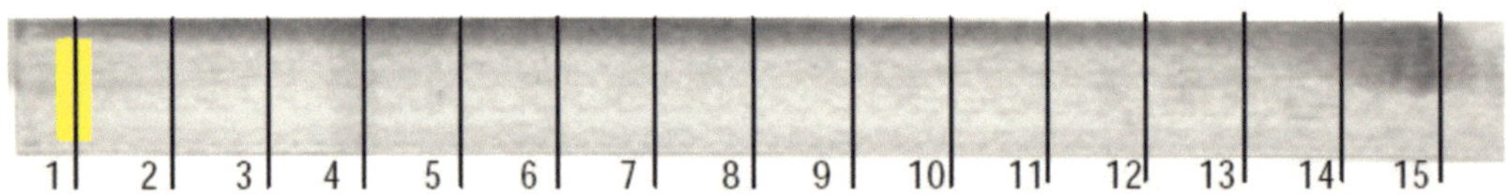

Figure 5. Lines showing the 15 regions along the workpiece where the surface roughness was measured

These data on surface roughness were naturally used in the neural networks, and they allow to verifying the magnitudes and efficiency of the system for prediction.

3.3 Statistical parameters used in the neural networks

Many parameters for monitoring fault in the grinding process have been studied. Two important parameters investigated by (Aguiar et al., 2002) and (Dotto et al., 2006) are DPO and DPKS. The DPO is defined as the standard deviation of the root mean square of acoustic emission signal multiplied by the maximum value of electric power in the grinding cycle or pass. The DPKS is the sum of the difference to the fourth power between cutting power and standard deviation of the cutting power multiplied by the standard deviation of the root mean square acoustic emission in the grinding cycle or pass. The DPO and DPKS are shown by equations 2 and 3 respectively.

$$DPO = std(AE)Max\left(Power\right) \tag{2}$$

$$DPKS = \{\sum_{i=1}^{i=m}[Power(i) - std(Power)]^4\}std(AE) \tag{3}$$

Where Max is the maximum value; Power the cutting power; AE the acoustic emission; std the standard deviation, m was set to 1024 in this work.
Thus, DPO and DPKS parameters were obtained from the data files stored during the acquisition. As these parameters were efficient at detecting grinding burn, they were tested as being the inputs of the neural networks in this research.

3.4 Neural network and input sets

Among several well known techniques of artificial intelligence are artificial neural networks, which basically consist of computational models analogous to the human brain whose main characteristic is the capability of learning (Aguiar et al, 2006). They are composed of many non-linear computational elements operating in parallel fashion. Neural networks, because of their massive nature, can perform computations at a higher rate. Due to their adaptive nature in using the learning process, neural networks can adapt to changes in the data and learn the characteristics of input signals (Kwak & Ha, 2004). In this work, the back-propagation algorithm of neural networks, which is one of the learning models, was used. Three hidden layers composed of 60, 40 and 20 neurons were found to be the best configuration for the neural networks studied. The following parameters were also found more suitable: sigmoid tangent activation function for the neurons of the hidden layers; linear activation function for the neurons of the last layer, downward gradient training algorithm; all data in the neural networks were normalized; training for 10000 epochs; square mean error value of 10-5; learning rate of 0.3; and momentum coefficient of 0.4. Table 1 shows three different sets used for the neural networks.

Set	Neural Network Configuration
1	AE, Cutting Power and Depth of Cut
2	AE, Cutting Power, DPO, DPKS and Depth of Cut
3	DPO and Depth of Cut

Table 1. Neural Network Configurations

4. Results and discussion

The measurements of surface roughness have shown an increase in magnitude as the depth of cut was increasing, mainly after the test with 35 μm depth of cut where grinding burn on the workpiece surface took place. In order to illustrate that increase in surface roughness, Figure 6 shows the mean values of surface roughness (μm) calculated for each depth of cut used for 15 regions along the workpiece (Figure 5) and three repetitions for each grinding condition. It can be observed the high standard deviation values more easily visible to those conditions of greater depth of cut, due mostly to encompass the mean of surface roughness along the workpiece. This factor can be better explained by the difference of surface roughness along the workpiece length; mainly in the tests where in certain region of the workpiece burn occurred. Thus, there was a great difference between the surface roughness values for the region where burn took place and the region with no burn.

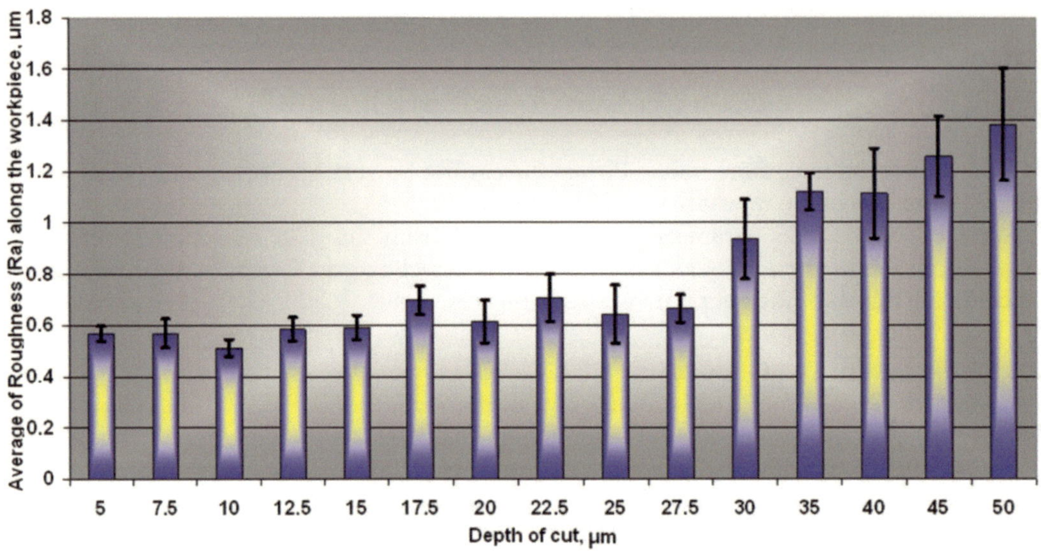

Figure 6. Average of Surface Roughness Ra (µm) along the workpiece in function of the depth of cut (µm) for 15 grinding runs.

It is important to point out that Figure 6 was built just for easily visualization of the surface roughness behaviour in relation to the depth of cut, provided the training and validation tests in the neural networks the surface roughness for each measured region was used.
From the grinding tests acoustic emission and cutting power vectors were extracted. Following each test, the root mean square of AE and filtering of both signals were obtained in order to check the success of the test as shown in Figure 7.

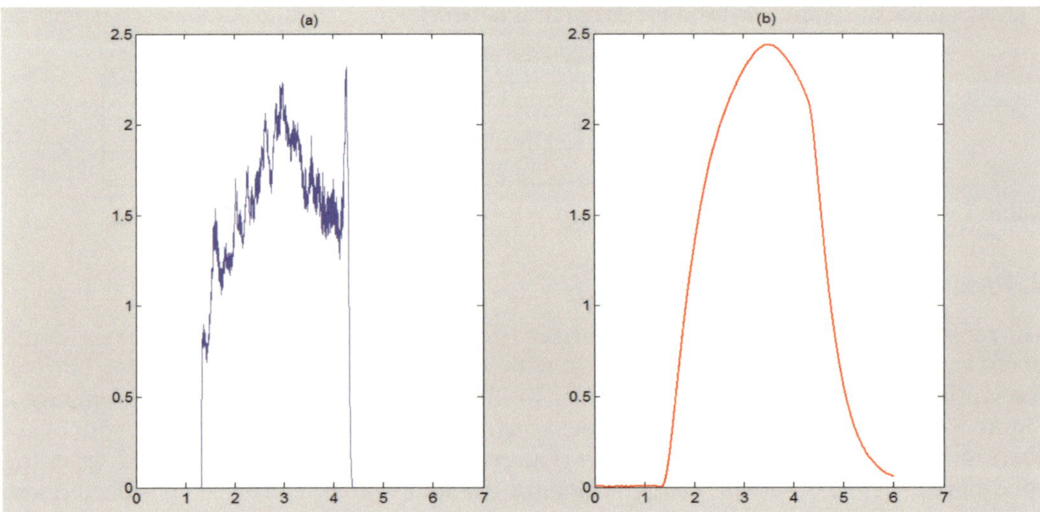

Figure 7. Acoustic Emission (a) and Cutting Power (b). Horizontal axis in seconds; Vertical axis in Volts.

These vectors were applied to the training of the neural networks, which provided surface roughness values and allowed the analysis of the system efficiency. The quantification of errors was done point by point through the computation of the absolute value of the difference between the measured values of surface roughness and the values given by the neural network. That was done for each of the 15 regions along the workpiece and for each workpiece.

Figure 8 shows the average values of square errors for all 15 regions along the workpiece length and 3 repetitions of each grinding condition as well as the variation of the error in each of those conditions. The lateral bar exhibits the three groups of errors found, that is, the lowest one the success, errors about 10^{-1} µm in the middle in green, and errors equal to or greater than 1 µm on the top in red.

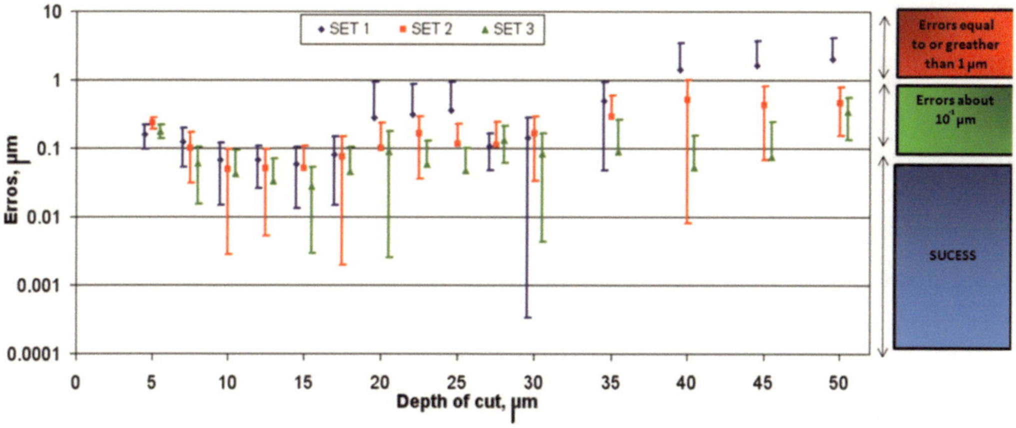

Figure 8. Average values of square errors, maximum and minimum errors values for each of 15 workpiece regions and 3 repetitions of each grinding condition.

It can be seen in Figure 8 the set number 1 presented error quite high to those greater depths of cut studied, overreaching the value of 1 µm for the last three grinding conditions. As mentioned previously, from the analysis of workpiece integrity it can be observed the grinding burn started at depth of cut of 35 µm. Thus, acoustic emission, cutting power and depth of cut have not responded very well as inputs to the neural network in the surface roughness prediction. However, the error observed does not necessarily mean that those parameters are not adequate to monitor other grinding control parameters, for instance, the grinding burn. The set number 2 has also presented greater errors in the last three grinding conditions but they have not overreached in average value of 1 µm.

Based on this information, the difference between the real measurements and the output of the neural networks can be classified in three groups: The first group for errors not greater than 10^{-2} µm, the second group for errors of about 10^{-1} µm, and the third group for errors equal to or greater than 1 µm.

As the surface roughness measurement device used in this work allows measurements with precision of two decimal digits, the last digit of this equipment is concerned with the uncertainty of the measurement. Thus, all the errors in the range of 10^{-2} µm or inferior were considered to be successful in the prediction of the surface roughness. This remark is reinforced by the tolerances applied to the grinding process in industry, which in general has tolerances in the range of 10^{-1} µm.

In the second group, errors of about 10^{-1} μm was thought as bad ones, but it would be mostly fine in industry depending on the type of part ground. Finally, the third group of errors is in the range of 1 μm or superior, which is classified as very bad ones.
Based on these analyses, we can see in Figure 8 that for many conditions, neural network had a hundred percent of success or almost that, mainly in the region from 5 μm of depth of cut to 30 μm, proving a total efficiency of the system for certain processes. These conditions are, for example, 10 μm, 12.5 μm, 15 μm and 17.5 μm for Set 1 and Set 2, and from 10 μm to 30 μm for Set 3.
It is important to point out that the great efficiency of the process occurred in the area of industrial application, since the largest rate of success was obtained exactly when burn on the workpiece doesn't succeed.
Comparing the values of surface roughness obtained in the tests and shown in Figure 6 it can be inferred that the studied process is optimized using 27.5μm depth of cut. This depth of cut represents the largest material removal rate employed before grinding burn takes place. Surface roughness values are similar to the grinding conditions before 27.5μm depth of cut.
Figure 9 was built in order to facilitate the comparison among the 3 input Sets and to show the general result of the values in all tested conditions.

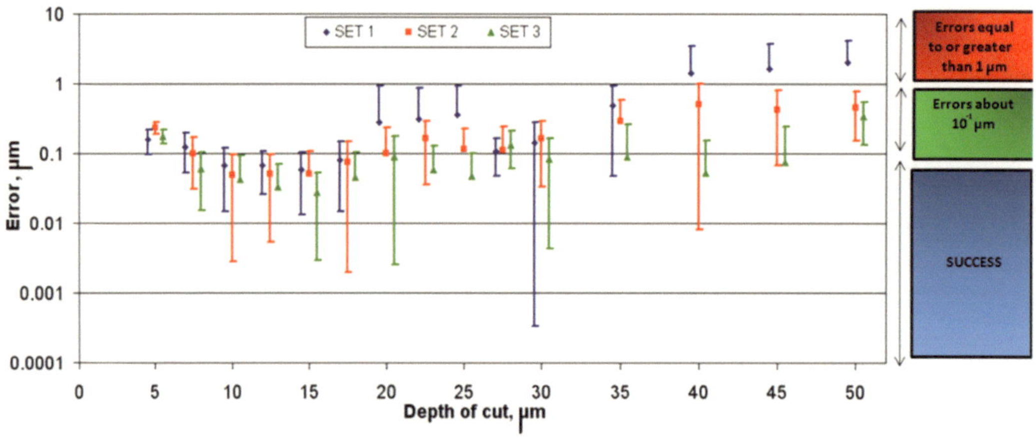

Figure 9. Percentage of errors and success of the neural network prediction for each set

Observing Figure 9 and other results from the neural network output as well as the comparison between the surface roughness values, the following remarks can be summarized.
From the sets utilized in the neural network training, the set number 3 (DPO and depth of cut) has presented the best output values for almost every grinding condition tested. This set of inputs has provided about 70% of success in the prediction of surface roughness analyzed region by region. In addition, it has not presented any error classified as bad one for the grinding process control.
The set number 1, composed by acoustic emission, cutting power and depth of cut, has presented satisfactory performance in the vast majority of the grinding condition tested, showing however pretty high errors mainly in the conditions of high depth of cut. Burning of the workpieces has occurred in these conditions, which can be understood as if this set is sensitive to others parameters besides surface roughness.

In the same way as occurred with set number 1, set number 2 (AE, cutting power, DPO, DPKS and depth of cut) also had good performance. The values of prediction for these two sets have alternated themselves, either for better or for worse when they are compared to each other. It is important to highlight that the set number 1 has showed higher errors than those seen in the set number 3 for some conditions. However, it has obtained good performance in the analysis of bad errors, showing only 3.1% for this kind of error.

5. Conclusions

The technique of surface roughness prediction has been developed using multi-sensor method with AE sensor and power meter for grinding process. Based on the results presented in this research, it can be concluded that acoustic emission and cutting power signals are very good input parameters to the neural network for surface roughness prediction of ground parts. Thus, the implementation of these signals is feasible in the control of grinding process in industry, provided that very low percentage of "bad" errors presented by the neural network was found. From the sets used in the training of the neural networks, and for the grinding conditions and methods employed, the set number 3 composed only by the DPO parameter and depth of cut has presented the best results. This set has reached about 70% of success in the prediction and none "bad" error, and for the condition where workpiece burn didn't occur success at prediction was almost 100%. In the comparison of errors generated by each set, set 3 was followed either by the set 1 or set 2. Because the level of errors found in the utilization of set number 3 as input to the neural network, the use of DPO parameter is highly attractive for predicting surface roughness in grinding. Therefore, an analog DPO output signal could be generated by an electronic circuitry with no need for further digital processing of AE and cutting power signals.

6. Acknowledgements

The authors want to express their gratitude to FAPESP -The State of Sao Paulo Research Foundation, and to IFM - The Institute Factory of Millennium for the financial support given to this research. Also, thanks go to the Grinding Laboratory - LUA and to Data Acquisition and Signal Processing Laboratory - LADAPS, both at FEB, Unesp, Bauru Campus, for their assets given to this work accomplishment.

7. References

Aguiar, P. R.; França T. V. & Bianchi E. C. (2006); Roughness and roundness prediction in grinding. *Proceedings of the 5th CIRP International Seminar on Intelligent Computation in Manufacturing Engineering (CIRP ICME ´06)*, 25-28 July, 2006, 183-188, Italy.

Aguiar, P.R.; Bianchi, E. C. & Oliveira, J. F. G. (2002); A method for burning detection in grinding process using acoustic emission and effective electrical power signal, *CIRP Journal of Manufacturing Systems*, 31, 2002, 253-257, Paris.

Bhushan B. (2001), Modern tribology handbook, *CRC Press LLC*, 2001.

Brinksmeier E., Tönshoff H. K., Czenkusch, C. & Heinzel, C.(1998), Modelling and optimization of grinding process, *Journal of Intelligent Manufacturing*, 9, 1998, 303-314.

Dornfeld, D. A. (1985), Manufacturing process monitoring and analysis using acoustic emission, *Journal Acoustic Emission*, 4, No. 2/3, 1985, S228-S231.

Dotto, F. R. L; Aguiar, P. R.; Bianchi, E. C. ; Serni P. J. A. & Thomazella R. (2006); Automatic system for thermal damage detection in manufacturing process with internet monitoring, *Journal of Brazilian Society of Mechanical Science & Engineering*, XXVIII, No. 2, 2006, 153-160.

Fredj, N. B. R., Amamou & Rezgui, M. A. (2002), Surface roughness prediction based upon experimental design and neural network models, *IEEE SMC*, 2002.

Hassui, A. & Diniz, A. E. (2003). Correlating surface roughness and vibration on plunge cylindrical grinding of steel, *International Journal of Machine Tools & Manufacture*, 43, 2003, 855-862.

Hundt, W.; Kuster F. & Rehsteiner, F. (1997); Model-based AE monitoring of the grinding process, *Annals of the CIRP*, 45, 1997, 243-247.

Hecker, R. L. & Liang, S. Y. (2003), Predictive modeling of surface roughness in grinding, *International Journal of Machine Tools & Manufacture*, 43, 2003, 755-761.

Kwak J. S., Sim S. B. & Jeong Y. D. (2006), An analysis of grinding power and surface roughness in external cylindrical grinding of hardened SCM440 steel using the response surface method, *International Journal of Machine Tools & Manufacture*, 46, 2006, 304-312.

Kwak J. S & Ha, M. K. (2004); Neural network approach for diagnosis of grinding operation by acoustic emission and power signals, *Journal of Materials Processing Technology*, 147, 2004, 65-71.

Lezanski, P. & Rafalowicz J. (1993), An intelligent monitoring system for cylindrical grinding, *CIRP* Ann., 42, 1993, 393–396.

Malkin S. (1989), Grinding technology - theory and applications of machining with abrasives, *Ellis Horwood Limited*, 1989, England.

Wang, J. Z.; Wang, L. S.; Li, G. F. & Zhou G. H. (2005). Prediction of surface roughness in cylindrical traverse grinding based on ALS algorithm, *Proceedings of the Fourth Internacional Conference on Machine Learning and Cybernetics*, 18-21 August 2005, 549-554, Guangzhou.

Wang, Z.; Willet, P.; Aguiar P. R. & Webster J. (2001); Neural network detection of grinding burn from acoustic emission, International. *Journal of Machine Tools & Manufacture*, 41, 2001, 283-309.

4

An Implementation of High Availability in Networked Robotic Systems

Florin Daniel Anton, Theodor Borangiu and Silvia Anton
University Politehnica of Bucharest
Romania

1. Introduction

Nowadays production flows are modular, each module in the enterprise being specialized and used to achieve a particular task. In many cases the modules are interconnected and materials are sequentially processed in each module resulting a final, unique product or assembly. One typical such production module is a flexible cell/system using multiple robots. In such complex enterprise environments, providing continuous services for applications is a key component of a successful implementing of robotized manufacturing. *High availability* (HA) is one of the components contributing to continuous service provision for applications, by masking or eliminating both planned and unplanned downtime of systems and applications. This is achieved by eliminating hardware and software *single points of failure* (SPOF).

The systems configured for high availability are a combination of hardware and software components configured to work together to ensure automated recovery in case of failure with a minimal acceptable downtime.

A high availability solution will ensure that the failure of any component of the solution - either hardware, software or system management, will not cause the application and its data to become permanently unavailable. High availability solutions should eliminate single points of failure through appropriate design, planning, hardware selection, software configuring, application control, carefully environment control and change management discipline.

In short, one can define high availability as the process of ensuring an application is available for use by duplicating and/or sharing hardware resources managed by a specialized software component. A high availability solution in robotized manufacturing provides automated failure detection, diagnosis, application recovery, and node (robot controller) re integration.

The chapter discusses the implementing of a high availability solution in a robotized manufacturing structure (cell, line). The solution is based on a High Availability Linux Cluster which is responsible for data availability and preservation on a NFS (Network File System) file system, and a second cluster – Fabrication Cluster (FC) which runs on Adept robot controllers, developed under the V+ programming language.

2. High availability versus fault tolerance

Based on the response time and response action to system detected failures, clusters and systems can be generally classified as:

- Fault-tolerant
- High availability

2.1 Fault-tolerant systems

The systems provided with *fault tolerance* are designed to operate virtually without interruption, regardless of the failure that may occur (except perhaps for a complete site going down due to a natural disaster). In such systems all components are at least duplicated for both software and hardware.
This means that all components, CPUs, memory, Ethernet cards, serial lines and disks have a special design and provide continuous service, even if one sub-component fails. Only special software solutions will run on fault tolerant hardware.
Such systems are very expensive and extremely specialized. Implementing a fault tolerant solution requires a lot of effort and a high degree of customization for all system components.
For environments where *no* downtime is acceptable (life critical systems), fault-tolerant equipment and solutions are required.

2.2 High availability systems

The systems configured for *high availability* are a combination of hardware and software components configured to work together to ensure automated recovery in case of failure with a minimal acceptable downtime.
In such industrial systems, the software involved detects problems in the robotized environment (production line, flexible manufacturing cell), and manages application survivability by restarting it on the same or on another available robot controller.
Thus, it is very important to eliminate all single points of failure in the manufacturing environment. For example, if a robot controller has only one network interface (connection), a second network interface (connection) should be provided in the same node to take over in case the primary interface providing the service fails.
Another important issue is to protect the data by mirroring and placing it on shared disk areas accessible from any machine in the cluster, directly or using the local area network.

3. High availability terms and concepts

For the purpose of designing and implementing a high-availability solution for networked robotic stations integrated in a manufacturing environment, the following terminology and concepts are introduced:
RMC: The Resource Monitoring and Control (RMC) is a function giving one the ability to monitor the state of system resources and respond when predefined thresholds are crossed, so that many routine tasks can be automatically performed.
Cluster: Loosely-coupled collection of independent systems (nodes - in this case robot controllers) organized in a network for the purpose of sharing resources and communicating with each other. A cluster defines relationships among cooperating systems, where peer cluster nodes provide the services offered by a cluster node should that node be unable to do so. These individual nodes are together responsible for maintaining the functionality of one or more applications in case of a failure of any cluster component.
There are two types of high availability clusters:

- Peer domain
- Managed domain

The general difference between these types of clusters is the relationship between the nodes.

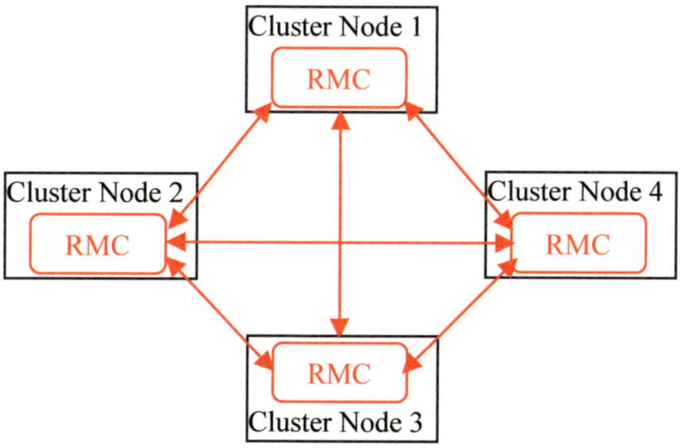

Fig. 1. Peer domain cluster topology.

In a *peer domain* (Fig. 1.), all nodes are considered equal and any node can monitor and control (or be monitored and controlled) by any other node (Harris et. al., 2004).
In a *management domain* (Fig. 2), a management node is aware of all nodes it is managing and all managed nodes are aware of their management server, but the nodes themselves know nothing about each other.

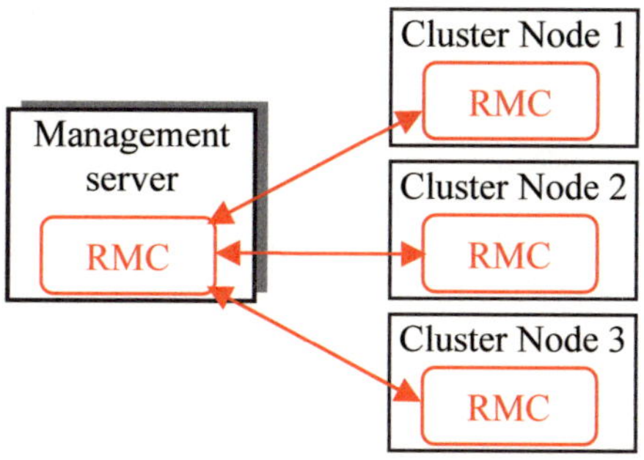

Fig. 2. Managed domain cluster topology.

Node: A robot controller that is defined as part of a cluster. Each node has a collection of resources (disks, file systems, IP addresses, and applications) that can be transferred to another node in the cluster in case the node or a component fails.
Clients: A client is a system that can access the application running on the cluster nodes over a local area network. Clients run a client application that connects to the server (node) where the application runs.
Topology: Contains basic cluster components nodes, networks, communication interfaces, communication devices, and communication adapters.
Resources: Logical components or entities that are being made highly available (for example, file systems, raw devices, service IP labels, and applications) by being moved from one node to another. All the resources that together form a highly available application or service are grouped in one resource group (RG). The cluster keeps the RG highly available

as a single entity that can be moved from node to node in the event of a component or node failure. Resource groups can be available from a single node or, in the case of concurrent applications, available simultaneously from multiple nodes. A cluster may host more than one resource group, thus allowing for efficient use of the cluster nodes.

Service IP label: A label matching to a service IP address and which is used for communications between clients and the node. A service IP label is part of a resource group, which means that the cluster will monitor it and keep it highly available.

IP address takeover: The process whereby an IP address is moved from one adapter to another adapter on the same logical network.

Resource takeover: This is the operation of transferring resources between nodes inside the cluster. If one component or node fails due to a hardware or operating system problem, its resource groups will be moved to the another node.

Failover: Represents the movement of a resource group from one active node to another node (backup node) in response to a failure on that active node.

Fallback: Represents the movement of a resource group back from the backup node to the previous node, when it becomes available. This movement is typically in response to the reintegration of the previously failed node.

Heartbeat packet: A packet sent between communication interfaces in the cluster, used to monitor the state of the cluster components - nodes, networks, adapters.

RSCT processes: (Reliable Scalable Cluster Technology). They consist of two processes (topology and group services) that monitor the state of the cluster and each node. The cluster manager receives event information generated by these processes and takes corresponding (response) actions in case of failure(s).

Group Leader (GL): The node with the highest IP as defined in one of the cluster networks (the first network available), that acts as the central repository for all topology and group data coming from the RSCT daemons concerning the state of the cluster.

Group leader backup: This is the node with the next highest IP address on the same arbitrarily chosen network, acting as a backup for the Group Leader; it will take over in the event that the Group Leader leaves the cluster.

Mayor: A node chosen by the RSCT Group Leader (the node with the next highest IP address after the GL Backup), if such exists; otherwise the Mayor node is the GL Backup itself. It is the Mayor's responsibility to inform other nodes of any changes in the cluster as determined by the Group Leader (GL).

Quorum: The notion of quorum is used to ensure that in case of loss of connectivity between two subsets of the peer domain only one subset considers itself as the peer domain. The quorum is defined as $n/2+1$, where n is the number of nodes defined in the peer domain. A peer domain cannot be turned online if less than the quorum of nodes can communicate.

SPOF: A single point of failure (SPOF) is any individual component integrated in a cluster which, in case of failure, renders the application unavailable for end users. Good design will remove single points of failure in the cluster - nodes, storage, networks. The implementation described here manages such single points of failure, as well as the resources required by the application.

4. RMC architecture and components design

The most important unit of a high availability cluster is the Resource Monitoring and Control (RMC) function, which monitors resources (selected by the user in accordance with the application) and performs actions in response to a defined condition.

The design of RMC architecture is presented for a multiple-resource production control system. The set of resources is represented by the command, control, communication, and operational components of networked robot controllers (station controllers) and robot terminals (station computers) integrated in the manufacturing cell.
The RMC subsystem to be defined is a generic cluster component that provides a scalable and reliable backbone to its clients with an interface to resources.

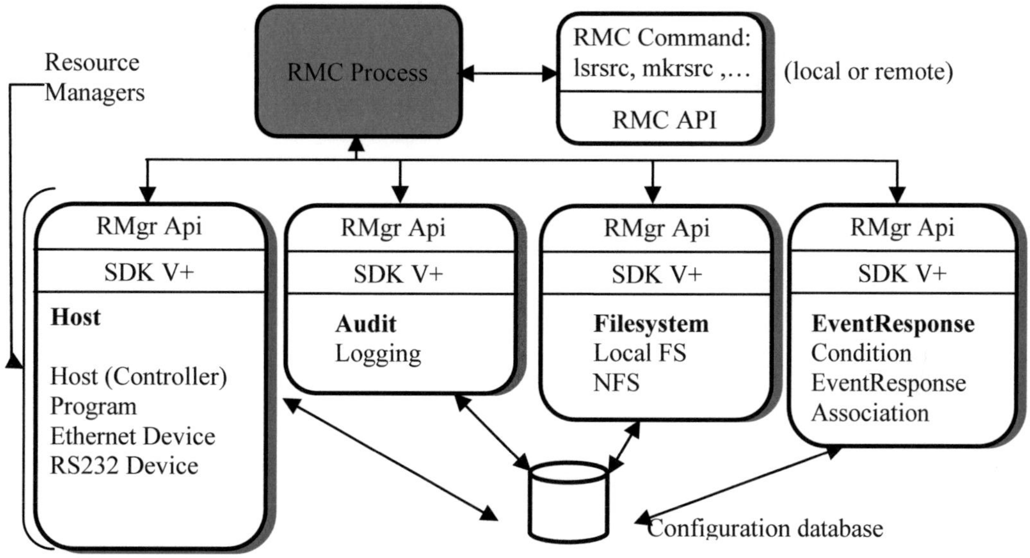

Fig. 3. The structure of the RMC subsystem.

The RMC has no knowledge of resource implementation, characteristics or features. The RMC subsystem therefore delegates to resource managers the actual execution of the actions the clients ask to perform. The architecture allows the following functions to be supported on clusters:

- Provide single monitoring and management infrastructure for clusters.
- Provide global access to subsystems and resources throughout the cluster.
- Support operations for configuring, monitoring, and controlling all cluster resources by RMC clients.
- Encapsulate all resource dependent operations.
- Provide a common access control mechanism across all resources.
- Support integration with other subsystems to achieve the highest levels of availability.

To support these functions, the structure of the RSCT Resource Monitoring and Control is not simple (see Fig. 3).
The RMC subsystem and RMC clients need not be in the same node; RMC provides a distributed service to its clients. The RMC clients can connect to the RMC process either locally or remotely using the RMC API i.e. Resource Monitoring and Control Application user Interface (Matsubara *et al.*, 2002).
Similarly, the RMC subsystem interacting with Resource Managers need not be in the same node. If they are on different nodes, the RMC subsystem will interact with local RMC subsystems located on the same node as the resource managers; then the local RMC process will forward the request between them.

Each resource manager is instantiated as one process. To avoid the multiplication of processes, a resource manager can handle several resource classes.
The commands of the Command Line Interface are V+ programs (V+ is the robot programming environment); the end-user can check and use them as samples for writing his own commands.

4.1 The Command Line Interface (CLI)

A RMC command line client can access all the resources within a cluster locally (A) and remotely (B) located (Fig. 4). The RMC command line interface (Fig. 5) is comprised of more than 50 commands (V+ programs): some components, such as the *Audit* resource manager, have only two commands, while others, such as *Event Response* resource manager, have 15 commands.

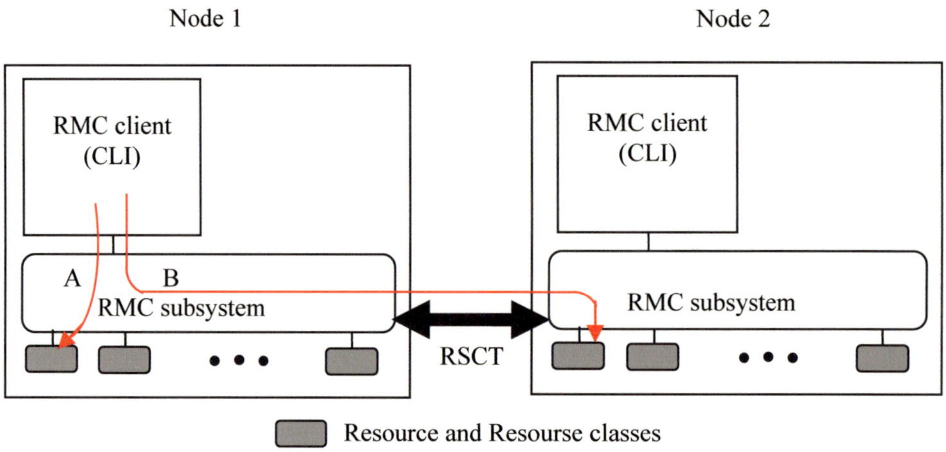

Fig. 4. The relationship between RMC Clients (CLI) and RMC subsystems.

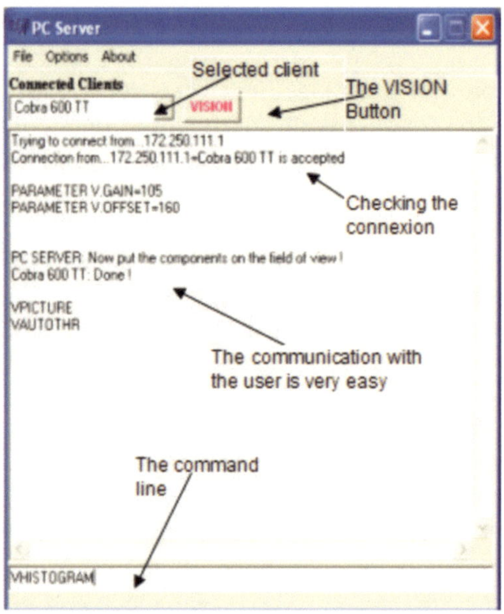

Fig. 5. The Command Line Interface.

Most RMC commands provide the user with a similar perception. Many commands have optional arguments to instruct the realization of the same functions. In addition to optional command arguments, RMC defines two environment variables:

- CT_CONTACT - define the communication target
- CT_MANAGEMENT_SCOPE - define the scope

These variables instruct most RMC commands to define their target and scope upon retrieving and setting values among cluster nodes.

Finally, RMC embeds the concept of *Resource Data Input*, which is a standardized file format for holding resource properties. Many RMC commands have a -f option that allows the end user to pass arguments to the command through a file instead of typing them on the command line. These files have the same Resource Data Input format for all commands so that the same file can be used for different commands. Two commands (**lsrsrcdef** - list resource definitions, and **lsactdef** - list actions definitions) can generate such a file so their output can be used as input for other commands.

4.2 The web-based Graphical User Interface (GUI)

The RMC graphical user interface is provided by the Web-based System Manager, and allow the user to configure the fabrication cluster (a HA cluster composed by industrial robots), and also acts as a teleoperation system.

The system allowing access to the web interface is composed by the following applications (Fig. 6):

The **Server Application (SA)**: Remote visual control and monitoring of multiple robot controllers from mobile and stationary matrix cameras.

- *Visual control*: the Server Application supports almost all V+ and AdeptVision program instructions and monitor commands. Robot training and control are interactive - menu-driven and acknowledged by image display in a VISION window. Some of the main functions available in this window are: choice of the physical camera and virtual cameras (multiple information counterparts of a single physical camera) and of the image buffers; selecting the display mode and resolution; histogram and average curve contrast analysis; selection of switches and parameters for virtual camera construction; display of vision system status; training and planning multiple ObjectFinder models for recognition and locating (Automated Visual Inspection - AVI & Guiding Vision for Robots - GVR); learning fingerprint models for collision-free grasping; editing, saving, loading and running V+ programs.
- *Monitoring*: a Monitoring/Treatment scheme can be defined for each Client/Station (the latter can be selected from a drop-down list of robot controllers connected to the server, by adding/removing them from the Client window). For each client a list of events and controller variables to be monitored according to a user-definable timing and precedence, and reacted at by user-definable actions/sequences can be specified in an Automatic Treatment Window.
- *Communication management*: the Server Application manages the communication with the robot controllers and the observation cameras, transfers real-time images from the cameras observing the robot workplace and production environment, reports status information, stores in a database and displays images taken by the robot camera via its controller. Finally, the SA outputs commands which are received from the eClients or acknowledges task execution.

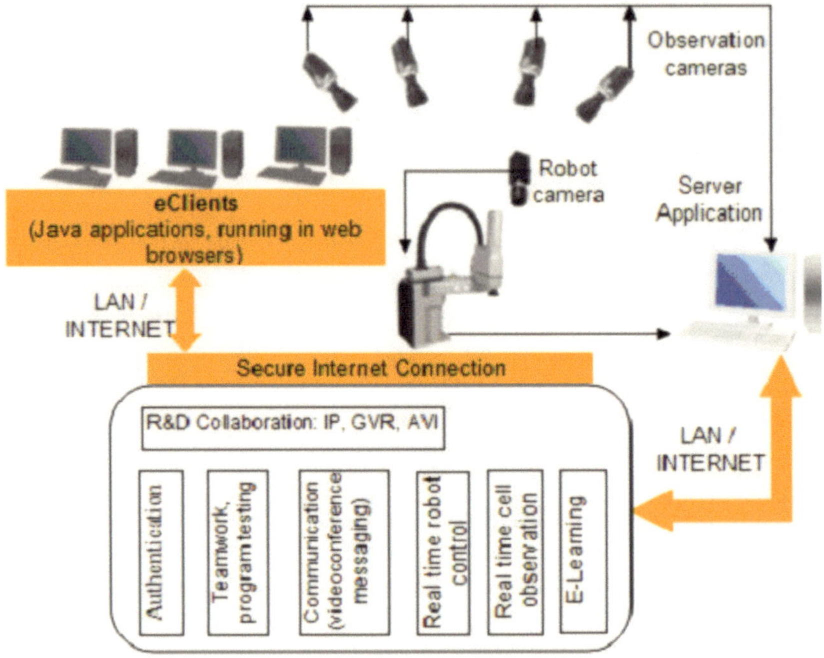

Fig. 6. The structure of the system which provides the web based interface.

The **eClients Applications (eCA)**: Java applications running in web browsers. They provide portal services and the connection of networked production agents: image data and RV program / report management; real-time robot control and cell / workplace observation. The eCA are composed by two applications:

- One application which has the function of retrieving images from the observation cameras (AXIS 214 PTZ), displaying them in real-time and also giving the user the possibility to change the orientation and zoom factor of the cameras.
- The second application is a VNC client.

The VNC viewer (Java client) is a web teleoperation application which can be executed into a web browser. The application connects to the Domino web server which makes a secure connection using a TCP/IP tunnel with a server having a private IP address, which cannot be accessed from internet but only using the Domino server. The private IP machine has a VNC server that exports the display, and also the teleoperation application (see Fig. 7). Using the exported display, the user can view and run the application as if the application runs on his own computer. The access is made using a username and a password, process managed by the Domino server.

For team training and application development, the system allows accessing related documents, presentations and multimedia materials, Web conferencing, instant messaging and teamwork support for efficient and security-oriented creation and use of group workspaces (students / trainers, researchers).

The Server and eClients Applications run on IBM PC workstations on which IBM Lotus software offerings are customized and integrated with other applications (error-free data and command transfer, V+ program editing, real time image display and refresh in multiple Client windows, replication functions, Client authentication, etc) in a virtual training and research laboratory across geographical boundaries.

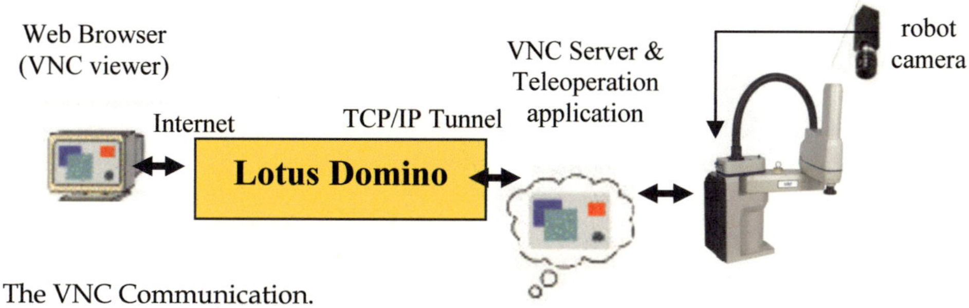

Fig. 7. The VNC Communication.

Lotus-oriented solutions have been considered for transferring messages, status reports, data and video camera images, interpreting them and accessing databases created by all partners. The final objective of the platform is to develop an E-Learning component allowing students to access and download technical documentation, create, test, debug and run RV and AVI programs, attend real-time laboratory demonstrations, and check their skills in proposed exercises.

Thus, IBM Lotus software unifies all Application modules, providing the necessary management and interconnecting tools for distributed industrial controllers, and the collaborative tools with back-end relational databases for team training and research.

To have access to the system, a user must have a username and a valid password to enter in the system. First the user must access the portal site using a java aware browser (like Internet Explorer, Opera, Firefox, with the JRE installed).

The portal which allows the web access is structured in two zones (Fig. 8):

- a public zone which contains all documentation, tutorials courses and others, needed by users to learn how to exploit the system; this part of the portal can be accessed by anyone;
- a private zone where the access is based on username and password. The private zone gives access to the eClients for teleoperation purposes.

After entering the correct username and password, the user is allowed in the system and has access to the teleoperation application which is a menu driven interface which allows him to interact with the system (see Fig. 9).

The web interface is also composed by two zones:

- An observation zone (right side in Fig. 9) where the user can monitor the jobs which are currently performed by the robots by using three networked cameras (due to the high resolution needed, only two cameras can be used at a time);
- A teleoperation zone where the user can interact with each robot.

The teleoperation application is composed by three windows (see Fig. 9):

- A camera control window, where the user can control all the camera parameters (tilt, pitch, zoom, focus and iris opening);
- A command window where the user can select the robot system which he wants to control, issue commands from the command line or activate the third window (Vision);
- A Vision window.

From the Vision window, vision commands can be issued by selecting the desired actions from the menus. The most important available functions are:

- selecting the physical and virtual cameras, and the virtual image buffers;
- selecting the display mode and the resolution;
- image acquisition;
- executing primary operations (histogram, thresholding, etc.);

- displaying the vision system status; training models;
- configuring switches and parameters for virtual camera set up.

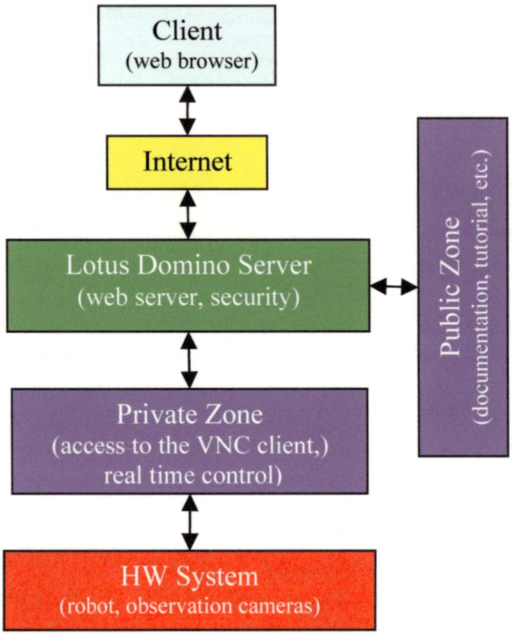

Fig. 8. The portal structure.

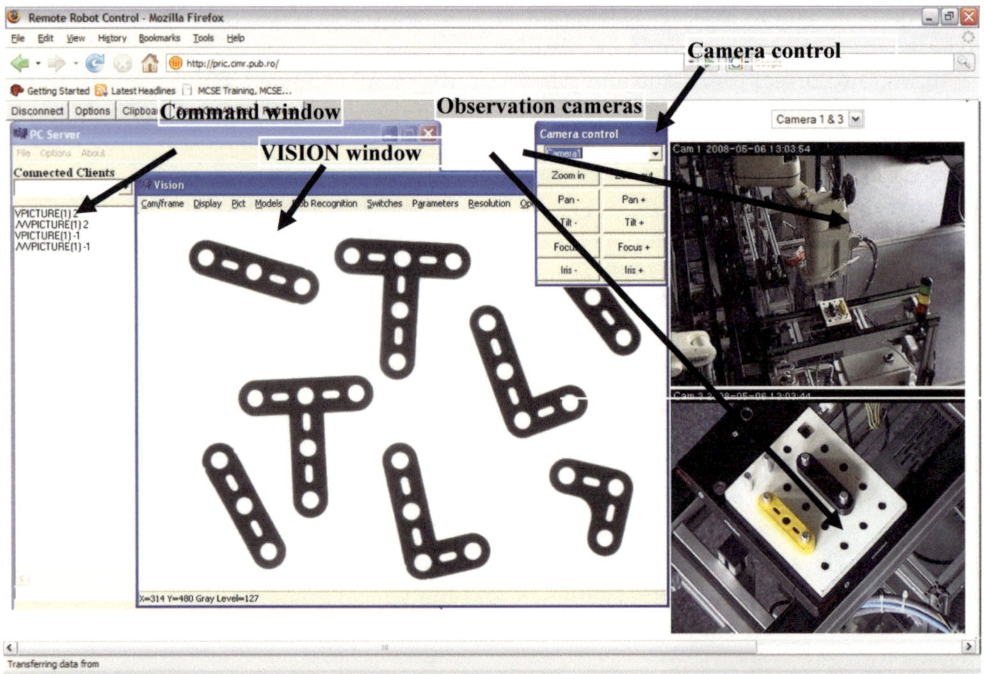

Fig. 9. The Web – based Graphical User Interface.

The advantage of the Vision window is that all commands can be issued using menus, but also the fact that the image acquired by the camera and sent to the server can now be accessed at pixel level. Another major advantage is that the training of part recognition and grasping models become a single-step process during which a unique window is used for parameters and constraints specification (Tomas Balibrea, *et al.*, 1997).

The client application can acquire full or partial images using the VGETPIC V+ operation and send them to the server (Adept Technology, 2001).

Captured image can be processed via the menus (filtering, binarization, convolution, morphing, etc.), saved in a common format and sent to a specialized image processing application. After processing, the image can be sent back to the client and integrated in the image processing module using the VPUTPIC V+ operation, for further use (an application using this mechanism was developed, and consists of a part identifying algorithm based on skeleton computation and matching.

In order to execute vision programs the user must set up the vision parameters such that the system can "see" the objects with the best image quality. The system has specialized functions to automatically establish these parameters or manually if some special settings are required.

After setting the parameters and loading the calibration camera-robot, the user can measure objects, apply filters, apply morphological operators, train and recognize objects.

The measurements include signature analysis; polar and linear offset signatures are computed and stored to be further used in applications. The skeleton computation is also included in the "Measurements" menu item (Borangiu, *et al.*, 2005).

An important feature of the system is the mechanism of training object recognition models and *multiple* grasping positions related to each model in order to provide alternative robot poses for collision-free object grasping at runtime.

Training of object models can be done in two ways:

- Creating a standard ObjectFinder model which is computed by a high level image processing module integrated in the system. The training procedure for this type of model is based on a sequence of steps which have been embedded in the application.
- Selecting a set of object features and storing them in a data structure which will further characterize each model (Borangiu, 2004).

After training and storing models, the user can write applications using these models for object search; the interface can be accessed from a great variety of places.

Therefore great care was taken to ensure optimal use in case of low bandwidth and poor quality hardware. High level of security and availability are key components; they were ensured by the selection of high quality technical devices and well-planned loading. As the portal must be prepared for an increasing number of users, the tool must be highly scalable. It needs to integrate contents and services and provide access for document repositories. User groups must be able to access personalised contents.

High-level availability must be also guaranteed. The system must be scalable according to load and requirements. The use of NLBS (*Network Load Balancing System*) provides a solution for an even load of the network. The portal user interface needs to be customized; templates must be created and uploaded. The authorisations and user groups must be defined.

The eClient application and Lotus Domino Server implement a security access policy to the virtual workspace. The access to the eClient application is granted based on the Domino defined ACL's (*Access Control Lists*), such that the user must specify a user name and a password in order to connect to the application. Two classes of privileges were defined:

1. A **user class** where the operator can observe images acquired from observation web cameras and images from the Vision system taken by multiple area cameras; he can also view the commands issued by the trainer and watch the results of the commands;
2. A **trainer class** where the operator is authorized to issue commands for every connected robot system, upload, download and modify programs. The trainer can also take pictures from an individual camera and use the specific vision tools to process that image. The observation cameras can also be moved and positioned in the desired location by the trainer. The trainer can give users full or partial permission to program testing.

The communication between users is realized by help of an integrated console (text mode) or using an Instant Messaging and Web Conferencing application (Sametime).

IBM Lotus Domino server software was used to combine enterprise-class messaging and calendar / scheduling capabilities with a robust platform for collaborative applications on a wide variety of operating systems. Designing the Lotus Domino Server made available three offerings: Domino Messaging Server (messaging only), Domino Utility Server (applications only), and Domino Enterprise Server (both messaging and applications) (Brooks, *et al.*, 2004).

Some of the most important Lotus Domino features were used:

- Encryption, signing, and authentication using the RSA public-key technology, which allows marking a document such that the recipient of this document can decisively find out that the document was unmodified during transmission.
- Access Control Lists (ACLs) determining who can access each database (application) and to what extent.
- Usage of Domino's new features to reduce network utilization. Network compression reduced the number of bytes sent during transactions by up to 50 percent. Connections across heavily loaded links such as WANs and XPCs saw the most benefit.
- Availability for the Windows NT and XP platforms, automatic fault recovery after shutdown and server restart without administrator intervention after the occurrence of an exception. Fault recovery used operating system resources, like message queues.

Because, the application eClient is accessed over the Internet, security represented a critical element of the design. Access to different levels of the application is controlled by xACLs (extended ACLs) to allow or disallow access. The existing database Access Control Lists (ACLs) and the new ACL file feature ensure that application-private databases remain secure. In addition, file protection documents for the Domino Web server used to serve the eClient (Java application) provide additional access control for files accessed via HTTP.

4.3 Resource, resource class, and attribute

This section provides information about the three terms used in RMC: resource, resource class, and attribute, as illustrated in Fig. 10.

Resource

The *resource* is a fundamental concept in the RMC architecture. A resource is an abstraction of an instance of a physical or logical entity that provides services to applications or system components. A system or cluster is composed of numerous resources of various types.

Each resource has the following characteristics:

- An *operational interface* used by its clients. This interface is the reason for the resource's existence. For example, the operational interface of a data file is given by the standard open, close, read, and write system calls.
- A set of persistent data values called *persistent attributes*, describing the characteristics or configuration of the resource (e.g. the file system name, data file name, and so on).

- A set of dynamic data values called *dynamic attributes,* reflecting the current state or other measurement values of the resource (e.g. the disk block usage of a file system).
- A handle that uniquely identifies the resource within the cluster. This value is unique across time and space and is called a *resource handle.*
- A *set of operations* that manipulate the state or configuration of the resource.

Resource class

A *resource class* is a collection of resources having similar characteristics. The resource class provides descriptive information about the properties and characteristics that are common to any resource within the resource class (for example, all the file system resources, files such as prog1.v2 and prog2.v2 belong to the file system resource class in RMC).

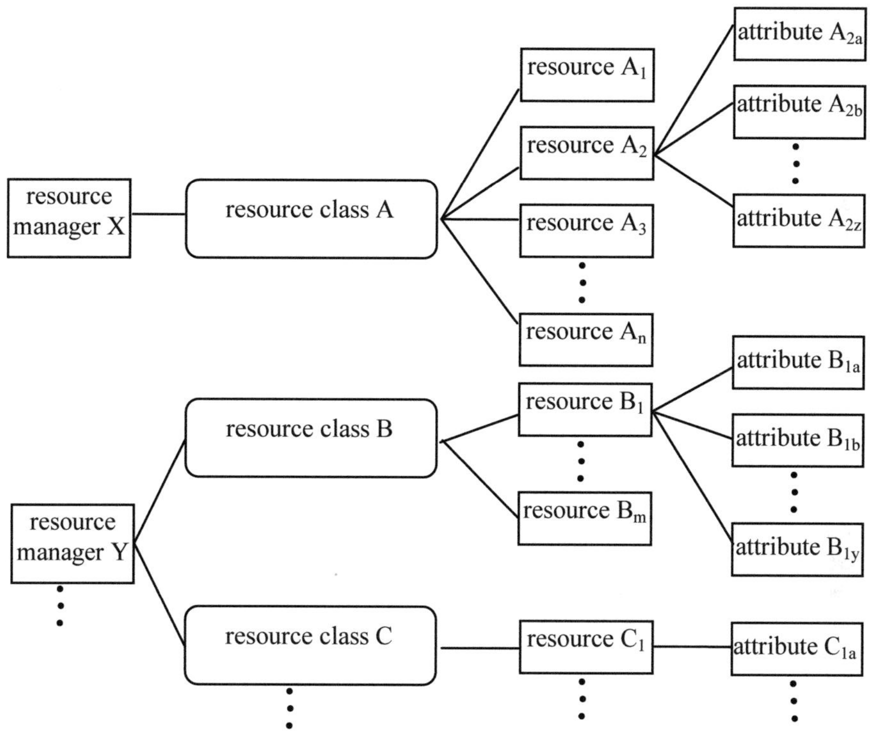

Fig. 10. Resource managers, resource classes, resources, and attributes.

Each resource class has the following characteristics:

- Descriptive information about the *class* and its *resources.*
- A list of the *resource handles* for all existing resources within the class.
- A set of persistent data values called *persistent attributes* that describe or control the operation of the resource class.
- A set of dynamic data values called *dynamic attributes* (e.g. every resource class supports a dynamic attribute that changes whenever the resource number in the class changes.
- An *access control list* (ACL) that defines permissions authorizing users to manipulate or query the resource class.
- A set of *operations* to *modify* or *query* the resource class.

Attribute

A resource class and a resource have several *attributes*. An attribute has a value and a unique name within the resource class or the resource.

A resource attribute is classified into either a *public* or a *private* property. The property is used as a hint for the RMC client for whom the attribute is to be presented to general users. Private attributes typically contain information which is not relevant to general users; therefore private attributes are hidden by default. However, one can display private attributes by specifying a flag in the command. Web-based System Manager does not support the display or modification of private resource attributes.

Attributes fall into two categories: *persistent* and *dynamic*. Each category has a different role, and is used differently by RMC clients:

- **Persistent attributes**: persistent attributes are configuration parameters for resources, and define the resource's characteristics (examples of persistent attributes: Device Name, IP Address, etc). For resource classes, persistent attributes describe or control the operations of the class. For the RMC client, the main usage of persistent resources is to specify which resources will be accessed. They are used as filters over the set of managed resources to select those on which the client wants to act.
- **Dynamic attributes**: Dynamic attributes reflect internal states or performance variables of resources and resource classes. One generally refers to dynamic attributes to define the monitoring condition of the resource to be monitored (for example, if you want to monitor a file system resource, you can monitor the disk space usage of the file system).

4.4 Resource managers

Each resource manager is the interface between the RMC subsystem and a specific aspect of the Adept Windows operating system instance it controls. All resource managers have the same architecture and interact with other RMC components. However, due to their specific nature they have different usage for the end user. Resource managers fall into four groups:

1. *Logging and debugging* (Audit resource manager) The Audit Log resource manager is used by other RMC components to log information about their actions, errors, and so on. The typical end user of Audit Log resource manager is a system administrator which scans these logs to see what is happening in the various RMC components. Unless a problem occurs, there is generally no need for the user to look at these logs.
2. *Configuration* (Configuration resource manager). The configuration resource manager is used by the system administrator to configure the system in a Peer Domain cluster. It is not used when RMC is configured in Standalone or Management Domain nodes.
3. *Reacting to events* (Event Response resource manager). The Event Response resource manager is the only resource manager directly used in normal operation conditions. The CLI allows the end user to define which events he wants to monitor in the nodes, and how the system should react to these events.
4. *Data monitoring* (Host resource manager, File system resource manager, Sensor resource manager). This group contains the file system resource manager and the Host resource manager. They can be seen by the end user as the containers of the objects and variables to monitor. However, it is important to note that these resource managers have no user interface. The end user cannot interact directly with them and has to pass through the Event Response resource manager to access the resources managed by these resource managers.

The only exception is the Sensor resource manager. It is the only resource manager that provides for expandability. Its user interface therefore contains four commands to let the end user define, modify or delete resources to be monitored. Once these resources are created, the normal interface to them is through the Event Response resource manager.

4.4.1 Event response resource manager

The Event Response resource manager (ERRM) plays the most important role to monitor systems using RMC. It uses several terms like *action, condition, expression, response* which will be discussed later.

The Event Response resource manager provides the system administrator with the ability to define a set of conditions to be monitored in the various nodes of the cluster, and to define actions to be taken in response to events (Lascu *et al.*, 2002; Lascu, 2005). These conditions are applied to dynamic properties of any resources of any resource manager in the cluster.

The Event Response resource manager provides a simple automation mechanism for implementing event driven actions. Basically, one can do the following actions:

- Define a condition composed of a resource property to be monitored and an expression that is evaluated periodically.
- Define a response that is composed of zero or several actions that consist of a command to be run and controls, such as to when and how the command is to be run.
- Associate one or more responses with a condition and activate the association.

In addition:

- Different users can employ the same conditions and responses, mix and match them to build different monitoring scenarios.
- Many predefined conditions and responses are shipped with the system and can be used as templates for the user who wants to define own monitoring scenarios.
- Any events that occur and any responses that are run are logged in the audit log. One can look at the audit log to check that the defined conditions were triggered, how the system reacted, and what was the result of the actions.

The Event Response resource manager provides one with both a CLI and a GUI (Web-based System Manager). It can then be decided whether you want to use the Web-based System Manager or the CLI. In a stand-alone environment and for setting up limited monitoring of the resources, the Web-based System Manager is the easiest tool to use, unless the user is already familiar with the CLI commands and arguments. However, for managing a cluster or developing sophisticated monitoring scenarios, the recommended tool is the CLI where the user can write his own CLI commands.

Expression

The expression is the means of making computations with data stored in attributes and audit log record fields. An expression is composed of variables, constants and operators, and returns a Boolean value: *true* or *false*.

The same syntax is used regardless of the following three environments; however, the expressions handle different types of variable:

- **ERRM conditions**: trigger events based on dynamic attributes of resources.
- **GUI or CLI**: filter resources based on variables representing persistent attributes of the resources.
- **Audit log expressions**: select audit records with variables being the name of a record field.

Condition

A condition applies to a resource class. It may contain a selection string, which is used to define which resource in the resource class should be monitored. It also contains a Boolean expression that must be satisfied by one (and only one) dynamic attribute of the resource to fire the condition. A condition has a name, so the user can refer to it once defined. A condition can also contain an optional rearm expression.

Fig. 11 shows the relationship between the monitored resource(s), conditions, events, and rearm events.

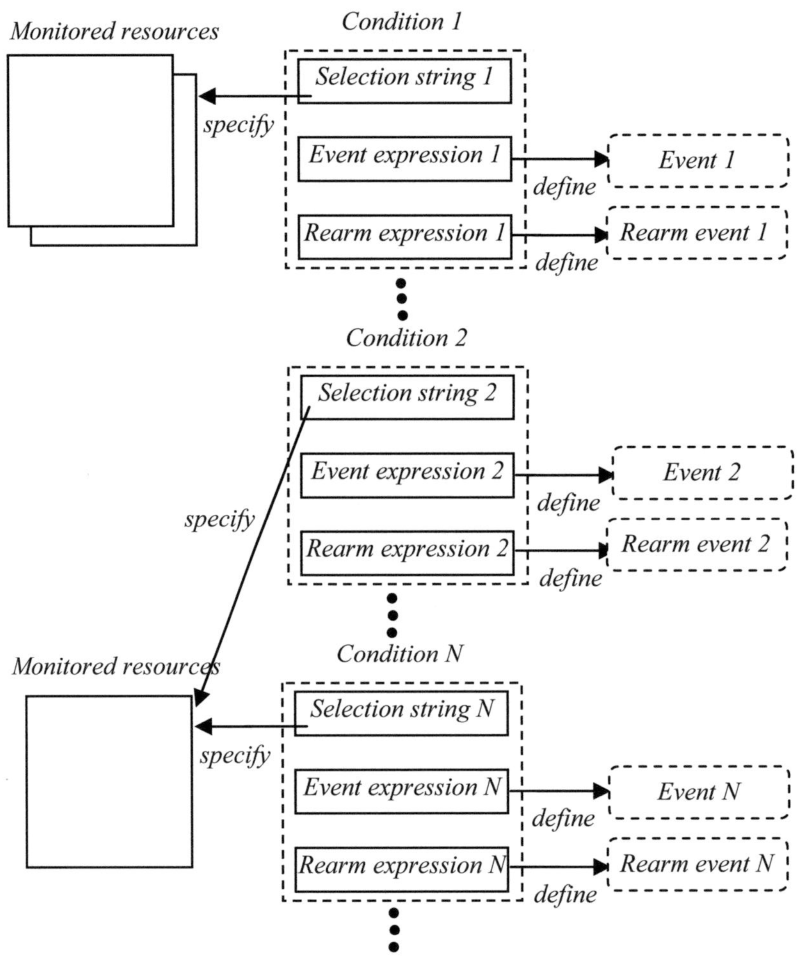

Fig. 11. Condition, expression, and events.

For example, to see if the file system uses more than 90 percent of its available disk space, a predefined condition called "disk space used" is provided with RMC. This condition uses the file system resource class and selects the local disk file system from all the file systems of the monitored nodes. It contains the expression "PercentTotUsed > 90" which applies to the PercentTotUsed attribute of the file system resource.

A condition may also contain a node group list that the event is monitored on. The default value for the node group list is NULL for all the nodes within the management scope. The default value for the management scope is local scope for the local node.

A condition can specify multiple monitored resources selected from a resource class by the selection string. A condition cannot select resources selected from multiple resource classes. Multiple conditions can specify the same resource(s); however, they must have different condition names. If multiple conditions that specify the same resource(s) have the same expression or the rearm expression, then they define the same event or the rearm event. However, you may want to specify different actions for them.

There are no objects defined as *event* and *rearm event*. If an event expression is evaluated as true, the result is called as an event. If a rearm expression is evaluated as true, the result is called as a rearm event.

Action

RMC comes with three *predefined actions*: logevent, notifyevent, and wallevent:

- The *logevent* action appends a formatted string to a user defined log file that can be read later on using the **alog** command.
- The *notifyevent* action sends the content of the event to the user.
- The *wallevent* event broadcasts a message to all users.

One can specify when an action should be executed from as follows:

- Only when the event defined by the associated condition occurs.
- Only when the rearm event defined by the associated condition occurs.
- Either the event or the rearm event defined by the associated condition occurs.

Finally, the user can also provide customized scripts as actions and the existing scripts can be used as examples or starting points for more complex operations. User defined scripts are executed in an environment where variables are set according to the event that would trigger the action. For example, the ERRM_RSRC_NAME variable would contain the name of the resource whose dynamic attribute change caused an event to fire.

Response

A *response* is a set of zero or more actions that will be performed by the system when an expression of a condition evaluates to true. A response contains the following information on how the actions are executed:

- Arguments for the actions.
- Time ranges (periods) to describe when the action can be run.
- Definitions of environment variables to set before executing the action.
- The expected return code of the action and whether the results of the action should be directed, for example, to the audit log.

A response can be defined with multiple actions. For example, the critical notification response calls the following three predefined actions:

- Logging an entry in the criticalEvents.log.
- Sending a message to the user.
- Warning all logged users with **wallevent**.

Association between a condition and a response

Conditions and responses can exist without being used and with nothing related to each other. Actions are part of responses and only defined to responses. Although it is possible that multiple responses have an action using the same name, these actions do not refer to the same object. One can associate a response with multiple actions.

To start observing the monitored resource, a condition must be associated with at least one response. A condition can be associated with multiple responses. However, each pair of a

condition and a response is considered a separate association. Once associated, the condition and response should be activated (started).
Fig. 12 illustrates the relationship between the conditions, the responses, and the actions. In this scheme, there are three associations (A, B, and C).

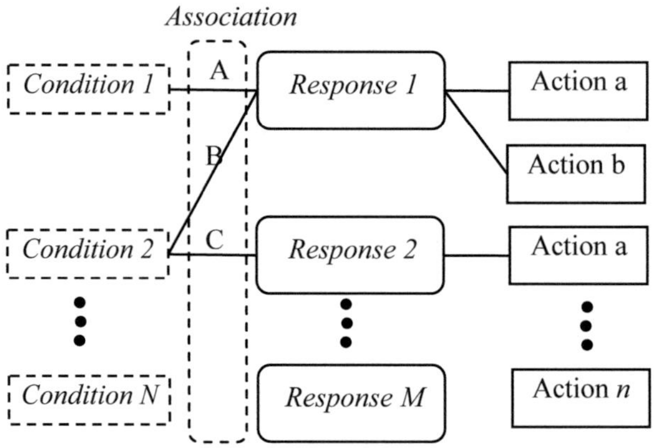

Fig. 12. Conditions, responses and actions.

An association has no name. The labels A, B, and C are for reference purposes. To refer to the specific association, one must specify the *condition name* and the *response name* that make the association. For example, one must specify the condition 1 and the response 1 to refer to the association A. Also, it must be clear that the same action name (in this example, action a) can be used in multiple responses, but these actions are different objects.
ERRM evaluates the defined conditions which are logical expressions based on the status of resources attributes; if the conditions are true a response is executed.

4.4.2 Audit log resource manager

The Audit Log resource manager provides the other RMC subsystems with a logging facility. It can be seen as a means to trace various actions taken by the cluster system.
Typically, these subsystems will add entries in the audit log corresponding to their normal behaviour or to error cases. Examples of information provided in the audit log are:

- Instances of starting and stopping monitoring
- Events
- Actions taken in response to events
- Results of these actions
- Subsystem errors and monitoring errors

Only subsystems can add records to the audit log; there is no user interface that allows the end user to add entries in the audit log.
Only two actions are offered by the Audit Log resource manager:

- **Lsaudrec**: List records in the audit log.
- **Rmaudrec**: Delete the specified records in the audit log.

4.4.3 Configuration resource manager

The configuration resource manager can create and manage a peer domain cluster. Thus, the configuration resource manager need not be used to use RMC on a stand-alone server or in a management domain.

4.4.4 File system resource manager

The File system resource manager manages only one resource class: the local file systems on the cluster node. It can be used to list the file systems, get their status, and retrieve values like the percentage of used space or *inodes*.

4.4.5 Host resource manager

The host resource manager provides the facility to monitor hardware devices and programs running on the cluster nodes. The monitored hardware includes disks, serial lines and network interface adapters. The program monitoring feature allows you to monitor when processes executing any arbitrary program are created or destroyed.

4.4.6 Sensor resource manager

The sensor resource manager allows the user to extend the functionality of RMC for monitoring parts of the Adept Windows robot controller operating system for which RMC does not provide a predefined resource class or attribute.
The sensor resource manager provides four commands to create, rename, or remove new resources called sensors. A *sensor* is a command that will be periodically executed, returning a value that can be monitored by defining a condition in the ERRM. By default, the period of execution is 60 seconds, but this period can be changed by the user to another value when creating the sensor.
If more than one returned value is needed, it is possible to define sensors returning multiple "property = value" pairs. This feature can be used, for example, to implement behaviours, such as notify user of the value of properties 2 and 3 when property 1 exceeds a threshold.

5. Solution implementing for networked robots

In order to implement the solution on a network of robot controllers, first a shared storage is needed, which must be reached by any controller from the cluster.
The file system from the storage is limited to NFS (*network file system*) by the operating system of the robot controllers (Adept Windows). Five Adept robot manipulators were considered, each one having its own multitasking controller.
For the proposed architecture, there is no option to use a directly connected shared storage, because Adept robot controllers do not support a Fiber Channel Host Bus Adapter (HBA). Also the storage must be high available, because it is a single point of failure for the Fabrication Cluster (FC), see Fig. 13.
Due to these constraints, the solution was to use a High Availability cluster to provide the shared storage option (NFS Cluster), and another cluster composed by Adept Controllers which will use the NFS service provided by the NFS Cluster (Fig. 13).
The NFS cluster is composed by two identical IBM xSeries 345 servers (2 processors at 2.4 GHz, 1GB RAM, and 75GB Disk space, two RSA 232 lines, two Network adapters, and two Fiber Channel HBA), and a DS4100 storage. The storage contains a Quorum volume used by the NFS cluster for communication between nodes, and a NFS volume which is exported by the NFS service which runs in the NFS cluster. The servers have each interface (network, serial, and HBA) duplicated to assure redundancy (Anton *et al.*, 2006; Borangiu *et al.*, 2006).
In order to detect the malfunctions of the NFS cluster, the servers communicate using heartbeat packets to ensure that the communication is established.

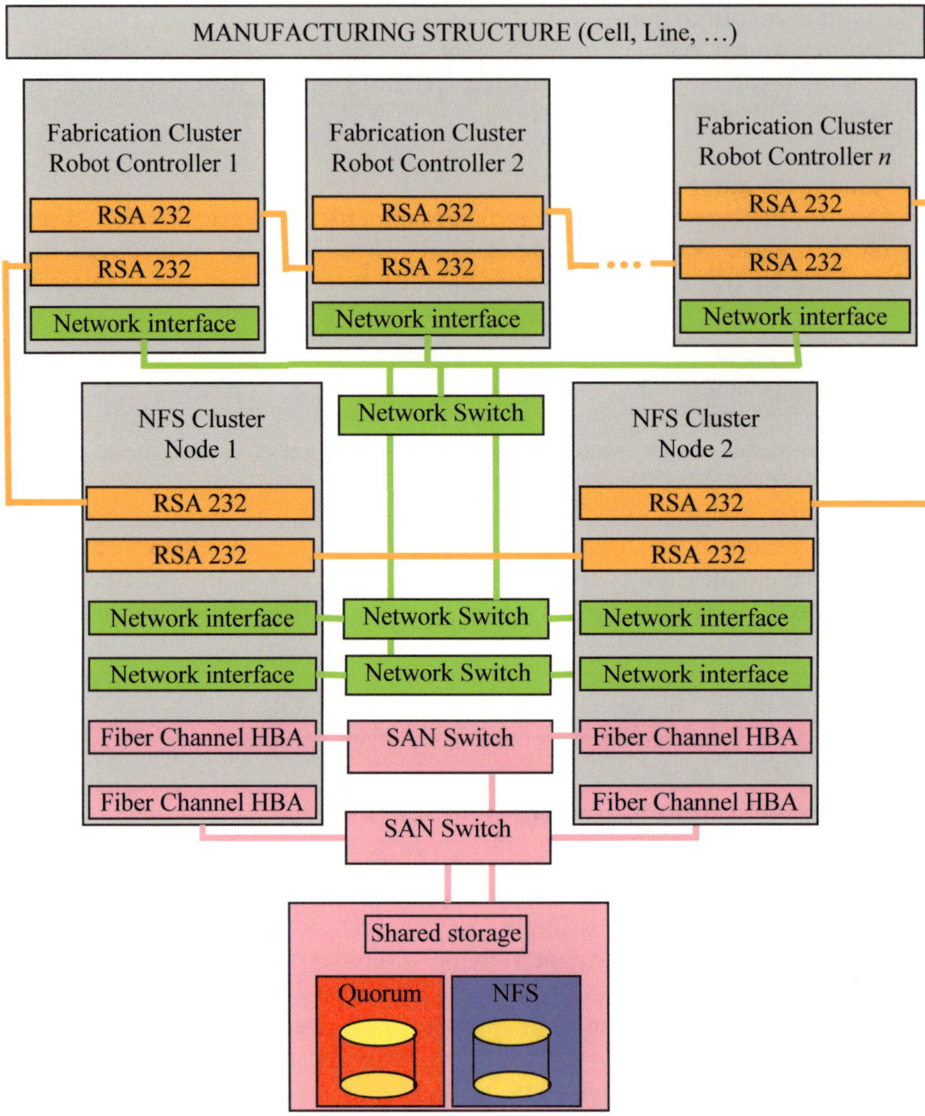

Fig. 13. Implementing the high availability solution for the networked robotic system.

There are three communication routes: the first route is the *Ethernet network*, the second is the *quorum volume* and the last communication route is the *serial line*. If the NFS cluster detects a malfunction of one of the nodes and if this node was the node which served the NFS service the cluster is reconfiguring as follows:

1. The server which is still running writes in the quorum volume which is taking the functions of the NFS server, then
2. Executes the resource takeover operation (the resource being the NFS volume) and mounts the NFS volume, then
3. Takes the IP of the other server (IP takeover) and
4. Starts the NFS service.

In this mode the Fabrication Cluster is not aware about the problems from the NFS cluster, because the NFS file system is further available.
The Fabrication Cluster can be composed by at least two robot controllers (nodes) - *group leader* and *group leader backup*. The nodes have resources like: robot manipulators (with attributes like: collision detection, current robot position, etc...), serial lines, Ethernet adapter, variables, programs, and NFS file system. The NFS file system is used to store programs, log files and status files. The programs are stored on NFS to make them available to all controllers, the log files are used to discover the causes of failure and the status files are used to know the last state of a controller.
In the event of a node failure, the production flow is interrupted. In this case, if there is a connection between the affected node and the group leader, the leader will be informed (if not, the heartbeat detects the failure) and the GL takes the necessary actions to remove the node from the cluster. The GL also reconfigures the cluster so the manufacturing process will continue. For example if one node cluster fails in a three-node cluster, the operations this node was doing will be reassigned to one of the remaining nodes.
The communication paths in the multiple-robot system are the *Ethernet network* and the *serial network*. The serial network is the last communication resort due to the low speed and also to the fact that it uses a set of Adept controllers to reach the destination (because it is a ring type connection). In this case the *ring network* will be down if more than one node fails.
Rebuilding the communication route in the network represents the last step required to restart the normal behave of the FC control system.
In case of malfunction of the communication network the following important cases can appear:

1. If the connection between the Switch and the Supervisor PC (the PC where the Server Application runs) is down the remote control will be lost, but the FMC will reconfigure as follows: the controller will use the Ethernet network for communication, and the controller with the first IP from the class will take the functions of the Supervisor PC. If the connexion is re-established the Supervisor PC makes a query, finds the replacing controller, transfers the databases and restarts the normal behaviour.
2. If the switch is not functioning, all the Ethernet connexions are lost, but the controllers will use the serial "network". The behaviour is like in the first case only that the web users can view the status from the Supervisor PC, including the images acquired by the observation cameras.
3. If a controller loses the Ethernet connexion, it will use one of the two serial lines to reach the Supervisor PC depending on the CPU time of the neighbours.

6. Conclusion

Fault-tolerance is provided to the cell communication system (Fig. 14), and therefore redundancy at both Station Controller level (a break down of a Robot Controller is detectable, the production tasks can be rescheduled to the remaining valid units for graceful degraded behaviour) and Station Computer level (replication of data bases for the IBM PC-type device terminals, reassignment of computers in case of break downs).
The fault tolerance solution presented in this paper is worth to be considered in environments where the production structure has the possibility to reconfigure, and where the manufacturing must assure a continuous production flow at batch level (job shop flow).
There are also some drawbacks in this solution, like the need of an additional NFS cluster. The spatial layout and configuring of robots must be done such that one robot will be able to

take the functions of another robot in case of failure. If this involves common workspaces, programming must be made with much care using robot synchronizations and monitoring continuously the current position of the manipulator.

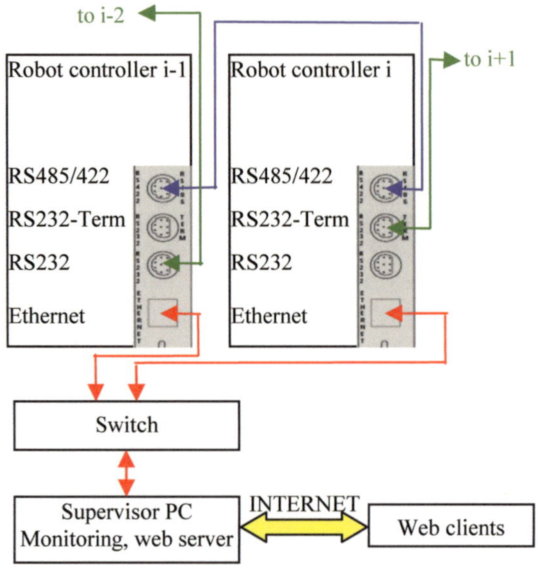

Fig. 14. Fault-tolerant communication architecture.

The advantage of the proposed solution is that the structure provides a continuous production flow with an insignificant downtime (during reconfiguration).
The solution was tested on a five-robot assembly cell located in the Robotics and IA Laboratory of the University Politehnica of Bucharest. The cell also includes two CNC milling machines and one Automatic Storage and Retrieval System, for raw material feeding and finite products storage.
During the tests, the robot network has detected a number of errors (end-effector collision with parts, communication errors, power failure, etc.) The robotic network has evaluated the particular situation, and the network was reconfigured and the abandoned applications were restarted in a time between 0.2 and 3 seconds. The network failure was also simulated during tests (Fig. 15).
One robot (R2) was disconnected from the Ethernet network, the heartbeat packet sent by the robot to the other cluster members has detected the malfunction and the robot has switched the communication using the serial line; this was done in 0.3 seconds after the Ethernet cable was removed. The communication between the affected robot and its neighbours was done using the serial lines, and the communication with other robots was done by routing the communication using the Ethernet line of the neighbours (R1 and R3). In this way the communication latency was reduced.
After the communication was re-established, the serial lines of the robot were disconnected. The robot has detected the communication failure, stopped the manipulation program and retracted the manipulator in a home position in the exterior of the working area in 0.8s.
The neighbours have sent the heartbeat packets using the serial lines and detected that they do not have any connection with the robot and announced the group leader (GL) which has

removed the robot from the cluster and reconfigured the cell (see Fig. 15). The neighbour R1 having the same working area as the affected robot R2 has loaded the values of the variables and the production program from the shared storage, and started production, continuing from the point where R2 has been stopped. The cell reconfiguration from the point where the serial lines were disconnected has taken 2.2 seconds.

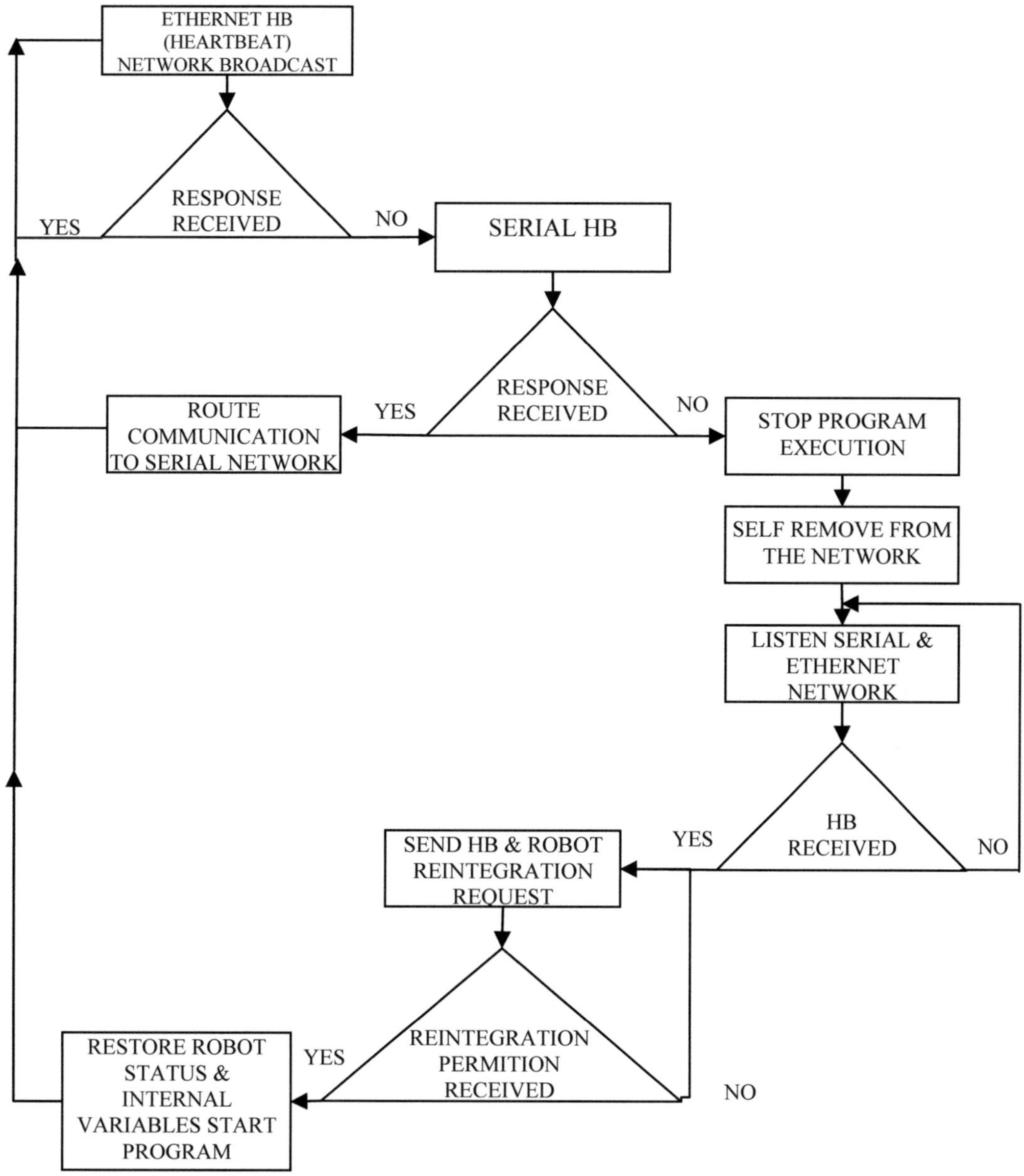

Fig. 15. Algorithm for network failure detection and reconfiguration.

Another communication test consisted in disconnecting both the serial lines and the Ethernet line at the same time, in this case the cluster tested the communication sequentially

and the cluster reconfiguration took 2.5 seconds. When configuring the cluster in order to test in parallel the communication lines, the reconfiguration took 2.3 seconds but the controllers processed a higher communication load.
The most unfavourable situation occurred when a robot manipulator was down; in this case the down time was greater because the application which was executed on that controller had to be transferred, reconfigured and restarted on another controller. Also if the controller still runs properly it will become group leader to facilitate the job of the previous GL (that has also a manipulation task to do).
The proposed solution is not entirely fault tolerant; however, in some situations it could be considered as a fault tolerant system due to the fact that even if a robot controller failed, the production continued in normal conditions.
The project is in the final stage of development and will be finished at the end of 2008. The research project will provide a portal solution for linking the existing fault tolerant pilot platform with multiple V+ industrial robot-vision controllers from Adept Technology located in different labs.

7. References

Adept Technology Inc., (2001). *AdeptVision Reference Guide Version 14.0 Part Number 00964-03000,* San Jose, Technical Publications.

Anton F., D., Borangiu, Th., Tunaru, S., Dogar, A., and S. Gheorghiu, 2006. Remote Monitoring and Control of a Robotized Fault Tolerant Workcell, *Proc. of the 12th IFAC Sympos. on Information Control Problems in Manufacturing INCOM'06,* Elsevier.

Borangiu, Th., (2004). Intelligent Image Processing in Robotics and Manufacturing, Romanian Academy Press, Bucharest.

Borangiu, Th., F.D. Anton, S. Tunaru and N.-A. Ivanescu, (2005). Integrated Vision Tools And Signature Analysis For Part Measurement And Recognition In Robotic Tasks, IFAC World Congress, Prague.

Borangiu, Th., Anton F., D., Tunaru, S., and A. Dogar, 2006. A Holonic Fault Tolerant Manufacturing Platform with Multiple Robots, Proc. of *15th Int. Workshop on Robotics in Alpe-Adria-Danube Region RAAD 2006.*

Brooks, K., R. Dhaliwal, K. O'Donnell, E. Stanton, K. Sumner and S. Van Herzele, (2004). *Lotus Domino 6.5.1 and Extended Products Integration Guide,* IBM RedBooks.

Harris, N., Armingaud, F., Belardi, M., Hunt, C., Lima, M., Malchisky Jr., W., Ruibal, J., R. and J. Taylor, 2004. *Linux Handbook: A guide to IBM Linux Solutions and Resources,* IBM Int. Technical Support Organization, 2nd Edition.

Lascu, O. et al, 2005. *Implementing High Availability Cluster Multi-Processing (HACMP) Cookbook,* IBM Int. Technical Support Organization, 1st Edition.

Lascu, O., Sayed, R., Carroll, S., Coleman, T., Haehnel, M., Klabenes, P., Quintero, D., Reyes, R., Jr., Tanaka, M., Truong, D., D., 2002. *An Introduction to Security in a CSM 1.3 for AIX 5L Environment,* IBM Int. Technical Support Organization, 1st Edition.

Matsubara, K., Blanchard, B., Nutt, P., Tokuyama, M., and T. Niijima, 2002. *A practical guide for Resource Monitoring and Control (RMC),* IBM Int. Technical Support Organization, 1st Edition.

Tomas Balibrea, L.M., L.A. Gonzales Contreras and M. Manu (1997). *Object Oriented Model of an Open Communication Architecture for Flexible Manufacturing Control,* Computer Science 1333 - Computer Aided System Theory, pp. 292-300, EUROCAST '97, Berlin.

5

Fault Diagnosis in Discrete Event Systems using Interpreted Petri Nets

Jesús Arámburo-Lizárraga, Antonio Ramírez-Treviño,
Ernesto López-Mellado and Elvia Ruiz-Beltrán
CINVESTAV Unidad Guadalajara
México

1. Introduction

Diagnosability property and fault detection schemes have been widely addressed on centralized approaches using the global model of the *Discrete Event System* (*DES*). Roughly speaking, diagnosability is the property of determining if using the system model is possible to detect and locate the faulty states in a finite number of steps. In the works (Sampath, et al., 1995) and (Sampath, et al., 1996), a method for modeling a *DES* using finite automata is proposed; based on this model, a diagnoser is derived. The cycles in the diagnoser are used to determine when the *DES* is diagnosable.

Recently, fault diagnosis of *DES* has been addressed through a distributed approach allowing breaking down the complexity when dealing with large and complex systems (Benveniste, et al., 2003; O. Contant, et al., 2004; Debouk, et al., 2000; Genc & Lafortune, 2003; Jiroveanu & Boel, 2003; Pencolé, 2004; Arámburo-Lizárraga, et al., 2005).

In (Debouk, et al., 2000) it is proposed a decentralized and modular approach to perform failure diagnosis based on Sampath's results (Sampath, et al., 1995). In (Contant, et al., 2004) and (Pencolé, 2004) the authors presented incremental algorithms to perform diagnosability analysis based on (Sampath, et al., 1995) in a distributed way; they consider systems whose components evolve by the occurrence of events; the parallel composition leads to a complete system model intractable. In (Genc & Lafortune, 2003) it is proposed a method that handles the reachability graph of the *PN* model in order to perform the analysis similarly to (Sampath, et al., 1995); based on design considerations the model is partitioned into two labelled *PN* and it is proven that the distributed diagnosis is equivalent to the centralized diagnosis; later, (Genc & Lafortune, 2005) extend the results to systems modeled by several labelled *PN* that share places, and present an algorithm to determine distributed diagnosis. In (Qiu & Kumar, 2005) it is studied the codiagnosability property, this property guarantees that any faults occurred in the system must be detected by at least one local diagnoser in a finite number of steps using the local information, besides, a notion of safe-codiagnosability is mentioned to capture the fact that the system has a safe specification while the system performance is tolerable. (Arámburo-Lizárraga, et al., 2005) proposes a methodology for designing reduced diagnosers and presents an algorithm to split a global model into a set of communicating sub-models for building distributed diagnosers. The diagnosers handle a system sub-model and every diagnoser has a set of communication events for detecting and locating the faults of the corresponding sub-model. (Arámburo-Lizárraga, et al., 2007) shows how to design low interaction distributed diagnosers reducing the communication among them and proposes a redundant distributed diagnoser scheme composed by a set of independent modules handling two kinds of redundancy (duplication or *TMR*).

This work considers the system modeled as an interpreted *PN* (*IPN*) allowing describing the system with partially observable states and events; the model includes the possible faults it may occur. In order to build such a model, this work presents a bottom-up modeling methodology in which the behavior of the system elements is decomposed into state variables; a range for each state variable must be settled. These ranges represent the possible values of state variables. Afterwards, these rages are coded into *IPN* (modules), where each value is represented by a different place. Then two composition operators for joining modules are used; the first one is named synchronic composition, which merges transitions according to certain rules. It is similar to the synchronic product presented in (Giua & DiCesare, 1994); the second one is named permissive composition, which uses selfloops for enabling transitions among modules; this allows constraining the model behavior into the actual system behavior. This new operator avoids the use of tuning phases of other modeling methods used in supervisory control (Giua & DiCesare, 1994). Based on the derived *IPN* model with the proposed methodology, the diagnosability property for *IPN* models is introduced. Roughly speaking, this property says that an *IPN* is diagnosable if it is possible to know both, when a faulty place is marked and which faulty place is marked. This property is closely related to the observability property (Aguirre-Salas, et al., 2002) and (Ramírez-Treviño, et al., 2003) and polynomial algorithms to test when an *IPN* is diagnosable are derived, avoiding the reachability analysis of other approaches. Also, a distributed diagnoser is presented; every distributed diagnoser uses the local information or communication among diagnosers for detecting and locating a system fault. The diagnosability property is preserved in the distributed architecture. Redundancy techniques could be applied to the distributed diagnosers to detect and locate a malfunction in the distributed diagnosers set.

The chapter is organized as follows: section 2 provides basic definitions of *PN*, *IPN* and the modeling methodology are presented. In section 3 the property of input-output diagnosability is defined and characterized, a diagnoser scheme devoted to detect and isolate failure states is also presented. Section 4 presents a procedure to build a reduced *IPN* model. Section 5 describes a method for model decomposition allowing interaction distributed diagnosers, also, it is presented a redundant scheme for reliable diagnosis applying redundancy to the distributed diagnosers. Finally, conclusions are given.

2. Basic notations and system modeling

2.1 Petri net basics

We consider systems modeled by Petri Nets and Interpreted Petri Nets. A Petri Net structure is a graph $G = (P, T, I, O)$ where: $P = \{p_1, p_2, ..., p_n\}$ and $T = \{t_1, t_2, ..., t_m\}$ are finite sets of nodes called respectively places and transitions, I (O): $P \times T \rightarrow \mathbb{Z}^+$ is a function representing the weighted arcs going from places to transitions (transitions to places), where $\mathbb{Z}^+$ is the set of nonnegative integers.

The symbol ${}^\bullet t_j$ $(t_j^\bullet)$ denotes the set of all places p_i such that $I(p_i,t_j)\neq 0$ $(O(p_i,t_j)\neq 0)$. Analogously, ${}^\bullet p_i$ $(p_i^\bullet)$ denotes the set of all transitions t_j such that $O(p_i,t_j)\neq 0$ $(I(p_i,t_j)\neq 0)$ and the incidence matrix of G is $C=[c_{ij}]$, where $c_{ij} = O(p_i,t_j) - I(p_i,t_j)$.

A marking function M: $P \rightarrow \mathbb{Z}^+$ represents the number of tokens (depicted as dots) residing inside each place. The marking of a *PN* is usually expressed as an n-entry vector.
A Petri Net system or Petri Net (*PN*) is the pair $N=(G,M_0)$, where G is a *PN* structure and M_0 is an initial token distribution. $R(G,M_0)$ is the set of all possible reachable markings from M_0 firing only enabled transitions. In a *PN* system, a transition t_j is enabled at marking M_k if $\forall p_i \in P$, $M_k(p_i) \geq I(p_i,t_j)$; an enabled transition t_j can be fired reaching a new marking M_{k+1} which can be computed as $M_{k+1} = M_k + Cv_k$, where $v_k(i)=0$, $i \neq j$, $v_k(j)=1$.
Interpreted Petri Nets (*IPN*) (Ramírez-Treviño, et al., 2003) is an extension to *PN* that allow to associate input and output signals to *PN* models. An *IPN* (Q, M_0) is an Interpreted Petri Net structure where $Q = (G, \Sigma, \lambda, \varphi)$ with an initial marking M_0, G is a *PN* structure, $\Sigma = \{a_1, a_2, \ldots, a_r\}$ is the input alphabet of the net, where a_i is an input symbol, λ: $T \rightarrow \Sigma \cup \{\varepsilon\}$ is a transition labelling function with the following constraint: $\forall t_j, t_k \in T$, $j \neq k$, if $\forall p_i$ $I(p_i,t_j) = I(p_i,t_k) \neq 0$ and both $\lambda(t_j) \neq \varepsilon$, $\lambda(t_k) \neq \varepsilon$, then $\lambda(t_j) \neq \lambda(t_k)$; ε represents an internal system event, and φ: $R(Q,M_0) \rightarrow (\mathbb{Z}^+)^q$ is an output function that associates to each marking an output vector, where q is the number of outputs. In this work φ is a $q \times n$ matrix. If the output symbol i is present (turned on) every time that $M(p_j) \geq 1$, then $\varphi(i,j)=1$, otherwise $\varphi(i,j)=0$.
A transition $t_j \in T$ of an *IPN* is enabled at marking M_k if $\forall p_i \in P$, $M_k(p_i) \geq I(p_i,t_j)$. If t_j is enabled at marking M_k, and $\lambda(t_j)$ is present, then t_j can be fired reaching M_{k+1}, i.e., $M_k \xrightarrow{t_j} M_{k+1}$; M_{k+1} can be computed using the state equation:

$$\begin{aligned} M_{k+1} &= M_k + Cv_k \\ y_k &= \varphi(M_k) \end{aligned} \tag{1}$$

where C and v_k are defined as in *PN* and $y_k \in (\mathbb{Z}^+)^q$ is the k-th output vector of the *IPN*.
The sequence $\sigma = t_i t_j \ldots t_k \ldots$ is a firing transition sequence of an *IPN* (Q,M_0) if $M_0 \xrightarrow{t_i} M_1 \xrightarrow{t_j} \ldots M_x \xrightarrow{t_k} \ldots$ According to functions λ and φ, transitions and places of an *IPN* (Q,M_0) are classified. If $\lambda(t_i) \neq \varepsilon$ the transition t_i is said to be manipulated. Otherwise it is non-manipulated. A place $p_i \in P$ is said to be measurable if the i-th column vector of φ is not null, i.e. $\varphi(\bullet,i) \neq 0$. Otherwise it is non-measurable. The following concepts are useful in the study of the diagnosability property.
The set $\pounds(Q,M_0) = \{\sigma = t_i t_j \ldots t_k \ldots \wedge M_0 \xrightarrow{t_i} M_1 \xrightarrow{t_j} \ldots M_x \xrightarrow{t_k} \ldots\}$ of all firing transition sequences is called the firing language of (Q,M_0).
A sequence of input-output symbols of (Q,M_0) is a sequence $\omega = (a_0,y_0)(a_1,y_1)\ldots(a_n,y_n)$, where $a_j \in \Sigma \cup \{\varepsilon\}$. The symbol a_{i+1} is the current *IPN* input when the output changes from y_i to y_{i+1}. It is assumed that $a_0 = \varepsilon$ and $y_0 = \varphi(M_0)$.
The firing transition sequence $\sigma \in \pounds(Q,M_0)$ whose firing actually generates ω is denoted by σ_ω. The set of all possible firing transition sequences that could generate the word ω is defined as $\Omega(\omega) = \{\sigma \mid \sigma \in \pounds(Q,M_0) \wedge \text{the firing of } \sigma \text{ produces } \omega\}$.
The set $\Lambda(Q,M_0) = \{\omega \mid \omega \text{ is a sequence of input-output symbols}\}$ denotes the set of all sequences of input-output symbols of (Q,M_0) and the set of all input-output sequences of length greater or equal than k will be denoted by $\Lambda^k(Q,M_0)$, i.e. $\Lambda^k(Q,M_0) = \{\omega \in \Lambda(Q,M_0) \mid |\omega| \geq k\}$ where $k \in \mathbb{N}$.

The set $\Lambda_B(Q,M_0) = \{\omega \in \Lambda(Q,M_0) \mid \sigma\in\Omega(\omega)$ such that $M_0 \xrightarrow{\sigma} M_j$ and M_j enables no transition, or when $M_j \xrightarrow{t_i}$ then $C(\bullet,t_i)=\vec{0}$ } denotes all input-output sequences leading to an ending marking in the *IPN* (markings enabling no transition or only self-loop transitions).

An *IPN* (Q, M_0) described by the state equation (1) is event-detectable iff the firing of any pair of transition t_i, t_j of (Q, M_0) can be distinguished from each other by the observation of the sequences of input-output symbols.

The following lemma (Rivera-Rangel, et al., 2005) gives a polynomial characterisation of event-detectable *IPN*.

Lemma 1: A live *IPN* given by (Q,M_0) is event-detectable if and only if:

1. $\forall t_i, t_j \in T$ such that $\lambda(t_i) = \lambda(t_j)$ or $\lambda(t_i) = \varepsilon$ it holds that $\varphi C(\bullet,t_i) \neq \varphi C(\bullet,t_j)$ and
2. $\forall t_k \in T$ it holds that $\varphi C(\bullet,t_k) \neq 0$.

2.2 System modeling

We work with the modeling methodology proposed in (Ramírez-Treviño, et.al.,2007). The methodology follows a modular bottom-up strategy. After identifying the system components, a set of state variables is assigned to every component, each state variable behavior is modeled by a *PN* model, herein named module. Then the set of modules are integrated into a single model according to the appropriate relationships achieved through two module composition operations. This methodology builds binary *IPN* modules to represent the behavior of each component of the identified *DES* and the relationships between them. The model captures the normal and faulty behavior of the individual components of the system. Below it is described the detailed steps of the methodology.

2.2.1 Modeling methodology

1. System components.- The system components must be identified and named. Afterwards, a finite set System_Components= $\{sc_1, sc_2,\ldots, sc_n\}$ of these names must be created. A system component could be a valve, a motor, a system resource, etc.
2. State variables.- For each system component, the different variables needed to represent its behavior must be chosen. In other words, the finite set State_Variables$_i$= $\{sv^i_1, sv^i_2,\ldots, sv^i_m\}$ associated to the system component $sc_i\in$ System_Components must be built. These variables could represent the position (for instance a valve position), velocity, voltage, etc. of each system component, or could represent a task descriptor (for instance a machine state). There exists at least one state variable for each system component.
3. Set of values.- For each state variable $sv^i_j \in$ State_Variables$_i$, the set $Value_{sv^i_j} = \{val_1^{ij}, val_2^{ij},\ldots, val_p^{ij}\}$ of possible values of sv^i_j must be stated. Necessary faulty values should also be considered in this set. For instance, the variable "valve_position" may take four values: "Open", "Closed", "ErrorOnOpen" and "ErrorOnClosed", or the variable "task_machine1" may take three values: "loading", "processing", and "unloading".
4. Codification.- The values in each set $Value_{sv^i_j}$ must be represented in terms of *PN* markings. This can be easily achieved if binary places are used. Thus, for each $sv^i_j \in$

State_Variables_i a set $P_{sv^i_j} = \{p^{ij}_1, p^{ij}_2, ..., p^{ij}_n\}$ of places such that $|Value_{sv^i_j}| = |P_{sv^i_j}|$ must be created. The marking of these places is binary and mutually exclusive. Then, $M(p^{ij}_z)=1$ means that the variable sv^i_j takes the value val^{ij}_z. Because of the existence of faulty values, the set of places can be partitioned into the subsets $P^F_{sv^i_j}$ and $P^N_{sv^i_j}$, representing the faulty and normal values respectively.

5. Event modeling.- For each pair of values val^{ij}_m, val^{ij}_n such that the state variable sv^i_j could change from value val^{ij}_m to a value val^{ij}_n, a transition t^{ij}_{mn} must be created. Then, one arc going from place p^{ij}_m to transition t^{ij}_{mn} and one arc going from transition t^{ij}_{mn} to place p^{ij}_n must be added.
6. Initial marking.- The initial marking is defined as: $M_0(p^{ij}_m)=1$ if the initial value of the variable sv^i_j is val^{ij}_m and $M_0(p^{ij}_m)=0$ otherwise.
7. Output.- The output of this algorithm is a set of isolated PN modules, each one modeling the behavior of a state variable sv^i_j.
8. Perform synchronous composition and permissive among modules *IPN* to obtain the global *IPN* model system (see Alcaraz-Mejía et al. 2003) and (Ramírez-Treviño et al. 2007).

In the *IPN* system model the sets of nodes are partitioned into faulty nodes (P^F, places coding faulty states, and T^F, transitions leading to faulty states) and normal functioning nodes (P^N and T^N); so $P = P^F \cup P^N$ and $T = T^F \cup T^N$. p^N_i denotes a place in P^N. Since $P^N \subseteq P$ then p^N_i also belongs to (Q, M_0). The set of risky places of (Q, M_0) is $P^R = {}^\bullet T^F$. The post-risk transition set of (Q, M_0) is $T^R = P^{R\bullet} \cap T^N$. (Q^N, M^N_0) denotes the embedded normal behavior of (Q, M_0), i.e., (Q^N, M^N_0) is the subnet induced by normal nodes.

Example 1. Consider the producer-consumer scheme depicted in figure 1.

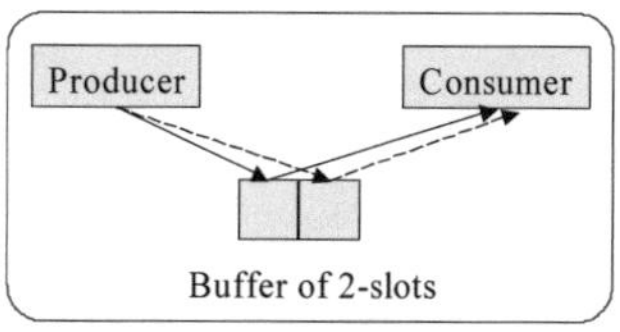

Fig. 1. Producer-Consumer with buffer of 2-slots scheme

The model consists of a producer unit (PU), a consumer unit (CU) and a buffer of 2-slots. The behavior of this system is the following. The producer unit PU creates and delivers products into the free buffer positions. The consumer unit CU retrieves products from the buffer when there is a product stored into a buffer slot. The producer unit PU could reach a faulty state from its producing state. Similarly, the consuming unit could reach a faulty state from its consuming state. We model each system component isolately (see figure 2), then the synchronous composition and permissive operations are applied to obtain the global system model of the producer-consumer scheme. In figure 3 is depicted the normal system behavior model; figure 4 includes the faulty behavior, in this case when PU stops its production and

CU stops its consumption. The places p_1, p_2, p_3 represent the normal PU behavior and p_{11} represents the faulty behavior. Places p_4, p_5, p_6 represent the normal CU behavior and p_{12} represents the faulty behavior. The places p_7, p_8, p_9 and p_{10} represent the 2- slots of the buffer. Function λ is defined as $\lambda(t_1)=a$, $\lambda(t_8)=b$; and $\lambda(t_i)=\varepsilon$ for others transitions. Measurable places are p_3, p_6, p_8, p_{10}, $P^R=\{p_3,p_6\}$, $T^R=\{t_1,t_8\}$, $T^F=\{t_9,t_{10}\}$ and $P^F=\{p_{11},p_{12}\}$.

3. Centralized diagnosability

The characterisation of input-output diagnosable *IPN* is based on the partition of $R(Q,M_0)$ into normal and faulty markings where all the faulty markings must be distinguishable from other reachable markings.

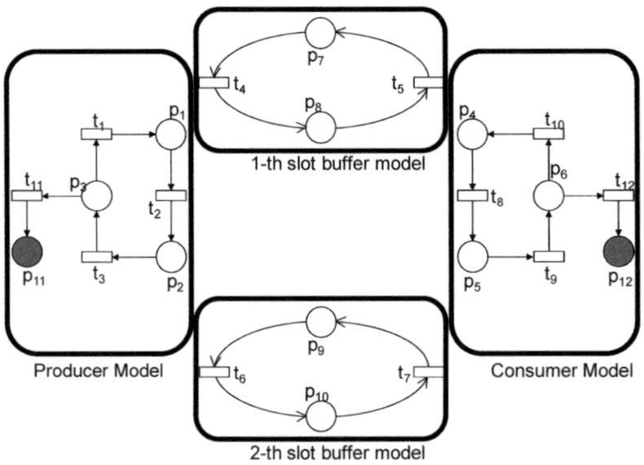

Fig. 2. *IPN* modules of the identified components

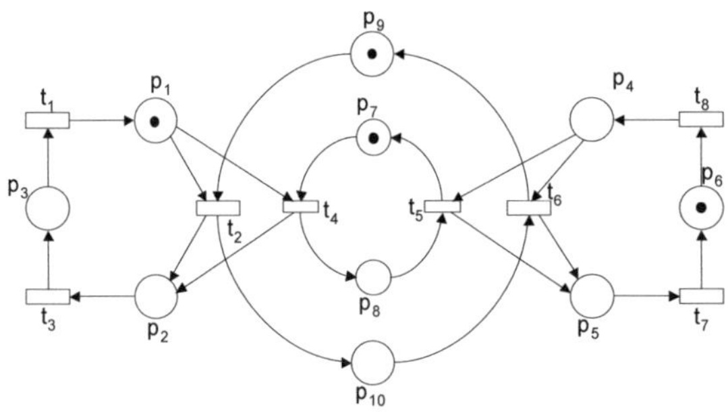

Fig. 3. Normal behavior of the *IPN* of the producer-consumer scheme

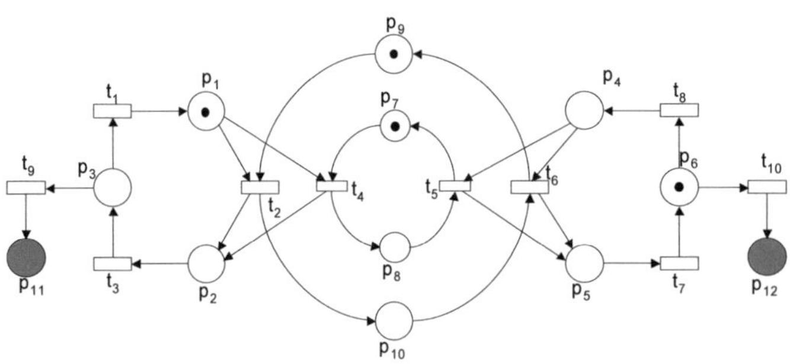

Fig. 4. Normal and Faulty behavior of the *IPN* of the producer-consumer scheme

Definition 1: An *IPN* given by (Q,M_0) is said to be input - output diagnosable in $k < \infty$ steps if any marking $M_f \in F$ is distinguishable from any other $M_k \in R(Q,M_0)$ using any word $\omega \in \Lambda^k(Q,M_f) \cup \Lambda_B(Q,M_f)$, where $F = \{M \mid \exists p_k \in P^F$ such that $M(p_k)>0, M \in R(Q,M_0)\}$.
The following result provides sufficient structural conditions for determining the input-output diagnosability of an *IPN* model.

Theorem 1: Let (Q,M_0) be a binary *IPN*, such that (Q^N, M_0^N) is live, strongly connected and event detectable. Let $\{X_1,...,X_T\}$ be the set of all T-semiflows of (Q,M_0). If $\forall\, p_i^N \in P^N$, $(p_i^N)^\bullet \cap T^F \neq \varnothing$ the following conditions hold:

1. $\forall r, \exists j\ X_r(j)\geq 1$, where $t_j \in (p_i^N)^\bullet - T^F$,
2. $\forall t_k \in (p_i^N)^\bullet - T^F, {}^\bullet(t_k)=\{\, p_i^N \}$ and $\lambda(t_k) \neq \varepsilon$.

then the *IPN* (Q,M_0) is input-output diagnosable.

Proof: Assume that (Q,M_0) meets all conditions of the theorem. Since (Q^N, M_0^N) is live, then it is live by places (see Desel & Esparza, 1995), i.e. $\forall p \in P$ there exists a marking M_k such that $M_k(p_i)>0$, previous observation is also valid for any $p_i^N \in P^N$, $(p_i^N)^\bullet \cap T^F \neq \varnothing$. Thus there exists σ_ω such that $M_0 \xrightarrow{\sigma_\omega} M_k$ and $M_k(p_i^N)=1$. The input-output symbol word generating σ_ω will be denoted by ω. Choose the longest transition firing sequence σ_ω not including transitions in T^F (could be the empty one). Such sequence must be marking a place $p_i^N \in P^N$, $(p_i^N)^\bullet \cap T^F \neq \varnothing$ since in the next step a faulty transition should be fired.
Since the initial marking is known, (Q^N, M_0^N) is event detectable, and σ_ω contains no faulty transitions, then the marking M_k such that $M_0 \xrightarrow{\sigma_\omega} M_k$, can be computed using the *IPN* state equation.
Since p_i^N has a token at M_k, then a transition $t_i^F \in (p_i^N)^\bullet \cap T^F$ could be fired reaching a faulty marking M_j. Thus, we will prove that it is possible to detect that M_j is reached.
Assume that such transition t_i^F at M_k. Thus $M_0 \xrightarrow{\sigma_\omega} M_k \xrightarrow{t_i^F} M_j$ and M_j adds a token into a place $p_j^F \in (p_i^N)^{\bullet\bullet} \cap P^F$ and removes a token from the place p_i^N. Since the use of the modeling methodology assures that $\lambda(t_i^F)=\varepsilon$ and $\varphi(M_k)=\varphi(M_j)$, then the firing of t_i^F cannot be detected using the current sequence of input-output symbols, i.e. it is not possible to determine when the faulty marking M_j was reached. In other words, the input-output symbol word ω generates both σ_ω and $\sigma_\omega t_i^F$. In order to detect when M_j is reached, we proceed as follows.
Since it holds that $\forall r, \exists j\ X_r(j)\geq 1$, $t_j \in T - T^F$, then no T-semiflow can be fired without firing a transition in $(p_i^N)^\bullet - T^F$. Moreover, since the number of transitions in (Q,M_0) is finite, then the length of a sequence attempting to fire a transition $(p_i^N)^\bullet$ should be finite. Then a transition $t_k \in (p_i^N)^\bullet \cap T^N$ must be attempted to fire after the firing of a finite sequence.
Since the current marking in the net is the faulty marking M_j, then t_k cannot be fired. Since the condition 2 indicates that $\lambda(t_k) \neq \varepsilon$, then we can detect when the symbol $\lambda(t_k)$ is given to

the system. Moreover, since ${}^{\bullet}(t_k)=\{p_i^N\}$, if t_k cannot be fired, then certainly that $M_j(p_i^N)=0$ and that $M_j(p_j^F)=1$, where $p_j^F \in (p_i^N)^{\bullet\bullet} \cap P^F$. Thus the faulty marking M_j is detected, i.e. the *IPN* is input-output diagnosable. □

3.1 Diagnosability test

Determining when an *IPN* is input-output diagnosable is reduced to the following tests.

1. Binarity of (Q,M_0). It is fulfilled because of the modeling methodology.
2. Liveness of modules can be tested efficiently; however the property is not preserved during arbitrary module composition operators. Thus some constraints in the application of synchronous and permissive compositions are introduced to guarantee liveness during composition. Such constraints may follow the rules introduced in (Koh & DiCesare, 1991) for module composition.
3. Event detectability and strongly connectedness are determined in polynomial time as well as detecting places $p_i^N \in P^N$ such that $(p_i^N)^{\bullet} \cap T^F \neq \emptyset$. Then condition 2 of previous theorem is efficiently tested.
4. Finally, condition 1 can be verified in polynomial time. In this case we need to check that there exists no T-semiflow that does not include transitions in $(p_i^N)^{\bullet}$, $p_i^N \in P^N$ such that $(p_i^N)^{\bullet} \cap T^F \neq \emptyset$. Thus we need to check that the following linear programming problems have no solutions.

$\forall p_i^N \in P^N$ such that $p_i^N \in P^N, (p_i^N)^{\bullet} \cap T^F \neq \emptyset$

$\nexists X$

s.t.

$CX=0$

$x(j)=0, \forall t_j \in (p_i^N)^{\bullet} - T^F$

There exist different strategies for constructing the diagnoser-model, we presented two different ways: a centralized and a distributed diagnoser. A centralized diagnoser model is composed by a copy of the normal system behavior of (Q,M_0). Also a reduced diagnoser model can be built. Also, this work handles a distributed diagnoser which is very useful for large and complex system besides incorporate reliability to the system diagnosis process through redundancy techniques applied to the distributed diagnoser model. The different diagnoser models are defined in the following sections.

3.2 Centralized diagnoser design

Diagnosability theorem given above provides the basis for designing an on-line diagnoser (see figure 5). Such diagnoser can detect a fault and locate faulty markings reached by an *IPN*. The proposed scheme for diagnosis (Ramírez-Treviño, et al., 2004) handles a copy of the normal behavior model which must evolve similarly to the system; the outputs of both the system and the model are compared and, when there is a difference, a procedure is started to compute the faulty marking.

Example 2. The *IPN* system model depicted in figure 3 represents the normal behavior model for the producer-consumer diagnoser. The initial marking M_0 represents that all the buffers are empty, the PU is waiting to deliver and the CU state is idle. Since this *IPN* is input-output diagnosable by theorem 1, then we can detect and locate the fault with the diagnoser. The system and its on-line diagnoser are depicted in figure 6, notice the diagnoser is a copy of the system. Now assume that the events represented by the sequence t_2t_3 are executed into the system, then the sequence t_2t_3 is fired in the diagnoser model. Thus both, the system and the diagnoser-model have the same output "producing" and "consuming". If the fault transition t_{11} is fired, then p_{11} is marked in the system and no system output change is detected. When the symbol of t_1, $\lambda(t_1)$ is given as input to the system, then the diagnoser-model evolves and its new output is "consuming", however the output of the system continues in "producing" and "consuming". Then the difference is the signal "producing"; thus the algorithm determines that p_{11} is marked, thus the fault is isolated.

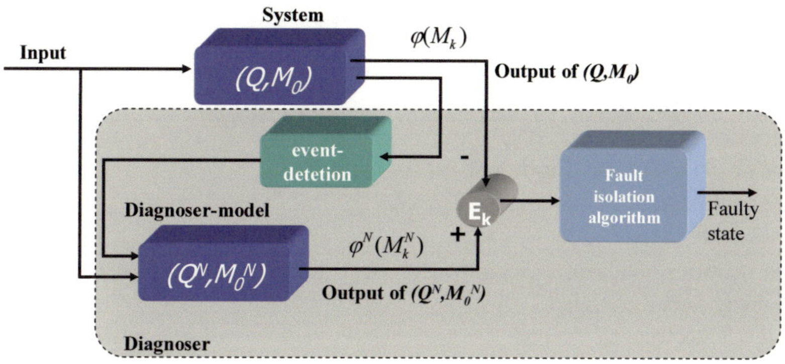

Fig. 5. On-line diagnose scheme

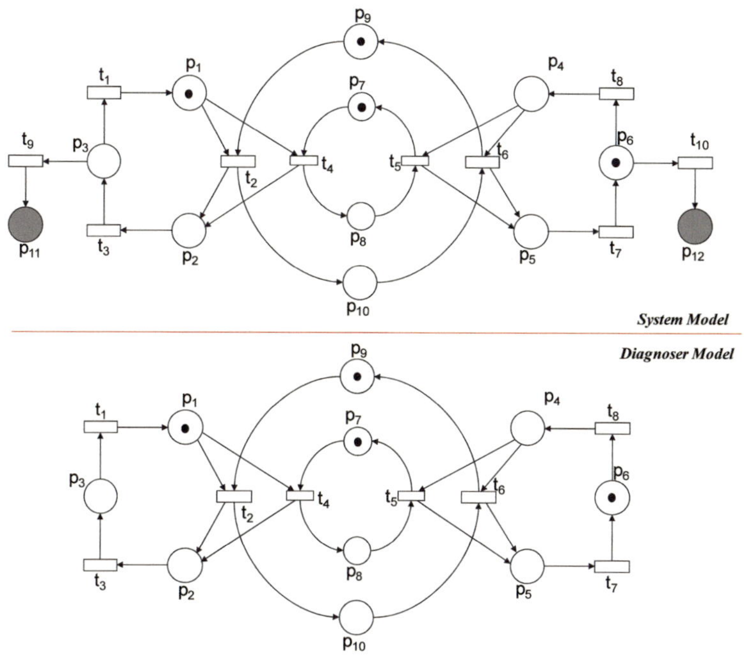

Fig. 6. On-line diagnoser model

4. Reduced diagnoser

4.1 Diagnoser model

Definition 2: The proposed diagnoser model structure for the system normal behavior (Q^N, M_0^N) is an *IPN* (Q^d, M_0^d) where the set of places is $P^d = \{p_d\}$ and the set of transitions is $T^d = T^N$, the incidence matrix C^d of (Q^d, M_0^d) is the following

$$C^d = B^T \varphi^N C^N \quad (2)$$

where B^T is a q×1 vector (q is the number of measurable places of $((Q^N, M_0^N)))$, φ^N is the output matrix of (Q^N, M_0^N), C^N is the incidence matrix of (Q^N, M_0^N).
The initial marking of the diagnoser model structure for one place is computed as:

$$M_0^d = B^T \varphi^N \left(M_0^N\right) \quad (3)$$

We propose a matrix B that is computed as follows:
Algorithm 1: Building B

Inputs: C-incidence matrix of an *IPN*,
 l - number of places in the diagnoser-model,
 q - number of measurable places in the *IPN*,
Outputs: A matrix B

1. The "base number" b should be computed. In this case $b=2max(abs(c_{ij}))+1$, where c_{ij} is an element of incidence matrix C.
2. Define a q × 1 vector.
3. $\begin{bmatrix} b^0 & b^1 & \cdots & b^{q-1} \end{bmatrix}$

This procedure computes matrix B.
According to the way in which B was constructed, all columns of C^d must different from zero and different from each other.
If a transition $t_i \in T - (T^R \cup T^F)$ is fired in (Q,M_0) then it is fired in (Q^d, M_0^d) (it is possible since these transitions are event detectable, thus the output system information is enough to detect when one of these transition is fired).
If a transition $t_j \in T^R$ is enabled in (Q_d, M_0^d) and $\lambda(t_j)$ is activated in (Q,M_0), then t_j must be fired in (Q^d, M_0^d). Thus, if t_j is not fired in (Q,M_0), then (Q,M_0) reached a faulty marking. In this case the output of the system and the output of the diagnoser are different from each other.

4.2 Error computation and fault isolation

Definition 3. Error computation. The k-th error is computed by the following equation:

$$e_k = M_k^d - B^T(\varphi M_k) \quad (5)$$

Notice that e_k is computed from the diagnoser-model output and not from the marking M_k. It means that the proposed diagnoser is using the system output and not internal system signals (those signals that are non measurable).
Definition 4. Fault isolation. When $e_k \neq 0$, an error is detected, then a faulty marking was reached. The mechanism used to find out the faulty marking is named fault isolation. This work proposes the following algorithm to accomplish this task.

Algorithm 2: Fault isolation

Inputs: M_k , M_k^d , e_k.
Outputs: p(faulty place), M_f(faulty marking)
Constants: C^d is the IPN diagnoser structure incidence matrix
i = index of the column of C^d such that $C^d(1,i)= e_k$

- $\forall p \in \cdot t_i, M_k(p)=0$
- $\forall p \in t_i \cdot, M_k(p)=0$
- $\forall p^F \in (^\bullet t_i)^{\bullet\bullet} \cap P^F, M_k(p^F)=1$
- $M_f = M_k$
- Return (p, M_f)

Example 3. Consider the system of the example 1. The *IPN* depicted in figure 4 represents the behavior of the system of example 1. Since this *IPN* is input-output diagnosable by theorem 1, then a diagnoser can be built for this system. In this case we will use the reduced structure presented in this section.

$$C^N = \begin{bmatrix} 1 & -1 & 0 & -1 & 0 & 0 & 0 & 0 \\ 0 & 1 & -1 & 1 & 0 & 0 & 0 & 0 \\ -1 & 0 & 1 & 0 & 0 & 0 & 0 & 0 \\ 0 & 0 & 0 & 0 & -1 & -1 & 0 & 1 \\ 0 & 0 & 0 & 0 & 1 & 1 & -1 & 0 \\ 0 & 0 & 0 & 0 & 0 & 0 & 1 & -1 \\ 0 & 0 & 0 & -1 & 1 & 0 & 0 & 0 \\ 0 & 0 & 0 & 1 & -1 & 0 & 0 & 0 \\ 0 & -1 & 0 & 0 & 0 & 1 & 0 & 0 \\ 0 & 1 & 0 & 0 & 0 & -1 & 0 & 0 \end{bmatrix}; \phi = \begin{bmatrix} 0 & 0 & 1 & 0 & 0 & 0 & 0 & 0 & 0 & 0 \\ 0 & 0 & 0 & 0 & 0 & 1 & 0 & 0 & 0 & 0 \\ 0 & 0 & 0 & 0 & 0 & 0 & 0 & 1 & 0 & 0 \\ 0 & 0 & 0 & 0 & 0 & 0 & 0 & 0 & 0 & 1 \end{bmatrix}$$

The base obtained to compute B is b=2*1+1=3; since we build B using algorithm 9. We obtain the following vector:

$$B^T = \begin{bmatrix} 1 & 3 & 9 & 27 \end{bmatrix}^T$$

Therefore C^d *is:*

$$C^d = \begin{bmatrix} -1 & 27 & 1 & 9 & -9 & -27 & 3 & -3 \end{bmatrix}$$

Hence, its associated *IPN* is depicted in figure 7.

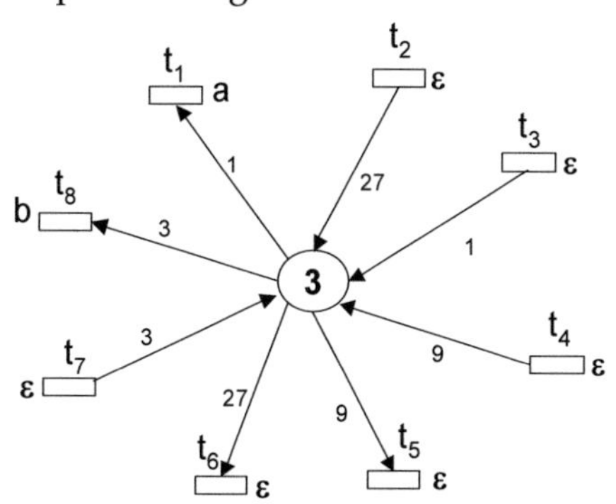

Fig. 7. The *IPN* reduced diagnoser-model

The initial marking of the diagnoser is M_0^d =[3]. In order to show how the diagnoser works, assume that the following sequence t_2t_3 is executed into the system, then this sequence is fired in the diagnoser. Thus the system output is $(\varphi(M_k))^T = [1 \;\; 1 \;\; 0 \;\; 1]$, and the marking of the *IPN* diagnoser is M_k^d =[3 1]. Then $e_k = M_k^d - (B^T)(\varphi M_k)$ = [31] - [31] = 0, thus the system is in a normal state. Now if the faulty transition t_9 is fired, then p_{11} is marked, however no change in the output system is detected. If the symbol of t_1 ($\lambda(t_1)$=a) is given as input to the *IPN* of the system model and *IPN* diagnoser, then the diagnoser evolves, and $M_{k+1}{}^d$ =[30]. Then e_k = –1 indicating the existence of an error. The fault isolation algorithm (algorithm 2) detects that the column 1 of C^d is equal to e_k, thus t_1 was not fired in the system. Then the same algorithm detects the faulty marking and determines that the faulty place p_{11} is marked.

5. Distributed diagnoser

5.1 Model distribution

In order to build a distributed diagnoser, the *IPN* model (Q, M_0) can be conveniently decomposed into *m* interacting modules where different modules share nodes (transitions and/or places).

Definition 5. Let (Q,M_0) be an *IPN*. A module μ_k = (N_k, Σ_k, λ_k, φ_k) is an *IPN* subnet of the global model (Q,M_0), where:

- $N_k = (T_k, P_k, I_k, O_k, I_k^C, O_k^C, M_{0k})$ where:
  - $T_k \subseteq T$,
  - $P_k = P_k^L \cup P_k^C$; $P_k^L \subseteq P$; P_k^C *represents the communication places among modules; this set is a copy of some places* P_l^L *that belongs to other modules,* $l \neq k$. P_k^C *is the minimal places that is required for the transitions of the module are event-detectable.* $M(P_k^C) = M(P_l^L)$.
  - $I_k(O_k)$: $P_k^L \times T_k \rightarrow Z^+$, s.t., $I_k(p_i,t_j) = I(p_i,t_j)$ $(O_k(p_i,t_j) = O(p_i,t_j))$, $\forall p_i \in P_k^L$ and $\forall t_j \in T_k$.
  - $I_k^C(O_k^C)$: $P_k^C \times T_l \rightarrow Z^+$, s.t., $I_k^C(p_i,t_j) = I(p_i,t_j)$ $(O_k^C(p_i,t_j) = O(p_i,t_j))$, $\forall p_i \in P_k^C$ and $\forall t_j \in T_l$, $l \neq k$. $I_k^C(O_k^C)$ *are the input(output) arcs to from* $p_i \in P_k^C$ *to transitions of other modules.*
  - $M_{0k} = M_0 \mid_{Pk}$
- $\Sigma_k = \{\alpha \in \Sigma \mid \exists t_i, t_i \in T_k, \lambda(t_i) = \alpha\}$
- $\lambda_k : T_k \rightarrow \Sigma_k \cup \{\varepsilon\}$, s.t. $\lambda_k(t_i) = \lambda(t_i)$ and $t_i \in T_k$
- $\varphi_k : R(\mu_k, M_{0k}) \rightarrow (Z^+)^{qk}$, q_k is restricted to the outputs associated to P_k.

Definition 6. Let (Q,M_0) be an *IPN*. A distribution DN_i of (Q,M_0) is a finite set of *m modules,* i.e., $DN_i=\{\mu_1, \mu_2,\ldots,\mu_m\}$. The distribution DN_i holds the following conditions:

1. $\bigcap_{k=1}^{m} P_k^L = \phi; \bigcup_{k=1}^{m} P_k^L = P$ (It is a partition over P_k^L).
2. $\bigcup_{k=1}^{m} T_k = T$ (It is not a partition)

The set of communication places $P_{com} = \bigcup_{k=1}^{m} P_k^C$ of a distribution represents the measurable places of each module $\mu_k \in DN_i$ needed to guarantee the event-detectability property of μ_k.

Definition 7. Let DN_i be a distribution of (Q,M_0) and $\sigma=t_it_j\ldots t_k\ldots$ be a sequence of £(Q,M_0). We define the natural projection PT_k of £(Q,M_0) over the languages of the modules $\mu_k \in DN_i$ of the following way:

$$PT_k: £(Q,M_0) \rightarrow £(\mu_k,M_{0k})$$

$$\forall t_i t_j \ldots t_s t_q \in £(Q,M_0)$$

$$PT_k(\varepsilon) = \varepsilon$$
$$PT_k(t_i t_j \ldots t_s t_q) = \begin{cases} PT_k(t_i t_j \ldots t_s) & \text{if } t_q \notin T_k \\ PT_k(t_i t_j \ldots t_s) t_q & \text{if } t_q \in T_k \end{cases}$$

There exists the input-output symbol projection over the input and output module symbols. Let $\omega = (a_0,y_0)(a_1,y_1)\ldots(a_n,y_n)$ be a sequence of symbols of (Q,M_0)
$P\Lambda_{\mu k}(\omega) = \left(\left(P_{IN\,k}\alpha_0, P_{OUT\,k}y_o\right), \left(P_{IN\,k}\alpha_1, P_{OUT\,k}y_1\right) \ldots \left(P_{IN\,k}\alpha_n, P_{OUT\,k}y_n\right)\right)$ where:

$$P_{INk}(\alpha_i) = \begin{cases} \varepsilon & \text{if } \alpha_i \notin \Sigma_k \\ \alpha_i & \text{if } \alpha_i \in \Sigma_k \end{cases} \quad \text{and } P_{OUTk}\left(y_i = \begin{bmatrix} y_i(1) \\ \vdots \\ y_i(q) \end{bmatrix} \right) = \begin{bmatrix} y_i'(1) \\ \vdots \\ y_i'(q) \end{bmatrix}$$

where $y_i'(s) = 0$ if the measurable place does not belong to μ_k

where $y_i'(s) = y_i(s)$ otherwise.

Example 4. Consider the *IPN* system model depicted in the figure 4. We partition the producer-consumer model to obtain a distributed model. The figure 8 depictes a distribution DN_i. For the sake of simplicity, we use in the example the same names for duplicated nodes (places or transitions) belonging to different modules. The distribution has three modules, i.e., $|DN_i| = 3$; $I_k^C(O_k^C)$ is represented by the dashed arcs. The module μ_1 has the transitions $T_1 = \{t_1,t_2,t_3,t_9\}$ and the place set $P_1 = \{p_1,p_2,p_3,p_{11}\} \cup \{p_{10}\}$. We are preserving the property of event detectability of common transitions duplicating measurable places, and using these copies in different modules.

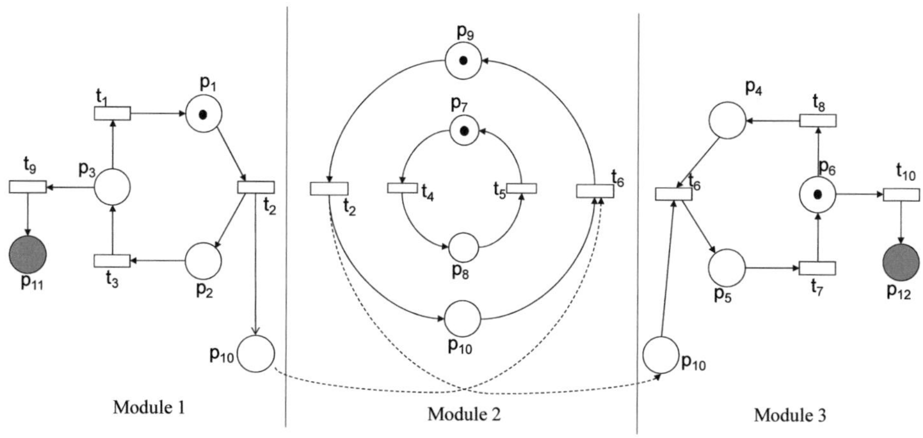

Fig. 8. Distributed Interpreted Petri Net Models

5.2 Distributed input-output diagnosability

The results of centralized diagnosability are applied to the modules.

A module is locally diagnosable if, for every local fault we can detect it only using local information, else it is conditionally diagnosable.

Definition 8. (Local Diagnosability) A module $\mu_n \in DN_i$ is said to be locally input-output diagnosable in $k < \infty$ steps if any faulty marking $M_{fn} \in R(\mu_n, M_{0n})$ is distinguishable from any other $M_{kn} \in R(\mu_n, M_{0n})$ using words ω_n, where $P\Lambda_{\mu n}(\omega) = \omega_n$.

Definition 9. (Conditional Diagnosability) A module $\mu_n \in DN_i$ is said to be conditional input-output diagnosable in $k < \infty$ steps if any faulty marking $M_{fn} \in R(\mu_n, M_{0n})$ is distinguishable from any other $M_{kn} \in R(\mu_n, M_{0n})$ using words ω_n and ω_m, where $P\Lambda_{\mu n}(\omega) = \omega_n$ and $P\Lambda_{\mu m}(\omega) = \omega_m$, ω_m denotes the set of all input-output sequences that lead to a marking a duplicate place in μ_m, where $\mu_n \neq \mu_m$.

5.3 Communication channels

Definition 10. The border transitions between two modules μ_k and μ_z is: $T^{kz}_{border} = \{t_i \mid t_i \in \{T_k \cap T_z\}\}$. This concept can be extended to several modules.
The communication channels between two modules are represented by $I_k{}^C(O_k{}^C)$. We assume that every module can communicate with every other module. The firing of a transition $t_j \in T_k$ may be local to the module μ_k and cause only a local marking change, or it may involve communication with another module if $t_j \in T^{kz}_{border}$. Every time a transition $t_j \in T^{kz}_{border}$ is fired in μ_k then a message is sent to every μ_z containing the same $t_j \in T^{kz}_{border}$ to put a token in some place $p_x \in P_z{}^C$ such that $p_x \in {}^{\bullet}t_j$. In (Lampson, 1993) it is proposed different ways to create protocols for implementing reliable messages.

5.4 Redundant diagnoser

Since distributed diagnosers leads to the use of several computers (*CPU*), then, redundancy can be introduced in the diagnosers. For instance, the Triple Modular Redundancy (*TMR*) can be used in this case.
Assume that a distribution $DN_i=\{\mu_1, \mu_2, \ldots, \mu_m\}$ was obtained and that it was distributed over m computers (see figure 9).

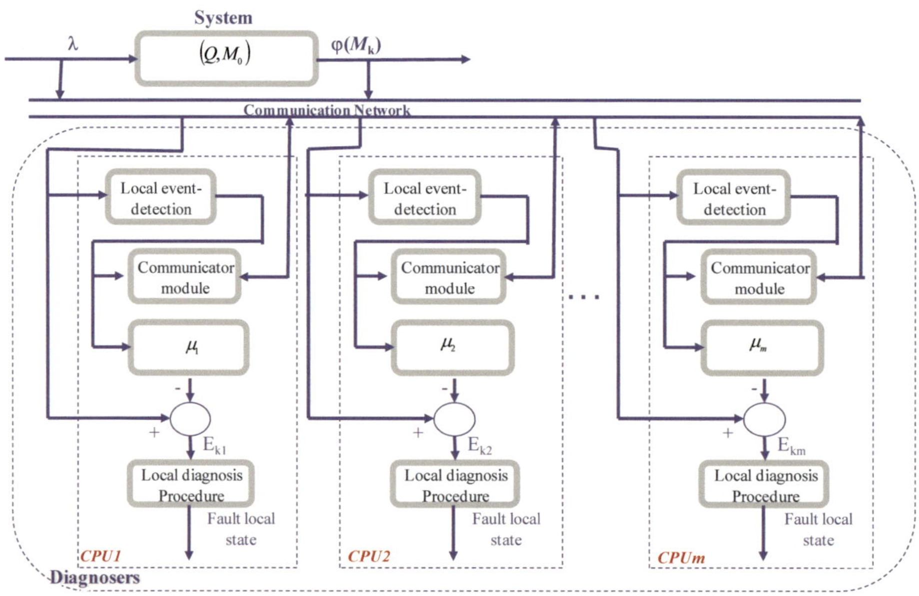

Fig. 9. On-Line Distributed Diagnoser

Then, a *TRM* scheme can be also applied to figure 9 (see figure 10), increasing the diagnoser reliability.

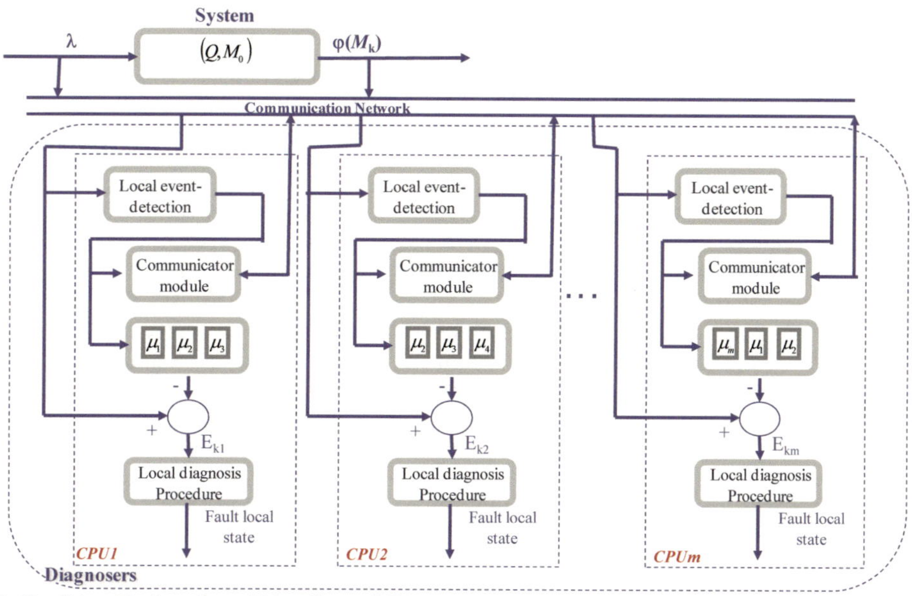

Fig. 10. On-Line Redundant Distributed Diagnoser

6. Conclusions

This chapter introduced the diagnosability property in *DES* modeled using *IPN*. It presented a structural characterization of this property using the T-semiflows of the *IPN*.

The approach herein presented exploits the *IPN* structure to determine when it is diagnosable, this approach leads to polynomial characterization of diagnosability.

Based on the *DES* model, three different types of diagnosers were presented. The first one was a centralized version, allowing to detect and locate faults. Sometimes, however, the system could very large; leading to large diagnoser models thus the other two diagnosers are designed to tackle this problem.

The second diagnoser is a reduced scheme. It uses one place; however the number of tokens could be large. The third diagnoser is a distributed one, were the diagnoser model is distributed over different computers. Adopting this approach the problems that appear in centralized versions are eliminated. Moreover, in this last case, a redundant diagnoser can be used for increasing the reliability of the distributed diagnosers.

7. References

Aguirre-Salas L., O. Begovich, A. Ramírez-Treviño (2002). "Observability in Interpreted Petri Nets using Sequence Invariants". Proc. of the IEEE Control and Decision Conference, pp. 3602-3607, Las Vegas, USA. December 2002.

Alcaraz-Mejía M., E. López-Mellado, Antonio Ramírez-Treviño, I. Rivera-Rangel (2003). "Petri Net Based Fault Diagnosis of Discrete Event Systems", Proc. of the IEEE International Conference on Systems, Man and Cybernetics. pp. 4730-4735, October 2003.

Arámburo-Lizárraga J., E. López-Mellado and A. Ramírez-Treviño (2005). "Distributed Fault Diagnosis using Petri Net Reduced Models". *Proc. of the IEEE Int. Conf. on Systems, Man and Cybernetics*. pp. 702-707, October 2005.

Arámburo-Lizárraga J., E. López-Mellado, and A. Ramírez-Treviño (2007). "Design of Low Interaction Distributed Diagnosers for Discrete Event Systems". Proc. of the 4th International Conference on Informatics in Control, Automation and Robotics. pp. 189-194. Angers, France. May 2007.

Arámburo-Lizárraga J., E. López-Mellado, A. Ramírez-Treviño and E. Ruiz-Betlrán (2007). "Reliable Distributed Fault Diagnosis using Redundant Diagnosers". Proc. of 1st IFAC Workshop on Dependable Control of Discrete Systems (DCDS'07). Cachan, France. June, 2007.

Benveniste A., S. Haar, E. Fabre and C. Jara (2003). "Distributed and Asynchronous Discrete Event Systems Diagnosis". *42nd IEEE Conference on Decision and Control.* 2003.

Contant O, S. Lafortune and D. Teneketzis (2004). "Diagnosis of modular discrete event systems". *7th Int. Workshop on Discrete Event Systems* Reims, France. September, 2004.

Debouk R, S. Lafortune and D. Teneketzis (2000). "Coordinated Decentralized Protocols for Failure Diagnosis of Discrete Event Systems", Kluwer Academic Publishers, Discrete Event Systems: Theory and Applications, vol. 10, pp. 33-79, 2000.

Desel J. and J. Esparza(1995). "Free Choice Petri Nets". Cambridge University Press, 1995.

Genc S. and S. Lafortune (2003). "Distributed Diagnosis of Discrete-Event Systems Using Petri Nets" *Proc. of the 24th. ATPN* pp. 316 - 336, June, 2003.

Genc S. and S. Lafortune (2005). "A Distributed Algorithm for On-Line Diagnosis of Place-Bordered Nets". 16th IFAC World Congress, Praha, Czech Republic, July 04-08, 2005.

Giua A. and F. DiCesare. "Petri Net Structural Analysis for Supervisory Control". IEEE Transactions on Robotics and Automation, Vol.10, No.2, pp. 185-195, April 1994.

Jalote P. (1994). "Fault Tolerance in distributed systems". Prentice Hall. 1994

Jiroveanu G. and R. K. Boel (2003). "A Distributed Approach for Fault Detection and Diagnosis based on Time Petri Nets". *Proc. of CESA.* Lille, France, July 2003.

Koh J, F. DiCesare(1991). "Transformation Methods for Generalized Petri Nets and Their Applications to Flexible Manufacturing Systems". IEEE Transactions on SMC. vol. 21, no. 6, pp. 1512-1522. 1991

Lampson B.W. Reliable Messages (1993). In S. Mullender, (ed), "Distributed Systems: Architecture and Implementation", chapter 10. Addison-Wesley, 1993.

Lefebvre D. and C. Delherm (2007). "Diagnosis of DES with Petri Net models". IEEE Transactions on Automation Science and Engineering, vol. 4, no. 1, pp. 114-118, 2007.

Pencolé Y. (2004) "Diagnosability analysis of distributed discrete event systems". *Proc. of the 15th International Workshop on Principles of Diagnosis.* Carcassonne, France. June 2004.

Qiu W. and R. Kumar (2005). "Decentralized Diagnosis of Event-Driven Systems for Safely Reacting to Failures". IFAC World Congress, Prague, July 2005.

Ramírez-Treviño A., I. Rivera-Rangel and E. López-Mellado (2003). "Observability of Discrete Event Systems Modeled by Interpreted Petri Nets". *IEEE Transactions on Robotics and Automation,* vol. 19, no. 4, pp. 557-565. August 2003.

Ramírez-Treviño A., E. Ruiz Beltrán, I. Rivera-Rangel, and E. López-Mellado (2004). A. Ramírez-Treviño, E. Ruiz Beltrán, I. Rivera-Rangel, E. López-Mellado. "Diagnosability of Discrete Event Systems. A Petri Net Based Approach". Proc. of the IEEE International Conference on Robotic and Automation. pp. 541-546, April 2004.

Ramírez-Treviño A., E. Ruiz Beltrán, I. Rivera-Rangel, E. López-Mellado (2007). "On-line Fault Diagnosis of Discrete Event Systems. A Petri Net Based Approach". *IEEE Transactions on Automation Science and Engineering.* Vo1. 4-1, pp. 31-39. January 2007.

Sampath M., R. Sengupta, S. Lafortune, K. Sinnamohideen, and D. Teneketzis (1995). "Diagnosability of discrete event systems", IEEE Transactions on Automatic Control, vol 40, no. 9, pp. 1555-1575, 1995.

Sampath M., R. Sengupta, S. Lafortune, K. Sinnamohideen, and D. Teneketzis (1996). "Failure Diagnosis Usign Discrete-Event Models", IEEE Transactions on Control System Technology, vol. 4, no.2, pp. 105-124. 1996.

6

Cognitive Approach to Control in Ill-structured Situation and the Problem of Risks

Abramova N.A., Avdeeva Z.K. and Kovriga S. V.
Institute of Control Science Russian Academy of Science
Russian Federation

1. Introduction

The decision theory studying control problems for systems of various natures (technical, biological, socio-economic) has become independent science branch during last decades. The decision theory actively uses methods of mathematics, psychology and informatics. The cognitive simulation is one of the new approaches of modern decision theory. The main purpose of this chapter is forming complete idea of cognitive approach actively developed now in the control science and interdisciplinary sciences (sociology, economy, etc.), further specification of basic concepts of this approach in the control science, as well as defining a class of control problems that are advisable to be solved by the cognitive simulation (simulation on the base of cognitive map).

The cognitive approach to simulation is directed to development of formal models and methods that support the intellectual process of problem solving owing to taking into account the cognitive abilities (perception, representation, cognition, understanding, explanation) of control agents. The basic technique of ill-structured problem solving is structurization of knowledge about object with further its environment and construction of a cognitive map (the static model of situation) and a dynamic model. The technique includes monitoring of dynamic of factors of the model (their tendencies), analysis of the model structure with the use of SWOT-approach, and modeling that permits to determine and solve ill-structured problems. This technique allows supporting of a vital control task that consists in goal setting of situation development, as far as solution of discovered problems turns into the system development control task. This technique application is useful when designing a strategy of development of social and economic objects.

The problem of risks for the results validity that arise due to the human factor in the cognitive approach is considered and two kinds of risks which we can pertinently treat as cognitive risks are exposed. On the example of a real cognitive map of a complex and ill-structured situation the analysis of some cognitive risks concerned with causal influence transitivity is carried out. Some explanatory mechanisms and criteria for the early detection of such risks are proposed. The issues important for further development of the cognitive approach to decision-making in the ill-structured situation control and especially of causal mapping techniques are highlighted.

The chapter includes the following topics:

1. Brief review of cognitive approach in control: evolution and some trends in its development. The problem of risks due to the human factor.

2. The models and methods of ill-structured problem solving on the basic construction and analysis of the cognitive map. Presenting the risks related with formalization of initial expert knowledge about ill-structured problem situation, by the example of cognitive map construction.
3. Analysis the experience of cognitive techniques application to control of socio-economic development and some trends of further development of cognitive approach.

2. Brief review of cognitive approach in control: evolution and some trends in its development. The problem of risks due to the human factor

2.1 Brief history of cognitive approach evolution

Origins of concept "cognitive map" lie in psychology (Tolman, 1946). Studying cognitive maps - subjective representations of spatial organization of outer world - has gained in fundamental importance in the framework of studying features of human perception of his surroundings. The cognitive map is the concept concerning cognitive processes related to gathering, representation, and processing of information on the environment during which an individual is not only a passive observer, but actively interacts with his environment. Forming cognitive maps by an individual is understood as the process consisting of series of psychological transformations. By means of these transformations, an individual gathers, stores, copies, recalls and manipulates information on relative locations and attributes of his spatial surroundings. This process is essential component of decision making for spatial behaviour. The psychological research is directed to a greater extent toward studying these very processes and their influence onto forming certain representations allowing an individual to act and make decisions in his environment.

In political science and sociology, the method of cognitive simulation was developed in 1960–1980 in papers of American researcher R. Axelrod and his colleagues from USA and Scandinavia (Heradstvein & Narvesen, 1978; Axelrod, 1976).

It should be noted that many authors also use the term "cognitive mapping". In any case, research of problem situation is based on model construction on the base of a cognitive map. The differences consist only in applied modifications of cognitive maps and methods of their formal processing.

In political science and sociology, "cognitive map" is not related to spatial orientation. It is interpreted as an individual's schematic representation of world image fragment relating to concrete problem situation. In this context, cognitive map is the way for representing thought structures directed toward a concrete problem and allowing to simulate politician's mentation during deliberation of action stimulating identification of further events (Heradstvein & Narvesen, 1978).

Construction and analysis of cognitive maps allow revealing causal structure of reasoning presented in political texts and, on this base, making conclusion on text author's vision of political situation, defining factors taken into account by politicians while making decisions.

Axelrod developed the method of cognitive simulation relying on ideas of psycho-logic (Abelson & Rosenberg, 1958), causal deduction (Shapiro & Bonham, 1973), as well as graph and decision theories (Axelrod, 1976). As the main problems of decision making, he distinguished the problems of situation explanation and test of hypothesis on situation arrangement, as well as prediction problems and problems of decision selection from a number of alternatives. On the base of these studies it was shown that in complex situations an individual is inclined to simplify situation representation, to not notice feedbacks, etc. As a consequence, long-term effects, interrelations between different problems, etc., are not taken into account while making decisions.

Axelrod placed the models constructed on the base of cognitive maps among normative models in that sense that they organize cognitive activity of an individual during the process of decision making. Just as any formalization, the cognitive map and methods of its analysis prescribe an individual his way of making decisions in complex situations.

The methods of cognitive simulation have found an application in collective working out and making decisions (Hart, 1976; Roberts, 1976). English scientist K. Eden developed general approach to construction of cognitive maps (Eden, 1988), resting on research in the field of decision making psychology, in particular, on the theory of personal constructs by Kelly (1955). Eden outlined importance of Kelly's statement to the effect that effectiveness of interaction in group of individuals making decisions essentially depends on each member's understanding of ways for situation interpretation by other group members.

Empirical studies of some authors (Eden, 1988; Axelrod, 1976; Hart, 1976; Holsti, 1976; Roberts, 1976; Shapiro & Bonham, 1973) shown that application of cognitive simulation methods allows increasing effectiveness of decision making in problem situation. Analyzing his own and other's cognitive maps, an individual can define problem representation more exactly, find contradictions, understands other individuals. In addition, the cognitive map is the convenient tool for changing settled stereotypes contributing to generation of new viewpoints.

At the same time, Axelrod notes a lack of formal methods for construction of cognitive maps directed toward reliability and interpretability of results of problem situation analysis (Axelrod, 1976).

As a rule, application of cognitive simulation methods in sociology and political science is aimed to revealing representations of an individual making decisions in various situations, resolution of subjective conflicts due to difference between problem representations and lack of mutual understanding among interacting individuals.

The development of cognitive simulation methods is substantially conditioned by need in analysis of ill-structured systems and situations including multitude of elements with different nature, as well as relationships between elements having both quantitative and qualitative nature. The concept "ill-structured problem (situation)" was introduced by G. Simon (1973).

The cognitive approach to analysis of ill-structured situations was proposed by Axelrod, (1976) and Roberts (1976). Axelrod to a greater extent was engaged into development of methodology, while Roberts - into development of mathematical tool. The primary precondition for rising this approach consists in boundedness of accurate model applicability to ill-structured system behaviour analysis, as well as making control decisions on resolution of problem situations.

For this approach, the subjective understanding and individual's representation of the controlled system parameters and relationships between them in form of cognitive map consists foundation for construction of ill-structured model of system or situation. Subsequent choice of formal methods for processing of representations reflected in the cognitive map depends on goals of analysis, as well as characteristic features of ill-structured system or situation.

Nowadays, the cognitive approach to analysis of ill-structured systems and situations is actively developed in Russia and abroad (Avdeeva et al., 2007; Kuznetsov et al., 2006; Kulba et al., 2004; Avdeeva et al., 2003; Chaib-draa, 2002; Maximov, 2001; Maximov & Kornoushenko, 2001; Kim, 2000; Huff, 1990; Kosko, 1986; Sawaragi et al, 1986; Heradstvein & Narvesen, 1978; Roberts, 1976). One of the typical trends of this development consists in looking for mechanisms joining various scientific areas of researching decision problems for control of ill-structured systems and situations.

The distinctive feature of cognitive approach consists in the following:

- process of solving of control problems is considered as cognitive human activity including application of formal models and methods as a part or stage of solution;
- another important stage is formalization of representation of ill-structured systems and situations, goals, interests and motivations of individuals involved into the problem solution processes.

2.2 Trends in development of the cognitive approach in decision-making

In theoretical researches on methods for searching and making decisions, such terms as cognitive approach, cognitive researches, cognitive modeling or cognitive mapping of complex objects, problems, and situations are used more and more often. It can explained by growing understanding of inevitable participation of people with their cognitive resources (and also restrictions) in the decision of practical problems of control, especially in case of complex and ill-structured situations (Abramova & Kovriga, 2008).

Recently the selective analysis of works proposed at International Conference "Cognitive analysis and situations evolution control" (CASC) has been carried out (Abramova, 2007b), with the following similar analysis of other publications using term "cognitive" in the context of solving the applied control problems. The conference has been held at the Institute of Control Sciences of Russian Academy of Sciences for the latter seven years, being focused on the integration of formal and cognitive (using resources of the person) problem solving methods.

The analysis has allowed seeing considerable distinction in understanding of what terms "cognitive approach" and "cognitive modeling" mean, in how the term "cognitive" operates, in scientific and applied problems being solved, in the formalization level, in considered cognitive aspects, in involved knowledge of the cognitive science. At all distinctions, it is possible to identify two basic directions in accordance with understanding of the cognitive approach in narrow and wide sense (in the context of decision-making and ill-structured situations).

Two directions of the cognitive approach. Today, the direction of the cognitive approach corresponding to narrow sense of the term is the most developed one. It means that various models of cognitive maps are applied as models for representation and creation of knowledge about ill-structured situations,. The most conservative branch of this direction focused on formal methods, at all does not consider specificity of human factors and features of structurization by the person of difficult situations et all; so the word "cognitive" bears purely nominative function of a label for models applied.

However as a whole it is relevant to speak about two trends in development of this direction. On the one hand, the positive tendency to larger account of such human-dependent stages as formalization of primary knowledge and representations of a problem situation, targets definition, etc is observed. On the other hand, the accepted models of knowledge and activity of the people solving practical problems (experts, analysts, decision-makers) are normative in relation to those people, and the justification of the models is defined by theorists at the level of common sense and traditions. Any knowledge of how people really think and what knowledge the cognitive science in this respect has, today are not usually involved today.

The cognitive approach in wide sense is not limited by the choice of cognitive maps as models for representing knowledges about complex objects and situations (Abramova & Novikov, 2006). The accent initially is made on human-dependent stages. Basically, the approach covers a wide spectrum of the models applied in decision-making for ill-structured problem situations. However, today the same tendencies in this direction dominates, as in the previous one.

Review (Abramova, 2007b) represents perspectives of more advanced development of the cognitive approach, with integration of formal methods and knowledge of psychology and the cognitive science as a whole. However, today bridging the gap appears to be difficult, at least, because of distinction in scientific languages. Among few exceptions it is useful to mention (Obal et al., 2007; Gabriel & Nyshadham, 2008; Vavilov, 2006).

2.3 The problem of risks due to the human factor

The problem of risks due to the human factor in the field of formal methods for searching and making decisions to control complex and ill-structured situations essentially consists in that because of inevitable and substantial human beings' participation in solving practical problems (at least, for formalization of primary representations) formal methods basically cannot provide validity of received decisions. Note that validity of method application results is understood here in wide intuitive sense as capability to rely upon these results in solving a specific practical problem. It is also possible to speak about validity of a method as its capability to yield valid results. Simplier speaking, such methods (which we refer to as subjective-formal ones) are basically risky concerning validity of their results.

The pragmatical importance of the given problem of risks obviously depends on how much significant are risks obtaining invalid results in solving practical problems. By present time theoretical, experimental and even practical knowledge is accumulated, leading to understanding or directly saying that human factors can be the significant source of risk for quality of results (Abramova, 2007a).

The most impressing is the research on "logic of failure", presented in D. Dörner's known book (Dörner, 1997). By means of vast computer-based simulation of how real people solve problems of the complex and ill- structured dynamic situation control, a number of typical errors is revealed which are inherent not only in dilettantes but also in experts at work with complex situations. These results lead to conclusion that some of the above errors should be expected for any kinds of cognitive maps whereas others should be characteristic of dynamic map modeling. Strictly speaking, in Dörner's experiments it is possible to admit that the risks of errors stem not only from natural ways of thinking but also from the accepted model for representation of knowledge about situations in the computer. However, Dörner finds evidences in favor of his theory in the analysis of known natural-thinking decisions concerning Chernobyl.

One more class of risk sources is revealed by O. Larichev (Larichev & Moshkovich, 1997). with his school in the traditional formal decision theory. It refers to methods and operations of receiving initial information from decision-makers (in other terms, knowledge acquisition) for subsequent processing with formal methods. Stemming in the accumulated psychological knowledge of more than 30 years, the inference has been drawn that "soft" qualitative measurements such as comparison, reference to a class, ordering are much more reliable, than appointment of subjective probabilities, quantitative estimates of criteria importance, weights, usefulnesses, etc. To increase reliability of such operations (in our terms, to decrease risk of human errors) the idea of psychologically correct (in other terms, cognitively valid) operations has been advanced by the school. From the above results it follows that validity of cognitive modeling should be various for different kinds of cognitive maps depending on kinds of estimates demanded from experts.

There is also a wide spectrum of psychological researches in the field of the limited rationality of the person, not relating to solving control problem activities, which evidencing to numerous types of risks in intellectual activity of a person.

To add evidences of practical significance of cognitive risk factors in decision-making, it is relevant to mention that some significant risk factors have been found out by these authors in the applied activity on safety-related software quality assurance including formal-method quality estimation and control. This practice along with further theoretical analysis have considerably expanded representations of the risk spectrum produced by theorists developing or choosing models for knowledge representation and activity of people solving practical problems, in comparison with the risks exposed by Larichev's school (Abramova, 2004; Abramova, 2006). For example, it has appeared that application of very natural, from the mathematical point of view, method of linear convolution of normalized estimates on partial quality indicators for estimation of a complex indicator creates paradoxical risk of loss of controllability on partial indicators.

Moreover, the analysis has confirmed presence of some general cognitive mechanisms of risk for diverse models of subject-matter experts' knowledge and related formal methods what serves evidence of reasonability of the general approach to the problem of risks due to the human factor in decision-making.

The subjective aspect of the problem of risks is that even more or less widely spread peaces of knowledge relative to risks and their sources mainly are not noticed by scientific community or, at the best, are underestimated. This ignorance is quite explainable theoretically with taking into account psychological risk factors (with the models presented in (Abramova, 2006; Abramova, 2007a). Thereby, there is a soil for theoretical development and practical application of subjective-formal methods, which cannot provide decision quality (adequacy, validity, reliability, safety) comprehensible for critical domains.

Among few works in the field of cognitive mapping, with recognizing not only the human factor influence, but also necessity of researches in this direction, it is worthwhile to note (Bouzdine-Chameeva, 2006), where validity of cognitive maps with internal validity between the data and the conceptualization of the data, including the definitions of concepts and influences, has been discussed, proceeding from the general ideas of content-analysis reliability.

Cognitive risks. Two kinds of risk factors. In diversity of risks due to the human factor in the considered field the special class is formed by cognitive risks which are explainable with taking into account factors (mechanisms) concerning cognitive sphere of a person. All risks mentioned above are referred to cognitive risks as well as a number of others, which are reported about in publications and private communications without doing them subject of the scientific analysis.

First of all, researchers are interested in the factors that, on the one hand, have regular nature (i.e. do not concern to dysfunctional cognition), and on the other hand, are hardly explainable from positions of common sense or even contradict it; so usually they are not assumed by the mathematicians developing working out normative models for expert knowledge representation or ignored if known.

Earlier cognitive processes have been analyzed in which subjects of intellectual activity (analysts, experts, decision makers and other staff who participates in searching and making decisions at collaboration with computers) turn to be under the influence of "ambient intelligence" due to imposed forms of thinking (Abramova & Kovriga, 2006; Abramova, 2006; Abramova, 2007a). This influence leads to dependence of decisions on theoretical beliefs of specialists on formal methods and computer-aided decision support systems and technologies. Therefore, results in risks of invalid decisions. Two kinds of risk factors explainable with the suggested models have been exposed which are pertinently considered as cognitive risks.

The risk factors psychologically influencing validity of expert methods during their application by experts belong the first-kind factors, or factors of direct action. Such factors can either objectively promote invalidity of results, or raise subjective confidence of experts in objective validity for the method application outcomes. One can tell that the latter represent themselves as factors of belief. Agents of these factors of influence are experts; just they appear in conditions which that may lead, eventually, to insufficiently valid (in the objective relation) outcomes.

The second-kind risk factors or factors of an indirect action psychologically influence upon validity of expert methods during their creation and justification. The agents of influence of such factors are the creators of methods, scientists and experts producing standards who, in turn, are subject to influence of scientific norms, paradigms, etc., that is the strongest factors of belief. Typical examples of the first-kind risk factors are the natural mechanisms of thinking with risk of errors, as in the case of the errors discovered by Dörner. Typical second-kind risk factors are psychologically incorrect models of knowledge of the experts, creating risk of unreliable (unstable or inconsistent) data from experts, with the models being supported by belief (often unconscious) in their validity.

Amongst a number of cognitive risk factors having been found out by these authors both in theoretical and experimental researches, as well as in practice, there are risks related to belief in universality of the principle of pairwise preference transitivity in decision-making (Abramova & Kovriga, 2006 ; Abramova & Korenyushkin, 2006). The principle means that from $a \succ b$ ("a is preferable over b") and $b \succ c$ it always follows $a \succ c$, though rationally reasoned violations of this principle are known (for example, preferences based on multicriteria comparisons).

From now onwards, the given work concerns one family of cognitive risks having been discovered quite recently. Risks take place in cognitive mapping based on causal maps. They are related to the property of causal influence transitivity and its violations. They have been admitted as a hypothesis by analogy to known rationally reasoned violations of transitivity of pairwise preferences and later found out in real causal maps.

3. Some methods of ill-structured problem solving on the basic a cognitive map. Some risks concerned with causal influence transitivity

For further turning to the cognitive approach in the narrow sense, it is necessary to clarify the concept of cognitive map in the meanings used in the decision-making context.

In this field of knowledge the concept of cognitive map is used more and more widely beginning from (Axelrod, 1976), but it takes various meanings, without saying about essentially differing concept of cognitive map in psychology. Recent years have brought out a number of reviews and articles with extensive reviews in which the diverse types of cognitive maps and other concept maps are compared, differing in substantial and formal interpretation, as well as in sights at their role in searching and making decision processes and control of complex objects and situations (in particular, Eden, 1988; Kremer, 1994 ; Mls, 2006;. Kuznetsov et al., 2006; Bouzdine-Chameeva, 2006; Abramova, 2007b; Avdeeva et al., 2007; Peña et al., 2007).

In this work term "cognitive map" refers to the family of models representing structure of causal (or, that is the same, cause-effect) influences of a mapped situation. Formally, the obligatory base of all models of the family is a directed graph, the nodes of which are associated with factors (or concepts) and arches are interpreted as direct causal influences (or causal relations, connections, links) between factors (Kuznetsov, Kilinich & Markovskii, 2006). Usually the obligatory base (that is the cognitive map in the narrow sense) is added with some parameters, such as an influence sign ("+" or "-") or influence intensity, and some or other interpretations both substantial, and mathematical are given to the map (Fig. 1).

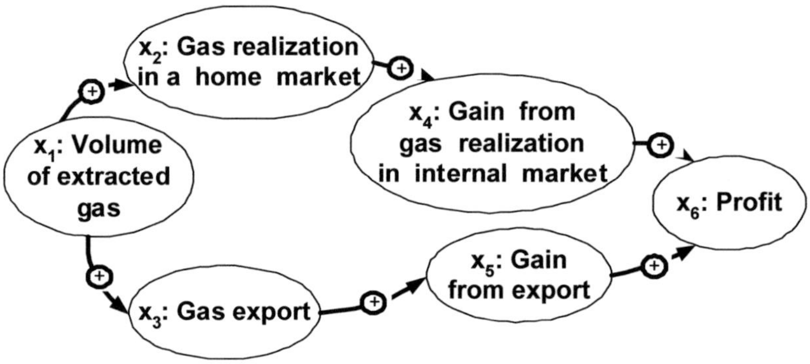

Fig. 1. Example of cognitive map

Various interpretations of nodes, arcs and weights on the arcs, as well as various functions defining influence of relations onto factors result in different modifications of cognitive maps and formal means for their analysis (Kuznetsov, Kilinich & Markovskii, 2006). Owing to multitude of cognitive map modifications, one can distinguish different types of models based on cognitive maps (in short, cognitive map models). Models of this family that are often referred to as causal maps or influence diagrams cover a wide spectrum of known types of models for cognitive mapping.

By type of used relationships, five types of cognitive maps are distinguished (Huff, 1990):

1. Maps that assess attention, association and importance of concepts: With these maps, the map maker searches for frequent use of related concepts as indicators of the strategic emphasis of a particular decision maker or organization, for example, and look for the association of these concepts with others to infer mental connection between important strategic themes. He also can make judgments about the complexity of these relationships or differences in the use of concepts;
2. Maps that show dimension of categories and cognitive taxonomies: The map maker investigates here more complex relationships between concepts. He can dichotomize concepts and construct hierarchical relationships among broad concepts and more specific subcategories. The maps of this type have been used to define the competitive

environment, and to explore the range and nature of choices perceived by decisions makers in a given setting;

3. Maps that show influence, causality and system dynamics (causal maps): These maps allow the map maker to focus on action, for instance, how the respondent explains the current situation in terms of previous events, and what changes are expected in the future. This kind of cognitive map is currently, has been, and is still, the most popular mapping method;
4. Maps that show the structure of argument and conclusion: This type of map attempts to show the logic behind conclusions and decisions to act. The map maker includes here causal beliefs, but looks more broadly at the text, as a whole, to show the cumulative impact of varied evidence and the links between longer chains of reasoning;
5. Maps that specify frames and perceptual codes: This approach supposes that cognition is highly conditioned by previous experience, and that experience is stored in memory as a set of structured expectations.

Applied practice shows that the maps of the third type are expedient for studying of ill-structured systems. Choice of method for structuring of ill-structured systems and situations in a form of set of factors and causal relations are conditioned by that the events and processes of operation and development of ill-structured systems include various events and trends defined by many factors, at that each of them in turn has an influence on some number of other factors. The networks of causal relations between them are formed (Kuznetsov et al., 2006; Dorner, 1997). Book of well-known German psychologist Dorner (1997) devoted to studying thought on control subject and analysis of causes of mistakes while resolving problem situations in operation and development of complex systems indicates that a momentary situation with its features is only actual state of system and its variables. One should not only understand events that happen, but also foresee events that will happen or may happen in the future, as well as suppose a way for situation changing depend on concrete interference. It requires structured knowledge, i.e. the knowledge of relationships and mutual influences of system variables. Dorner notes that in ideal case this knowledge is represented in the form of "mathematical functions", but in case when the latter cannot be constructed one can apply diagrams of causal relations allowing to reconstruct various assumptions (hypotheses) in mind of the subject of control, at that not in the form of "causal chains", but in the form of "causal networks".

According to Kuznetsov, Kilinich & Markovskii (2006), the situation analysis problems on the base of cognitive maps can be divided into two types - static and dynamic. Static analysis or influence analysis is analysis of considered situation via studying structure of mutual influence in a cognitive map. The influence analysis chooses factors with the strongest influence onto goal factors that are the factors with values to be changed. The dynamic analysis lies in the base of generating possible scenarios of situation development with time. Thus, abilities of solution of analysis and control problems are defined by type of used models, static or dynamic.

As a rule, two types of mathematical tools are used for carrying out these kinds of analysis - linear dynamic systems and fuzzy mathematics.

Nowadays, the research abroad is mainly devoted to the development of models on the base of fuzzy mathematics and static models using different mathematical tools (Chaib-draa, 2002; Kim, 2000; Kosko, 1986; Sawaragi et al, 1986; Heradstvein & Narvesen, 1978).

In Russia, along with the development of methods for static analysis of cognitive maps, special efforts are directed to research of ill-structured systems and situations using linear dynamic models that are presented in this paper by works of Russian scientists (Avdeeva et al., 2007; Avdeeva, 2006; Kulba et al., 2004; Avdeeva et al., 2003; Maximov & Kornoushenko, 2001; Maximov, 2001).

In linear dynamic model based on cognitive map, factor is formally defined as variable taking values from some numerical scale.

The model construction is based on the cognitive structurization of knowledge about object and its environment.

The purpose of structuring consists in revealing of the most essential factors describe a "boundary" layer of interaction between object and external environment, as well as establishing of qualitative relationships between cause and effect between them.

The cognitive structurization is finished by formalization of singled out knowledge that consists in generalization of essential information into a system of basic factors.

Analysis of a graph model of a situation associated with a cognitive map allows to reveal the structural properties of a situation. The basis of the model is a weighed digraph G =(X, A), where X is a set of nodes that biuniquely corresponds to the set of basic factors, A is a set of arcs reflecting the fact of direct influence of factors. Each arc connecting some factor xi with some factor x_j has the weight a_{ij} with the sign depicting the sign of influence of the factor xi on the factor x_j, and the absolute value of a_{ij} depicting the strength of the influence. Thus, the cognitive map can be examined as the connectivity matrix A_g of the graph G.

When constructing a cognitive map , the set of basic factors, X, is grouped in blocks relevant to external environment, X^{ext}, and an internal environment, $X^{int} = X \backslash X^{ext}$. Besides determination of factors and influence between them the vector of initial factor trends, $X^{ext}(0) \cup X^{int}(0)$, is established.

Change of factor value with time is given by Kulba et al. (2004), Maximov (2001), Maximov & Kornoushenko (2001)

$$x_i(t+1) = x_i(t) + \sum_{j \in I_i} a_{ij}\left(x_j(t) - x_j(t-1)\right), \quad i = 1,...,N \tag{1}$$

where $x_i(t+1)$ and $x_i(t)$ are the values of i-th factor at instants t+1 and t, respectively, $x_j(t) - x_j(t-1) = \Delta x_j(t)$ is the increment of factor x_j at instant t characterizing the rate of change (trend) of x_j, a_{ij} is the weight of the factor x_j influence onto factor x_i, I_i is the set consisting of numbers of factors directly effecting the factor x_i.

With knowledge of the initial situation state X (0) and accepting x (t) =0 for all t <0, the state of a situation in self-development (without applied control) at any moment *t* can be characterized by the vector of factor values

$$X(t) = (E_N + A + A^2 + \ldots + A^t)\, X(0) \tag{2}$$

Estimation of the sum of this series can be obtained only if the graph G adjacency matrix A_g is stable. Then all elements of this series approach to finite limits at unlimited increase of t.

To determine the transitive closure, it is sufficient to consider N terms in a power series of matrix B, where N is the order of matrix B, i.e. the number of basic factors in the situation cognitive map. Then the transitive closure of the matrix B is estimated by the matrix

$$Q = E_N + A + A^2 + \ldots + A^N \cong \left(E_N - A\right)^{-1} \tag{3}$$

Matrix Q characterizes the direct and mediated influences effecting each factor.
Let $U(0)=(u_1(0),...,u_p(0))$ be the vector of pulse control effecting at the moment t=0 the factors X. Then, with knowledge of the initial state of situation X(0), the state of the situation at any moment can characterized by the factor value vector

$$X(t)=QX(0)+QBU \quad (4)$$

where $u = (u_1,...,u_p)^T$ is the external input vector and B - (0,1) – is the (pxn)-matrix with nonzero elements specifying numbers of corrected coordinates of the initial state X (0).
The dynamic model is constructed to obtain a new knowledge of structure and dynamics of considering situations. On the basis of that model, one can carry out the scenario research with use of methods of computer modeling of self-development and controlled development of ill-structured situation (Maximov, 2001; Maximov & Kornoushenko, 2001; Avdeeva, et al., 2003; Makarenko, et al., 2004).
Our research group focuses special attention on the following:

1. searching and development of structuring methods aimed at construction of cognitive maps;
2. increasing technological effectiveness of scientific and instrumental support for solving practical control problems:
3. approach to revealing and prohibition of the semantic errors and risks of formalization (before application formal method to analysis of ill-structured situation).

The problem of risks outlined below in section 3.1 represents advanced approach in wide sense, with taking cognitive mapping as the representative example.
On the basis (1)-(4) has developed the following methods and approaches have been developed (Avdeeva et al., 2007; Avdeeva, 2006; Avdeeva et al., 2003; Maximov & Kornoushenko, 2001):

- method of structure and goal analysis of ill-structured system development (section 3.3);
- SWOT-analysis on the basis of analysis of the model structure (section 3.2);
- approach and methods for solving ill-structured problems and deriving scenarios of ill-structured system development; (section 3.3);
- approach to studying conflict situations generated by contradictions in interests of subjects influencing development of considered system and other.

At that, the problem is defined as discrepancy between current condition of ill-structured system and its dynamic and desired condition that is given by the subject of control. Complex application of the mentioned methods allows carrying out static and dynamic analysis while studying ill-structured systems. The socio-economic systems (SEO) constitute a typical class of ill-structured systems that are expedient for application of cognitive simulation for development problem solving.
But it is necessary to take into account that such methods essentially depend on techniques of revealing of the factors and definition of their interrelations describing situations under research. Construction of the cognitive map reflecting representation of complex system development includes forming of the conceptual scheme of the situation (the description of the subject domains defining complex system development); choice or revealing of the important factors (detailing of the conceptual scheme); definition of interrelations between the factors.
Despite of popularity of such methods, now there is no coordination in the literature concerning a way of revealing of the important factors influencing and defining the

situation under research. Methods of a cognitive map construction can be useful as to understanding cognitive processes of the persons participating in decision-making, and as a basis for control of active research of complex situations (Avdeeva, 2006). The analysis of works of scientists (Axelrod, 1976; Hodgkinson, Maule & Bown, 2004; Avdeeva, 2006, and other), actively applied cognitive methods for research of complex system, shows, that

- revealing of factors and interrelations by means of a content-analysis of documents, in particular, in the first work, devoted to a cognitive map application for support of decision-making, the author results bases of such analysis of stenograms of session of politicians;
- revealing of factors and interrelations by means of the analysis of expert opinions thus are involved experts on the subject domains;
- revealing of factors and interrelations by means of the analysis of quantitative data, for example, regression analysis.

The choice of a method for cognitive map construction depends on availability of data and also on the purposes of researchers. The listed above methods are not devoid of disadvantages. For example, the main problem in the documentary approach consists in that information is represented differently. Thus, the stenograms of debate of English politicians were represented as well structured material. Other documentary sources can unlikely be considered as well structured. One of the main disadvantages of the expert methods is insufficient validity of revealing of factors and their interrelations. The analysis of quantitative data assumes that the revealed factors and interrelations are impartial. But to get some factor in sight, the quantitative information about it and its influence should be collected.

On the basis of long-term experience of practical application of cognitive methods in the analysis of problem situations by the example of development of social and economic objects, our research group has developed techniques of carrying out strategic meetings and heuristic schemes of a cognitive map construction on the basis of PEST and SWOT (Strengths, Weaknesses, Opportunities, Threats) ideologies. This group of methods is directed onto structurization and coordination of primary representations of problem situation of non-uniform collective of experts.

As a whole, application of such techniques is complicated by lack of checking adequacy criteria for cognitive models as well as presence of reliability problem for results of application of expert methods.

3.1 Some risks concerned with causal influence transitivity and their analysis.

The example description. The facts. In fig. 2 the fragment of a real cognitive map slightly simplified to demonstrate action of risks is presented. Influence in pair of factors (3,2) at the verbal level is interpreted as follows: "increase in access of manufacturers to gas export pipelines (with other things being equal) causes increase in volume of extracted gas". This influence is positive (in mathematical sense) that means the same direction of changes of factors. Positive influence in pair (4,2) is verbalized similarly. Influence in pair (2,1) is negative: "increase in volume of extracted gas (with other things being equal) causes decrease in deficiency of gas in the country".

All three influences, as well as the set of factors significant for the investigated situation of dynamics of the market of gas in the country, are established by the expert. (Substantially, this map corresponds to a situation when there are stocks of gas and manufacturers have resources for increase gas production in volume but their access to means for its delivery to consumers is limited.)

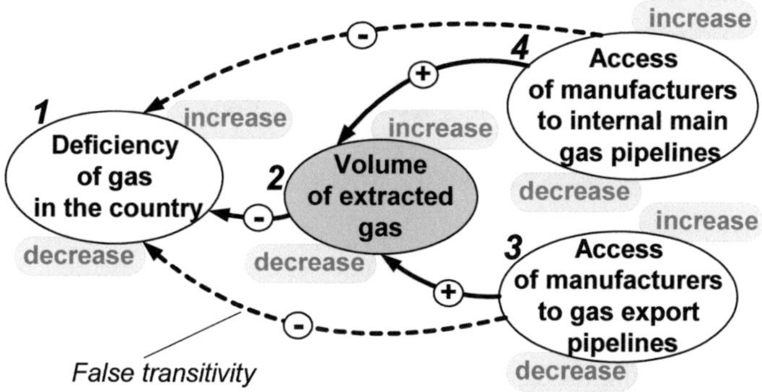

Fig. 2. An initial fragment of a cognitive map with false transitivity

According to formal model of causal influences and intuitive logic, from positive influence $3\xrightarrow{+}2$ and negative influence $2\xrightarrow{-}1$ follows transitive negative (in mathematical sense) influence $3\xrightarrow{-}1$; influence $4\xrightarrow{-}1$ is deduced similarly.

However later the expert has noticed that "logically deduced" influence $3\xrightarrow{-}1$ is absent in reality: thereby false transitivity of influences takes place in the map. The analysis of expert knowledge of the situation leads to the following correction (fig. 3).

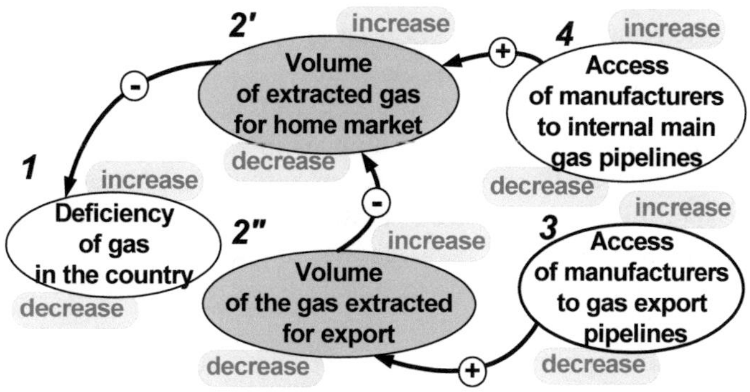

Fig. 3. The corrected fragment of a cognitive map

It is worth while to underline that at such refinement replacements of influence $3\xrightarrow{+}2$ with $3\xrightarrow{+}2''$, $4\xrightarrow{+}2$ with $4\xrightarrow{+}2'$ and $2\xrightarrow{-}1$ with $2'\xrightarrow{-}1$, in essence have not changed expert's interpretation of influences: only the form of representation of knowledge has changed only. However, in this way the chain $3\xrightarrow{+}2\xrightarrow{-}1$ generating false influence $3\xrightarrow{-}1$ has disappeared.

Essentially different situation occurs with introduction of additional negative influence $2''\xrightarrow{-}2'$. This influence means that at increase of access of manufacturers to export gas pipelines (with other things being equal) it is possible to increase volume of the gas extracted for export not only by means of increase in volume of extraction, but also by simple «valve switching». Thus growth of volume of gas for export is made at the expense of decrease in volume of gas for home market. In this case knowledge is entered into a map, well known to experts, but not represented within the frame of initial system of concepts (factors).

As a result of correction new transitive influences $3 \xrightarrow{-} 2'$, $3 \xrightarrow{+} 1$ have appeared. Instead of positive (in substantial sense) situation in the map of fig. 1, when it is possible to reduce deficiency of gas at the expense of access of manufacturers both to internal, and to external gas pipelines, more complicated and more realistic situation comes up in the map of fig. 3.

Along with positive (as a matter of fact) transitive influence $4 \xrightarrow{-} 1$, negative (as a matter of fact) influence $3 \xrightarrow{+} 1$ takes place, and their proportion at the decision of the problem of gas deficiency in the country demands comparative estimation of influences.

The invalidity of the fragment of cognitive map in fig. 2 relative to reality, or in other words, error in cognitive mapping is obvious.

The above practical example along with some other ones serve as actual evidences of the fact that in cognitive modeling of complex situations there are possible cases of erroneous inferences by transitivity, i.e. false transitivity.

The analysis of cognitive risks. In the above example the false transitivity can be explained with cumulative action of two modes of risk factors in the course of cognitive mapping: assumption of causal influence transitivity as the universal principle and disproportion of extension of concepts.

First, in each of the three direct influences between factors of the initial map having been set by an expert, there is disproportion of extension of concepts with excess of extension of concept 2, which denotes the influence receiver in pairs (3,2), (4,2) and the influence source in pair (2,1).

Denote that in cognitive mapping it is traditional to speak about factors (concepts) as causes and effects. However we prefer, at least in the analysis, to speak about sources and receivers of influences , because, in modeling of complex and ill-structured situations substantial cause-effect interpretations of individual influences in a map quite often happen more or less difficult. (There is such a situation in the above example.) Moreover, in the theoretical analysis which we spend it is more exact to distinguish "factors" and "concepts of factors" (that is concepts designating factors). It is relevant to speak about factors at the analysis of situation content, and more pertinently to speak about concepts of factors when it is a question of the logic analysis of concept extensions.

Excess of extension of concepts in some direct influences informally means that it would be possible to take concepts with smaller extension for mapping the same substantial cause-effect influences. Just this action has been made at correction.

It is hardly admissible to consider such disproportions with excess of extension of concepts as errors because they are typical for the conceptualization of complex and ill-structured situations. This is evidenced both with practice of cognitive mapping and with informal reasonings of experts on such situations. Therefore we consider such disproportions as cognitive risks. They are natural for carrying to first-kind risk factors , the carriers of which are experts and which objectively reduce validity of cognitive modeling in complicated situations.

Causal influence transitivity is accepted as the universal principle by theorists. It leads to an automatic inference of indirect influences by transitivity at the formal simulation of a given map. Thus the given assumption should be considered as the second-kind risk factor.

Some criteria for early detection of risks. Let us more formally define some criteria which could help an expert to monitor risks and make the decision on possibility to correct disproportional concept extensions in case of false influence transitivity detection. In the definition, the fact, which has been found out in practice, is taken into account: the same (as a matter of fact) causal influence can be represented in a cognitive map in different forms so that we speak about different representations of the influence. Let we have factors A, B_1 (represented with the same name concepts), which are linked by direct causal influence $B_1 \to A$, and let there exists (is found by an expert) factor B_2 such that replacement of representation of influence $B_1 \to A$ with $B_2 \to A$ does not change the influence substantially, and herewith

$$\mathcal{V}_{B_1} \supset \mathcal{V}_{B_2},$$

where $\mathcal{V}_{B_i}$, $i = 1,2$, stands for the extension of the corresponding concept, and the relation between extensions is treated as usual inclusion or, that is the same, verbally: "c_1 has smaller extension than c_2". Then the factor B_2 is **more proportional in its concept extension** than B_1 as the source in the direct influence on A, and factor B_1 is **extensionally excessive** in this influence.

The proposed expert criterion of extensional proportionality, $K^S(B,A)$ is applicable to any pair of factors of a cognitive map, connected by direct influence. It allows to estimate whether factor-source of influence B is extensionally proportional to influence (or set of influences) on the receiver being modeled with link (B, A).

For example, factor 2 in fig. 2 is extensionally excessive in the influence (2,1), according to $K^S(2,1)$. The criterion of extensional proportionality for the influence receiver $K^D(A,B)$ is formulated and applied similarly, though in case of many influences onto one factor it is less informative at risk detection and error correction.

Summary. The problem of risks due to the human factor in the field of formal methods of searching and making decisions in the control of complex and ill-structured situations is considered as the general problem for diverse models of subject-matter experts' knowledge and related formal methods (Abramova & Novikov, 2006; Abramova, 2007a).

Earlier the idea about productivity of the uniform approach to the problem for diverse models of experts' knowledge, solved problems and formal methods has been stated, and some theoretical and empirical evidences in favor of this idea have been found (Abramova, 2007a).

The idea has found the reinforcement and further development at current studying risks concerned to causal influence transitivity in cognitive modeling, with carrying out analogies to risks in decision-making based on pairwise preferences. It is enough to say that just the analogy of principles of transitivity of paired preferences and causal influences has led to a hypothesis on possible violation of the axiom of causal influence transitivity, i.e. to risk of false transitivity at formal cognitive mapping what has been confirmed in practice.

However, in few practical situations of violating the axiom of causal influence transitivity which we have analyzed by now, violations took place in cases when detection of false transitivity was perceived by experts as a signal of an error with necessity of cognitive map correction, and herewith correction was successful. Nevertheless, by analogy situations are conceivable when modeling of expert knowledge in terms of the chosen general formal model of cognitive maps appears impossible without separate violations of the axiom of influence transitivity, and the violations are estimated by subject-matter experts not as a signal of an error but as limitation of the formal model.

The question on existence or possibility of such situations in practical cognitive maps of the complex and ill-structured situations is opened by today. Further research on the empirical material of practical cognitive maps of complicated situations is planned with the aim to revealing such situations, their practical importance, and also rational experts' reasoning in such situations.

3.2 SWOT-analysis on the basis of analysis of the model structure

Generally, SWOT-analysis is expert determination of strength and weaknesses of SEO, opportunities and threats of its environment, and estimation of their interaction. Results of SWOT-analysis are represented as a matrix called "Window of opportunities".

The mathematical procedure of generation of matrix "window of opportunities" on the basis of analysis of structural features of SEO development cognitive map has been worked out. Thus, there is no need in regular attraction of experts with all accompanying procedures.

The essence of the procedure of SWOT-analysis is the following.

Analyzing a situation of SEO development, it is possible to put forward various hypotheses about desirable dynamics of any factor of the model. So, the parameter "attitude to factor dynamics" (AFD) is brought in for each factor of the model. If dynamics of a factor is positive (negative) AFD is equal 1 (–1). If it is difficult to evaluate the factor dynamics its AFD is equal to 0. The set of AFD vector on some set of model factors reflects desirable change of a situation in SEO.

Let's designate through $R^{ext}(X^{ext})$, $R^{int}(X^{int})$ vectors of AFD of factors of the external and internal environment, where $X^{ext} \cup X^{int} = X$; $X^{ext}(0)$, $X^{int}(0)$ - vectors of initial trends accordingly.

While the situation evolves each factor is being influenced not only by "neighbouring" factors, but also by more "distant" ones and these indirect influences are transferred through chains of the appropriate factors and graph arcs that connect them. Set of influences both direct, and indirect to which each factor in a situation is subject to, is described with the use of concept of transitive closure of a cognitive map of the situation (matrix Q (3)).

When constructing a matrix "Window of opportunities", opportunities and threats of the environment, strengths, and weaknesses of SEO are determined on the basis of observation of the dynamics of model factors and estimations of their integrated influence on desirable dynamics of factors $R^{ext}(X^{ext})$ и $R^{int}(X^{int})$. The significance of strengths and weaknesses of SEO is determined as well.

Let us introduce the basic definitions.

Definition 1. If the initial trend of the internal environment factor, $x_i^{int}(0)$, is negative, i.e. does not correspond to a desirable direction of change (AFD), the given factor is regarded as a weakness of functioning and development of SEO, otherwise (the trend is favourable) - as a strength of object. The weaknesses determine internal threatening trends to SEO development, and strengths - internal favourable trends.

Using a terminology of SWOT-analysis, we shall designate X^{st} – a subset of factors the strengths determine of SEO, X^{w} – a subset of the factor–weaknesses of SEO, $X^{st} \cup X^{w} \subset X^{int}$.

Definition 2. The initial factor trend from X influences positively on desirable dynamics of the factor from X if the following equality holds true

$$\mathrm{sign}(x_i(0)q_{ij}) = r_j(x_j),$$

where q_{ij} – (i,j) element of a transitive closure matrix Q, which determines integrated influence of i factor on the j factor; q_{ij}=0 if x_i does not influence x_j.

If AFD of some factors are given equal to zero ($r_j(x_j) = 0$) such factors are excluded from the analysis (integrated influences on them of other factors are not taken into account).
The following definitions follow from definition 2.

Definition 3. The factor of an environment x_i^{ext} is neutral for X^{int}, if the initial trend of this factor does not influence ($q_{ij}=0$) the desirable dynamics of all factors of the internal environment of SEO, X^{int}.

Definition 4. The factor of environment x_i^{ext} characterizes the opportunity for SEO development if the factor is not neutral and its initial trends does not negatively influence (through the appropriate integrated influences) the desirable dynamics of all factors of internal environment of SEO, $R^{int}(X^{int})$. In other words the initial trend of factor x_i^{ext} promotes SEO development in a desirable direction.

Definition 5. The trend of the environment factor x_i^{ext} threatens the SEO development, if it negatively influences (through the appropriate integrated influences) desirable dynamics even of one factor of internal environment $x_j^{int} \in X^{int}$.

Using a terminology of SWOT-analysis, we shall designate X^{op} – a subset of factors-opportunities for SEO development, X^{th} – a subset of factors–threats to SEO development, $X^{op} \cup X^{th} \subset X^{ext}$.

Definition 6. The internal environment factor, x_i^{int} ,is neutral for X^{ext}, if the initial trend of the factor does not influence ($q_{ij}=0$) the desirable dynamics of the environment factors X^{ext}.

Definition 7. The internal environment factor, x_i^{int}, promotes strengthening of the opportunity of environment x_j^{op} if x_i^{int} is not neutral and its initial trend favourably influences (through the appropriate integrated influence) the desirable dynamics of the factor x_j^{op}. Otherwise x_i^{int} promotes decrease of the opportunity of environment.

Definition 8. The internal environment factor x_i^{int} promotes parrying of threats of environment x_j^{th} if x_i^{int} is not neutral and its initial trend favorably influences (through the appropriate integrated influence) the desirable dynamics of the factor x_j^{th}. Otherwise x_i^{int} promotes strengthening of threats of environment.

On the basis of definitions 1-8 SWOT-analysis comes to the following stages:

1. Building of cognitive map of SEO development with extraction of external X^{ext} and internal blocks of factors. Vector of initial trends of factors $X^{ext}(0)$ and $X^{int}(0)$ is set.
2. AFD for each factor $R^{ext}(X^{ext})$, $R^{int}(X^{int})$ is set.
3. Strengths and weaknesses for each object (X^{st} and X^{w}) are found on the basis of the vector $X^{int}(0)$.
4. 4. Matrix of transitive closure Q (2) is used to build

- matrix "Window of opportunities ext-int" on the basis of which opportunities X^{op} and threats X^{th} of environment and their importance (how great is their influence on factors of internal environment) are determined, $X^{op} \cup X^{th} \subset X^{ext}$;
- matrix "Window of opportunities int-ext" with the purpose of determination of internal opportunities of SEO that can neutralize the threats of environment X^{th}, and also the problems connected with possible negative influence of SEO on environment X^{ext};
- matrix "Window of opportunities op-th". The analysis of interferences between opportunities and threats allows to reveal opportunities which promote parrying of threats;
- matrix "Window of opportunities st-w" for revealing the latent internal opportunities allowing to remove weaknesses of SEO due to advantages.

As a result of the analysis all factors are being grouped into the following classes: S (Strengths), W (Weaknesses), O (Opportunities), T (Threats). Factors inside of each class are being ranked according to the force of their favourable (unfavourable) influence on the factors of another class. This procedure lets us estimate the importance of strengths and weaknesses, opportunities and threats for SEO development.

3.3 Structure and goal analysis of SEO development

When setting the goals of a SEO development a decision maker doesn't always manage to trace if the goals he has set are inconsistent, i.e. reaching of a goal will prevent from reaching of another one. Inconsistency of goals can also be influenced by the chosen ways of their reaching.

Thus, it is very important to reveal the contradictions already at the stage of goal setting.

The technique of the structure and goal analysis of SEO development (Avdeeva et al., 2003) allows to determine integrated (direct and all possible indirect) influences of one factor on the other and due to it to reveal inconsistencies between goal and control factors. The structure and goal analysis also allows to determine the most effective controls.

The goal of a situation development is described by a subset of goal factors of cognitive model. That means that the vector of goals of a situation development is a vector of values of goal factors (fixed goal), or a vector of directions of change of these values (unfixed goal).

Definition 9. The fixed vector goal includes vector of goals Y^* and preset values of trends of change of each goal Y_i^*.

The fixed vector goal is a point in m-dimensional space of trends of goal change. In other words, goal is a vector of some "ideal" values of trends of goal factors change.

Definition 10. The unfixed vector goal includes vector of goals Y^* and directions of favourable trends of change of its coordinates according to their AFDs.

Vector of favourable trends is a vector of interests of a decision maker (analyst). Restrictions are not imposed on the value of favourable change of goal factors (the more - the better).

Thus, the structural analysis of cognitive model of a situation development under control consists of the following stages:

Stage 1 - analysis of goals (coordinates of a vector of goals) on mutual consistency in order to answer the question "whether the vector of goals (fixed or unfixed) is inconsistent, i.e. whether the reaching of any of goals (coordinates in a vector of the given goals) will prevent from reaching of other goals?"

Let $Y = \{y_1 ... y_m\}$ be a set of goal factors and r(Y) - a vector of desirable AFDs.

Definition 11. Vector of goals *Y* is consistent if

$$r_j r_k = \mathrm{sign}(q_{ik}) \text{ for any } y_i, y_k \in Y. \quad (5)$$

where q_{ik} - (i, k)-th element of matrix Q (3).

If (5) is fulfilled for goal factors y_i, y_k they refer to as consistent, otherwise these factors are inconsistent.

When the consistent vector of goals is formed the desirable integrated change of any of goal factors will not result in undesirable integrated change of other goal factors in a vector of goals.

Stage 2 - check of a consistency of the set of control factors with the given vector of goals, i.e. whether the change of the value of any control factor (with the help of the appropriate control) will promote reaching of some goals in a vector of goals and at the same time prevent from reaching of other goals of a vector of goals.

Definition 12. Vector of control factors is consistent with a vector of goals *Y*, if for each coordinate of a vector of control actions $U=(u_1...u_p)$ it is possible to determine such sign, that for a resulting sign vector sign (U) it will be fair:

$$r_s = sign(q_{st})sign(u_t) \text{ for any } u_t \in U, y_s \in Y \qquad (6)$$

When control factors are consistent with the vector of goals and (6) is fulfilled, any change of control factors according to a vector sign (U) will not cause the change of any coordinate of a vector of goals Y in undesirable direction. Let U*(0) be a vector of control actions the signs of which are selected according to (6), and $|U^*(0)|$ - vector $U^*(0)$ in which all coordinates are replaced with their absolute values. The concepts entered above allow to formulate the following statement.

Statement. If the selected vector of goals Y is consistent and the set of control factors is coordinated with a vector of goals it is possible to choose such vector of control actions U for which it will be fair

$$|U^*_1(0)| \le |U^*_2(0)| \rightarrow Y(U^*_1(0)) \le Y(U^*_2(0))$$

where $Y(U^*_i(0))$ is a vector of changes of goal factors caused by activation of vector of control actions $U^*_i(0), i=(1...m)$, i.e. property of "domination" by modules of control transfers into property of "domination" by results of their influence on goal factors.

In other words, more "intensive" control (with large absolute values of coordinates) will cause more "intensive" changes of coordinates of the goal vector in desirable directions.

The mentioned definitions are used in situation analysis and modelling. Thus, violation of conditions of consistency of the selected vector of goals can help the analyst to understand the interaction of goal factors and set his vector of goals "more correctly", conforming to the situation. Analysis of the vector of selected control factors on consistency with a vector of goals will allow to resign inconsistent control actions and, on the contrary, to more actively use "advantageous" control factors, the change of which according to control actions affecting them will result in great favourable changes of goal factors.

Stage 3 - estimation of efficiency of influence of control factors on all coordinates of the vector of goals. Such estimation is useful when choosing the most effective control factors the changes of which with the help of the selected control actions will provide the purposeful development of a situation.

Formally, the parameter of efficiency $E(u_k)$ of the control factor u_k (i.e. the maximal positive effect from the change of u_k) is determined as absolute value of the sum of coefficients of influence of the given control factor u_k on the goal factors multiplied by AFDs of the goal factors, i.e.

$$E(u_k) = \left| \sum_{i=1}^{m} r_i q_{ki} \right|$$

where r_i is the AFD of the goal factor y_i,

q_{ki} - (k,i)-th element of matrix Q.

Really, the maximal positive effect Δy from realization of control u_k on the factor x_k is estimated as

$$\Delta y = \left(\sum_{i=1}^{m} r_i q_{ki} \right) u_k$$

where the sign of action g_k coincides with the sign of the sum $\sum_{i=1}^{m} r_i q_{ki}$, and its value is equal to 1.

On the basis of results of SWOT- and Structure and goal analysis we form the set of vectors of goals of SEO and vectors of controls that reflect possible strategy of secure SEO development. On the basis of modelling controlled development SEO general strategy of SEO purposeful development is worked out in view of trends in environment.

The combination of the stages described above enables one to diagnose and define problems in SEO development and find ways of its solution and form well-founded goals and strategy of SEO development.

3.4 Approach to deriving of strategy for problems solving

In general, the control of socio-economic system can be represented as construction of strategy for the system development, defining the main goals and general directions for their reaching, and its implementation.

Revealing the system development problems influencing negatively achievement of strategic control goals is one of the key stages of construction of strategy for socio-economic system control.

This section presents the general scheme of method for forming solution strategy for ill-structured problems on the base of linear dynamic models on the base of cognitive map in regard to socio-economic systems (Fig.4).

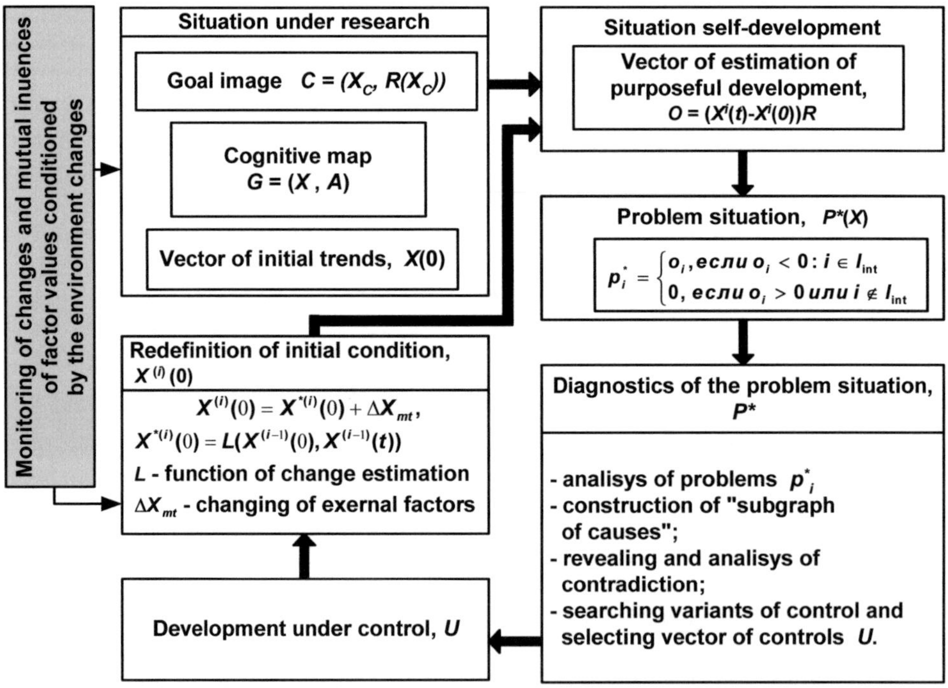

Fig. 4. General scheme of method for forming strategy for ill-structured problems solution

The control problem consists in transfer of socio-economic system into one of the states corresponding to goal image. At that, the proposed approach allows determining the system state in both values of model factors and rates of factor changes.

In the variant of approach described in this paper the state of socio-economic system is defined by rates of change of model factors. Correspondingly, the dynamics of change in simulated situation is analyzed on the basis of factor trends.

The goal image of socio-economic system defines desirable directions of changing of the system state from the position of control subject. Formally, it is represented as

$$C = (X^C, R(X^C)) \tag{7}$$

where X^C is the subset of goal factors, $X^C \subseteq X$ (X is the set of model factors), $R(X^C)$ is the vector of estimations of factor dynamics defining desirable directions for changing of goal factors:

$$R(x_i^c) = \begin{cases} +1, \text{ if acceleration of } x_i^c \text{ changing rates} \\ \quad \text{is desirable,} \\ -1, \text{ otherwise.} \end{cases}$$

The strategy for socio-economic system development problems solution consists of w strategic steps that define the sequence of system state changing

$$S^0 \to S^1 \to S^2 \ldots \to S^m \to S^C$$

where S^0 is the initial state, S^C is the goal state (corresponding to goal image (7)), $S^i \to S^{i+1}$ is the strategic step, at which the problem is revealed, and, on the base of its analysis, from the set of model factors X, the subset of local goals (goal factors) and the subset of controls (control factors) are selected. At that, changing of control factors results in desirable changing of goal factors.

Each strategic step $S^i \to S^{i+1}$ includes the following:

Revealing the problem on the base of self-development simulation for initial state at i-th step (Maximov, 2001; Maximov & Kornoushenko, 2001). As a result, the problem becomes defined more precisely in form of the subset P^i of factors, which changes do not correspond to the goal image.

Diagnostics of the problem via construction of "subgraph of causes" and structure and goal analysis (Avdeeva, 2006; Avdeeva et al., 2003, Maximov & Kornoushenko, 2001) with the purpose of extracting the subset of local non-contradictory factors Y^i from P^i and searching variants of control (subsets of control factors) U_j^i facilitating change of Y^i in desired direction.

Non-contradiction of the goal factors means that the desired change of any goal factor from Y^i does not result in undesirable change of other factors form Y^i.

Simulation of the controlled system development consisting the base for forming various scenarios of the controlled development for resolving the problem applying obtained variants of control U_j^i and carrying out the comparative appraisal of scenarios with the purpose of selecting control being optimal for this strategic step (Avdeeva et al., 2007; Avdeeva, 2006; Avdeeva et al., 2003; Maximov, 2001).

Redefinition of initial condition for the next strategic step via transformation $L(S^{i*})$ of the resulting state factor values S^{i*} at current step (Avdeeva, 2006) taking into account results of

monitoring (control) of changes and mutual influences of factor values conditioned by the environment changes (Avdeeva et al., 2007; Avdeeva, 2006; Avdeeva et al., 2003).
At the next step, the cycle is repeated since new initial conditions can result in new problems preventing the desired development of socio-economic system.
The strategy construction process is finished with reaching satisfactory result that consists in reaching nearest approach to the desired system state corresponding to the goal image in conditions of given constraints (invariability of model structure, limitations on control resources, etc.) (Avdeeva, 2006). This means that, starting from some strategic step, estimation of purposeful development of the system given as functional of goal achievement degree does not change.

4. Conclusions

Experience of applying of various models and methods on the basis of cognitive approach (in Russia and abroad), as well as increasing interest of practical specialists to developments in this directions show expediency of development of this approach in control. At that, we should note some unsolved (or partially solved) problems. The problem of risks due to the human factor in the field of formal methods of searching and making decisions in the control of complex and ill-structured situations is considered as the general problem for diverse models of subject-matter experts' knowledge and related formal methods (Abramova & Novikov, 2006; Abramova, 2007a).
Let us outline some directions of research planned by the authors within the framework of development of cognitive approach.

- **Development of theoretical principles, methods, and technologies for constructing models on the base of cognitive approach while studying ill-structured systems and situations.** This direction supposes forming the main principles and system of criteria directed toward the following:
- increasing coordination and mutual understanding between participants of the process of resolving complex problem situation.
- increasing formalization authenticity for initial knowledge (representation) of the problem situation.

The general conceptual scheme for control of model construction process for ill-structured system (situation) is developed. The criteria of appropriate transition from initial representation of ill-structured system (situation) in form of the cognitive map to one or another mathematical model defining further formal processing of initial representations are formed. Applicability of formal model is estimated reasoning from features and specificity of considered system (situation).
The approach to formalization of initial representations of ill-structured problem in form of collective cognitive map with the purpose of generalization and agreement of different representations between problem bearers, which are competent in various object areas of knowledge. Solution of this problem rests upon the developed methods of conceptual structuring (Avdeeva, 2006; Avdeeva et al., 2003), as well as criteria and particular technologies of forming and agreement of collective concepts (Abramova et al., 1999).
The established problem of risks caused by human factor is now under research directed onto the following:

- generalizing results of empirical verification of hypotheses about control subjects put into typical methods of decision making support;

- making recommendations on estimation of adequacy of subject-formal methods for solving problems of control in ill-structured situations with describing the methods in a language using principles proposed at the first stage with selective empiric check;
- developing conception of cognitive simulator for users of simulation systems on the base of cognitive maps taking into account cognitive features and typical risks caused by human factor in the life circle of the considered subject-formal methods;
- making recommendations on revealing and blocking risks while applying the methods on the base of cognitive maps while solving practical control problems with selective empiric check of recommendations.
- **Development of instrumental tools for support of intellectual activity of subject in control of development of ill-structured systems and situations.** By now, the software analytical system (STRICE - Strategy Intelligent Creation Environment) implementing the function of model construction on the base of cognitive maps, scenario simulation and comparative estimation of scenarios has been developed (Avdeeva et al., 2007, Avdeeva, 2006, Avdeeva et al., 2003). The modular architecture of the developed system allows to build it up with other tools for solving various control problems, as well as interaction with modern information and analytical systems (for example, systems of gathering and analysis of information, ERP-systems).

Further development of this software system is aimed at the development of interface providing soft intellectual control of purposeful process of generation of formalized knowledge of subject (individual and collective).

5. References

Abelson, R.P. & Rosenberg, M.J. (1958). Symbolic psycho-logic: A model of attitudinal cognition. *J. Behavioural Science*, №3, (1958), (1–13).

Abramova, N.A. (2004). About some myths in software quality estimation, *J. Reliability*, No.1, (Jan. 2004), (pp. 38-63), ISSN:1729-2646.

Abramova, N.A. (2006). A subject of intellectual activity under cognitive control of ambient intelligence, *Proceedings of 9th IFAC Symposium on Automated Systems Based on Human Skills and Knowledge*, pp. 73-78, Nancy, France, 2006.

Abramova, N.A. (2007a). On the problem of risks due to the human factor in expert methods and information technologies. *J. Control Sciences*, №2, (Mar/Apr. 2007), (11-21), ISSN: 1819-3161.

Abramova, N.A. (2007b). On the development of cognitive approach to control of ill-structured objects and situations, *Proceedings of 7th Intern. Conf. "Cognitive analysis and Situations evolution Control"*, pp.9-15 , ISBN:978-5-91450-009-9, Moscow, november 2007, Institute of Control Sciences, Moscow.

Abramova, N.A. & Kovriga, S.V. (2006). On risks related to errors of experts and analysts, *J. Control Sciences*, №6, (Nov./Dec. 2007), (60-67), ISSN: 1819-3161.

Abramova, N.A. ; Kovriga, S.V. (2008). Cognitive Approach to Decition-maiking in Ill-Structured Situation Control and the Problem of Risks, *Proceedings of Conference on Human System Interaction - HIS'2008*, pp. 83-88, ISBN:1-4244-1543-8, Krakow, Poland, May 25-27, 2008.

Abramova, N. ; Korenyushkin, A. (2006). Another approach to intransitivity of preferences, *Proceedings of 22nd European Conference on Operational Research,* p. 226, Prague, Czech republic, july 2007.

Abramova, N.A. ; Novikov, D.A. (2006). Evolution of representations about the human factor in control science, In: *Human factor in control sciences,* N. Abramova, D. Novikov, and K. Hinsberg (ed.), (5-54), URSS, ISBN: 5-484-00391-1, Moscow.

Avdeeva, Z. (2006) Revealing of threats of complex systems development on the basis of cognitive approach, Modeling and analysis of safety and risk in complex systems, *Proceeding of the seventh international scientific school MA SR – 2007,* pp. . 478-483, ISBN: 978-5-8088-0266, Saint-Petersburg, Russia, Sept. 2007, SPb.:SUAI.

Avdeeva, Z. ; Kovriga, S. & other. (2007). Cognitive approach in control. *J. Control Sciences,* №3, (May/Jun. 2007), (2-8), ISSN: 1819-3161.

Avdeeva, Z., Kovriga, S. & Makarenko, D. (2007). Cognitive approach to problem solving of social and economic object development. *Proceedings of 4th International Conference on Informatics in Control, Automation and Robotics,* pp. 432–435, Angers, France, may 2007.

Avdeeva, Z., Kovriga, S. & other. (2003). Goal setting and structure and goal analysis of complex systems and situations. *Proceedings of the 8th IFAC Symposium on Automated Systems Based on Human Skill and Knowledge,* pp. 247-252, Göteborg, Sweden, September 2003.

Axelrod, R. (1976). The cognitive mapping approach to decision making. In: *Structure of Decision. The Cognitive Maps of Political Elites,* R. Axelrod (ed.), (3-18), ISBN: 069107578-6, Princeton University Press, Princeton.

Chaib-draa, B. Causal maps: theory, implementation, and practical applications in multiagent environments. *IEEE Trans. on Knowledge and Data Engineering,* Vol.14, №6, (november/december 2002), (1201-1217), ISSN: 1041-4347.

Bouzdine-Chameeva, T. (2006). An application of causal mapping technique ANCOM-2 in management studies, *Proceedings on the 6th Global Conference on Business & Economics,* pp. 11-21, ISBN: 0-9742114-6-x, USA, oktober 2006, Gutman Conference Center, USA.

Dorner, D. (1997). *The Logic of Failure: recognizing and avoiding error in complex situations,* Basic Books, ISBN: 978-0201479485, Massachusetts.

Eden, C. (1988). Cognitive mapping. *Eur. J. of Operational Res.,* Vol.36, №1, (1988), (1–13), ISSN: 0377-2217.

Gabriel, I. J. ; Nyshadham, E. (2008). A Cognitive Map of People's Online Risk Perceptions and Attitudes: An Empirical Study, *Proceedings of the 41st Hawaii International Conf. on System Sciences,* pp. 1-10, Big Island. 2008.

Hart, J. (1976). Comparative Cognition: Politics of International Control of the Oceans. In: *Structure of Decision. The Cognitive Maps of Political Elites,* R. Axelrod (ed.), (180-221), ISBN: 069107578-6, Princeton University Press, Princeton.

Heradstvein, D. & Narvesen, O. (1978). Psychological constraints on decision-making. A discussion of cognitive approaches: operational code and cognitive map. *J. of Cooperation and Conflict.* Vol.13, №2, (june1978), (77 – 92), ISSN: 0010-8367.

Hodgkinson, G.; Maule, A. & Bown, N. (2004). Causal Cognitive Mapping in the Organizational Strategy Field: A Comparison of Alternative Elicitation Procedures, *Organizational Research Methods*, Vol. 7, No. 1, (Jan. 2004), (pp. 3-26), ISSN: 1094-4281, SAGE Publications

Holsti, O. (1976). Foreign policy formation viewed cognitively. In: *Structure of Decision. The Cognitive Maps of Political Elites*, R. Axelrod (ed.), (18-55), ISBN: 069107578-6, Princeton University Press, Princeton.

Huff, A. (1990). Mapping strategic thought. In: *Mapping strategic thought* A. Huff (ed.), (pp. 11-49), Wiley, Chichester.

Kelly, G.A. (1955). *The psychology of personal constructs. Vol.1: A theory of personality*. Norton, New York.

Kim, D. (2000). A simulation method of cognitive maps. *Proceedings of 1st International Conference on Systems Thinking in Management.*, pp. 294-299, 2000, Geelong, Australia, Deakins University, Geelong.

Kosko, B. (1986). Fuzzy cognitive maps. In: *Int. J. of Man-Machine Studies*, №1, (1986), (65–75), ISSN 0020-7373.

Kremer, R. (1994). Concept mapping: informal to formal, *in Proc. of the Second International Conference on Conceptual Structures*, pp. 45-55, ISBN 3-540-58328-9, College Park, Maryland, USA, Aug. 1994, Springer, Maryland.

Kulba, V.V.; Kononov, D.A. & other. (2004). *Methods for forming development scenario for socio-economic systems*. SYNTEG, Moscow.

Kuznetsov, O.P.; Kilinich, A.A. & Markovskii, A.V. (2006). *Influence analysis in control of ill-structured situations based on cognitive maps*. In: *Human factor in control sciences*, N. Abramova, D. Novikov, and K. Hinsberg (ed.), (311-344), URSS, ISBN: 5-484-00391-1, Moscow.

Larichev, O.; Moshkovich H. (1997). *Verbal decision analysis for unstructured problems*. Kluwer Academic Publishers, Boston.

Maximov, V. & Kornoushenko, E. (2001). Analytical basics of construction the graph and computer models for complicated situations. *Proceedings of the 10th IFAC Symposium on Information Control Problems in Manufacturing*, Vienna, Austria, september 2001, Vienna University of Technology.

Maximov, V. (2001). Cognitive analysis and situation modelling. *Proceedings of the 8th IFAC Conference on "Social Stability: The Challenge of Technology Development"*, Vienna, Austria, september 2001, Vienna University of Technology.

Mls, K. (2004). From concept mapping to qualitative modeling in cognitive research, *Proceedings of the 1st International Conference on Concept Mapping*, pp. 159-163, Pamplona, Spain. 2004.

Obal, L., Judge R. & Ryan T. (2007). The Influence of Cognitive Mapping, Learning Styles and Performance Expectations on Learning Effectiveness, *Proceedings of AIS SIG-ED IAIM Conference*, pp. 1-20, Montreal, Canada. 2007.

Peña, A.; Sossa, H. & Gutiérrez, A. (2007) Cognitive maps: an overview and their application for student modeling. *J. Comp. y Sist.*, Vol.10, №3, (Jan./Mar. 2007), (230-250), ISSN: 1405-5546.

Roberts, F. (1976). *Discrete mathematical models with applications to social, biological and environmental problems.* Prentice Hall, ISBN: ISBN-13: 978-0132141710, New Jersey.

Sawaragi, T.; Iwai, S. & other. (1986). An integration of qualitative causal knowledge for user-oriented decision support, In: *Control Theory and Advanced Technology,* №2, (1986), (451-482), ISSN: 0911-0704.

Shapiro, M.J. & Bonham, G.M. (1973). Cognitive processes and foreign policy decision-making, In: *International Studies Quarterly,* №17, (1973) , (147–174), ISSN: 0020-8833.

Simon, H. (1973). The Structure of Ill-structured Problems, In: *Artificial Intelligence,* №4, (1973), (181–202).

Tolman, E.C. (1946). Cognitive maps in rats and men, In: *Psychological Review,* № 55, (1946), (189–208).

Vavilov, S.V. (2006). Psychological Space of Managerial Decisions, *J. Sociological investigations,* no. 5, (May 2006) ,(pp 93-102), ISSN: 0132-1625.

7

Attentional Selection for Action in Mobile Robots

Pilar Bachiller, Pablo Bustos and Luis J. Manso
Universidad de Extremadura
Spain

1. Introduction

During the last few years attention has become an important issue in machine vision. Studies of attentional mechanisms in biological vision have inspired many computational models (Tsotsos et al., 1995; Itti & Koch, 2000; Frintrop et al., 2005; Torralba et al., 2006; Navalpakkan & Itti, 2006). Most of them follow the assumption of limited capacity associated to the role of attention from psychological proposals (Broadbent, 1958; Laberge, 1995). These theories hypothesize that the visual system has limited processing capacity and that attention acts as a filter selecting the information that should be processed. This assumption has been criticized by many authors who affirm that the human perceptual system processing capacity is enormous (Neumann et al., 1986; Allport, 1987). From this point of view, a stage selecting the information to be processed is not needed. Instead, they claim the role of attention from the perspective of selection for action (Allport, 1987). According to this new conception, the function of attention is to avoid behavioural disorganization by selecting the appropriate information to drive task execution. Such a notion of attention is very interesting in robotics, where the aim is to build autonomous robots that interact with complex environments, keeping multiple behavioural objectives. Attentional selection for action can guide robot behaviours by focusing on relevant visual targets while avoiding distracting elements. Moreover, it can be conceived as a coordination mechanism, since stimuli selection allows serializing the actions of, potentially, multiple active behaviours.

To exploit these ideas, a visual attention system based on the selection for action theory has been developed. The system is a central component of a control architecture from which complex behaviours emerge according to different attention-action links. It has been designed and tested on a mobile robot endowed with a stereo vision head. Figure 1 shows the proposed control model. Sensory-motor abilities of the robot are divided into two groups that lead to two subsystems: the visual attention system, which includes the mechanisms that give rise to the selection of visual information, and the set of high-level behaviours that use visual information to accomplish their goals. Both subsystems are connected to the motor control system, which is in charge of effectively executing motor responses generated by the other two subsystems.

Each high-level behaviour modulates the visual system in a specific way in order to get the necessary visual information. The incoming flow of information affects high-level behaviours guiding their actions, so the attentional system also modulates the behavioural system. This double modulation results in two control loops running in parallel, affecting each other and working in cooperation to get a common objective. The control loop associated to the visual system controls ocular movements that lead to the fixation of a visual target. The other loop controls the sensory-motor activity of high-level behaviours.

The attentional nature of the proposed system favours the simultaneous execution of multiple behaviours. In our model, action selection is solved by selecting a visual target. Fixation of a focus of attention ensures that only actions compatible with the selected information take place, solving this way the problem of coordination among behaviours. The sensory selection matter can be treated easily in comparison to the action selection problem. Actions could be contradictory, making more difficult the process of selecting the most suitable one. However, stimuli are selected according to the relevance of their properties in the current situation. At any given time and for any situation, the stimulus having the most suitable properties can always be found. It results in a sensory selection that provides the necessary information to carry out the most appropriate action for that situation. Extending this idea, attention acts as a means of action serializing, which is produced by the ordered execution of actions associated to the sequence of selected stimuli. The specific interleaving among actions comes from a time relation that links internal system requirements and external world features.

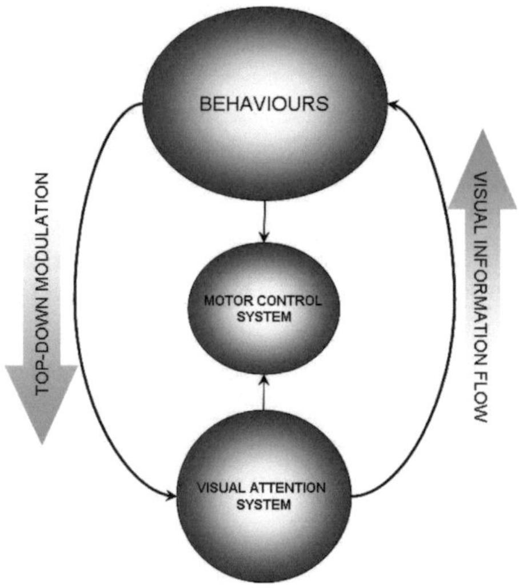

Fig. 1. Attention-based control model

2. The visual attention system

The proposed attention system has been modelled as a set of components collaborating to select, fix and track visual targets according to different task requirements. Components are organized in different processing levels as shown in figure 2. The low-level components are related to image acquisition, motor control, as well as computation and maintenance of regions of interest (ROI's). Components of intermediate level extract sets of ROI features related to "what" (appearance information) and "how" (spatial information) issues. These features are used by high-level components, called target selectors (TS), to drive attention according to certain top-down behavioural specifications. Attention control is not centralized, but distributed among several target selectors. Each of them drives attention from different top-down specifications to focus on different types of visual targets. At any given time, overt attention is driven by one TS, while the rest attend covertly to their corresponding targets. The frequency of overt control of attention of each TS is modulated by high-level behavioural units according to their information requirements. The fixation of a selected target is accomplished by two independent camera movements: a saccadic and tracking movement in one of the cameras and a vergence movement in the other. This allows controlling attention from monocular information while keeping stable binocular fixation.

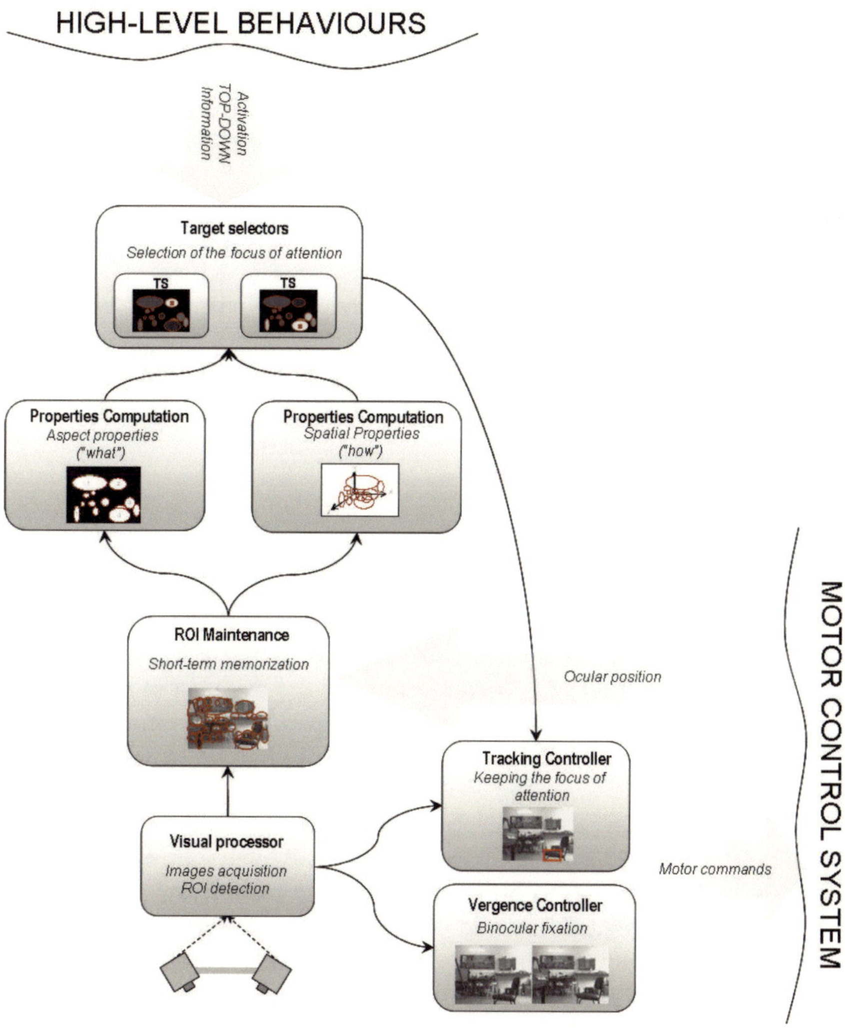

Fig. 1. Architecture of the proposed visual attention system

The whole process constitutes a perception-action cycle that begins with image acquisition and detection of regions of interest and provides a motor response as a result of the selection of a focus of attention. The rest of this section details the different processing stages of this cycle.

2.1 ROI detection and maintenance

In its first stage, system detects image regions that exhibit highly informative contents. To get stable results, the detection process must be invariant to different transformations, such as different scales or illumination changes. Several interest point detectors have been proposed in the literature (Lindeberg, 1998; Lowe, 1999; Mikolajczyk & Schmid, 2001). Among the existing approaches, the Harris-Laplace method, proposed by Mikolajczyk and Schmid (Mikolajczyk & Schmid, 2001), shows outstanding results regarding to important scale changes as well as different rotations, translations and changes of illumination. This method is highly suitable for our purpose. However, looking for a reduction in processing time, we propose a variant of the Harris-Laplace interest point detector. Our proposal consists in applying the Harris-Laplace method to the multiscale prism resulting from a centred section of the scale space representation (figure 3). This new representation exhibits a Cartesian topology that approaches the retinal structure of the human eye (Bandera & Scott, 1989).

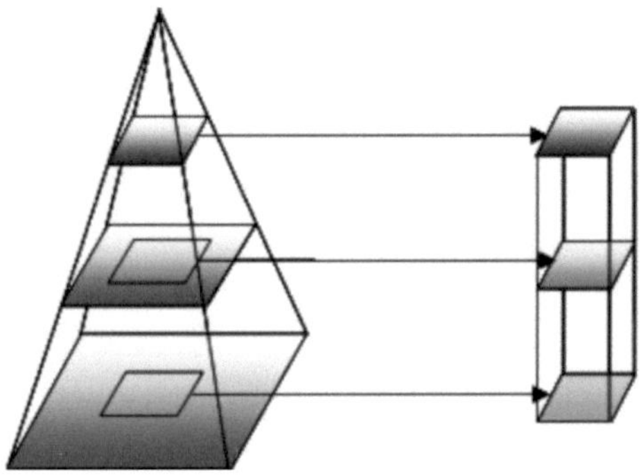

Fig. 3. Multiscale prism

The multiscale prism keeps an eccentricity-dependent resolution structure. With this representation, foveal areas are represented at every resolution level, whereas peripheral areas only appear at the higher levels. As a consequence, Harris-Laplace method detects regions of any size at the fovea and big-size regions at the periphery (figure 4). This provides a significant reduction of processing time at the expense of some loss of visual information. Nevertheless, non-detected regions correspond to less informative areas of the periphery. They can be recovered at any moment placing the fovea onto these areas through camera movements.

Regions detected by the above process are integrated in a map built on the camera reference frame. This map serves as a short-term memory that maintains information about recently perceived regions of interest. Each region is indexed by its angular position. This allows keeping information about regions situated outside the current visual field, so attention can quickly shift to previously seen regions.

At every processing cycle, new detected regions are stored in this memory and the previously stored ones are updated according to the new visual information of the scene. The updating process consists in searching every stored region in the neighbourhood of its position inside the current scale space. The spatial search allows locating regions whose size changed due to, for example, a translation movement of the robot.

Fig. 4. Harris-Laplace regions from multiscale pyramid (upper images) and prism (lower images)

Each region maintains an attribute of permanence time that is increased or decreased depending on the success of its position updating. Regions whose permanence time is under a threshold are considered forgotten and are thus removed from the memory map. For remaining regions, other attributes, such as raw image window, 2D velocity and attention time, are updated. The group of attributes computed during this stage forms a first set of properties that are extended in later stages of the system. To get different pathways of processing that allow computing properties in a separate and independent way, regions keep an identification number. This identification value allows integrating subsets of separately computed properties that come from the same original region.

2.2 Computation of high-level properties

At this stage, processing flow is divided into two subsystems that compute high-level properties of each region of interest. These two subsystems are an analogous of the "*what*" and "*how*" systems proposed in neuroscience (Milner & Goodale, 1995). This division allows a specialization of functions, dedicating specific resources to each subsystem and sharing what is common from low-level processes. The "*what*" system computes properties related to the visual appearance of regions. The "*how*" system extracts spatial information of each visual region, such as position, orientation and movement.

The "*what*" system

Properties computed by the "*what*" system must provide region categorization from its visual appearance. To get consistent results in this categorization process, properties must be invariant to transformation of the point of view, changes of illumination, etc. Histogram-based descriptors, like SIFT (Lowe, 2004), RIFT and Spin (Lazebnik et al., 2005), provide stability against different image deformations. Experimental results obtained from several comparative studies (Mikolajczyk & Schmid, 2005) show better discriminative power of this kind of descriptors in relation to traditional descriptors, like banks of filters and differential invariants. Based on these results and considering a lower computational cost, we use RIFT and Spin descriptors (Lazebnik et al., 2005) to represent aspect properties of regions. Both descriptors are invariant to scale and rotation changes and extract information about gradient orientations and image intensities, respectively.

The "*how*" system

The group of properties extracted by the second subsystem is composed of features of regions that describe their action possibilities. They must provide information about how to interact with each region. In our current implementation, properties computed by this component are: 3D position of the region computed using stereo matching; spatial motion from region tracking and 3D position; and planarity and plane orientation of the overall region gathered by homography estimation.

2.3 Selection of the focus of attention

The selection of the focus of attention is accomplished by multiple control components (target selectors) that individually determine the region to get the focus of attention following specific criteria. Distributed control of attention provides strong advantages compared to centralized control. Firstly, it provides a clearer and simpler design of the selection process. Secondly, it admits the coexistence of different types of visual targets, which is a key aspect of any task-driven attentional system. In this sense, we hypothesize that it is necessary to separate the way properties of regions are integrated for attentional selection according to behavioural objectives. For instance, in a generic navigation task such as going somewhere by following a predetermined set of landmarks, two types of visual targets can be distinguished: landmarks guiding navigation and potential obstacles. Integration of the different sets of properties characterizing both targets could turn any distracting area for the described task into a salient region, so defining global selection criteria including the two targets could be unfeasible. Moreover, although an effective integration of different kind of targets was achieved, the system can not guarantee to provide the necessary sequence of visual fixations that allow properly distributing attention time among them. If we put these ideas in the context of robot surveillance, a more important question emerges: how are reactive and deliberative abilities integrated? If attentional control is centralized, this issue becomes complicated. It is due to the fact that unexpected and foreseen things, such as moving and static objects, could be defined by contradictory properties. As a consequence, a centralized control strategy would cause either an excessive alerting state or a passive response to unsafe situations.

Individual selection process

Each target selector selects a focus of attention whose properties keep the greatest correspondence with its selection criteria. For this purpose, it computes a saliency map that represents the relevancy of each region according to its top-down specifications. This map acts as a control surface whose maxima match with candidate visual regions to get the focus of attention.

There are several alternatives for computing the saliency map. The most representative models that integrates top-down and bottom-up influences in attentional control employ a weighting process where features are modulated by multiplicative factors representing top-down specifications (Frintrop et al., 2005; Navalpakkan & Itti, 2006). This integration method presents some limitations that can be observed in figure 5. This figure shows the behaviour of a weighting method for the selection of a visual target on the basis of two properties (P_1 and P_2). For any value of P_1 and P_2, saliency is computed as:

$$saliency = P_1 w_1 + P_2 w_2 \quad (1)$$

where w_1 and w_2 are top-down weights associated to P_1 and P_2, respectively.
Assuming that both properties are equally relevant in the selection process, weights w_1 and w_2 should have the same value to express the desired target description. Hence, there is no distinction between considering simultaneous or separate matching with several properties, which limits the definition of potential selection criteria. Moreover, as it can be observed in figure 5, the resulting saliency distribution in the feature space (a) would cause some ambiguity in the selection process. Thus, as shown in (b), any horizontal line across the plane in (a) gathers stimuli with equal saliency even though they correspond to very different candidate regions.

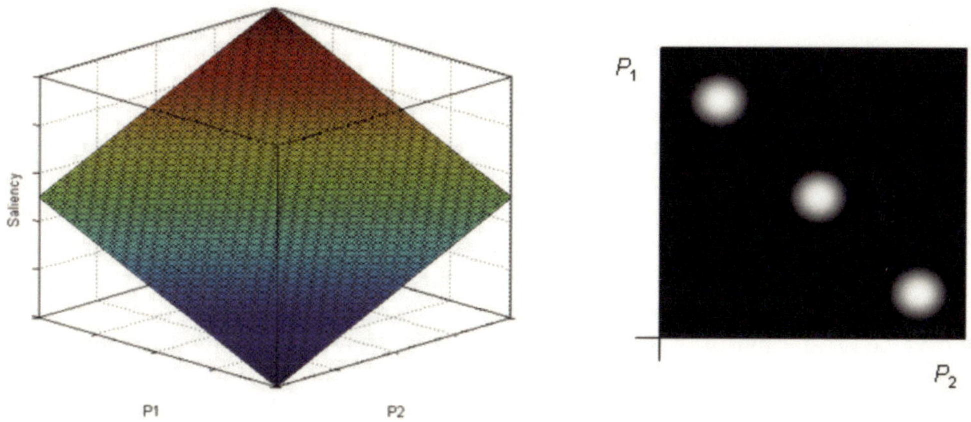

(a) Saliency distrbution in the feature space of P_1 and P_2 (b) Different stimuli with the same saliency value

Fig. 5. Saliency computation using a weighting method (see the text above for details)

The proposed selection process overcomes the limitations described above by using fuzzy techniques. Fuzzy logic (Zadeh, 1965) provides a framework of development that is very suitable for our purpose, mainly for three reasons:

- Properties used in the selection process present some degree of uncertainty and imprecision. Fuzzy logic allows dealing with this problem by treating each property as a linguistic variable.
- Selection criteria can be easily defined using fuzzy rules that associate matching of properties to saliency levels.
- Output of every rule can be fused together using a fuzzy inference process that integrates all selection criteria to provide a final saliency value.

Following these ideas, the saliency map is obtained from a set of fuzzy rules that associate, for every visual region, the fulfilment of certain selection criteria (antecedents) to different saliency levels (consequents). Each premise of a rule antecedent is expressed by a linguistic label on a region property. These labels give rise to fuzzy sets defined in the domain of the properties that are relevant for the selection of the focus of attention. Thus, each selection criterion is expressed by a rule antecedent that determines matching between properties and associated fuzzy sets. The consequent part indicates the saliency value that is assigned to a region that fulfils that selection criterion. Since selection priority depends on the degree of fulfilment of antecedents and not directly on property values, we use a zero-order Takagi-

Sugeno model (Takagi & Sugeno, 1985). Hence, rule outputs are expressed through constant values that allow ordering regions according to the corresponding selection criteria. Thus, a rule of the proposed system is written as:

IF P_1 **is** A_i **AND** P_2 is B_j … **AND** P_n **is** X_k **THEN** *saliency* $= s_{ij..k}$

where P_1, P_2, …, P_n are properties of regions, A_i, B_j, …, X_k fuzzy sets associated to those properties and $s_{ij...k}$ a constant value expressing a saliency level.

To assign specific values to each rule output, two types of rules are distinguished in our system: exclusion rules and selection rules. Strictly negatives values (considering a 0 value as negative) are associated with exclusion rules, whose antecedents define regions that have no interest in the selection process and, therefore, should not be attended. Positive values are assigned to selection rules in such a way that rules that express greater fulfilment of top-down specification will take greater saliency values than those that define less important regions.

To illustrate these ideas, a fuzzy system for an obstacle selector is shown below. In this example, robot must detect potential obstacles situated in the trajectory to a target position. Three properties are considered for determining the obstacle quality of any visual region:

- *Relative Depth (RD)*: it is defined as the relation between region depth and target depth. Both depth values are computed at a reference system situated in the ground position of the robot with its Z axis fixed in the straight trajectory to the target. This relation quantifies the proximity degree of a region to the robot, to the target or to both of them. A value greater than 1 of this property can be considered as a region situated outside a potential trajectory to the target. Three fuzzy sets are considered for this property (figure 6): *NearR, NearT* and *Far*. *NearR* considers depths that are nearer to the robot than to the target, *NearT* is related to regions situated in a nearer depth to the target than to the robot and *Far* is associated to regions situated behind the target.
- *Deviation (Dv)*: it quantifies the proximity degree of a region to the straight line between the robot and the target. It is measured by the distance between the central point of the region and that straight line. This property is partitioned in three fuzzy sets (figure 7): *Low, Medium* and *High.*
- *Height (He)*: Properties described above consider ground parallel distances between a region and the robot. However, those regions situated high enough from the ground should not be selected as obstacles, since they are not interfering areas in the way to the target. For this reason, height of regions is included as the third property that allows defining selection criteria of our system. This property allows deciding whether to include or not regions as potential obstacles, so only two fuzzy sets are considered (figure 8): *Low* and *High.*

Once fuzzy partitions have been determined, the rule system is established by defining a fuzzy rule for each combination of fuzzy sets associated to the group of properties. For the current example, it results in a system with 18 rules, i.e., one rule for each combination of fuzzy sets of *He, Dv* and *RD*. Table 1 shows a simplified representation of the whole rule system. Each inner cell corresponds to a fuzzy rule whose antecedent is formed by the conjunction of three premises: the premise about *Height* (*He*) of the superior column, the premise about *Deviation* (*Dv*) of the inferior column and the premise about *Relative Depth* (*RD*) of the corresponding row. Inside each cell, the consequent value of the associated rule is shown. For instance, the first column of inner cells represents the following three rules:

IF *He* **is** *Low* **AND** *Dv* is *Low* **AND** *RD* **is** *NearR* **THEN** *saliency* =s_{111}
IF *He* **is** *Low* **AND** *Dv* is *Low* **AND** *RD* **is** *NearT* **THEN** *saliency* =s_{112}
IF *He* **is** *Low* **AND** *Dv* is *Low* **AND** *RD* **is** *Far* **THEN** *saliency* =s_{113}

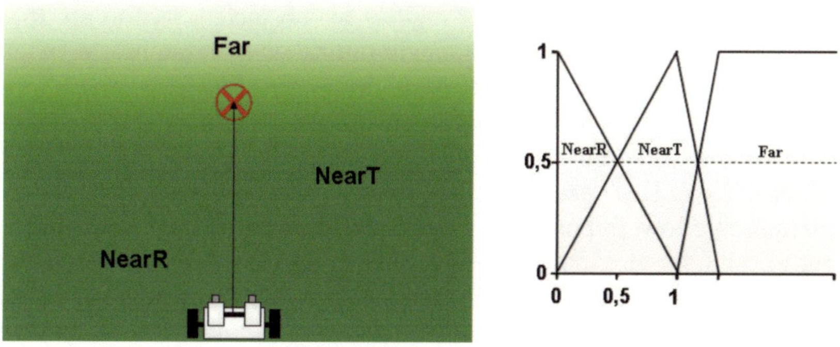

Fig. 6. Fuzzy sets of the *Relative Depth* property

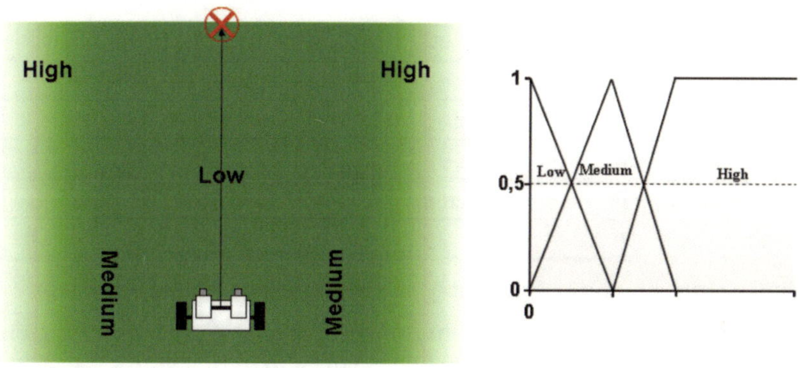

Fig. 7. Fuzzy sets of the *Deviation* property

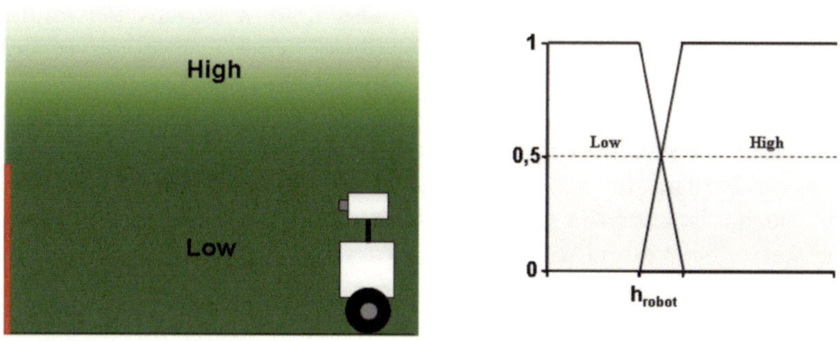

Fig. 8. Fuzzy sets of the *Height* property

Consequent values are set according to the excluding/selecting criteria of each rule. In the above example, every rule containing a premise with the form "*He* is *High*", "*Dv* is *High*" or "*RD* is *Far*" is considered an exclusion rule, since any region characterized by those properties does not interfere in the trajectory to the target. This group of regions should not be selected as obstacles, which is expressed by assigning a negative saliency value to consequents of the corresponding rules (consequents in red of table 1). Remaining rules are selection rules whose positive output values must impose a saliency order. The specification of these output values is carried out by ordering rules from the most to the least adequate in the selection process. The most important regions for obstacle selection are those situated nearest to the robot. This implies that the maximum saliency value has to be associated to regions fulfilling the hypothesis "*He* is *Low* AND *Dv* is *Low* AND *RD* is *NearR*". On the other hand, regions at near positions to the straight line between the robot and the target are more important than those situated far away from the approaching line. It means that regions that meet the premise "*Dv* is *Low*" are prioritized over those that do not fulfil the premise. All these considerations allow imposing the appropriate order for consequents of selection rules, leading to the following output relations: $s_{111} > s_{112} > s_{121} > s_{122}$. According to these relations, rule consequents can be established. Also, other considerations, such as the relative relevance of one rule in relation to others, can be taken into account in order to provide additional running conditions.

		He					
		Low			*High*		
RD	*NearR*	s_{111}	s_{121}	s_{131}	s_{211}	s_{221}	s_{231}
	NearT	s_{112}	s_{122}	s_{132}	s_{212}	s_{222}	s_{232}
	Far	s_{113}	s_{123}	s_{133}	s_{213}	s_{223}	s_{233}
		Low	*Medium*	*High*	*Low*	*Medium*	*High*
		Dv					

Table 1. Tabular representation of the fuzzy rule system of an obstacle selector

Once the set of rules is defined, the system is operative to evaluate visual regions according to their properties. This evaluation results in a saliency map that stores the saliency value of each region. Through this map, system discards regions with negative saliency and orders remaining regions from the most to the least relevant. Final selection is achieved by choosing those regions whose saliency differ less than a certain percentage from the maximum saliency value. This makes possible to decide whether to obtain several candidate regions to get the focus of attention or to select only the most salient one.

Inhibition of return

When a target selector obtains several candidate regions where focus of attention should be directed to, a mechanism for distributing attention time among all of them must be included. The mechanism used is an inhibition of return (IR) process that is applied as the final selection step of each target selector. IR process allows displacing the focus of attention from currently attended region to the next most salient one throughout time. For this purpose, it maintains an inhibition map that represents the attentional fixation degree of each area of the visual field. According to both inhibition and saliency maps a winning region is finally selected to become the next focus of attention.

Global selection process

The simultaneous running of multiple target selectors requires including a global selector that decides which individually selected region gains the overt focus of attention at each moment.

Target selectors attend covertly to their selected regions. They request the global selector to take overt control of attention at a certain frequency that is modulated by high-level behavioural units. This frequency depends on the information requirements of the corresponding behaviour, so, at any moment, several target selectors could try to get the overt control of attention. To deal with this situation, global selector maintains a time stamp for each active target selector that indicates when to cede control to that individual selector. Every so often, the global selector analyses the time stamp of every target selector. The selector with the oldest mark is then chosen for driving the overt control of attention. If several selectors share the oldest time stamp, the one with the highest frequency acquires motor control. Frequencies of individual selectors can be interpreted as alerting levels that allow keeping a higher or lower attention degree on the corresponding target. In this sense, the described strategy gives priority to those selectors with the highest alerting level that require faster control responses.

2.4 Binocular fixation of the focus of attention

Overt control of attention is achieved by binocular fixation, centring the selected visual target in both images of the stereo pair. This process keeps attention focused on the target until the next one is selected.

Our proposal for binocular control is to divide the stereo fixation into two independent camera movements: a saccadic and tracking movement in one of the cameras and an asymmetric vergence movement in the other one. This separation results in a master-slave configuration that simplifies the global control. Moreover, it allows fixating visual targets with unknown spatial properties, so binocular focusing on the target can be achieved although attention is guided by monocular information (Enright, 1998).

Tracking control

In our master-slave configuration, the master or dominant camera is in charge of fixating and tracking the focus of attention. For this purpose, a hierarchical correlation-based method is employed using a multiscale prism representation of the target. This representation keeps information about target and its near areas, allowing false matching regions to be discarded according to target neighbourhood.

The proposed tracking method starts locating positions of the highest resolution level that exhibit high similarity to the target. If more than one position is found, a new comparison is done at the next resolution level. This comparison discards regions whose neighbouring areas do not match with proximal areas around target. If more than one similar region is found again, the comparison process is repeated at the next resolution level. Thus, the method proceeds by an ascendant search of the target position in the scale space until a winning region is obtained.

Figure 8 shows two frames of a sequence of tracking obtained while robot approaches a target. In this scene, three similar regions appear in the visual field. The one situated at the right constitutes the visual target. Images show the application of the hierarchical tracking method to both situations. For each level, regions of high similarity to the target are highlighted in green. Similar regions obtained from previous level that are discarded at a given level are marked in red. As it can be observed in this figure, the method employs an ascendant searching strategy that is maintained until a unique position of high similarity is found.

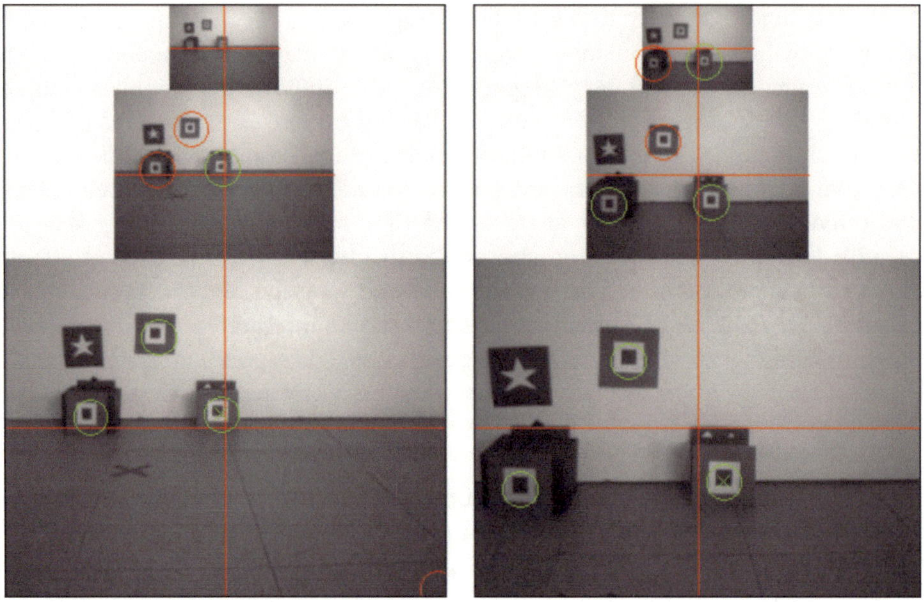

Fig. 8. Focus of attention tracking control

Vergence control

The function of an asymmetric vergence movement is to centre the visual target situated at the fovea of the dominant camera in the image of the slave camera. Hence, vergence control can be treated as a problem of maximization of the correlation coefficient measured between the central window of the dominant camera and the epipolar homologous window in the other one. Now then, the right size of that central window is dependent on the properties of the visual world, which makes it necessary to compute a new parameter: the size of the correlation window. To solve this problem, we introduce a hierarchical control method that carries out a descendent search in the scale space. This method is the inverse process used in tracking control. The reason for using a new strategy in vergence control is that the unknown size of vergence area impedes a reliable ascendant location of vergence position. In contrast, a descendent search provides both extension and position of the vergence area.

To deal with different potential extension of vergence area, the central image window of the dominant camera is represented at different resolution levels using a multiscale prism. Initial search is done at the lowest resolution level. It consists of locating the position inside the widest central image window that maximizes correlation with the corresponding image patch of the dominant camera. If a position of high similarity is found, the next level tries to find a more precise vergence position by searching in a neighbouring window of the previously obtained position. If no initial position can be found, the complete search is repeated at the lower level. This procedure is applied for each level until a final vergence position is obtained in the highest resolution level.

Figure 9 shows the proposed vergence method working in two real situations. Left images in (a) and (b) show the multiscale pyramidal representation of the image in the dominant camera. Central prism is delimited by the red squares of each level. Right images in (a) and (b) depict the multiscale representation of the image in the vergence camera. The search window within each level is represented by a red rectangle. As shown in this figure, computing vergence position (marked with an "X") at every resolution level allows selecting the search window of the subsequent level. If a given level produces no result, the process maintains the width of the search window in the subsequent one, adapting its value to the corresponding resolution. This behaviour can be observed in the second level of image (b). Once the whole process is completed, vergence position is obtained in the highest resolution level.

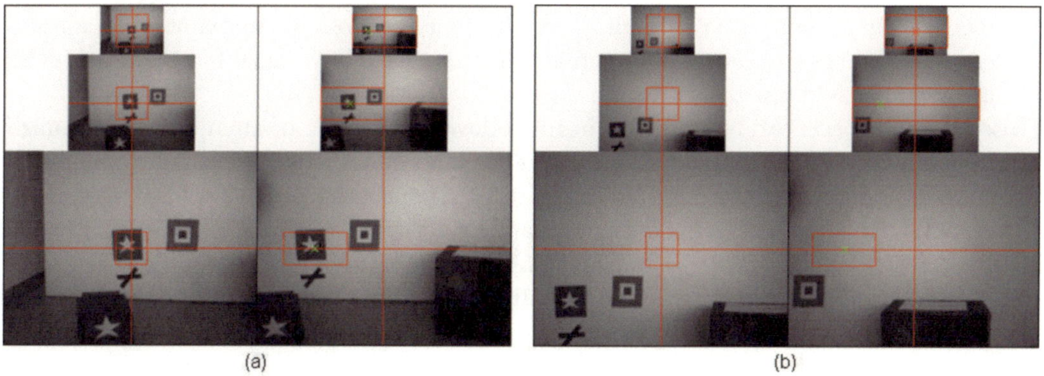

Fig. 9. Vergence control in real situations

3. Attention-based control

Intelligent robot control is carried out through pre-established links between the set of high-level behaviours and the visual attention system. Each high-level behaviour starts up an attentional control mechanism that provides a coherent visual behaviour according to its task-goals. This control mechanism is a specific target selector that responds to high-level activation and modulation by supplying the appropriate flow of visual information.

Links between attentional control components and high-level behaviours provide the necessary organization of the group of actions that solve a certain task. The attention system always guarantees the selection of a visual region that constitutes the unique visual input of every high-level behavioural unit. This sensory selection results in an effective execution of a subset of actions from the total of potential actions associated to active behaviours. The final sequence of actions, which is produced as result of the serialized attentional access to visual stimuli, forms a control dynamic that gives rise to the desired compound behaviour.

The proposed visual system models several kinds of attention that provide different high-level control dynamics:

- Bottom-up attention: selection of a focus of attention depends on properties of visual regions in the scene. If any region maintains properties that agree with selection criteria of active target selectors, the focus of attention shifts to that region. In this sense, the system provides a bottom-up control on non-searched stimuli, allowing the robot to react appropriately to unexpected events.
- Top-down attention: attention system is modulated from high-level processes through the activation of some target selectors that provide the appropriate visual behaviour for

the current situation. From this point of view, the acting context has a remarkable influence on the selection of a focus of attention, allowing global control to centre on task-relevant visual stimuli of the environment.

- Overt attention: visual attention ensures the selection of a focus of attention that acts as the unique source of visual information for high-level control components. This selection process allows high-level behaviours to produce appropriate reactions to the attended stimulus, avoiding any conflict among behaviours.
- Covert attention: coexistence of multiple target selectors allows keeping a "mental" focusing on several visual stimuli. It implies that it is possible to maintain an alerting state that allows displacing attention at any moment to any visual target for achieving different behavioural goals.

These control aspects can be observed in the following example of attention-based control for solving a task of navigation following a set of landmarks. The system is formed by the components and connections depicted in figure 10. Each behavioural component defines an attentional dynamic by activating and modulating a specific target selector. Three high-level behaviours take part in this system: *LANDMARK-GUIDED NAVIGATION, GO TO POINT* and *EXPLORE*. They modulate three attentional selectors of landmarks, obstacles and unattended regions, respectively.

Two situations are considered in this system. The first one occurs when the location of the landmark that the robot must approach is known. In this case, active behaviours are *LANDMARK-GUIDED NAVIGATION* and *GO TO POINT*. The first one activates a landmark selector and sends the aspect information of the landmark to be located. Attentional frequency on landmark is modulated by the *NAVIGATION* behaviour with an inversely proportional value to distance between landmark and robot. Thus, the less the distance to target position is, the greater the degree of attention on landmark is dedicated. When the *NAVIGATION* behaviour receives visual information that match up with the expected landmark, it sends to the *GO TO POINT* behaviour the target position in order to make the robot reach that position. To accomplish its goals, this last behaviour is connected to an obstacle selector, which allows shifting attention to regions situated close to the trajectory between the robot and the landmark. The *GO TO POINT* behaviour interprets incoming visual information as the nearest region that could interfere in the approach to landmark. Thus, it performs different actions according to the position of the attended region and the goal position. Firstly, if both positions are near enough, it is considered that the attended region is part of the target area and, consequently, it responds by approaching that region. However, if there is some distance between the target and the selected region, it is assumed that attention is focused on an obstacle that must be avoided. To get low reaction times in view of unsafe situations, obstacle selector is modulated with an activation frequency that is proportional to robot velocity.

The second situation that is considered arises when either *NAVIGATION* behaviour gets no initial information about landmark or when, after locating it, no new information is received for a long time. To deal with these situations, *NAVIGATION* behaviour deactivates *GO TO POINT* behaviour and activates *EXPLORATION*. The function of this new behaviour is to get visual access to regions of the environment that have not been recently attended. This function is accomplished through its connection to a selector of unattended regions of the visual system. This selector maintains a purely bottom-up dynamic that, together with the inhibition of return mechanism, leads to the visual exploration of the scene. Simultaneous execution of landmark and unattended region selectors give rise to the localization of the landmark that is being sought as soon as it appears in the visual field. As a consequence, landmark gains the focus of attention and system returns to the first described situation. Once the goal position is reached, *NAVIGATION* behaviour reinitiates the whole process by sending to its target selector the description of a new landmark.

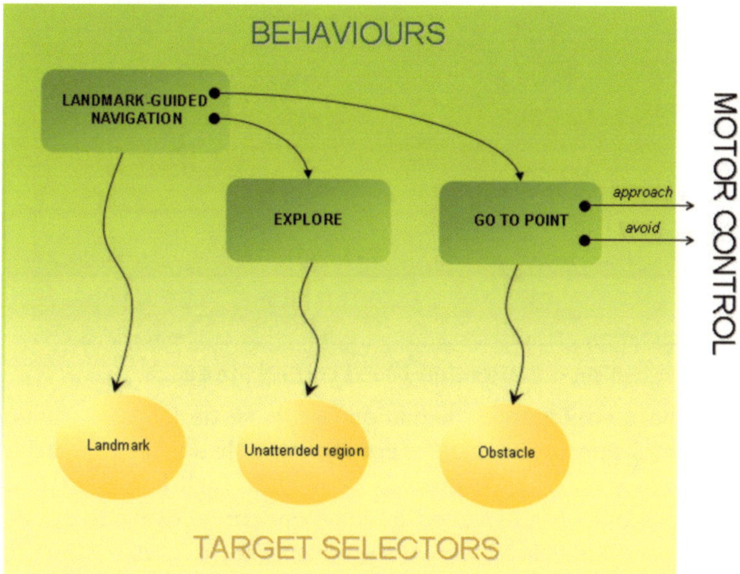

Fig. 10. Control components of a landmark-guided navigation task

The two situations described above are solved using a multi-target attentional control that exhibits several kinds of attention, providing different high-level control dynamics. On one hand, the system provides top-down attentional control on landmarks, allowing high-level processes to decide whether to approach a target position or to explore the environment to locate that position. When a landmark is located, the focus of attention alternates between landmark and potential obstacles, keeping overt control on one of the targets and covert control on the other one. This allows robot to react appropriately to obstacles while landmark position can be quickly recovered. In the exploring situation, searching is carried out through a bottom-up control where no target specifications are considered, so exploration depends exclusively on external world features. As in the previous situation, simultaneous covert control in landmark selector during exploration allows an immediate focusing on target once it is visualized.

4. Hardware and software architectures

Our attention-based control model has been tested in a five degrees of freedom mobile robot (three d.o.f. for the head and two for the base) that is endowed with a stereo vision head. It is equipped with digital PID controlled servos and two ISight Firewire cameras (Figure 11).

This model of robot has been developed in our Laboratory and is widely used for prototyping and algorithm testing.

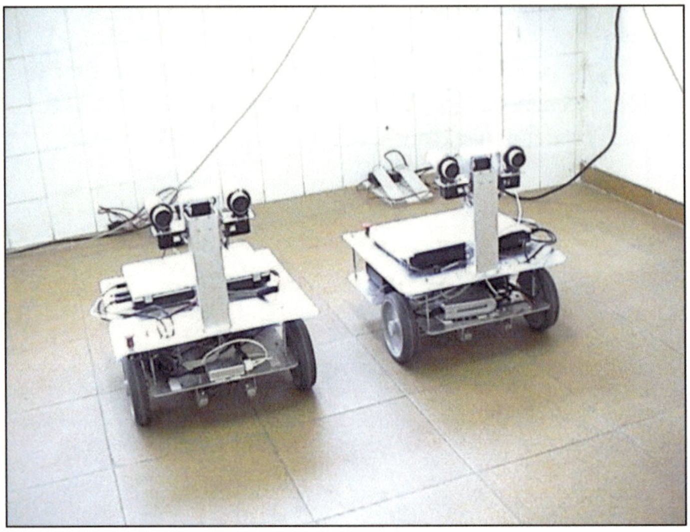

Fig. 11.Robots use for testing our attention-based control approach

To obtain a real-time modular implementation, we have designed a component-oriented software architecture that exhibits an efficient and flexible structure. Every component of the attentional system depicted in figure 2, as well as high-level behaviours and other support elements, has been implemented as independent software components following the principles of components-based software engineering (Szyperski, 1998; He et al., 2005). The Internet Communication Engine (Ice) middleware platform (ZeroC, 2007) has been used for connection among components. Thus, each component has been written as an independent C++ application that includes Ice objects and proxies to communicate with other components. The resulting architecture is characterized by a flexible structure, a high degree of reusability of its elements and real-time running capabilities (Bachiller et al., 2007).

5. Experiments

The attention-based control model has been tested through several real experiments using our testing mobile robot for solving autonomous navigation tasks. In this section, results of two experiments are presented. Several frames of the navigation sequences are shown. For every frame, two different views of the corresponding scene are presented. The first one is a view from the robot represented by the two images of the stereo pair. Through these images, the focus of attention that is selected at every moment can be appreciated. The second one is a general view obtained from a camera situated in the ceiling of the room. In this second group of images, the attended area is highlighted in red to facilitate the association between the two sets of images.

The control system employed in these experiments is the same that was described in section 3. Navigation is guided by a landmark or a sequence of landmarks that the robot must locate and approach while potential obstacles in the way are avoided.

5.1 Experiment 1: approaching a landmark with detection and avoidance of obstacles

In this first experiment, the robot must approach a landmark (star) situated in a wall in front of it. Two boxes stand in the way to the target position, so the robot must avoid them by going through the free space between them. On the floor, behind these two obstacles, several planar objects have been placed. The robot has to determine that these objects do not interfere in the trajectory to the landmark, since it can drive over them. Planarity and orientation properties of regions provide this distinction between obstacles and ground areas. Detection of ground areas is depicted in the sequence of images of the stereo pair by green regions. This process is only carried out for regions situated near to the focus of attention, so not all the planar regions placed on the floor are detected as ground regions.

Figures 12-15 show the result of this experiment. Initially (a), the robot is attending the landmark, causing the activation of the *GO TO POINT* behaviour. Consequently, attentional control responds by fixating the focus of attention on the first obstacle (b), which allows behavioural control to produce suitable avoiding actions (c-e). Proximity to the second obstacle gives rise to a shift of attention towards the corresponding area (f). This attentional change causes variations in the steering of the robot allowing the avoidance of this second obstacle (f and g). Once the robot is appropriately oriented, attention shifts again to regions near the target and new approaching actions follow (h). After some time, landmark selector recovers attentional control producing the updating of the target position (i). This updating causes new changes of attention that directs to proximal areas between the robot and the landmark. Specifically, attention focuses on plane regions of the floor (j), which allows high-level processes to detect them as ground regions. This detection gives rise to the correct treatment of the new situation that results in advancing movements towards target position (j-l). After several instants, attention is again centred on the landmark (m) making the robot to approach it in a more precise way (n-o) and finally reach the goal position (p).

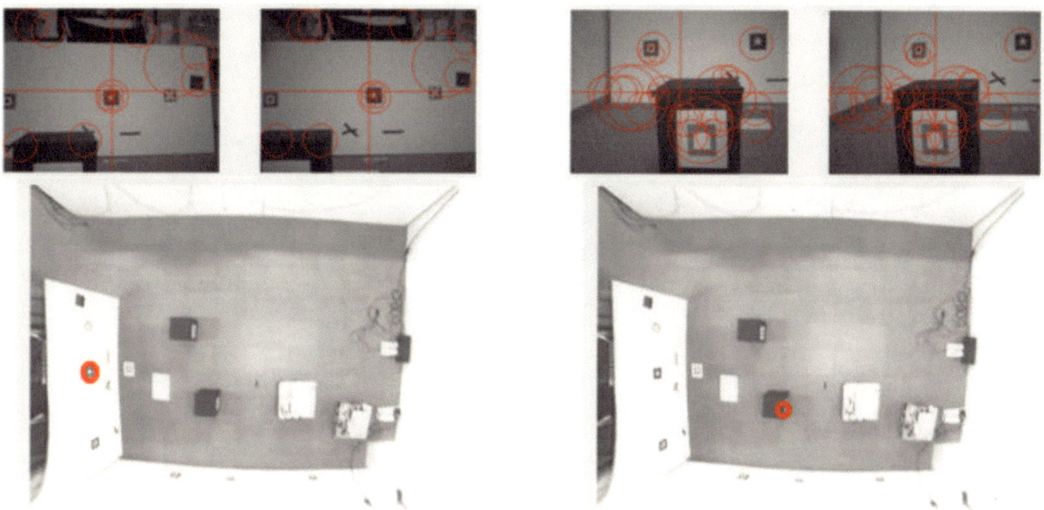

(a) (b)

Fig. 12. Experiment of navigation avoiding obstacle (first part)

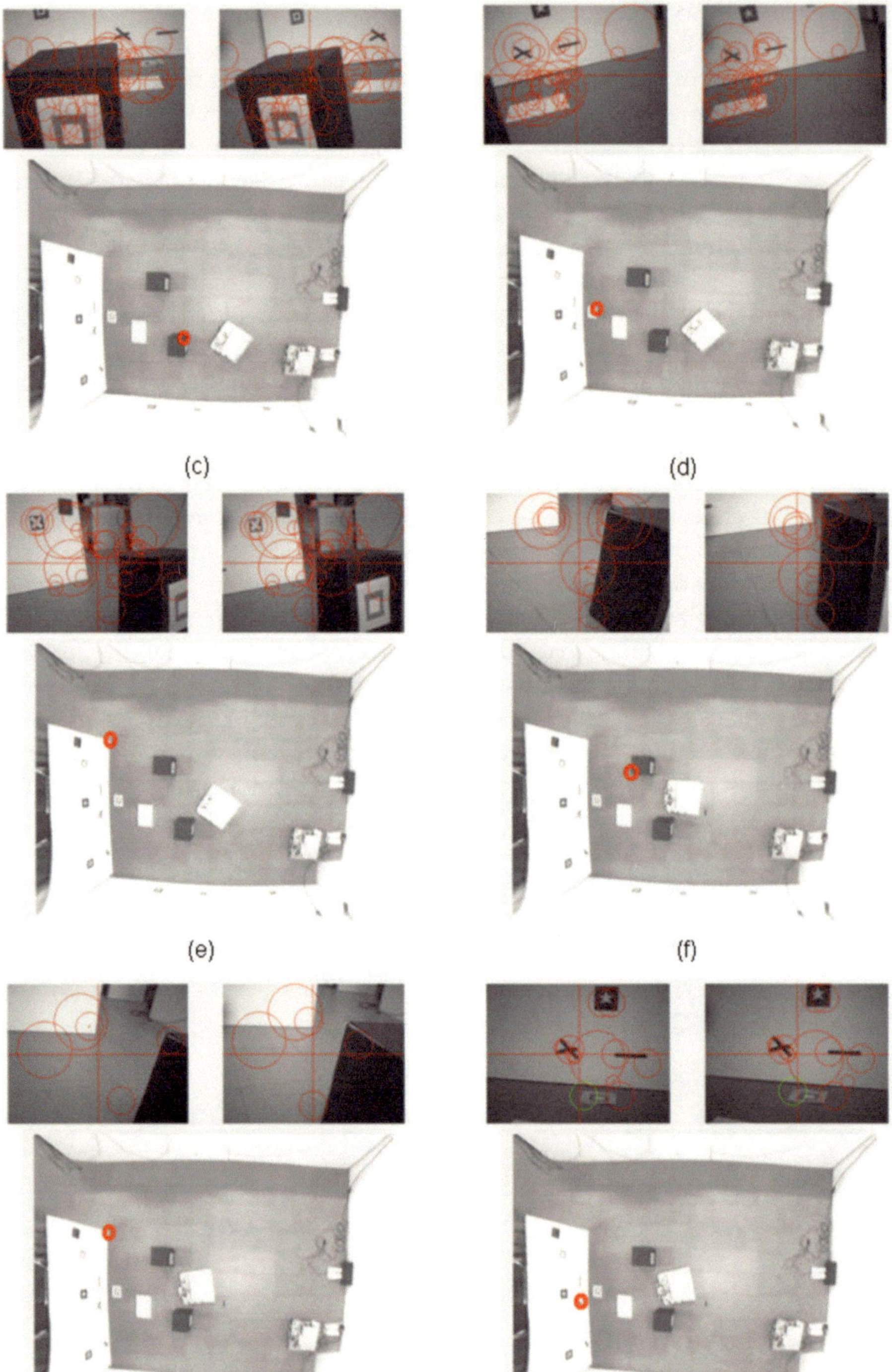

Fig. 13. Experiment of navigation avoiding obstacle (second part)

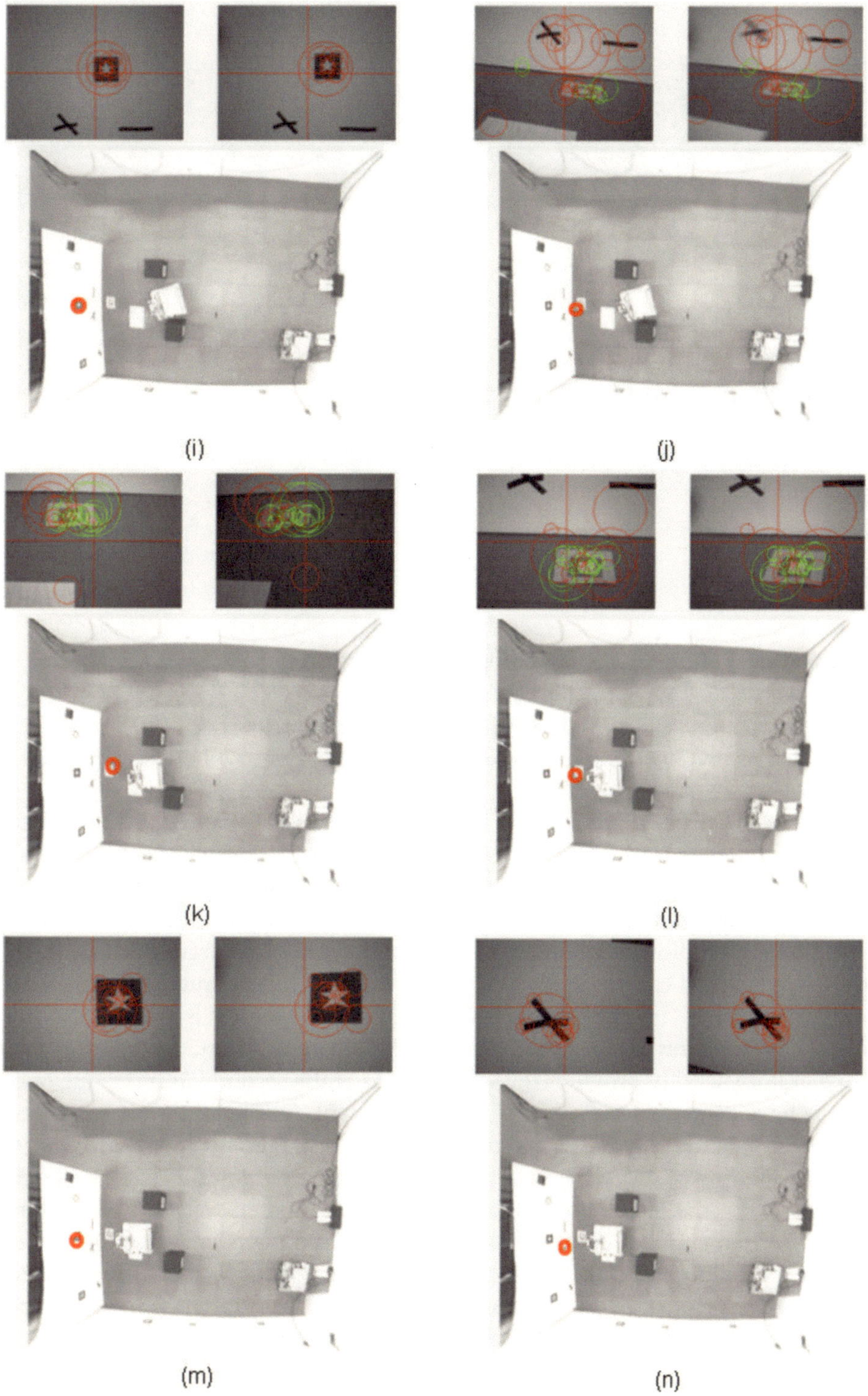

Fig. 14. Experiment of navigation avoiding obstacle (third part)

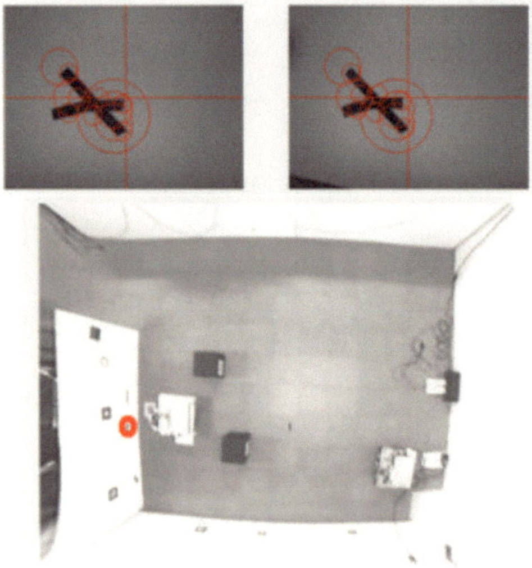

(o)

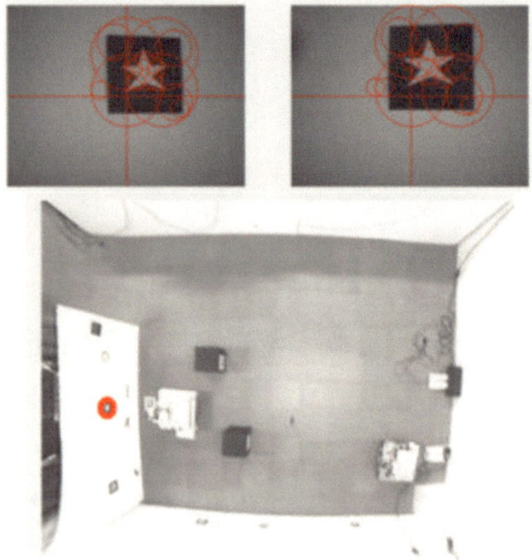

(p)

Fig. 15. Experiment of navigation avoiding obstacle (fourth part)

5.2 Experiment 2: navigation following a sequence of landmarks

The aim of this second experiment is to test the robot behaviour when navigation is guided by a sequence of landmarks.

The scenario of this experiment consists of two landmarks that the robot must sequentially reach. First landmark is formed by three concentric squares of different colours. The second one is the same landmark of the previous experiment. Figures 16-18 show results obtained in this test. At the beginning of the experiment (a), attention is fixated on the first landmark. Absence of obstacles allows a continuous advance towards the target position (b-d). Once this position is reached (d), the *NAVIGATION* behaviour reprograms landmark selector to locate the second landmark. Since there is no visual region that fulfils appearance properties of this new landmark, an exploratory behaviour is activated. Attentional control associated to this exploratory activity provides the visual access of the second landmark (g), once other visual areas have been explored (e-f). Since that moment, first landmark is conceived as an obstacle that robot must avoid for reaching the new target position. Obstacle selector provides this interpretation of the situation by centring attention on regions associated to the first landmark (h). Once the robot is appropriately oriented to avoid collisioning with the obstacle (i-j), attentional control focuses on regions situated near to the target. As a consequence, behavioural control produces approaching actions (k-l), which are interrupted when landmark selector recovers attentional control (m). After updating the target position, new approaching actions follow (n-o), allowing the robot to eventually reach the desired position (p).

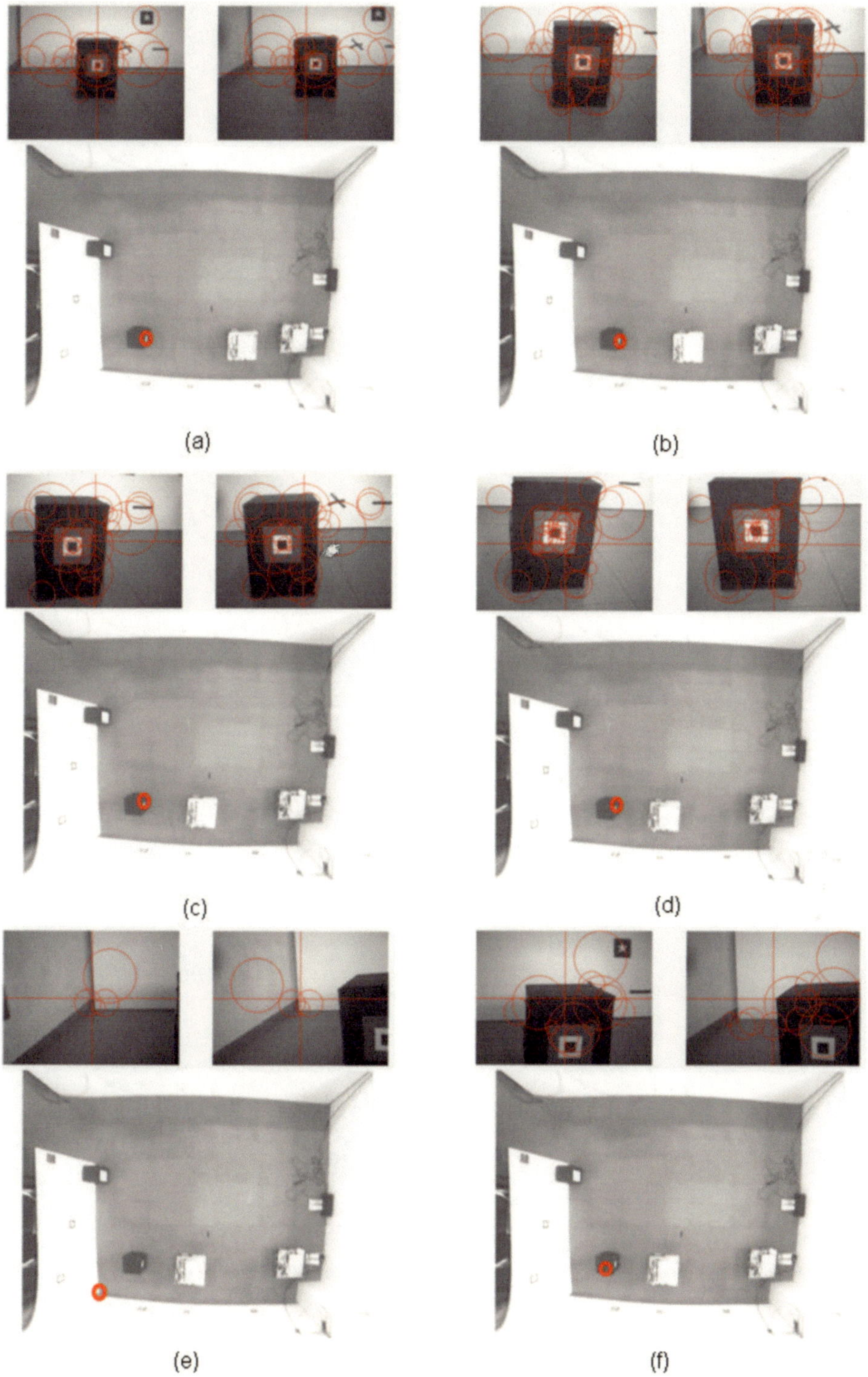

Fig. 16. Experiment of navigation following a sequence of landmarks (first part)

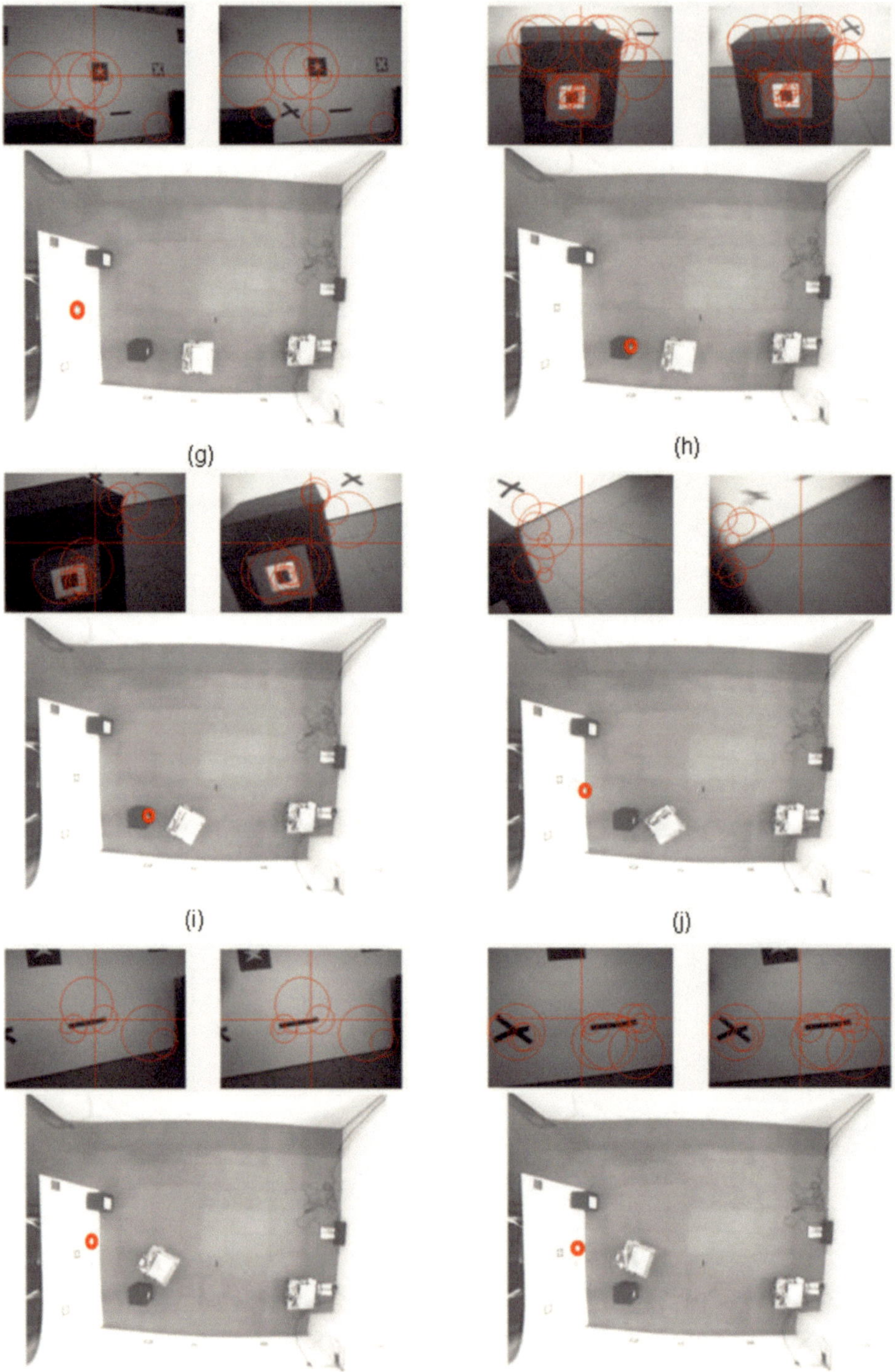

Fig. 17. Experiment of navigation following a sequence of landmarks (second part)

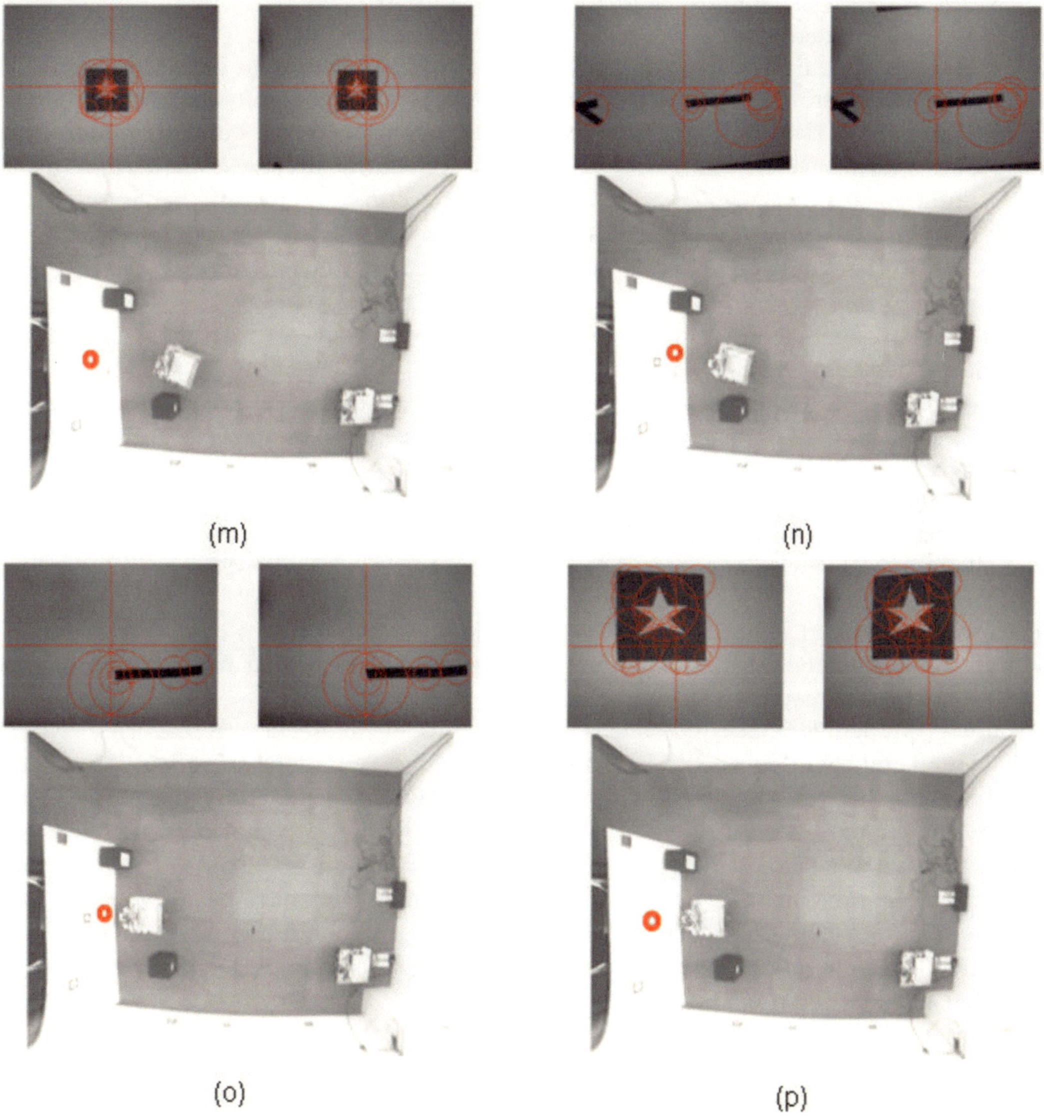

Fig. 18. Experiment of navigation following a sequence of landmarks (third part)

6. Conclusions

In this chapter, a novel computational model of visual attention based on the selection for action theory has been presented. In our system, attention is conceived as an intermediary between visual perception and action control, solving two fundamental behavioural questions:

- *Where to look?* Attention drives perceptual processes according to action requirements. It selects relevant visual information for action execution.
- *What to do?* Attentional selection limits potential actions than can be performed in a given situation. In this sense, actions are modulated according to perceptual result of attentional control.

These ideas give rise to what we have called attention-based control model. The proposed global system is a behavioural architecture that uses attention as a connection means between perception and action. In this model, behaviours are control units that link attention and action. They modulate the attention system according to their particular goals and generate actions consistent with the selected focus of attention.

Links between attention and action become effective through several properties of the visual attention system:

- **Top-down specifications**: attention is modulated from high-level behaviours according to their acting requirements. Hence, attentional selection is strongly influenced by action.
- **Distributed control**: attentional control is distributed between multiple individual control units that keep several visual targets simultaneously. This control scheme allows attention to be modulated from multiple behaviours with different information requirements.
- **Overt and covert attention**: the system provides overt and covert control of attention, allowing attentional selection to alternate among several visual targets according to external world features and internal robot requirements.

All these properties make possible to define proper interacting dynamics between behavioural and attentional controls that result in a global compound behaviour. Moreover, they provide the necessary coordination among individual behaviours that leads to an effective autonomous control.

The whole system has been designed as a component-based software architecture that provides a real-time modular implementation. Real experiments have been conducted in a mobile robot for solving navigation tasks. Results show good performance of the system in real situations, providing initial groundings for deeper explorations of links between attention and action in robot control.

7. References

Allport, A. (1987). Selection for action: some behavioural and neurophysiological considerations of attention and action, In: *Perspective on perception and action*, H. Heuer and A. Sanders (Ed.), Erlbaum

Bachiller, P.; Bustos, P.; Cañas, J.M. & Royo R. (2007). An experiment in distributed visual attention, *Proceedings of the 4th International Conference on Informatics in Control, Automation and Robotics*

Bandera, C. & Scott, P. (1989). Foveal machine vision systems, *IEEE International Conference on Systems, Man and Cybernetics*, pp. 596-599

Broadbent, D.E. (1958). *Perception and communication*, Pergamon Press, New York

Enright, J. (1998). Monocularly programmed human saccades during vergence changes?, *Journal of Physiology*, Vol. 512, pp. 235-250

Frintrop, S.; Backer, G. & Rome E. (2005). Goal-directed search with a top-down modulated computational attention system, *Lecture Notes in Computer Science*, W.G. Kropatsch, R. Sablatnig and A. Hanbury (Ed.), Vol. 3663, pp. 117-124, Springer

He, J.; Li, X. & Liu, Z. (2005). Component-based software engineering: the need to link methods and their theories, *Proceedings of ICTAC 2005, Lecture Notes in Computer Science*, Vol. 3722, pp. 70-95

Itti, L. & Koch, C. (2000). A saliency-based search mechanism for overt and covert shifts of visual attention, *Vision Research*, Vol. 40, pp. 1489-1506

LaBerge, D. (1995). *Attentional processing*, Harvard University Press.

Lazebnik, K.; Schmid, C. & Ponce, J. (2005). A sparse texture representation using local affine regions, *IEEE Transactions on Pattern Analysis and Machine Intelligence*, Vol. 27, pp. 1265-1278

Lindeberg, T. (1998). Feature detection with automatic scale selection, *International Journal of Computer Vision*, Vol. 30, No. 2, pp. 77-116

Lowe, D. (1999). Object recognition from local scale-invariant features, *The Proceedings of the Seventh IEEE International Conference on Computer Vision*, Vol. 2, pp. 1150-1157

Lowe, D. (2004). Distinctive image features from scale-invariant keypoints, *International Journal of Computer Vision*, Vol. 60, pp. 91-110

Mikolajczyk, K. & Schmid, C. (2001). Indexing based on scale invariant interest points, *Proceedings of the Eighth IEEE International Conference on Computer Vision*, Vol. 1, pp. 525-531

Mikolajczyk, K. & Schmid, C. (2005). A performance evaluation of local descriptors, IEEE Transactions on Pattern Analysis and Machine Intelligece, Vol. 27, pp. 1615-1630

Milner, A.D. & Goodale, M.A. (1995). *The visual brain in action*, Oxford University Press.

Navalpakkan, V. & Itti, L. (2006). An integrated model of top-down and bottom-up attention for optimizing detection speed, *Proceedings of the 2006 IEEE Computer Society Conference on Computer Vision and Pattern Recognition*, IEEE Computer Society, pp. 2049-2056.

Neumann, O.; van der Hejiden, A.H.C. & Allport, A. (1986). Visual selective attention: introductory remarks, *Psychological Research*, Vol. 48, pp. 185-188

Szyperski, C. (1998). *Component software: beyond object-oriented programming*, ACM Press/Addison-Wesley Publishing Co.

Takagi, T. & Sugeno, M. (1985). Fuzzy identification of systems and its applications to modeling and control, *IEEE Transactions on Systems, Man and Cybernetics*, Vol. 15, pp.116-132

Torralba, A.; Oliva, A.; Castelhano, M. S. & Henderson, J. M. (2006). Contextual guidance of eyes movements and attention in real-world scenes: the role of global features in object search, *Psychological Review*, Vol. 113, No. 4, pp. 766-786

Tsotsos, J.; Culhane, S.M.; Winky, W.; Lay, Y.; Davis, N. & Nuflo, F. (1995). Modeling visual attention via selective tuning model, *Artificial Intelligence*, Vol. 78, No. 1-2, pp. 507-545

Zadeh, L. (1965). Fuzzy sets, *Information and Control*, Vol. 8, pp. 338-353
ZeroC (2007). Internet communication engine, URL: *http://www.zeroc.com/ice.html*

8

Estimation of Tire-Road Forces and Vehicle Sideslip Angle

Guillaume Baffet[1], Ali Charara[2] and Daniel Lechner[3]
[1]*HEUDIASYC laboratory – CNRS (during this study)*
[1]*IRCCYN Laboratory, Ecole des Mines de Nantes – CNRS (Today)*
[2]*HEUDIASYC Laboratory, Universite de technologie de Compiegne – CNRS*
[3]*INRETS-MA laboratory*
France

1. Introduction

This chapter focuses on the estimation of car dynamic variables for the improvement of vehicle safety, handling characteristics and comfort. More specifically, a new estimation process is proposed to estimate longitudinal/lateral tire-road forces, velocity, sideslip angle and wheel cornering stiffness. One main contribution of the process is that it uses measurements from currently available standard sensors (yaw rate, longitudinal/lateral accelerations, steering angle and angular wheel velocities). Another contribution is that the process provides robust sideslip angle estimations with respect to cornering stiffness changes (or tire-road friction variations). Finally, the estimation process is applied and compared to real experimental data, notably sideslip and wheel force measurements.
The last few years have seen the emergence in cars of active security systems to reduce dangerous situations for drivers. Among these active security systems, Anti-lock Braking Systems (ABS) and Electronic Stability Programs (ESP) significantly reduce the number of road accidents. However, these systems may be improved if the dynamic potential of a car is well known. For example, information on tire-road friction means a better definition of potential trajectories, and therefore a better management of vehicle controls. Nowadays, certain fundamental data relating to vehicle-dynamics are not measurable in a standard car for both technical and economic reasons. As a consequence, dynamic variables such as tire forces and sideslip angle must be observed or estimated.
Vehicle-dynamic estimation has been widely discussed in the literature, e.g. (Kiencke & Nielsen, 2000), (Ungoren et al., 2004), (Lechner, 2002), (Stephant et al., 2006), (Baffet et al, 2006a). The vehicle-road system is usually modeled by combining a vehicle model with a tire-force model in one block. One particularity of this study is that it separates the estimation modeling into two blocks (shown in Figure 1), where the first block concerns the car body dynamic while the second is devoted to the tire-road interface dynamic.
The first block contains an Extended Kalman Filter (denoted as *O1,4w*) constructed with a four-wheel vehicle model and a random walk force model. The first observer *O1,4w* estimates longitudinal/lateral tire forces, velocity and yaw rate, which are inputs to the observer in the second block (denoted as *O2,LAM*). This second observer is developed from a sideslip angle model and a linear adaptive force model.

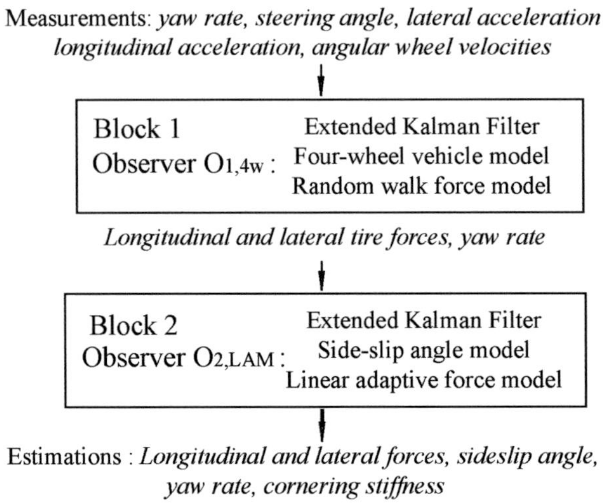

Figure 1. Estimation process. Observers *O1,4w* and *O2,LAM*

Some studies have described observers which take road friction variations into account (Lakehal-ayat et al, 2006), (Rabhi et al, 2005), (Ray, 1997). In the works of (Lakehal-ayat et al, 2006) road friction is considered as a disturbance. Alternatively, as in (Rabhi et al, 2005), the tire-force parameters are identified with an observer, while in (Ray, 1997) tire forces are modeled with an random walk model. In this study a linear adaptive tire force model is proposed (in block 2) with an eye to studying road friction variations.

The rest of the paper is organized as follows. The second section describes the vehicle model and the observer *O1,4w* (block 1). The third section presents the sideslip angle and cornering stiffness observer (*O2,LAM* in block 2). In the fourth section an observability analysis is performed. The fifth section provides experimental results: the two observers are evaluated with respect to sideslip angle and tire-force measurements. Finally, concluding remarks are given in section 6.

2. Block 1: observer for tire-road force

This section describes the first observer *O1,4w* constructed from a four-wheel vehicle model (see Figure 2), where ψ is the yaw angle, β the center of gravity sideslip angle, *Vg* the center of gravity velocity, and *L1* and *L2* the distance from the vehicle center of gravity to the front and rear axles respectively. *Fx,y,i,j* are the longitudinal and lateral tire-road forces, *δ1,2* are the front left and right steering angles respectively, and *E* is the vehicle track (lateral distance from wheel to wheel).

In order to develop an observable system (notably in the case of null steering angles), rear longitudinal forces are neglected relative to the front longitudinal forces. The simplified equation for yaw acceleration (four-wheel vehicle model) can be formulated as the following dynamic relationship (*O1,4w* model):

$$\ddot{\psi} = \frac{1}{I_z}\begin{bmatrix} L_1[F_{y11}\cos(\delta_1) + F_{y12}\cos(\delta_2) + F_{x11}\sin(\delta_1) + F_{x12}\sin(\delta_2)] \\ - L_2[F_{y21} + F_{y22}] \\ E/_2[F_{y11}\sin(\delta_1) - F_{y12}\sin(\delta_2) + F_{x12}\cos(\delta_2) - F_{x11}\cos(\delta_1)] \end{bmatrix} \tag{1}$$

where m the vehicle mass and Iz the yaw moment of inertia. The different force evolutions are modeled with a random walk model:

$$\dot{F}_{xij}, \dot{F}_{yij} = [0,0], \quad i = 1,2, \quad j = 1,2 \tag{2}$$

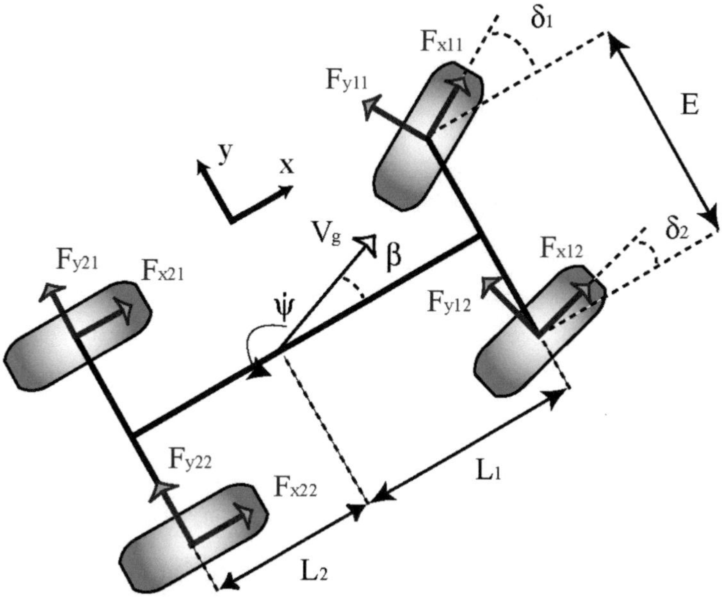

Figure 2. Four-wheel vehicle model

The measurement vector Y and the measurement model are:

$$\begin{aligned}
&Y = [\dot{\psi}, \gamma_y, \gamma_x] = [Y_1, Y_2, Y_3] \\
&Y_1 = \dot{\psi} \\
&Y_2 = \frac{1}{m}[F_{y11}\cos(\delta_1) + F_{y12}\cos(\delta_2) + F_{y21} + F_{y22} + F_{x11}\sin(\delta_1) + F_{x12}\sin(\delta_2)] \\
&Y_3 = \frac{1}{m}[-F_{y11}\sin(\delta_1) - F_{y12}\sin(\delta_2) + F_{x11}\cos(\delta_1) + F_{x12}\cos(\delta_2)]
\end{aligned} \tag{3}$$

where γx and γy are the longitudinal and lateral accelerations respectively.

The *O1,4w* system (association between equations (1), random walk force equation (2), and the measurement equations (3)) is not observable in the case where *Fy21* and *Fy22* are state vector components. For example, in equation (1, 2, 3) there is no relation allowing the rear lateral forces *Fy21* and *Fy22* to be differentiated in the sum *(Fy21+Fy22)*: as a consequence only the sum *(Fy2=Fy21+Fy22)* is observable. Moreover, when driving in a straight line, yaw rate is small, $\delta 1$ and $\delta 2$ are approximately null, and hence there is no significant knowledge in equation (1, 2, 3) differentiating *Fy11* and *Fy12* in the sum *(Fy11+Fy12)*, so only the sum *(Fy1=Fy11+Fy12)* is observable. These observations lead us to develop the *O1,4w* system with a state vector composed of force sums:

$$X = [\dot{\psi}, F_{y1}, F_{y2}, F_{x1}] \tag{4}$$

where *Fx1* is the sum of front longitudinal forces *(Fx1=Fx11+Fx12)*. Tire forces and force sums are associated according to the dispersion of vertical forces:

$$\begin{aligned}
F_{x11} &= \frac{F_{z11}}{F_{z11}+F_{z12}}F_{x1}, \quad F_{x12} = \frac{F_{z12}}{F_{z11}+F_{z12}}F_{x1} \\
F_{y11} &= \frac{F_{z11}}{F_{z11}+F_{z12}}F_{y1}, \quad F_{y12} = \frac{F_{z12}}{F_{z11}+F_{z12}}F_{y1} \\
F_{y21} &= \frac{F_{z21}}{F_{z21}+F_{z22}}F_{y2}, \quad F_{y22} = \frac{F_{z22}}{F_{z21}+F_{z22}}F_{y2}
\end{aligned} \tag{5}$$

where *Fzij* are the vertical forces. These are calculated, neglecting roll and suspension movements, with the following load transfer model:

$$\begin{aligned}
F_{z11} &= \frac{L_2 mg - h_{cog} m\gamma_x}{2(L_1+L_2)} - \frac{L_2 h_{cog} m\gamma_y}{E(L_1+L_2)} \\
F_{z12} &= \frac{L_2 mg - h_{cog} m\gamma_x}{2(L_1+L_2)} + \frac{L_2 h_{cog} m\gamma_y}{E(L_1+L_2)} \\
F_{z21} &= \frac{L_1 mg + h_{cog} m\gamma_x}{2(L_1+L_2)} - \frac{L_1 h_{cog} m\gamma_y}{E(L_1+L_2)} \\
F_{z22} &= \frac{L_1 mg + h_{cog} m\gamma_x}{2(L_1+L_2)} + \frac{L_1 h_{cog} m\gamma_y}{E(L_1+L_2)}
\end{aligned} \tag{6}$$

hcog being the center of gravity height and *g* the gravitational constant. The superposition principle means that the load transfer model assumes the assumption of independent longitudinal and lateral acceleration contributions (Lechner, 2002). The input vectors *U* of the *O1,4w* observer corresponds to:

$$U = [\delta_1, \delta_2, F_{z11}, F_{z12}, F_{z21}, F_{z22}] \tag{7}$$

As regards the vertical force inputs, these are calculated from lateral and longitudinal accelerations with the load transfer model.

3. Block 2: observer for sideslip angle and cornering stiffness

This section presents the observer *O2,LAM* constructed from a sideslip angle model and a tire-force model. The sideslip angle model is based on the single-track model (Segel, 1956), with neglected rear longitudinal force:

$$\dot{\beta} = \frac{F_{x1}\sin(\delta-\beta) + F_{y1}\cos(\delta-\beta) + F_{y2}\cos(\beta)}{mV_g} - \dot{\psi} \tag{8}$$

Rear and front sideslip angles are calculated as:

$$\beta_1 = \delta - \beta - L_1 \frac{\dot{\psi}}{V_g}, \quad \beta_2 = -\beta + L_2 \frac{\dot{\psi}}{V_g} \tag{9}$$

where δ is the mean of front steering angles.
The dynamic of the tire-road contact is usually formulated by modeling the tire-force as a function of the slip between tire and road (Pacejka & Bakker, 1991), (Kiencke & Nielsen, 2000), (Canudas-De-Wit et al 2003). Figure 3 illustrates different lateral tire-force models (linear, linear adaptive and Burckhardt for various road surfaces (Kiencke & Nielsen, 2000). In this study lateral wheel slips are assumed to be equal to the wheel sideslip angles.

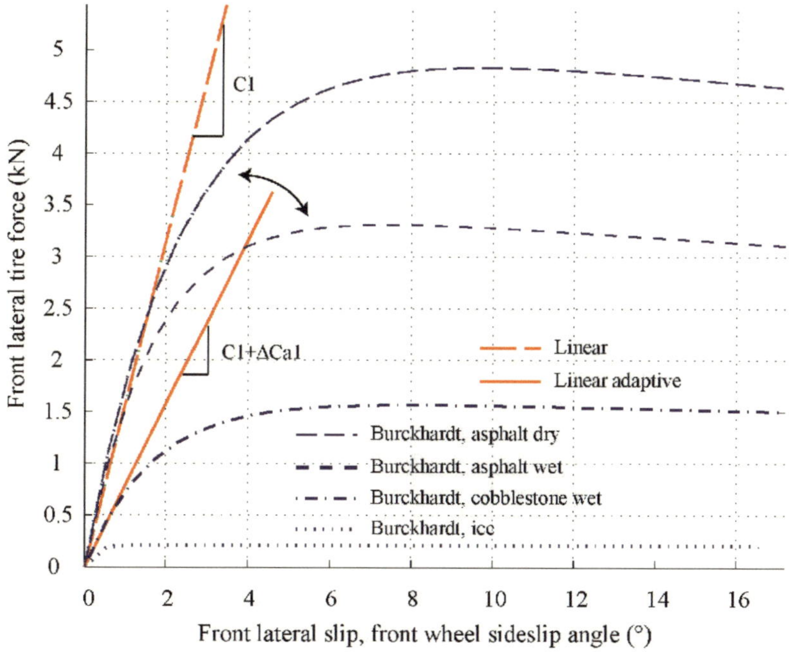

Figure 3. Lateral tire force models: linear, linear adaptive, Burckhardt for various road surfaces

In current driving situations, lateral tire forces may be considered linear with respect to sideslip angle (linear model):

$$F_{yi}(\beta_i) = C_i \beta_i, \quad i = 1,2, \tag{10}$$

where Ci is the wheel cornering stiffness, a parameter closely related to tire-road friction.
When road friction changes or when the nonlinear tire domain is reached, "real" wheel cornering stiffness varies. In order a take the wheel cornering stiffness variations into account, we propose an adaptive tire-force model (known as the linear adaptive tire-force model, illustrated in Figure 3. This model is based on the linear model at which a readjustment variable ΔCai is added to correct wheel cornering stiffness errors:

$$F_{yi}(\beta_i) = (C_i + \Delta C_{ai}) \beta_i, \quad i = 1,2, \tag{11}$$

The variable ΔCai is included in the state vector of the *O2,LAM* observer and it evolution equation is formulated according to a random walk model. Input *U'*, state *X'* and measurement *Y'* are chosen as:

$$\begin{aligned} U' &= [u_1', u_2', u_3', u_4'] = [\delta, \dot{\psi}, V_g, F_{x1}], \\ X' &= [x_1', x_2', x_3'] = [\beta,\ \Delta C_{a1}, \Delta C_{a2}], \\ Y' &= [y_1', y_2', y_3'] = [F_{y1}, F_{y2}, \gamma_y]. \end{aligned} \tag{12}$$

The measurement model is

$$\begin{aligned} y_1' &= (C_1 + x_2')\beta_1 \\ y_2' &= (C_2 + x_3')\beta_2 \\ y_3' &= \frac{1}{m}((C_1 + x_2')\beta_1 \cos(u_1') + (C_2 + x_3')\beta_2 + u_4'\sin(u_1')) \end{aligned} \tag{13}$$

where

$$\begin{aligned} \beta_1 &= u_1' - x_1' - L_1 \, u_2'/u_3' \\ \beta_2 &= -x_1' + L_2 \, u_2'/u_3' \end{aligned} \tag{14}$$

Given the state estimation denoted as

$$X' = [x_1', x_2', x_3'] \tag{15}$$

the state evolution model of *O2,LAM* is:

$$\begin{aligned} \dot{x}_1' &= \frac{1}{m u_3}[u_4 \sin(u_1 - x_1') + F_{yw1aux}\cos(u_1' - x_1') + F_{yw2aux}\cos(x_1') - u_2'] \\ \dot{x}_2' &= 0 \\ \dot{x}_3' &= 0 \end{aligned} \tag{16}$$

where the auxiliary variables *Fyw1,aux* and *Fyw2,aux* are calculated as:

$$\begin{aligned} F_{yw1,aux} &= (C_1 + x_2')(u_1' - x_1' - L_1 \, u_2'/u_3') \\ F_{yw2,aux} &= (C_2 + x_3')(-x_1' + L_2 \, u_2'/u_3') \end{aligned} \tag{17}$$

4. Estimation method

The different observers (*O1,4w*, *O2,LAM*) were developed according to an extended Kalman filter method. In 1960 R. E. Kalman published a paper describing a recursive solution to the discrete-data linear filtering problem (Kalman, 1960). Since this publication, Kalman's method, usually known as the "Extended Kalman Filter", has been the object of extensive

search and numerous applications. For example, in (Mohinder & Angus, 1993), Mohinder and Angus present a broad overview of Kalman filtering.
This paragraph describes an EKF algorithm. *bs,k*, *be,k* and *bm,k* represent measurement noise at time *tk* for the measurements, inputs and state evolution model respectively. This noise is assumed to be Gaussian, white and centered. *Qs*, *Qe* and *Qm* are the noise variance-covariance matrices for *bs,k*, *be,k* and *bm,k*, respectively. The discrete form of models is:

$$\begin{aligned}
&\mathrm{F}(\mathrm{X_k}, U_k^*) = X_k + \int_k^{k+1} f(X_k, U_k^*)dt \\
&X_{k+1} = \mathrm{F}(\mathrm{X_k}, U_k^*) + b_{m,k} \\
&Y_k = h(\mathrm{X_k}, U_k^*) + b_{s,k} \\
&U_k^* = U_k + b_{e,k}
\end{aligned} \tag{18}$$

where f and h are the evolution and measurement functions. X_k^- and X_k^+ are state prediction and estimation vectors, respectively, at time *tk*. The first step of the EKF is to linearize the evolution equation around the estimated state and input:

$$\begin{aligned}
A_k &= \frac{\partial F}{\partial X}\left(X_k^+, U_k^*\right) \\
B_k &= \frac{\partial F}{\partial U}\left(X_k^+, U_k^*\right)
\end{aligned} \tag{19}$$

The second step is the prediction of the next state, from the previous state and measured input:

$$X_{k+1}^- = F\left(X_k^+, U_k^*\right) \tag{20}$$

The covariance matrix of state estimation uncertainty is then:

$$P_{k+1}^- = A_k P_k^+ A_k^\perp + B_k Q_e B_k^\perp + Q_m \tag{21}$$

The third step is to calculate the Kalman gain matrix from the linearization of the measurement matrix:

$$\begin{aligned}
C_k &= \frac{\partial h}{\partial X}\left(X_{k+1}^-, U_k^*\right) \\
B_k &= \frac{\partial F}{\partial U}\left(X_{k+1}^-, U_k^*\right) \\
D_k &= \frac{\partial h}{\partial U}\left(X_{k+1}^-, U_k^*\right)
\end{aligned} \tag{22}$$

The following intermediate variables are used:

$$\begin{aligned}
R_k &= C_k P_{k+1}^- C_k^\perp + D_k Q_e D_k^\perp \\
S_k &= B_k Q_e D_k^\perp \\
T_k &= P_{k+1}^- C_k^\perp + S_k
\end{aligned} \tag{23}$$

and the Kalman gain matrix is:

$$K_k = T_k(R_k + Q_s + C_k S_k + S_k^{\perp} C_k^{\perp})^{-1} \tag{24}$$

The estimation step is to correct the state vector in line with measurement errors:

$$X_{k+1}^{+} = X_{k+1}^{-} + K_k \left[Y_{k+1} - h\left(X_{k+1}^{-}, U_{k+1}^{*} \right) \right] \tag{25}$$

Finally, the covariance matrix of state estimation uncertainty becomes:

$$P_{k+1}^{+} = P_{k+1}^{-} - K_k (C_k P_{k+1}^{-} + S_k^{\perp}) \tag{26}$$

5. Observability

From the two vehicle-road systems (*O1,4w, O2,LAM*), two observability functions were calculated. The two systems are nonlinear, so the observability definition is local and uses the Lie derivative (Nijmeijer & Van der Schaft 1990).
The Lie derivative of *hi* function, at *p+1* order, is defined as:

$$L_f^{p+1} h_i = \frac{\partial L_f^p h_i(X)}{\partial X} f(X,U) \tag{27}$$

With

$$L_f^1 h_i(X) = \frac{\partial h_i(X)}{\partial X} f(X,U) \tag{28}$$

The observability function *oi* corresponding to the measurement function *hi* is defined as:

$$o_i = \begin{pmatrix} dh_i(X) \\ dL_f^1 h_i(X) \\ \dots \\ dL_f^{(n-1)} h_i(X) \end{pmatrix} \tag{29}$$

where n is the dimension of X vector and d is the operator:

$$dh_i = \left(\frac{\partial h_i}{\partial x_1}, \dots, \frac{\partial h_i}{\partial x_6} \right) \tag{30}$$

The observability function of the system is calculated as:

$$O = \left(o_1, \dots, o_p \right)^{\perp} \tag{31}$$

where p is the dimension of the Y vector. Figure 4 illustrates observability analysis of the two systems for an experimental test, presented in section 6. Ranks of the two observability functions were 4 (for *O1,4w*) and 3 (for *O2,LAM*) (state dimensions) throughout the test, and consequently the state of the two systems were locally observable.

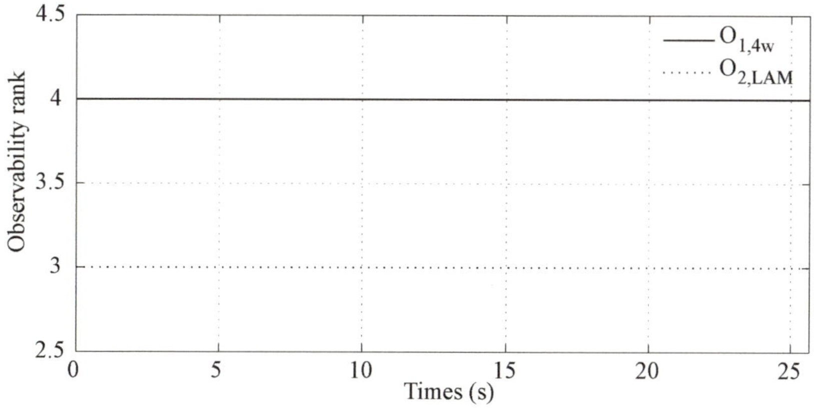

Figure 4. Ranks of the two observability functions for systems O1,4w and O2,LAM, during an experimental test (slalom)

6. Experimental results

The experimental vehicle (see Figure 5) is a Peugeot 307 equipped with a number of sensors including GPS, accelerometer, odometer, gyrometer, steering angle, correvit and dynamometric hubs. Among these sensors, the correvit (a non-contact optical sensor) gives measurements of rear sideslip angle and vehicle velocity, while the dynamometric hubs are wheel-force transducers.

Figure 5. Laboratory's experimental vehicle

This study uses an experimental test representative of both longitudinal and lateral dynamic behaviors. The vehicle trajectory and the acceleration diagram are shown in Figure 6. During the test, the vehicle first accelerated up to $\gamma x \approx 0.3g$, then negotiated a slalom at an approximate velocity of $12\ m/s$ ($-0.6g < \gamma y < 0.6g$), before finally decelerating to $\gamma x \approx 0.7g$.

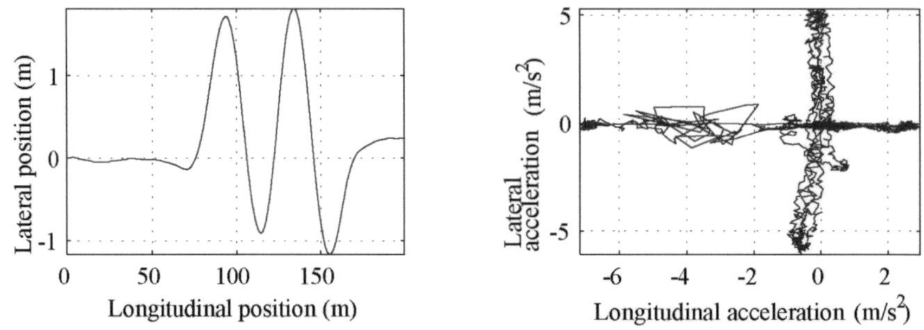

Figure 6. Experimental test, vehicle positions, acceleration diagram

The results are presented in two forms: figures of estimations/measurements and tables of normalized errors. The normalized error εz for an estimation z is defined in (Stephant et al, 2006) as

$$\varepsilon_z = 100 \frac{\left| z - z_{measurement} \right|}{\left| \max(z_{measurement}) \right|} \tag{32}$$

6.1 Block 1: observer O1,4w results

Figure 7 and table 1 present *O1,4w* observer results.

Estimates	Max	Mean	Std
$\dot{\psi}$	24.6°/s	0.4 %	2.6 %
Fy1	5816 N	3.1 %	4.0 %
Fy2	3782 N	2.9 %	5.4 %
Fx1	9305 N	3.1 %	4.1 %

Table 1. Maximum absolute values, *O1,4w* normalized mean errors and normalized standard deviation (Std)

The state estimations were initialized using the maximum values for the measurements during the test (for instance, the estimation of the front lateral force *Fy1* was set to *5155 N*). In spite of these false initializations the estimations converge quickly to the measured values, showing the good convergence properties of the observer. Moreover, the *O1,4w* observer produces satisfactory estimations close to measurements (normalized mean and standard deviations errors are less than *7 %*). These good experimental results confirm that the observer approach may be appropriate for the estimation of tire-forces.

6.2 Block 2: observer O2,LAM results

During the test, (*Fx1, Fy1, Fy2, Vg*) inputs of *O2,LAM* were originally those from the *O1,4w* observer. In order to demonstrate the improvement provided by the observer using the *linear adaptive force model* (*O2,LAM*, equation 11), another observer constructed with a *linear fixed force model* is used in comparison (denoted *Orl*, equation 10, described in (Baffet et al, 2006b). The robustness of the two observers is tested with respect to tire-road friction variations by performing the tests with different cornering stiffness parameters (*[C1, C2]*0.5, 1, 1.5*). The observers were evaluated for the same test presented in section 6.

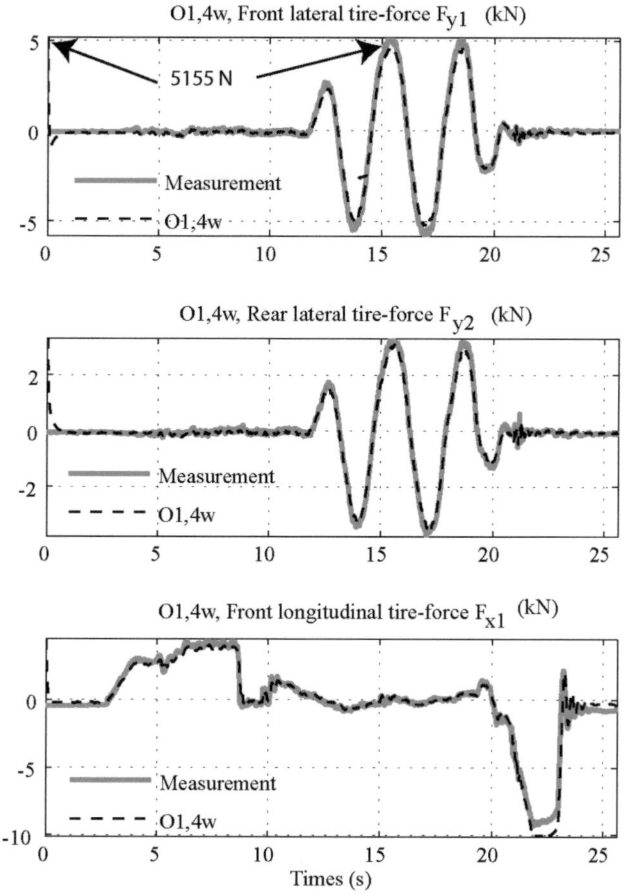

Figure 7. Experimental test. O1,4w results in comparison with measurements.Figure 8 shows the estimation results of observer *Orl* for rear sideslip angle. Observer *Orl* gives good results when cornering stiffnesses are approximately known (*[C_1, C_2]*1*). However, this observer is not robust when cornering stiffnesses change (*[C_1, C_2]*0.5, 2*).

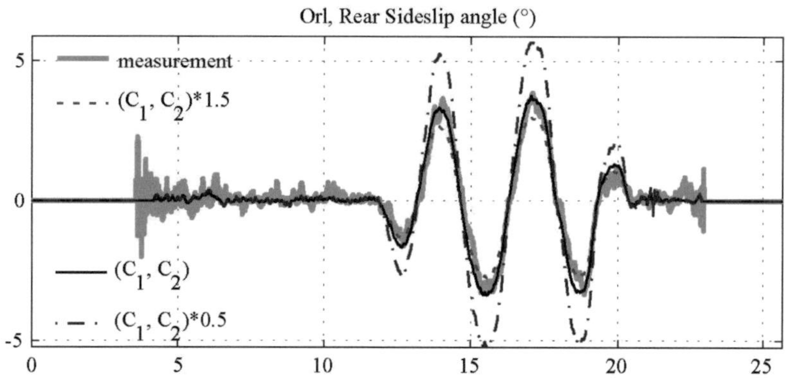

Figure 8. Observer Orl using a fixed linear force model, rear sideslip angle estimations with different cornering stiffness settings.

Figure 9 and table 2 show estimation results for the adaptive observer *O2,LAM*. The performance robustness of *O2,LAM* is very good, since sideslip angle is well estimated irrespective of cornering stiffness settings. This result is confirmed by the normalized mean errors (Table 2) which are approximately constant (about *7 %*). The front and rear cornering stiffness estimations (Ci + ΔCi) converge quickly to the same values after the beginning of the slalom at *12 s*.

O2,LAM	0.5(C1,C2)	(C1,C2)	1.5(C1,C2)
Max	3.0°	3.0°	3.0°
Mean	7.4 %	7.0 %	7.2 %

Table 2. Observer *O2,LAM*, rear sideslip angle estimation results, maximum absolute value, normalized mean errors.

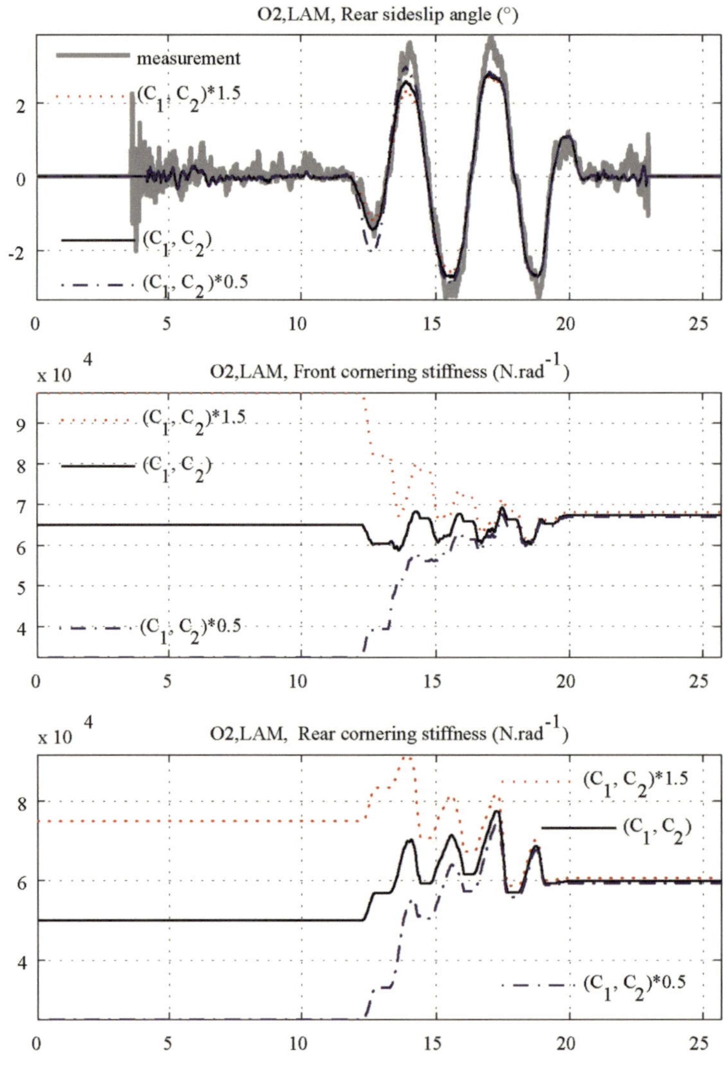

Figure 9. O2,LAM adaptive observer, Sideslip angle estimation results, Front and rear cornering stiffness estimations Ci+ ΔCi, with different cornering stiffness settings

6. Conclusions

This study deals with two vehicle-dynamic observers constructed for use in a two-block estimation process. Block 1 mainly estimates tire-forces (without an explicit tire-force model), while block 2 calculates sideslip angle and corrects cornering stiffnesses (with an adaptive tire-force model).

The first observer *O1,4w* (block 1), an extended Kalman Filter, is constructed with a random walk force model. The experimental evaluations of *O1,4w* are satisfactory, showing excellent estimations close to the measurements and good convergence properties.

The second observer *O2,LAM* (block 2), developed with an adaptive tire-force model, was evaluated for different cornering stiffness settings and was compared with an observer constructed with a fixed tire-force model (Orl). Results show that Orl is not robust when cornering stiffness parameters change, whereas *O2,LAM* gives excellent estimations of the sideslip angle. This result justifies the use of an adaptive tire-force model to take into account road friction changes.

The different results show the potential of the two-block estimation process. The first block has the advantage of providing satisfactory force estimations without a tire-force model, whereas the second block provides robust sideslip angle estimations with respect to cornering stiffness changes (or tire-road friction variations).

Future studies will improve vehicle-road models, notably for the calculation of the front/rear sideslip angles, in order to widen validity domains for observers. Subsequent vehicle-road models will take into account roll, vertical dynamics and vehicle-tire elasto-kinematics. Moreover, experimental tests will be performed, notably on different road surfaces and in critical driving situations (strong understeering and oversteering).

7. References

Baffet, G.; Stephant, J. & Charara, A. (2006). Vehicle Sideslip Angle and Lateral Tire-Force Estimations in Standard and Critical Driving Situations: Simulations and Experiments. *Proceedings of the 8th International Symposium on Advanced Vehicle Control AVEC2006*, pp 41-45, Taipei Taiwan

Baffet, G.; Stephant, J. & Charara, A. (2006). Sideslip angle lateral tire force and road friction estimation in simulations and experiments. *Proceedings of the IEEE conference on control application CCA*, pp. 903-908, Munich, Germany

Bolzern, P.; Cheli, F.; Falciola, G. & Resta, F. (1999). Estimation of the nonlinear suspension tyre cornering forces from experimental road test data. *Vehicle system dynamics.* Vol.31, 23-34

Canudas-De-Wit, C.; Tsiotras, P.; Velenis, E.; Basset, M. & Gissinger, G. (2003). Dynamic friction models for road/tire longitudinal interaction. *Vehicle System Dynamics,* Vol.39, 189-226

Kalman, R.E. (1960). A New Approach to Linear Filtering and Prediction Problems. *Transactions of the ASME - PUBLISHER of Basic Engineering,* Vol.82, 35-45

Kiencke U. & Nielsen, L. (2000). *Automotive control system.* Springer

Lakehal-ayat, M.; Tseng, H.E.; Mao, Y. & Karidas, J. (2006). Disturbance Observer for Lateral Velocity Estimation. *Proceedings of the 8th International Symposium on Advanced Vehicle Control AVEC2006,* pp 889-894, Taipei Taiwan

Lechner, D. (2002). *Analyse du comportement dynamique des vehicules routiers legers: developpement d'une methodologie appliquee a la securite primaire.* Ph.D. dissertation Ecole Centrale de Lyon, France

Mohinder, S.G. & Angus, P.A. (1993) *Kalman filtering theory and practice.* Prentice hall.

Nijmeijer, H. & Van der Schaft, A.J. (1990) *Nonlinear Dynamical Control Systems.* Springer-Verlag

Pacejka, H.B. & Bakker, E. (1991). The magic formula tyre model. *Int. colloq. on tyre models for vehicle dynamics analysis,* 1-18

Rabhi, A. ; M'Sirdi, N.K. ; Zbiri, N. & Delanne, Y (2005). Vehicle-road interaction modelling for estimation of contact forces. *Vehicle System Dynamics,* Vol.43, 403-411

Ray, L. (1997). Nonlinear Tire Force Estimation and Road Friction Identification : Simulation and Experiments. *Automatica,* Vol.33, 1819-1833

Segel, M.L. (1956). Theorical prediction and experimental substantiation of the response of the automobile to steering control. *Automobile division of the institut of mechanical engineers,* Vol.7, 310-330

Stephant, J.; Charara, A. & Meizel, D. (2006). Evaluation of a sliding mode observer for vehicle sideslip angle. *Control Engineering Practice,* Vol. 15, 803-812

Ungoren, A.Y.; Peng, H. & Tseng, H.E. (2004). A study on lateral speed estimation methods. *Int. J. Vehicle Autonomous Systems,* Vol.2, 126-144

9

A New Load Adjustment Approach for Job-Shops

Zied Bahroun[1], Mourad Fendouli[1] and Jean-Pierre Campagne[2]
[1]Laboratory LIP2, Faculty of Sciences of Tunis,
[2]Laboratory LIESP, University of Lyon, INSA-Lyon,
[1]Tunisia
[2]France

1. Introduction

This chapter presents a middle-term new **load adjustment approach** by overlapping for a set of jobs in a **job-shop** context, **guaranteeing the existence of a limited capacity schedule without scheduling** under the assumption of pre-emptive tasks.
The proposed approach is based on the exploitation of the tasks scheduling time segments overlapping and on the distribution of the job's margins between tasks in a just in time context. First, we present a literature review concerning load adjustment approaches. Second, we introduce the overlapping load adjustment approach. Third, we present an original heuristic to use this approach in the case of job-shops organized firms. After that, we present the associated scheduling approach. Finally, we will discuss a more general use of this approach and the possible extensions.
Generally, the production planning is made in a hierarchical way in two planning and scheduling decision levels. In the first step, we decide which products to supply, in which quantities and delays, in the second step, we adjust load to the capacity and schedule the tasks on the machines.
There are three main and classical load adjustment approaches. First of all, we have the placement. This approach consists in calculating a detailed tasks schedule. A new task is integrated in the planning if we find a hole in the planning which is bigger than the duration of this task. This approach estimates only one schedule which can be destroyed by any disturbance. The second approach is the periodic and cumulative approach. It consists in calculating the cumulative load and capacity for a latest loading and for each period. This approach does not guarantee the existence of a scheduling solution because it does not take into consideration the ready dates constraints. The third approach is the periodic and non-cumulative approach. It consists in assigning tasks to periods and comparing period by period the available and the required capacities. This method estimates only the solutions in which the tasks are fixed in a specific period.
The problem of sequencing decisions in production planning and scheduling was studied by (Dauzere-Peres & Laserre, 1999). They think that it is better in some cases to integrate the scheduling decision level in the lot sizing decision level and propose an iterative approach for planning and scheduling production. Some researchers integrated the scheduling and capacity constraints in their lot sizing model (see for example, Fleishmann & Meyr, 1997).

We can also find a survey on lot sizing and scheduling in (Drexl & Kimms, 1997). Nevertheless, a single machine is generally considered in most of these works which are difficultly applicable in a real situation.
(Hillion & Proth, 1994) and (Gunther, 1987) studied the problem of finite capacity planning. (Hillion & Proth, 1994) considered the problem of a finite capacity flow control in a multi-stage/multi-product environment. (Gunther, 1987) developed two heuristics for the lot sizing in the context of finite capacity. (Néron et al., 2001) developed an approach for solving hybrid flow shop problem using energetic reasoning. This approach has some similarities in the concept with our approach.
The problem of finite capacity planning in MRP based systems has also been studied. In order to obtain capacity-based production plans, (Pandey et al., 2000; Wuttipornpun & Yenradee, 2004) proposed a finite capacity material requirement planning system. (Grubbström & Huynh, 2006) considered multi-level, multi-stage capacity-constrained production-inventory systems in a MRP context.
In comparison with all these approaches, the overlapping load adjustment approach allows to distinguish two main phases. The first phase consists in establishing a long or mid-term production planning where the feasibility is ensured without scheduling and tasks placement, which allows us to characterize a set of feasible scheduling solutions. The scheduling will be done only in the second phase.

2. The overlapping load adjustment approach

The time scale is divided into time periods. Each task of a job has got a processing time, requires one or more resources and has to be realized during a scheduling time segment associated with one or more consecutive periods. The scheduling time segments of consecutive tasks of the same job cannot overlap. From now on and throughout this paper a lapse of time called here lapse, designates a succession of a number of n consecutive periods. Let (a,b) be a lapse composed of a succession of periods which are limited by the periods a and b including them. The shortest lapses are composed of only one period, for instance (a,a). Such a lapse (a,a) is called a basic lapse. The longest lapse is noted (1,H) in which number 1 is associated with the first period of the planning time frame and the letter H the last one. From such a planning time frame, the total number of different lapses is equal to $H^*(H+1)/2$.This number is of course to be multiplied by the number of existing processors. The sub-lapse of a lapse is a subset of one or more consecutive periods of this lapse. For instance, the sub-lapses of [1,3] are [1,1], [2,2], [3,3], [1,2] and [2,3]. Every lapse containing a lapse [a,b] is called the over-lapse of [a,b]. For instance, [1,3] is an over-lapse of [1,1]. Each lapse is characterized by:

- An accumulated capacity: sum of the capacities of each period included in this lapse.
- A direct capacity requirement: sum of the capacities required by the tasks whose scheduling time segment is exactly equal to this lapse.
- An accumulated capacity requirement: sum of the capacities required by the tasks whose scheduling time segment is fully included in this lapse. It is the sum of the direct capacity requirements of this lapse and its sub-lapses.

(Dillenseger, 1993) sets the following proposition out: for any lapse, its accumulated capacity requirement must be equal or smaller than its accumulated capacity. He proves that it is a necessary and sufficient condition for the existence of a loading solution of the set of tasks (within the limits of their scheduling time segments and considering the capacity levels), according to pre-emptive possibility.

Let's consider the following example of a production plan composed of 7 jobs which will be treated on a single processor with ready and due dates as shown in Table 1 below. The considered period for this example is the week composed of five days (the day is the unit time):

Job	processing time (days)	Ready date (beginning of the week)	Due date (ending of the week)
A	3	4	4
B	1	2	4
C	2	2	4
D	2	3	3
E	2	2	2
F	1	2	3
G	4	3	4
H	2	1	3
I	2	1	4
J	1	1	1

Table 1. Example 1

Figure 1 represents a capacity requirement planning (CRP) corresponding to the example of Table 1. For instance, task F needs a load of 1 day and have a time scheduling segment composed of weeks 2 and 3 (it means that this task should be scheduled and produced in any time inside the weeks 2 and 3). The capacity of a period (the week in this example) is 5 days. The margin of task F is so equal to 9 days (10 days of weeks 2 and 3 minus its load of 1 day). The associated planning feasibility control graph (PFCG) is shown in Figure 1. For each lapse of weeks, we calculate the direct capacity requirement, the accumulated capacity requirement and the remaining capacity. For instance, the lapse [3,4] composed of weeks 3 and 4 has a direct capacity requirement equal to 4 days (task G), an accumulated capacity requirement equal to 9 days (tasks with a scheduling time segment included in the lapse: A, D and G) and a remaining capacity equal to (10-9) days. The planning feasibility control graph proves the feasibility of this set of jobs (all the remaining capacities for all the lapses are positive).

Firstly, this load adjustment approach was applied to plan the activities of a make-to-order company in a mono-level context (Dillenseger, 1993). This approach was applied then to a flow-shop composed of m processors (Bahroun, 2000a), to a generalised flow-shop (Bahroun, 2000b) and for the cyclic production context (Bahroun, 1999).

3. Application for job-shop organized firms

Let us consider N jobs with their due and ready dates. Each job is composed of one to m tasks realized on one to m processors with a certain order which is not necessarily the same for all the jobs (Figure 2). We suppose that these jobs are the results of a products supply calculation in an M.R.P. based system for instance. **We aim at adjusting the load resulted** by these jobs in a **finite capacity way** by adapting the overlapping load adjustment approach to the job shop case.

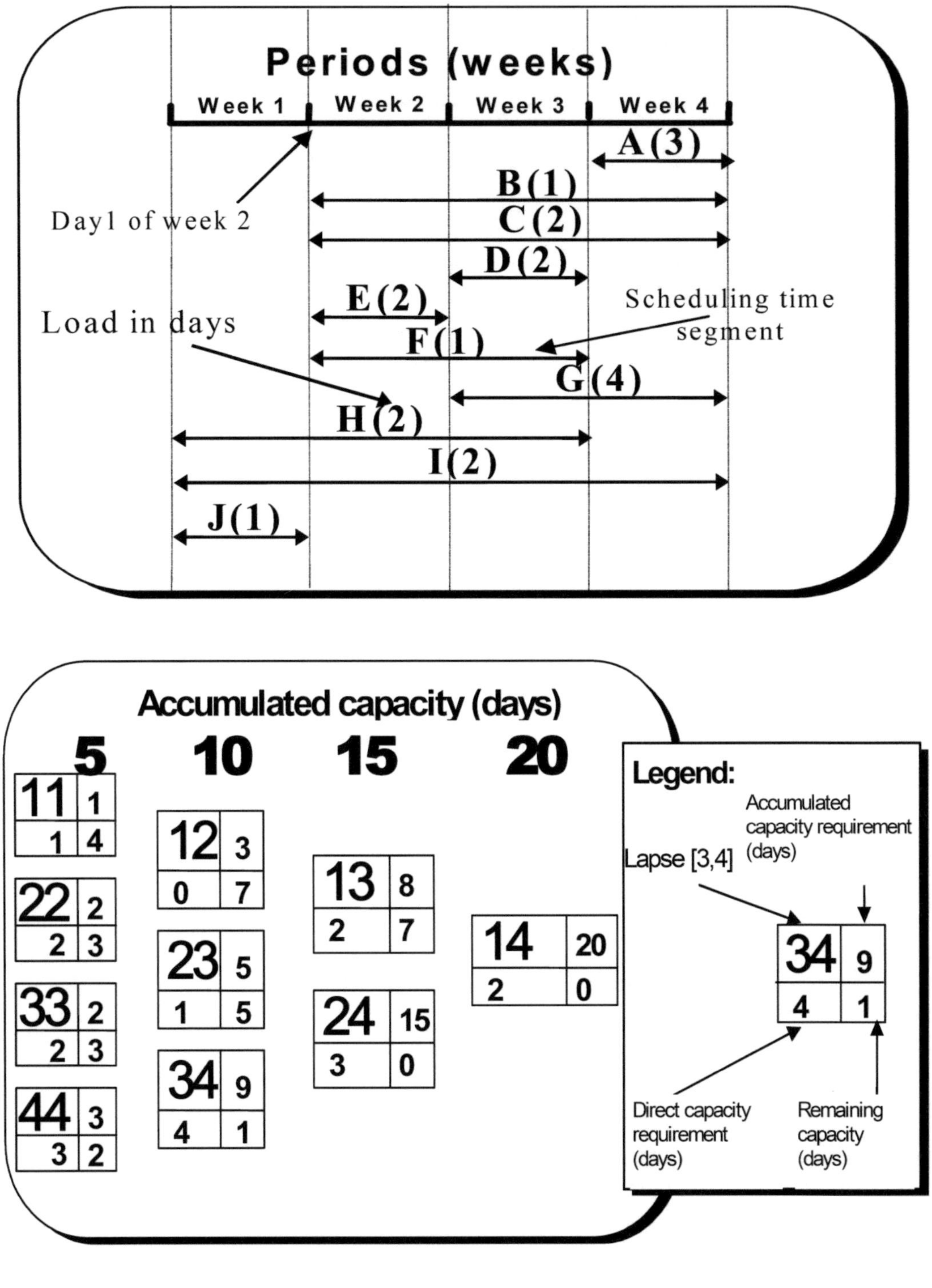

Fig. 1. Capacity requirements planning (CRP) and Planning feasibility control graph (PFCG)

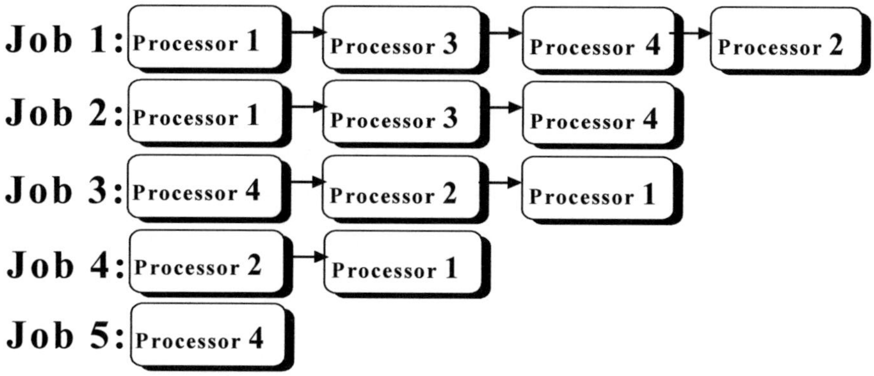

Fig. 2. Example of a job shop

First, we will calculate the scheduling time segment of each task, considering the due and the ready dates of their job, their precedence constraint and the capacity constraints. After that, we will try to exploit the existing margins. They will be distributed on the different job's tasks and will be assigned with priority to the tasks corresponding to overloaded processors. We have developed a heuristic which tries to share out judiciously the job's margins on their tasks. For this, we calculate a latest loading on all the processors without margins (we only assign the margins of jobs with a unique task). We classify in the load decreasing order the processors. We then assign all the margins to the most loaded processor and after that we keep only the necessary margins to validate the processor loading (for any lapse, its accumulated capacity requirement must be equal or smaller than its accumulated capacity) and transfer the unused margins to the next processor accordingly to the load classification. Then, we reiterate the same treatment to the following processor until reaching the last processor.

We define the following parameters:

- N = number of jobs.
- m = number of processors.
- Ma_i = the global margin of job i.
- p_{ij} = processing time of the task corresponding to job i on processor j.
- p_{ij} = 0 if there is not a task of job i on processor j.
- r_i = release or ready date of job i.
- d_i = due date of job i.
- b_{ij} = beginning of the scheduling time segment of the task corresponding to job i on processor j.
- e_{ij} = ending of the scheduling time segment of the task corresponding to job i on processor J.
- $ACCP[a,b]_j$ = accumulated capacity of the lapse [a,b] for processor j.
- $ACCPR[a,b]_j$ = accumulated capacity requirement of the lapse [a,b] for processor j.
- $RC[a,b]_j$ = remaining capacity of the lapse [a,b] on processor j.
- d_p = duration of an elementary period.

We note $\lceil x \rceil$ the smallest integer which is greater than or equal to x and $\lfloor x \rfloor$ the biggest integer which is smaller than or equal to x.

Our approach is based on five main steps (we will illustrate our approach with the example of Table 2):

Job	Processing order	Process. time on proc. 1 (day)	Process. time on proc. 2 (days)	Process. time on proc. 3 (days)	ready date (beginning of the period)	due date (ending of the period)	Global margins in periods (weeks)
A	P1→P2→P3	**3**	**2**	**2**	1	6	**3**
B	P1→P2→P3	**2**	**3**	**1**	1	5	**2**
C	P1	**3**	0	0	3	5	**2**
D	P2	0	**2**	0	2	4	**2**
E	P3	0	0	**2**	4	6	**2**
F	P3→P2→P1	**3**	**3**	**2**	2	5	**1**
G	P3→P2→P1	**1**	**1**	**1**	2	6	**2**
H	P3→P1→P2	**2**	**2**	**2**	3	5	**0**
I	P3→P1→P2	**1**	**2**	**2**	3	5	**0**
J	P2→P1	**3**	**3**	0	2	5	**2**
K	P2→P3	0	**3**	**4**	3	6	**2**
L	P3→P1	**2**	0	**3**	4	6	**1**

Table 2. Example 2

1st step :

We calculate the global job's margins:

$$Ma_i = d_i - r_i - \left(\sum_{j=1}^{m} \lceil (p_{ij} / d_p) \rceil\right) + 1 \quad (1)$$

For our example, we consider a production system composed of three processors (P1, P2 and P3) and a set of jobs (Table 1, d_p = 5 days). We calculate the global margins using the last formula and we obtain the results reported in the last column of Table 2. After that, we determine the scheduling time segment of each task according to a latest loading without margins (we assign only the margins of jobs with a unique task like jobs C, D and E. In fact, these margins will be used only on a unique processor and will not be distributed on several processors). For instance, for job A, the last task on processor P3 will have a scheduling segment that ends at the end of week 6 and will begin so, at the beginning of the same week (because the processing time of this task is inferior to a week), the task number 2 for the same job A on processor P2 will have a scheduling segment that begins and ends at week 5. The first task of Job A on processor P1 will have a scheduling segment that ends and begins at week 4. Job C is composed of only one task, so we will assign its margin and the time scheduling segment will begin at week 3 and end at week 5. We calculate the scheduling segments of the other jobs in the same way and we obtain the capacity requirement planning of P1, P2 and P3 as shown in Figures 3 and 4.

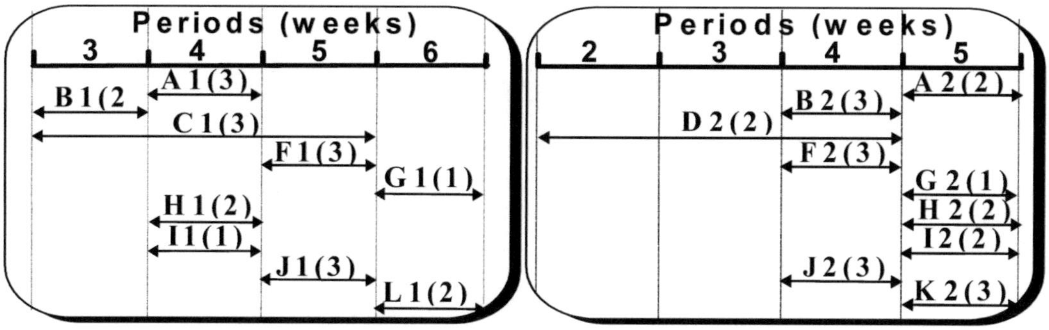

Fig. 3. Capacity requirements planning (CRP) of P1 and P2 before treating

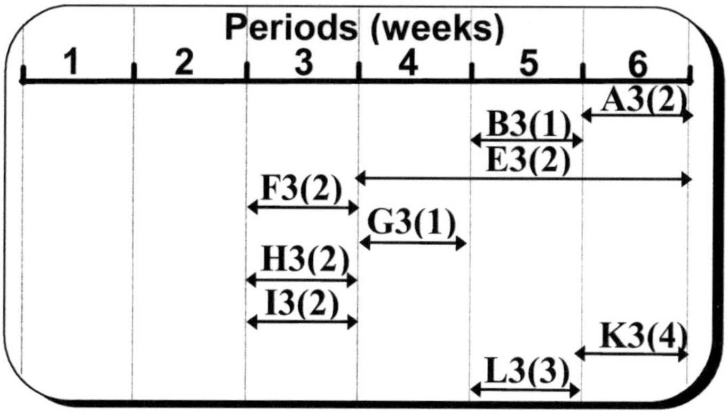

Fig. 4. Capacity requirements planning (CRP) of P3 before treating

After that, we calculate the total load and the load by period for each processor as below:

Processor	Total load in days	Number of concerned Periods (weeks)	Load/period
P1	20	4	5
P2	21	4	5,25
P3	19	4	4,75

Table 3. Calculation of the processor load

We classify and treat the processor in the decreasing order of the load/period: P2, P1, P3.

2nd step:

We assign all the global margins to the processor P2. Then, we calculate for each task the beginning and the ending periods of the scheduling time segment of this task:

$\forall$ i and for a processor j:

e_{ij} remains the same

$$b_{ij} = e_{ji} - Ma_i - \lceil (p_{ij} / d_p) \rceil + 1 \qquad (2)$$

If we calculate the beginning time of the second processor of our example, we can generate the corresponding capacity requirement planning (CRP) and the planning feasibility control graph (PFGC):

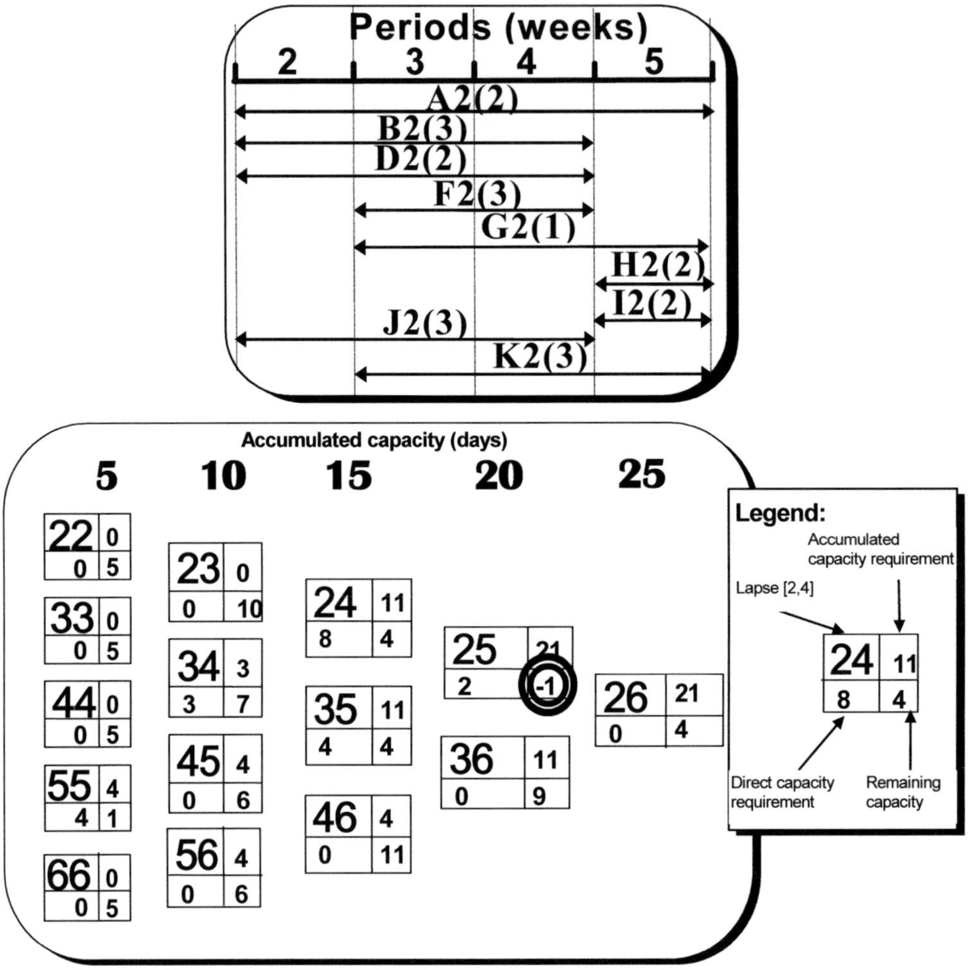

Fig. 5. CRP and PFCG of the processor P2 after assigning all the margins

<u>3rd Step:</u>

We then check the feasibility condition. If we find problems in certain lapses (even with all available margins), we should lengthen the scheduling time segment of some tasks. If for a lapse [a, b] in the processor j, the validation condition is not verified:

- We consider in the duration increasing order all the tasks that $b_{ij} \geq a$ et $e_{ij}= b$.
- In this list, we begin to treat the first task in the kth position in the list with a load equal or superior to the overloading in the lapse which allows us to delay the minimum number of tasks.
- We proceed progressively to the lengthening of the scheduling time segment of these tasks, period by period until the verification of the feasibility condition or arrival to the end of the list.
- If we arrive to the end of the list we try with tasks positioned in the $(k-1)^{th}$, $(k-2)^{th}$ … position until the verification of the feasibility condition.

In our example (Figure 5), we remark that we have an overloading in the lapse [2,5] which obliges us to delay of one period, one of the tasks included in this lapse and which finishes in period 5 (A2, G2, H2, I2 or K2 as we can see in CRP of Figure 5). We choose in this example to lengthen the scheduling time segment of task G2 from the lapse [3,5] to the lapse [3,6].

Remark: If we do not accept to delay jobs, the lapses with negative remaining capacity indicate where we must increase the capacity by using for example overtime or interims. We can also introduce the notion of jobs priority for choosing which tasks must be delayed.

4th Step:

Now, we will try to regain margins. We begin with the tasks corresponding to jobs with weak global margins. Tasks of jobs without margins are assigned to the elementary lapses ([1,1], [2,2] etc.), those corresponding to jobs with one period margin are assigned to the lapses of the second column of the feasibility control graph, those corresponding to jobs with k periods margins are assigned to the column number k etc. Our treatment begins with the lapses of the second column because the corresponding jobs have only one period global margin and we must preserve these precious margins to validate the other processors and use the margins from jobs that have important global margins.

A transfer of a task from the lapse [a,b] to the lapse [a+w,b], adds load to all the over-lapses of [a+w,b] which are not initially over-lapses of [a,b]. The transferred load must be equal or smaller than the remaining capacity on these lapses for maintaining the validation condition.

The proposed approach for this transfer tries to transfer the maximum number of tasks and tries to match in the best way the transferred load in regard to the remaining capacity. Consequently, we construct the set of tasks which can be transferred, and we classify this set in the increasing order of their load. We transfer the tasks one by one in this order while the sum of their load is smaller than the remaining capacity. Then, we take the last task transferred and we try to change it by another task from the remaining tasks of the set and which matches better the remaining capacity. If two tasks have the same load we can choose for example the task corresponding to a product with a greater carrying cost. If we take the example of PFCG in Figure 5, we begin with the lapses of the second column. If we try, for instance, to regain margins from the tasks corresponding to the lapse [3,4], we should transfer the maximum number of tasks to the lapse [4,4]. We can transfer task F2 because the minimum of the remaining capacity of the over-lapses of [4,4] which are not over-lapses of [3,4] (the lapses [4,4], [4,5] and [4,6]) is 5 and it is greater than the load of the task F2. Then, we pass to lapses [4,5], [5,6], and next to the lapses of the third column (for the task G2, we succeed to regain 2 periods, the scheduling time is shortened from the lapse [3,6] to lapse [5,6]). We reiterate this treatment until arriving to the last column. We obtain after treatment of the processor P2 the capacity requirement planning of Figure 6.

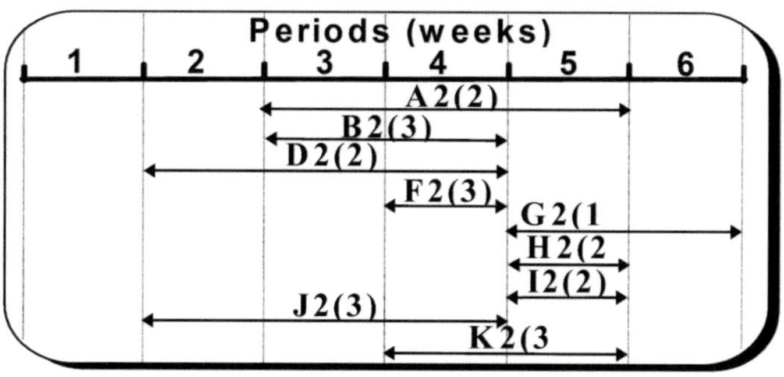

Fig. 6. C.R.P. of processor P2 after treatment

We can generalize this step for a processor j:

$\forall$ the lapse [x,y] / x= a and y= a+w with $1 \le w \le H$ and $1 \le a \le H-w$:

- Construct the set A = {set of tasks / b_{ij} = a and e_{ij} = a+ w} (The set of tasks that their scheduling time segment can be shortened, this set will be ordered in the increasing order of their load).
- Iterate then for f = w -1 $\rightarrow$ 0
  - ➢ Cap = Min (RC_j [a+1+z, a+w+n]) For n = 0 $\rightarrow$ H - a - w, z = f $\rightarrow$ 0 (we calculate the maximum load that can be transferred).
  - ➢ Transfer the maximum number of tasks from A (using the approach described in the precedent page) that the sum of their processing times is inferior or equal to Cap. Let C be the set of these tasks transferred and Q the sum of their processing time:

    $\forall$ the task ij $\in$ C, we put b_{ij} = a+1 + f

    For n = 0 $\rightarrow$ H - a - w and for z = f $\rightarrow$ 0, we do:

    RC_j [a+1+z, a+w+n] = RC_j [a+1+z, a+w +n] – Q (We update the new time scheduling segment of the tasks and the remaining capacity of the concerned lapses).
- We update the beginning and the ending of the scheduling time segments of the other tasks on the other processors as follow:
  - ➢ If a task u precedes task j of the same job i on the processor j, we move its scheduling time segment in a manner that the ending time becomes equal to the beginning time of the task j. We effectuate the same treatment until arriving to the first task.
  - ➢ If a task u follows task j of the same job i on the processor j, we update in a symmetrical manner the scheduling time segment of this task and all the other tasks up to the last one.

5th step:

We assign all the unused margins to the next processor (in this case the processor P1). We calculate the scheduling time segments of the tasks corresponding to this processor using the following formulae:

$\forall$ i, for a processor j :

$$b_{ji} = e_{ji} - M'a_i - \lceil (p_{ij} / d_p) \rceil + 1 \qquad (3)$$

Where $M'a_i$ is the remaining margin.

We can assign margins for a task i on a processor j only if the precedent tasks of the same job are not already treated. We obtain the C.R.P. and the control feasibility graph of the processor P1 of figure 7.

We check then the feasibility condition. If for a lapse [a,b] in the processor j, the validation condition is not verified, we should lengthen the scheduling time segment of the tasks as below:

- We consider in the duration increasing order the list of all the tasks included in the lapse [a,b] and that b_{ij} = a or e_{ij}= b.
- In order to treat the minimum number of tasks, we begin to treat in this list the first task in the kth position in the list which has a load equal or superior to the overloading in the lapse.
- We try to regain a margin for this task by shortening the scheduling time segment of the precedent task of the same job in a processor already treated.

- We try this for all the tasks positioned in the (k+1)th position in the list until succeeding or arriving to the end of the list.
- If we arrive to the end of the list, we try with tasks positioned in the (k-1)th, (k-2)th … position until succeeding or arriving to the beginning of the list.
- We must lengthen the scheduling time segment of as many tasks as necessary to validate the feasibility of the problematic lapse.
- If the remaining capacity continues to be negative, we reiterate the treatment of the tasks in the same order but by trying in this case to move completely if possible the precedent task of the same job in the past which allows us to lengthen the scheduling time segment of the tasks of this processor.

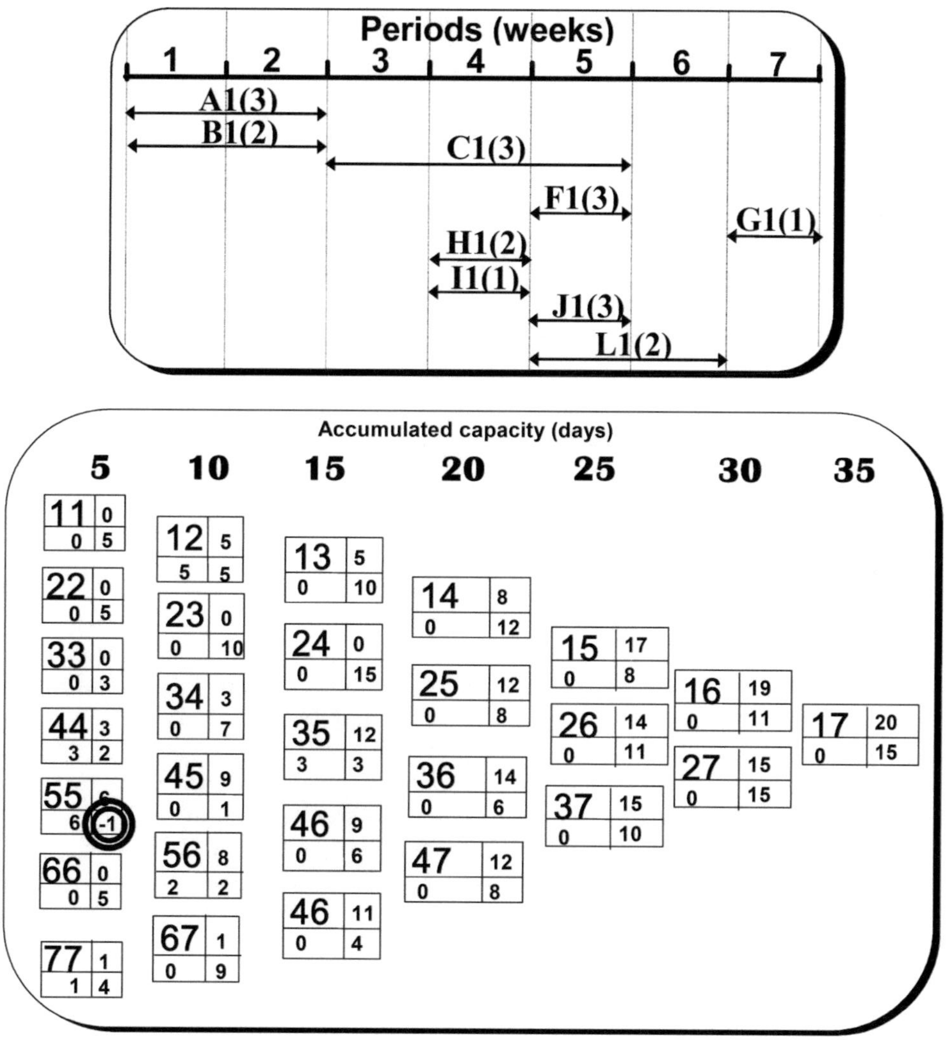

Fig. 7. CRP and PFCG of processor P1 after assigning all the available margins

If we do not succeed, we treat the tasks in the same order by trying to delay the due date on a minimum number of jobs or by increasing the capacity of the incriminate lapses.

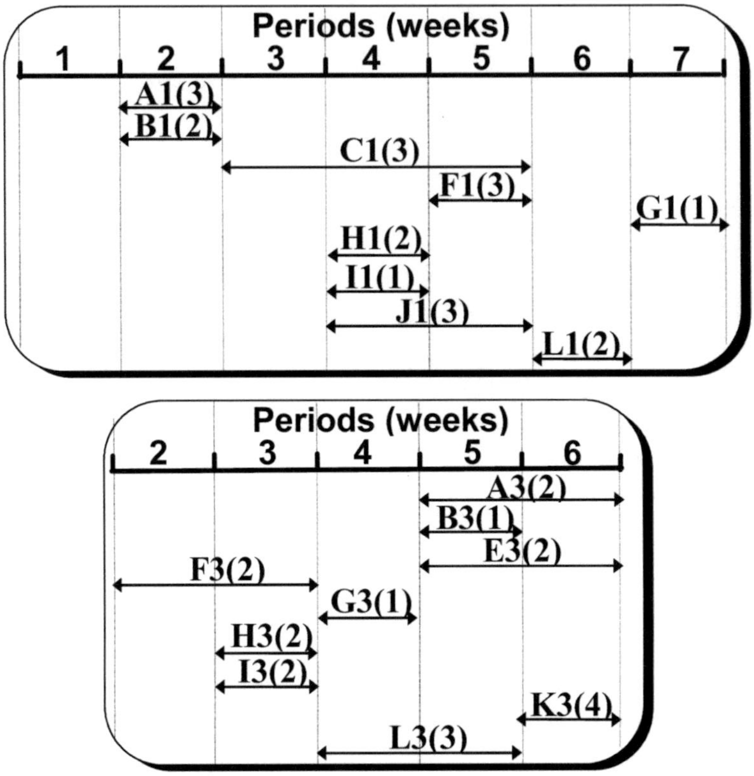

Fig. 8. C.R.P. of processor P1 and P3 after treatment

If we take our example into consideration, the control feasibility graph of the processor P1 indicates that we have a problem in the lapse [5,5] (Figure 7): the accumulated capacity requirement is 6 days and the available capacity is 5 days, so we should lengthen the scheduling time segment of one of the tasks F1 or J1. If we lengthen the task in the right, the due date order will be delayed (F1 or J1 are the last task of the jobs F and J), so we will try to shorten or move the scheduling time of F2 or J2 in the processor P2. F2 can not be shortened (the scheduling time segment is one period). J2 can be shortened, the scheduling time segment will be shortened to the lapse [2,3]. We verify that the control feasibility graph of P2 remains valid. The scheduling time segment of J1 will become [4,5]. The control feasibility graph of P1 becomes valid and we can try to regain margins like done for the processor P2 in the fourth step. We obtain the C.R.P. of processor 1 as shown in figure 8.

Then, we apply step 5 for the last processor P3 and we obtain the C.R.P. as shown in Figure 8. We update the beginning and the end of the scheduling time segments of the other tasks on the other processors as explained in step 4 and we obtain the final CRP of processors P1 and P2 (figure 9).

4. Scheduling

In case where we admit to interrupt at least one task by period, the overlapping load adjustment approach furnishes a necessary and sufficient condition for the existence of a feasible scheduling solution.

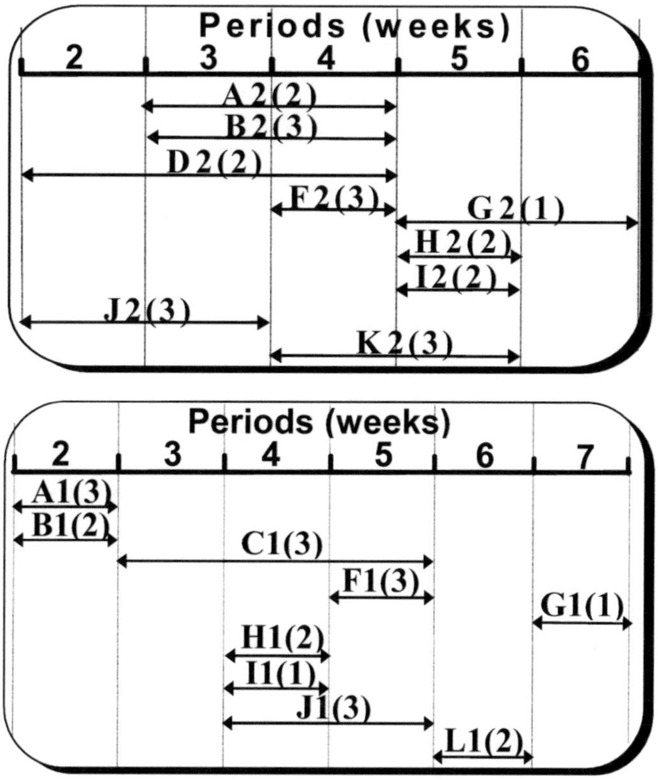

Fig. 9. Final C.R.P. of processors P2 and P1 after treatment

The scheduling will be made in a real time manner. Indeed, in the end of each task the responsible will choose the next task between all the tasks that can be loaded and so on. This load adjustment approach will be coupled with a scheduling tool which can function as described below. If we take our example of Figure 1, the real time scheduling can be made as follow:

- In the beginning of week 1 we can choose to begin the task J, H or I (Figure 10), we choose for instance task J, then tasks I and H.
- We arrive at the beginning of day 1 of week 2 and we see that we can choose between three tasks: B, C, E or F (Figure 10). We choose for example tasks B and C and we obtain the partial scheduling of figure 10.
- We are in the beginning of day 4 of week 2 and we can choose between task E or F. The scheduling tool will inform the user if he can really choose one task and during how much time without breaking the feasibility condition. The user in this example can not choose task F because the task must be completed at the end of week 2 so the user is obliged in this case to choose E and after that F.
- We are at the beginning of day 2 of week 3, we can choose between task D and G. If the user chooses for example task G, the scheduling tool will inform him that the task G can be scheduled for only two days until the end of the third day of the week because there is the task D with a load of 2 days that must be scheduled on week 3. So, if the user decides to choose G, he must interrupt G after two days, schedule D and after that, continue with the task G, but the user can choose to schedule D and G without

interruption. We suppose that the user chooses task D and after that G. Finally, the user will schedule the last task A and we obtain the final scheduling of Figure 11.

In our opinion, this original approach of scheduling presents many advantages in comparison with the automatic calculation of a schedule. First, the approach is really dynamic, each decision is taken in the last moment and we do not produce plans which will be out of date. Moreover, the user can have its own reasons to choose one task or another. A system that proposes and does not impose but exposes the consequences of each choice in regard to the scheduling which allows the user to decide with full knowledge of the facts and integrates his own criteria. This characteristic can promote the scheduling performance, and allow in all cases the responsibility of the user and his comprehension of the system.

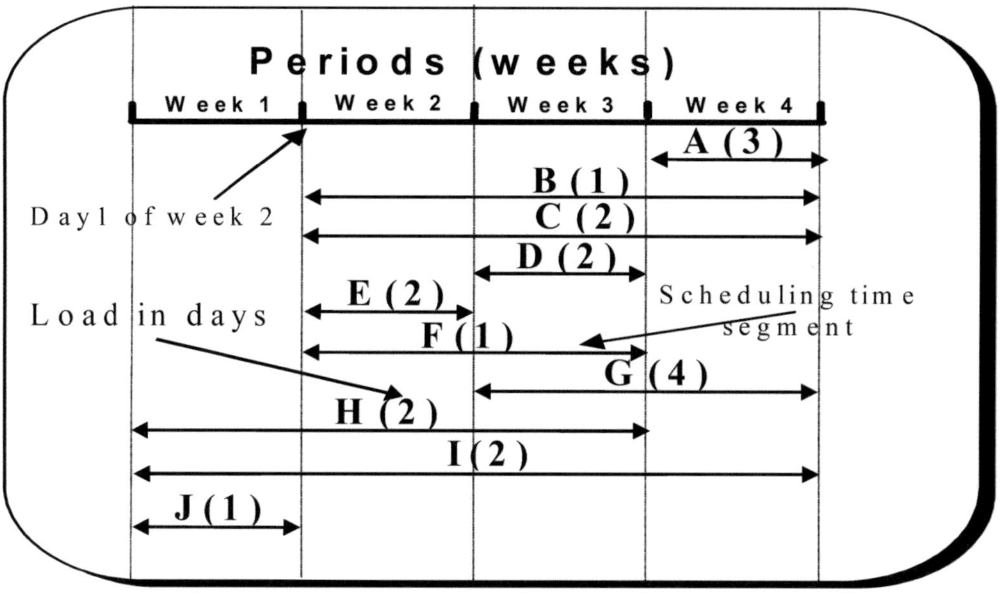

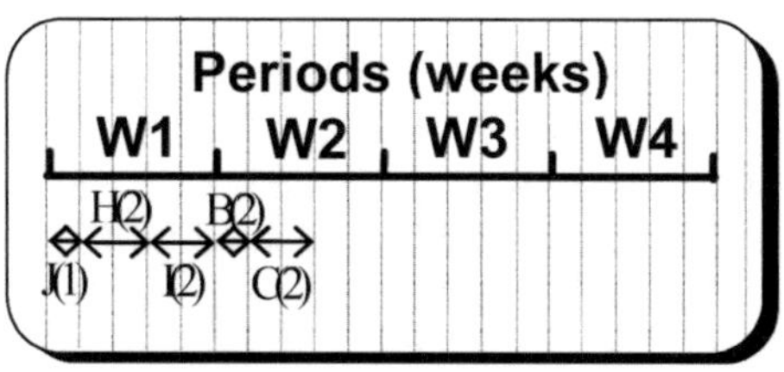

Fig. 10. the partial scheduling

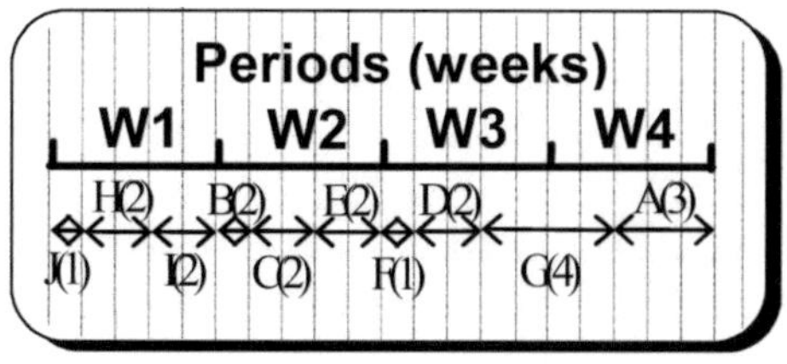

Fig. 11. the final scheduling

5. Experimentation

An experimentation has been carried out on a set of examples. We will describe the conditions and the results of this experimentation. We have constructed 35 examples in a job shop context where we have varied the number of jobs (6 to 16), the number of processors (2 to 4) and the size of the period (4 to 7 days). The total load is always near or superior to the capacity. We compared our approach with the classic approach of placement. For each example, we apply our approach and the placement approach. This placement is applied in two steps: first an earliest placement to determine the earliest due dates and second a latest placement using job's due dates as the maximum between the requested due dates and the earliest ones.

The results of the experimentation are summarized in Table 4. We will use for this these notations:

NMO: Number of Manufacturing Orders.

NPr: Number of processors.

PS: Period size in days.

NPe: Number of periods.

PTO: Processing time of the overlapping approach which represents the duration in days between the beginning of the first task of the M.O. until the ending of the last task of the last M.O.

PTP: Processing time of the placement approach which represents the duration in days between the beginning of the first task of the M.O. until the ending of the last task of the last M.O for the placement approach.

DPT: Difference in % for the processing time between the two approaches.

ORO: Occupation rate of the processors in % for the overlapping approach. It is in fact an average of the occupation rate of all the processors.

ORP: Occupation rate of the processors in % for the placement approach.

DOR: Difference in % for the occupation rate between the two approaches.

NDO: Number of delayed orders for the overlapping approach.

NDP: Number of delayed orders for the placement approach.

DDO: difference in the delayed orders between the two approaches.

First, we have compared the two approaches in terms of processing times (figures 12 and 13). We call the processing time the duration in days between the beginning of the first task on the first processor and the end of the last task on the last processor. The difference is about 4.5 % between the two approaches.

Second, we have compared the two approaches in terms of occupation rate (figures 14 and 15). We remark that there is a difference of about -4.3 % in the occupation rate for the two approaches. This fact is normal because with the first approach we try to keep the maximum of freedom in a middle term without giving a detailed production schedule.

Third, we have compared the two approaches in terms of number of delayed jobs (figures 16 and 17). The average number of delayed jobs with the overlapping approach is about 5.1 whereas it is equal to 4.7 with the placement approach. We observe that the results given by the two approaches are very close even if the placement gives only one scheduling solution when the overlapping approach gives a set of scheduling solutions.

All these facts prove really the efficiency of our approach. However, we should experiment and compare our approach with other approaches and with a biggest number of examples.

Example	NMO	NPr	PS	NPe	PTO	PTP	DPT	ORO	ORP	DOR	NDO	NDP	DDO
1	12	3	5	6	30	28	7	80	87	-7	2	2	0
2	14	4	5	7	30	25	17	70	74	-4	0	2	-2
3	9	3	5	6	25	13	48	54	88	-34	0	0	0
4	6	2	5	7	25	20	20	57	77	-20	0	0	0
5	16	3	5	14	70	59	16	91	92	-1	12	13	-1
6	6	3	5	8	40	40	0	78	91	-13	2	3	-1
7	6	3	4	9	36	40	-11	87	98	-11	2	3	-1
8	12	3	7	6	28	28	0	79	78	1	0	0	0
9	14	4	7	7	42	27	36	48	93	-45	0	0	0
10	6	3	6	7	30	40	-33	93	92	1	0	1	-1
11	6	2	7	7	35	26	26	47	54	-7	0	0	0
12	16	3	7	10	70	59	16	91	92	-1	8	9	-1
13	6	3	7	7	42	40	5	67	93	-26	0	0	0
14	8	3	7	7	28	44	-57	84	57	27	0	0	0
15	12	3	5	14	70	71	-1	87	84	3	11	9	2
16	12	3	7	10	70	71	-1	91	84	7	8	6	2
17	16	3	5	18	90	96	-7	100	89	11	14	14	0
18	16	3	7	14	98	96	2	92	90	2	10	12	-2
19	12	3	5	23	110	112	-2	85	79	6	11	11	0
20	12	3	7	16	112	112	0	85	79	6	12	12	0
21	6	2	5	12	60	55	8	89	100	-11	4	4	0
22	6	2	7	9	63	60	5	87	90	-3	4	4	0
23	14	4	5	22	125	100	20	70	68	2	14	13	1
24	14	4	7	13	91	101	-11	72	88	-16	12	10	2
25	9	3	5	7	35	31	11	78	85	-7	4	3	1
26	9	3	7	7	49	31	37	60	85	-25	4	2	2
27	10	3	5	10	45	46	-2	76	73	3	6	4	2
28	10	3	7	8	56	46	18	100	75	25	5	2	3
29	8	3	4	8	28	36	-29	84	73	11	4	1	3
30	8	3	6	7	30	36	-20	90	78	12	0	0	0
31	14	4	4	7	28	24	14	69	89	-20	1	0	1
32	12	3	4	7	28	26	7	79	88	-9	5	3	2
33	12	3	6	6	30	29	3	77	82	-5	1	0	1
34	16	3	4	16	64	59	8	90	92	-2	13	12	1
35	16	3	6	16	66	59	11	90	92	-2	10	11	-1
Average	10,9	3,0	5.7	10,2	53,7	51,0	4,5	79,3	83,7	-4,3	5,1	4,7	0,4

Table 4. Results of the experimentation

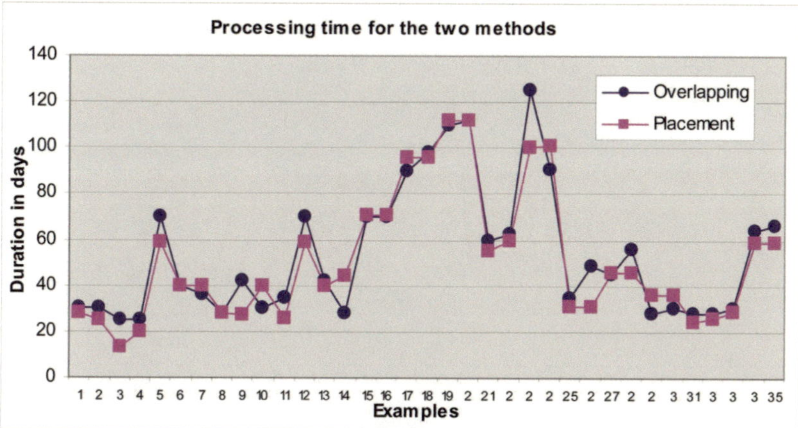

Fig. 12.The processing times obtained with the overlapping and placement approaches.

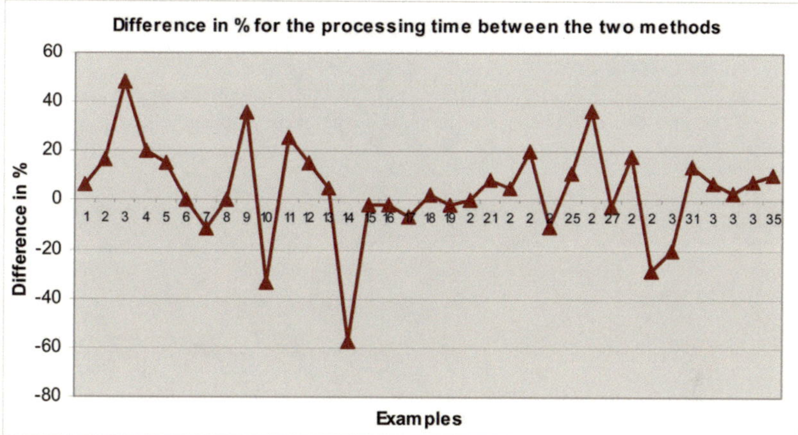

Fig. 13. The difference in % for the processing time between the two methods.

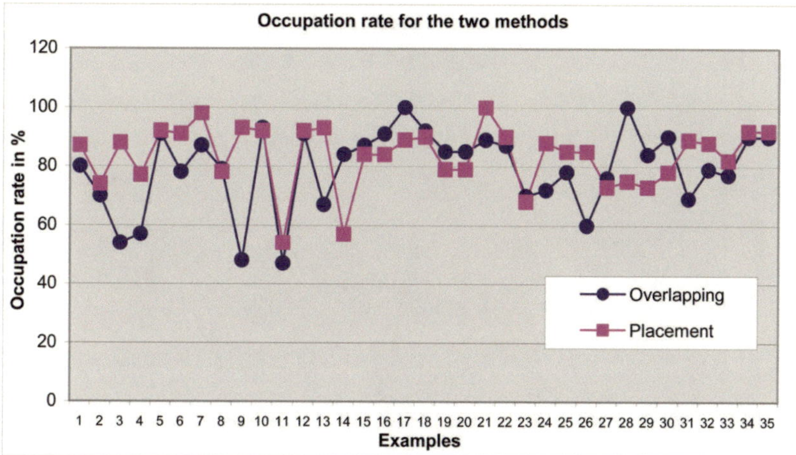

Fig. 14. The occupation rates obtained with the overlapping and placement approaches.

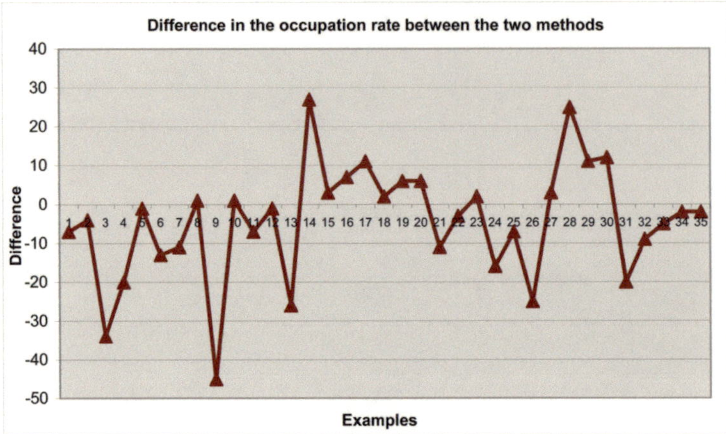

Fig. 15. The difference in the occupation rate between the two methods.

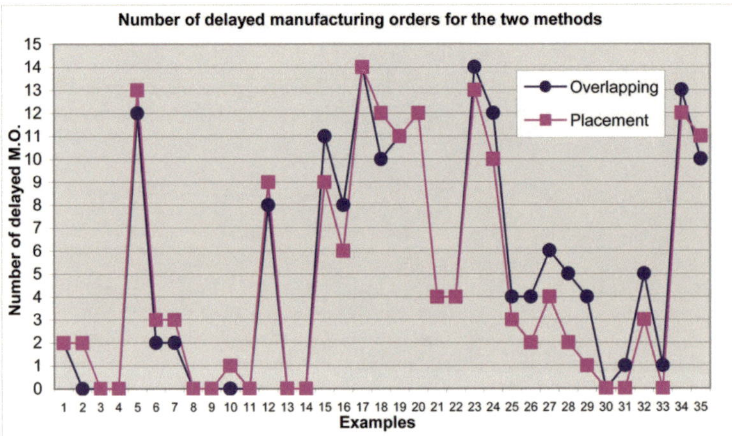

Fig. 16. The number of delayed orders for the overlapping and placement approaches.

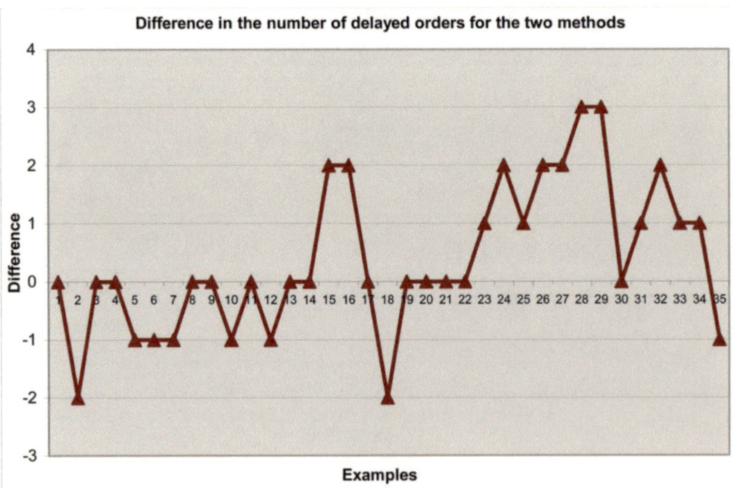

Fig. 17. The difference in the number of delayed orders for the two methods.

6. Conclusion

The most common approach used in production planning remains MRP II (Manufacturing Resource Planning). The proposed approach in this paper constitutes an alternative to the traditional load adjustment approaches used in the CRP (Capacity Requirement Planning) modules in software based on MRP II philosophy.

The new heuristic presented in this paper, in comparison with the usual middle and/or long-term planning and scheduling approaches, has the following advantages:

- not setting a long-term tasks scheduling to assure that the planning can be properly carried out;
- exploiting and distributing judiciously the job's margins on their tasks and trying to respect the just-in-time principles;
- permitting the postponement of the final scheduling job's problems until the short term at the order release phase and/or the scheduling phase;
- delaying, if necessary, the due dates of some jobs or increasing the capacity in some lapses for guaranteeing in every case the feasibility of the production planning.

We can extend and improve our work by studying the possibility of introducing the overlapping of the scheduling time segments of consecutive tasks. We can also improve our heuristic in order to use it in the case where the tasks are not preemptive.

7. References

Bahroun, Z., Jebali, D., Baptiste, P. & Campagne, J-P. & Moalla, M. (1999). Extension of the overlapping production planning and application for the cyclic delivery context, in *IEPM '99 Industrial Engineering and Production Management*, Glasgow.

Bahroun, Z., Campagne, J-P. & Moalla, M. (2000a). The overlapping production planning: A new approach of a limited capacity management. *International Journal of Production Economics*, 64, 21-36.

Bahroun, Z., Campagne, J-P. & Moalla, M. (2000b). Une nouvelle approche de planification à capacité finie pour les ateliers flow-shop. *Journal Européen des Systèmes Automatisés*, 5, 567-598.

Dauzere-Peres, S., & Lassere, J-B. (2002). On the importance of scheduling decisions in production planning and scheduling. *International Transactions in Operational Research*, 9, 6, 779-793.

Dillenseger, F. (1993). Conception d'un système de planification à moyen terme pour fabrications à la commande. PhD thesis, INSA Lyon, France.

Drexl, A. & Kimms, A. (1997). Lot sizing and scheduling - Survey and extensions. *European Journal of Operation Research*, 99, 221-235.

Fleischmann, B. & Meyr, H. (1997). The general lot sizing and scheduling problem. *Operation Research Spektrum*, 19, 1, 11-21.

Grubbström, R.W. & Huynh, T.T.T. (2006). Multi-level, multi-stage capacity-constrained production-inventory systems in discrete time with non-zero lead times using MRP theory. *International Journal of Production Economics*, 101, 53-62.

Gunther, H.O. (1987). Planning lot sizes and capacity requirements in a single stage production system. *European Journal of Operational Research*, 31, 1, 223-231.

Hillion, H. & Proth, J-M. (1994). Finite capacity flow control in a multi-stage/multi-product environment. *International Journal of production Research*, 32, 5, 1119-1136.

Néron, E., Baptiste, P. & Gupta, J.N.D. (2001). Solving Hybrid Flow Shop problem using energetic reasoning and global operations. *Omega*, 29, 501-511.
Pandey, P.C., Yenradee, P. & Archariyapruek, S. (2005). A finite capacity material requirement planning system. *Production Planning & Control*, 2, 113-121.
Wuttipornpun, T. & Yenradee, P. (2004). Development of finite capacity material requirement planning system for assembly operations. *Production Planning & Control*, 15, 534-549.

10

Discovering Strategic Behaviors in Multi-Agent Scenarios by Ontology-Driven Mining

Davide Bacciu[1], Andrea Bellandi[1], Barbara Furletti[1],
Valerio Grossi[2] and Andrea Romei[2]
[1]*IMT Institute for Advanced Studies Lucca*
[2]*Department Of Computer Science – University of Pisa*
Italy

1. Introduction

Providing human users with a structured insight into extensive data collections is a problem that, in the latter years, has gathered increasing attention by the scientific community. An even more daring and ambitious research challenge is the recent attempt to address the same problem within the scope of the so-called multi-agent systems. In this context, multiple autonomous and heterogeneous entities, i.e. the agents, populate the environment performing actions and taking decisions based on internal policies and their knowledge of the world.

The courses of action of the agents and their interaction with the environment and with themselves generates extensive amounts of information. Underneath such flat data collections lies fundamental knowledge concerning strategies and common behaviors emerging from the agent actions. The advantage of extracting such "strategic" knowledge is twofold. On the one hand, it allows users a deeper insight into the multi-agent system, permitting to discover interesting patterns of actions that can be identified as interesting behaviors in the particular problem at hand. Consider, for instance, a multi-agent system simulating negotiation/selling phases in a particular market (Viamonte et al., 2006): the detection of interesting patterns of actions can bring to light specific market strategies that can be exploited to optimize revenues. On the other hand, the identified strategic knowledge can be exploited by the agents themselves to improve their performance, for instance by updating their internal representation of the world and their behavioral policies.

In order to extract such "strategic" knowledge, in this work we take a data mining approach using Association analysis, enriching it with the expressive power and the flexibility of ontologies. Association rules can effectively tackle both aspects of strategic behavior extraction, since they both provide a human understandable representation means for the extracted knowledge and an action rule base that can be used to supply agents with procedural and high-level knowledge concerning the identified strategies.

Ontologies offer a structured description of the domain knowledge while maintaining data and their representation separated. An ontology refers to an "engineering artifact", consisting of a specific vocabulary containing the terms used to describe a certain domain, and a set of explicit assumptions regarding the meaning of vocabulary words. This set of

assumptions has usually the form of a logical theory, where terms correspond to unary or binary predicates, respectively called concepts and relations. For our purposes the logics supplies the formal language, the semantics and a set of proof procedures for representing the various aspects of the knowledge and for reasoning about it, in order to drive the data mining process.
In general, data mining consists of a series of transformation steps, starting from data preprocessing up to post-processing, for converting raw data into useful information. The ontology drives the association rules extraction process intervening in the pre and post processing phases. Complex categories of domain-specific constraints are specified during the preprocessing phase, while during the post-processing phase, the extracted association rules are further filtered for providing better and more fitting results.
Recent works have introduced the use of association rules for imitation learning in robotic agents within a reactive paradigm (Hellström, 2003), as well as in learning simple grasping and picking behaviors in agricultural robots (Zou et al., 2006), while (Makio et al., 2007) exploited association analysis to discover basic motor skills from human motion capture data. However, here we are not interested in determining low level motor behaviors in single agents, e.g. fight or fly reactions; rather, we seek to discover high level strategies underlying the actions of complex multi-agents systems. Within this respect, the introduction of ontologies for constraint-based multi-level association rule mining induces advantages in terms of the human interpretability of the strategies resulting from the Association analysis. Moreover, ontologies offer a simple mean for providing the agents with a representation of the extracted strategies that can be used, for instance, as high level primitives in the planning step of the agents' control hierarchy.
The rest of the chapter is organized as follows: Section 2 introduces the data mining concepts that are used thorough the paper and describes related works in the area of behavior mining in agent-based systems. Section 3 formalizes the proposed model, while in Section 4 we describe the behavior mining approach and we present an example related to the simulated robotic soccer league. Section 5 concludes the paper with final considerations and future developments.

2. Background

The following subsections describe the background knowledge as well as the main concepts of Data Mining and Association analysis, together with a brief survey of the application of association rules to Behavior Mining in Agent-based Systems. Moreover we introduce the ontology theory that is used in the definition of the proposed model.

2.1 Constrained association rule mining

Knowledge Discovery in Databases (KDD) (Tan et al., 2005) is an emerging field that covers a wide range of applicative domains, several models for representing extracted patterns and models, and a large number of algorithms for data preprocessing, model extraction and model reasoning. KDD addresses the task of extracting "meaningful information" from large volumes of data. *Data Mining* (DM) is an important part of the *knowledge discovery process*, whose aims is to convert raw data into useful information. DM consists of a series of transformation steps, from data *pre-processing* to *post-processing*. Pre-processing reduces and transforms raw input data into an appropriate format for the mining step. Post-processing ensures that only valid and useful results are retained.

Association analysis is a central task in DM. It aims at finding strongly correlated sets of events from large databases. In 1993, (Agrawal et al., 1993) introduced association rules for discovering regularities among products in large scale transaction data recorded by point-of-sale (POS) systems in supermarkets. For example, the rule {onions, vegetables} → {beef} indicates that if a customer buys onions and vegetables together, he/she is likely to buy beef as well. Such information can be used as the basis for decisions about marketing activities, such as promotional pricing or product placements. In addition to the above example from market basket analysis, association rules are employed today in many application areas including web mining, intrusion detection, bioinformatics and so on.

Nowadays, one of the most important open issue in association rules mining is due to the large number of extracted rules. In fact, it is not unusual for the collection of rules to be too large for being handled directly by an analyst. Moreover, often one would like to be able to focus the mining process on a particular portion of the search space or, in other words, it is important to have some additional expressiveness in order to define what is interesting and what is not. In order to solve these problems, frameworks for constrained association rule mining have been developed.

In the remaining of this section, we formalize the problem of association rules mining, briefly introducing the Apriori algorithm; finally, we review the constraints-based association rule mining problem.

Definition. Let $I = \{i_1, \ldots, i_n\}$ be a set of literals, called *items*. Each item has some predefined attributes (e.g. price, type, etc.). Let D be a set of *transactions*, which form the database that is going to be disclosed. Each transaction $t \in D$ is an *itemset*, such that $t \subseteq I$. A unique identifier (TID) is associated with each transaction. A transaction t supports X, i. e. a set of items in I, if $X \subseteq t$. An itemset X has *support* $s = supp(X)$ if s% of the transactions supports X. If $|X| = k$, then X is called a *k-itemset*.

An *association rule* (AR) is an implication of the form $X \rightarrow Y$, where $X \subset I$, $Y \subset I$ and $X \cap Y = \emptyset$. X is said *antecedent* (or *body*) and Y *consequent* (or *head*). An AR has a *support* s on D, if s% of the transactions in D contain $X \cup Y$. An association rule has a *confidence* c on D if the c% of the transactions in D containing X, also include Y. In formulas:

- $supp(X \rightarrow Y) = supp(X \cup Y) = |X \cup Y| / |D|$;
- $conf(X \rightarrow Y) = supp(X \cup Y) / supp(X)$.

It is worth noticing that, while support is a measure of the frequency of a rule, confidence is a measure of the strength of the relation between sets of items.

An itemset is said *frequent*, if its support is higher than a user-specified minimum support. A rule is said *strong* if its support and its confidence are higher than a user-specified minimum support and a user-specified minimum confidence, respectively.

A constraint on itemsets is a function $C_I{:}2^I \rightarrow \{true, false\}$. An itemset I satisfies a constraint C, if and only if $C_I(I) = true$. Similarly, a rule constraint is a function $C_R{:}R \rightarrow \{true, false\}$, where R is the set of association rules. In general, given a conjunction of constraints, the constrained association rule mining problem requires to compute all the rules that satisfy the constraints.

Algorithms for AR mining. Association rule mining algorithms scan the database of transactions and calculate the support and confidence of the candidate rules to determine whether they are significant or not. Their computational process is composed of two steps: (1) given a minimum support, find all frequent itemsets; (2) given a set of frequent itemsets and a minimum confidence, find all the association rules that are strong. Step (1) is, usually,

harder than rule generation and the main difficulty of itemsets generation comes from the large size of the search space.
Apriori has been the first frequent itemset algorithm to be proposed (Agrawal et al., 1994). It exploits a bottom-up, level-wise exploration of the lattice of frequent itemsets. During a first scan of the dataset, all the frequent singletons, denoted with L_1, are found. Next, L_1 is used to find the set of frequent 2-itemsets L_2, and so on until all the frequent patterns have been discovered.
Each iteration is composed of two steps: *candidate generation* and *support counting*. During the first step, a collection of possibly frequent itemsets C_k of length k is generated, and their actual support is calculated with a single scan over the dataset during the second step. In principle, it would be easy to generate all the possible k-itemsets at each iteration, given that I is known. As a result, we will have an extremely large collection of itemsets that have to pass the support counting phase. The Apriori algorithm exploits a pruning of the search space in order to reduce the number of candidates before each iteration. Pruning is based on the concept that, if $X \in C_k$ has a subset of length $k - 1$, that does not belong to L_{k-1} i.e. the subset is not frequent, X cannot be frequent as well, and it can be removed from C_k.
A survey of constraints-based AR mining. The fundamental idea inspiring constrained AR mining is to introduce a set of constraints that association rules should satisfy. The analyst can use a conjunction of constraints for specifying the properties of the patterns of interest. As a trivial example, let us focus on market basket analysis. Lets suppose that each item is associated to a fixed price, and that we are interested only on those co-occuring set of items such that their average price is higher than a given threshold. KDD systems should be able to exploit such constraints to speedup the knowledge extraction process.
The most important classes of constraints and their properties are shown in Fig. 1. For a complete and formal definition please refer to (Bonchi & Lucchese, 2007).
The first interesting class is the *anti-monotone* (Agrawal et al., 1993), which was already introduced with the Apriori algorithm (i.e. the minimum frequency is an anti-monotone constraint). It simply states that, if a predicate holds for a set S, then it holds for any subset of S. At the opposite, the *monotone* (Bucila et al., 2002) constraints state that, if a predicate holds for a set S, it holds also for any superset of S. For example, the constraint $max(S.price) > 100$ is monotone, while $max(S.price) \leq 100$ is anti-monotone.
The *succinct* constraint (Han et. al., 1999) states that the decision concerning the satisfaction of a predicate can be determined based on the single element of the S set. For instance, the max constraints defined above are both succinct, since a single element of S can be used as a comparator. The $avg(S)$ constraint is not succinct.
The class of *convertible anti-monotone* (resp. *convertible monotone*) (Pei & Han, 2000) includes constraints C_{CAM} (resp. C_{CM}) that define an item ordering relation such that whenever an itemset S satisfies C_{CAM} (resp. violates C_{CM}), so does any prefix of S. For instance, $mean(S.price) > 100$ is a convertible anti-monotone constraint, with respect to the descending price order. In fact, $C_{CAM}(S) \Rightarrow C_{CAM}(T)$, where S is a prefix of T.
The class of *loose anti-monotone* constraints, introduced by (Bonchi & Lucchese, 2005), simply states that, given an itemset S with $|S| > 2$, a constraint is loose anti-monotone (denoted C_{LAM}) if: $C_{LAM}(S) \Rightarrow \exists i \in S: C_{LAM}(S \setminus \{i\})$. It is trivial to show that $variance(S.price) < 100$ is a loose anti-monotone constraint.

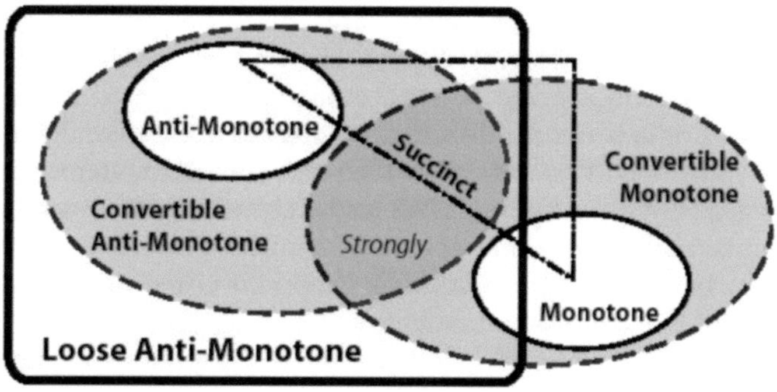

Fig. 1. Characterization of the classes of constraints.

For each of these classes, there exists a specialized algorithm, which is able to take advantage of its peculiar properties. Techniques for constraint-driven pattern discovery can also be classified, depending on the KDD phase in which they are evaluated, in the following categories:

- *pre-processing constraints* are evaluated during the pre-processing phase, restricting the source data to the instances that can generate only patterns satisfying the constraint;
- *mining constraints* are directly integrated into the mining algorithm (e.g. Apriori) used for extracting the rules;
- *post-processing constraints* are evaluated for filtering out patterns after the mining algorithm.

(Wojciechowski & Zakrzewicz, 2002) focuses on improving the efficiency of constraint-based frequent pattern mining, by using dataset filtering techniques that conceptually transform a given data mining task into an equivalent one operating on a smaller dataset.

A first preliminary effort to propose a general framework for constrained frequent itemsets mining is (Bonchi & Lucchese, 2007), in which the authors present the design of *ExAMiner^{GEN}*, a general Apriori-like algorithm, which is able to exploit all the possible kinds of constraints presented above. In particular, the authors suggest a data reduction technique for obtaining advantages from the conjunction of monotone and anti-monotone constraints. They observe that a transaction that does not satisfy the monotone constraint CM, can be removed since none of its subsets will satisfy CM either, and therefore the transaction cannot support any valid itemset. This data reduction, in turn, lowers the support of other frequent but invalid itemsets, thus reducing the search space and improving anti-monotone pruning techniques. This virtuous circle is encoded level-wise in the Apriori algorithm.

Constraints can also improve the comprehension of the extracted rules, with the aid of a concept hierarchy. A concept hierarchy is a multilevel organization of the various entities or concepts defined in a particular domain. For example, in the market basket analysis a concept hierarchy has the form of an item taxonomy describing the "is-a" relationships among the items sold in a store, e.g. *"milk"* is a kind of *"food"* and *"DVD"* is a kind of *"home electronics"* equipment. Mining on concept hierarchies is also called constraint-based multilevel DM and association rules extracted with the aid of a taxonomy are also called *multi-level*.

The main advantages of incorporating concept hierarchies into association analysis are: (i) items at lower levels may not have enough support to appear in any frequent itemsets and (ii) rules found at lower levels of a hierarchy tend to be too specific and may not be interesting as rules at higher levels.

Methods to define and integrate item constraints on a concept hierarchy are originally introduced in (Agrawal & Srikant, 1995) and (Fu & Han, 1995). Originally, multilevel constraints are evaluated during the pre-processing phase, generalizing items at bottom level to higher levels of the hierarchy before applying Apriori. Basically, each transaction *t*, is replaced with its "extended transaction", which contains all the items in *t* along with their corresponding ancestors. (Srikant et al., 1997) and (Han et al., 1999) can be seen as the first attempts to integrate both constraint-based and multilevel mining directly into the rule extraction algorithm, employing anti-monotone proprieties existing across levels: "if a high-level item is infrequent, none of its descendants can be frequent".

In (Goethals & Van den Bussche, 2003) the authors integrate the evaluation of constraints into the mining process, so that the amount of computation is proportional to what the user gets, limiting the exponential generation of patterns and rules. The constraints considered in this work are Boolean combinations of atomic conditions, e.g. *Body(i)* (or *Head(i)*) specifies that a certain item *i* occurs in the body (respectively, in the head) of a rule. Another interesting work on integrating constraints evaluation in the mining process is represented by DualMiner (Bucila et al., 2003). This system implements a new algorithm for finding frequent itemsets, that efficiently prunes its search space by employing both monotone and anti-monotone constraints.

Other remarkable constraint-based optimizations can be found in (Gallo et al., 2005) and (Esposito et al., 2006). The former proposes an "incremental" approach exploiting the results of previous queries in order to reduce the response time to new queries. In this context the term *incremental* describes the fact that the outcome of a new query starts from a previous result. The proposed algorithm is able to deal with context-dependent constraints. A similar idea is used in (Esposito et al., 2006), where a new query is rewritten in terms of union and intersection of the result sets of other previously executed and materialized queries. The authors propose a two-step approach. Firstly, they find conditions to apply various typologies of constraints (essentially on items properties), then, by using an optimizer, they recognize equivalent queries avoiding repeated heavy computations.

Finally, (Wang et al., 2003) explores *support constraints*, that specify what minimum support is required for each itemsets, so that only the necessary itemsets are generated. The three constraints $supp_1(B_1, B_3) \geq 0.2$, $supp_2(B_3) \geq 0.4$ and $supp_3(B_2) \geq 0.6$ are examples of support constraints where each B_i, called *bin*, is a set of disjoint items which, taken in isolation, do not need to satisfy any minimum support requirement.

In this chapter, we focus on pre-processing and post-processing multi-level constraints. However, we plan to develop further our model by integrating constraints evaluation directly into the mining phase, taking advantage of the peculiar properties of each constraint.

2.2 Behavior mining in agent-based systems

In last twenty years, data mining has been extensively applied to provide users with a deep insight into data collections, supplying powerful tools for extracting high-level knowledge from extensive, often poorly structured, information sources. Recently (Cao et al., 2007) attention has started to shift from the classical view of data mining as a tool for transferring knowledge from raw data to human agents (i.e. the users), to a novel interpretation where data mining serves to transfer knowledge from data collections to intelligent artificial agents, under the form of strategic patters of actions, i.e. behaviors.

At the beginning, most of the works focused on learning single agent behaviors from demonstrational data: in (Hellström, 2003), for instance, a robotic system based on the reactive paradigm that learns complex behaviors by imitation is presented. Through a stimuli/response mapping, the system determines the appropriate robot action as a direct function of its sensory inputs. This mapping is provided by a set of association rules that are extracted from sensory/action data gathered as the human user exemplifies the behaviors that have to be learned by the robotic agent. Similarly, (Zou et al., 2006) use association rules to learn picking behaviors for agriculture robots. Other recent works focused on data mining as a tool for extracting motor primitives from motion data: (Makio et al., 2007) discusses how to discover motor skills by mining human motion capture data to extract association rules and frequent patterns that describe the dependencies among the body parts and the relative movements. In (Xu et al., 2007), on the other hand, association rules are used to discover and describe knowledge of chewing behavior in an object oriented framework. (Mori et al., 2005) collects information on the household behavior of people in daily life: by means of association rules, they discover frequent combinations of events, called episodes, that are used to predict and support the daily actions of the user.

The models described so far relate to single agent scenarios (either robotic or purely software), but much of the recent works have been devoted to multi-agent systems. These ones differ from single agent systems in the sense that there is no global control and globally consistent knowledge and, since data and control are distributed, multi-agent models have the inherent advantages of distributed systems, such as scalability, fault-tolerance and parallelism. Notice that the term "multi-agent" refers to two aspects of the problem: the former and most straightforward one relates to the fact that the system itself is, indeed, composed of multiple agents. The latter aspect relates to the fact that the data used in the mining process is produced by multiple sources, i.e. by multiple human or artificial agents. This is the case of several agent based models with application to information retrieval, computational economics and robotics. In (Viamonte et al., 2006), for instance, is presented a Multi-Agent Market Simulator where the agents behavior follows the customers and seller profiles extracted by mining information from the market segment under analysis. (Bezek et al., 2006) presents an algorithm for extracting strategies from a RoboCup soccer scenario comprising several interacting agents with different roles and behaviors. In particular, the model extracts association rules from the sequences of basic multi-agent actions, showing how the constructed rules capture basic soccer strategies such as offensive moves and defense schemes.

More generic models, such as the Agent Academy framework (Mitkas et al., 2004) support design and instantiation of agent communities as well as ontology definition. This system integrates an agent training procedure that is a mechanism for embedding rule-based reasoning capabilities and application-specific knowledge into agents. This is realized by means of a data-mining module that processes available behavioral knowledge in order to extract decision models for the agents. Another general purpose multi-agent framework is (Kaya et al., 2005). It uses association rule mining to estimate agent actions, maintaining an historic database of the past agent actions that is processed to extract association rules that are, in turn, used to update the action selection strategies of the agents. With a similar intent, (Dudek, 2007) proposes the APS (Analysis of Past States) framework, where web browsing agents alternate between work phases, during which actions are performed and information concerning the user preferences are recorded, and stand-by phases, where the agents are idle and the historic information is mined to extract new user profiles under the form of association rules.

In general, there is an increasing trend of integrating strategic behavioral knowledge into agents by mining repositories containing data collected either exogenously to the system (e.g. by human exemplification) or by analyzing historic data of the agents themselves. Such knowledge extraction mechanism requires tools capable of extracting significant behavioral patterns from large data collections. Additionally, the representation of such behavioral patterns should allow cross interpretability of the data mining results both from the machine and the user perspectives. We advise that association rules can play a key role in this sense, serving two key purposes: as knowledge representation means they provide humans with insight into data collections; as an action rule base, they can be used to supply virtual agents with procedural and high-level knowledge that can effectively be exploited for increasing the agents' performance. In particular, we will focus on how the introduction of ontologies in the association mining process can successfully improve the interpretability of the association analysis's results both from the point of view of the user and from the agents' perspective.

2.3 Ontology

The ontology may be defined as the *study of being as such*. The construction of the word itself has remote origins and may be seen as an attempt to modernize the classical debate about metaphysics. One of the first exponent of this new "discipline" was Aristotle that, in (Aristotle, 350 B.C.), presents a first work that can be resembled to an "ontology". He provides a list of categories that represent an inventory of what there is, in terms of the most general kinds of entities. Later on, the Latin term "ontologia" appeared for the first time in the circles of German protestants around the 1600 (Göckelin, 1980) (Lorhard, 1613), while the first English appearance goes back to 1721 in the Baileys dictionary in which it was defined as "an Account of being in the Abstract". The first use of the term ontology in the computer and information science literature occurs instead in 1967, in a work on the foundations of data modeling by S. H. Mealy (Mealy, 1967).

From the scientific point of view, the ontology is, in its first approximation, a table of categories in which every type of entity is captured by some node within a hierarchical tree.

While the philosopher-ontologist has only the goal of establishing the truth about the domain in question, in the world of the information systems the ontology is a software (or formal language) designed with a specific set of uses and computational environments. In this field, the ontologies are rapidly developing (from 1990 up to now) thanks to their focus on classification and their abilities in representing the domains, in sharing the knowledge and in keeping separate the domain knowledge from the operational one. Furthermore, they play a key role in the Semantic Web by providing a source of shared and precisely defined terms that can be used for describing the resources. Reasoning over such descriptions is essential for accessibility purposes, automating processes and discovering new knowledge.

One of the modern ontology definition given by (Maedche, 2003) is reported below. An ontology O is defined as

$$O = \{C, R, A^O\},$$

where:

1. C is a set whose elements are called concepts.

2. $R \subseteq (C \times C)$ is a set whose elements are called relations. For $r = (c_1, c_2) \in R$, one may write $r(c_1) = c_2$.
3. A^O is a set of axioms on O.

To cope with the lexical level, the notion of a lexicon is introduced. For an ontology structure $O = \{C, R, A^O\}$ a lexicon L is defined as

$$L = \{L^C, L^R, F, G\},$$

where:

1. L^C is a set whose elements are called lexical entries for concepts.
2. L^R is a set whose elements are called lexical entries for relations.
3. $F \subseteq L^C \times C$ is a reference for concepts such that

$$F(l_C) = \{c \in C: (l_C,c) \in F\} \text{ for all } l_C \in L^C,$$

$$F^{-1}(c) = \{l_C \in L^C: (l_C,c) \in F\} \text{ for all } c \in C.$$

4. $G \subseteq L^R \times R$ is a reference for concepts such that

$$G(l_R) = \{r \in R: (l_R,r) \in G\} \text{ for all } l_R \in L^R,$$

$$G^{-1}(r) = \{l_R \in L^R: (l_R,r) \in G\} \text{ for all } r \in R.$$

We note that a hierarchical structure can be explicitly defined in terms of R.

Finally, the mapping from elements in the lexicon to elements in the ontology structure corresponds to the notion of an interpretation in declarative semantics (Genesereth and Nilsson, 1987). Such a (semantic) interpretation associates elements of a language to elements of a conceptualisation. Based on the above definitions, an ontology can be formally defined as the couple $<O, L>$ where O is an ontology structure and L is a corresponding lexicon. The ontology structure O plays the role of an explicit specification of a conceptualisation of some domain while the lexicon L provides the agreed vocabulary to communicate about this conceptualisation.

The following example may be useful for explaining the above definitions.

Let $O = \{C, R, A^O\}$ be an ontology structure such that $C = \{c_1, c_2\}$ and $R = \{r\}$, where $r(c_1) = c_2$. Suppose that $A^O = \varnothing$. Also, let $L = \{L^C, L^R, F, C\}$ be a corresponding lexicon such that $L^C = \{Mouse, Input\_device\}$, $L^R = \{is\_a\}$, $F(Mouse) = c_1$, $F(Input\_device) = c_2$ and $G(is\_a) = r$.

Fig. 2 depicts the elementary ontology described above. The top of the figure shows the ontology structure O. It consists of two concepts c_1, c_2 and one relation r that relates them. This corresponds to a conceptualisation of a domain of interest without any lexical reference. The latter is provided by the lexicon L depicted at the bottom of the figure. L provides the lexical references for O by means of F and G. F and G map lexical reference strings to the concepts and relations defined in O, respectively. For instance, for $r \in R$ one may consider using $G^{-1}(r)$ to get the lexical reference, i.e. "*is_a*", corresponding to r and vice versa.

The Web Ontology Language (OWL) is the current standard provided by the World Wide Web Consortium (W3C) for defining and instantiating Web ontologies. An OWL ontology can include descriptions of classes, properties and their instances. The OWL language provides three increasingly expressive sublanguages:

- OWL Lite supports a classification hierarchy and simple constraint features.

- OWL DL is more expressive than OWL Lite without losing computational completeness (all entailments are guaranteed to be computed) and decidability (all computations will finish in finite time) of reasoning systems. Its name is due to its correspondence with description logics (Baader, 2003), a field of research that has studied a particular decidable fragment of first order logic.
- OWL Full is the most expressive with no computational guarantees.

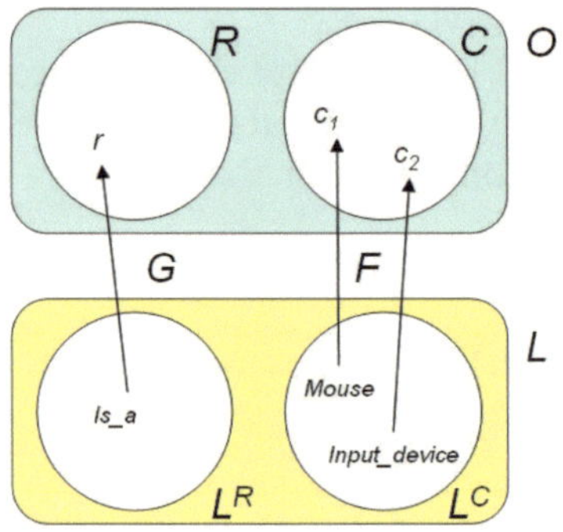

Fig. 2. Graphical representation of an ontology.

We focus mainly on the first two variants of OWL because OWL-Full has a nonstandard semantics that makes the language undecidable and therefore difficult to implement. OWL comes with several syntaxes, all of which are rather verbose. Hence, in this chapter we use the standard DL syntax. For a full DL syntax description, please refer to (Baader, 2003).

The main building blocks of an OWL ontology are (i) concepts (or classes), representing sets of objects, (ii) roles (or properties), representing relationships between objects, and (iii) individuals, representing specific objects. Furthermore they are composed of two parts: intensional and extensional. The former part consists of a *TBox* and an *RBox*, and contains knowledge about concepts (i.e. classes) and the complex relations between them (i.e. roles). The latter one consists of an *ABox*, containing knowledge about entities and the way they are related to the classes as well as roles from the intensional part. In the following is an example of DL formalization (according to the syntax defined in (Baader, 2003)) related to the previous ontology fragment. We add a property specifying the number of keys of each input device

$$
\begin{aligned}
TBox = ((Mouse \subseteq Input\_device) \cap (Input\_device \subseteq Thing) \\
\cap (=1\ hasKeysNumber) \subseteq Input\_device \\
\cap (\forall\ hasKeysNumber.integer)).
\end{aligned}
$$

All the possible ontology instances constitute the *ABox* which is interlinked with the intensional knowledge. An example of *ABox* related to the previous *TBox* is the following:

$$
\begin{aligned}
ABox = \{Logitech\_device : Input\_device, \\
hasKeysNumber = 88, \\
Optical\_IBM\_Mouse : Mouse, \\
hasKeysNumber = 2\}
\end{aligned}
$$

Given such an ontology, the OWL formal semantics specifies how to derive its logical consequences, i.e. facts not literally present in the ontology, but entailed by the semantics. The semantics of OWL DL is fairly standard by DL standards. An example is in Fig. 3.
An interpretation $I = (\Delta^I, \cdot^I)$ is a tuple where Δ^I, the domain of discourse, is the union of two disjoint sets $\Delta_O{}^I$, (the object domain) and $\Delta_D{}^I$ (the data domain), and I is the interpretation function that gives meaning to the entities defined in the ontology. I maps each OWL class C to a subset $C^I \subseteq \Delta_O{}^I$, each object property P_{Obj} to a binary relation $P^I{}_{Obj} \subseteq (\Delta_O{}^I \times \Delta_O{}^I)$, each datatype property P_{Data} to a binary relation $P^I{}_{Data} \subseteq (\Delta_O{}^I \times \Delta_D{}^I)$. The complete definition can be found in the OWL W3C Recommendation[1].

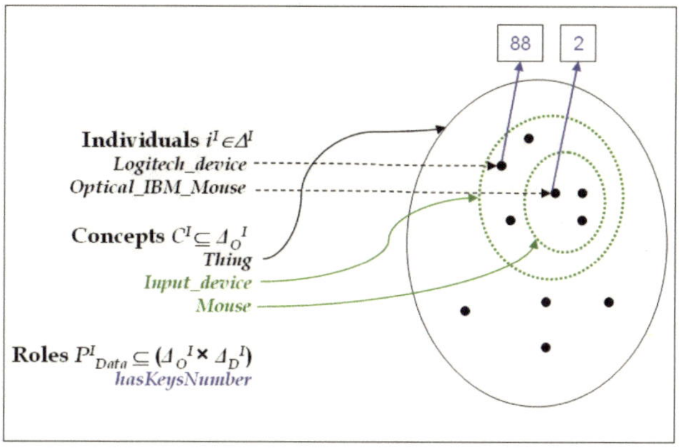

Fig. 3. Example of well defined (model theoretic) semantics.

Concepts and relations represent the explicit knowledge of an ontology, nevertheless another kind of knowledge (i.e. the implicit knowledge) can be "deduced" starting from the known facts. Its existence is implied by or inferred from observable behavior or performance by using the rule inference mechanism. Rules are an extension of the logical core formalism, which can still be interpreted logically. The simplest variant of such rules are expressions of the form $C \Rightarrow D$ where C and D are concepts. The meaning of such a rule is "if an individual is proved to be an instance of C, then derive that it is also an instance of D". Operationally, the semantics of a finite set R of rules can be described by a forward reasoning process. Starting with an initial knowledge base K, a series of knowledge bases $K^{(0)}$, $K^{(1)}$, ... is constructed, where $K^{(0)} = K$ and $K^{(i+1)}$ is obtained from $K^{(i)}$ by adding a new assertion $D(a)$ whenever R contains a rule $C \Rightarrow D$ such that $K^{(i)} \leq C(a)$ holds, but $K^{(i)}$ does not contain $D(a)$.
Semantic Web Rule Language (SWRL) is the W3C proposal for the rule language based on a combination of the OWL DL and OWL Lite sublanguages with the Unary/Binary Datalog RuleML sublanguages of the Rule Markup Language[2]. SWRL allows users to write Horn-like rules that can be expressed in terms of OWL concepts and that can reason about OWL individuals (Horrocks, 2004). According to the SWRL syntax and referring to the previous example of TBox and ABox, the inference rule for deducing that an ontology element is an instance of Mouse stated that is an Input_device and has 2 keys, is reported below:

$$\forall x,y \ \mathit{Input\_device}(?x) \wedge \mathit{hasKeysNumber}(?x,?y) \wedge \mathit{equalTo}(?y, 2) \rightarrow \mathit{Mouse}(?x),$$

where equalTo(., .) is a SWRL built-in function.

[1] The W3C webpage: http://www.w3c.org/.

[2] Rule Markup Language is specified at www.ruleml.org.

3. The system

This section describes in detail our framework and the process for extracting multi-level association rules by using the ontological support (Bellandi et al., 2008).
Our scenario consists of the set of components shown in Fig. 4.

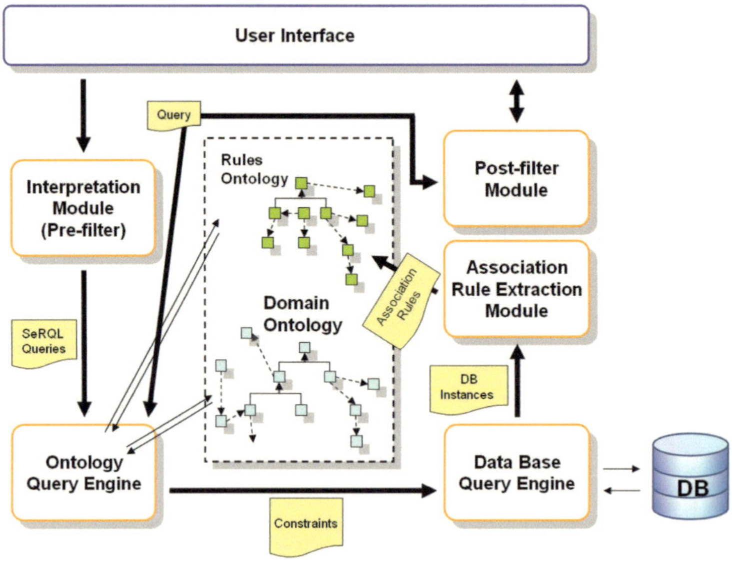

Fig. 4. Architecture of the system.

As in the most common system architectures, our framework is composed of a set of "modules" and a set of repositories. There is a repository for the metadata, i.e. the ontologies, and a repository for the transactions, i.e. the DataBase (DB). The former repository contains the **Domain Ontology** (DO) and the **Rule Ontology** (RO). The DO describes the domain of interest, i.e. all the items that compose the transactions and that are stored in the DB. The RO, instead, is used for storing the association rules extracted during the process. It is different from the DO because it models the structure of an association rule according to its definition (see Section 2.1). The DO is the description model for the collection of data and is not designed for containing actual instances. The RO, instead, is used both for describing and containing the inference rules.
The **User Interface** provides mechanisms for interacting with the system and for visualizing the results. The user can provide the system with two types of input: a set of user constraints for specifying what items must be considered in the discovering process, and another one for filtering the extracted association rules. The user specifies constraints by using the syntax presented in the Table 1. It includes both pruning constraints, used to filter a set of non-interesting items, and abstraction constraints, which permit a generalization of the item on a concept of the ontology. We informally describe each constraint by means of an example.
By using pruning constraints, one can specify a set of items which can be excluded from the input transactions set, and, as a consequence, from the extracted rules. This kind of constraints refers either to a single item, or to an ontology concept, and can include a condition expressed on a set of ontology properties. There exist two kinds of pruning constraints:

$\mathcal{I}$ is the set of items $(i_1, i_2, ... i_n \in \mathcal{I})$.
$\mathcal{C}$ is the set of the concepts of the ontology $(c_1, c_2, ... c_n \in \mathcal{C})$.
$\mathcal{P}_c$ is the set of the properties of the concept $c \in \mathcal{C}$ $(p_1, p_2, ... p_n \in \mathcal{P}_c)$.
$cond_c$ is a Description Logic expression.
ALL represents all the instances defined in the ontology.

A constraint is defined on $\mathcal{I}$, $\mathcal{C}$ and $\mathcal{P_C}$ in the following form:

1. **Pruning Constraints.** A pruning constraint is of one of the following forms:
   (a) $prune(e)$, where $e \in \mathcal{I} \cup \mathcal{C} \cup \{ALL\}$.
   (b) $prune_{cond_c}(c)$, where $c \in \mathcal{C} \cup \{ALL\}$.
2. **Abstraction Constraints.** An abstraction constraint is of one of the following forms:
   (a) $abstract(e, c)$, where $e \in \mathcal{I} \cup \mathcal{C}$, $c \in \mathcal{C}$ and c is a super-concept of e.
   (b) $abstract_{cond_{c_1}}(c_1, c_2)$ where $c_1 \in \mathcal{C} \cup \{ALL\}$, $c_2 \in \mathcal{C}$ and c_2 is a super-concept of c_1.
   (c) $abstract^{l}_{cond_e}(e)$, where $e \in \mathcal{I} \cup \mathcal{C} \cup \{ALL\}$, and l is a non-negative integer indicating the level of the hierarchy; $cond$ can be unspecified.

Table 1. Syntax of the Pruning and Abstraction constraints.

- The constraint *prune*(e) excludes a single item or all the items belonging to an ontology concept expressed by "e". As one could expect, removing an ontology concept $c \in C$ also implies excluding all the items belonging to the descendent concepts of c.
- The constraint $prune_{cond}$(c) is similar to *prune*(e), except for uninteresting items that are selected according to a condition defined on a set of properties of c. The condition can also refer to a property of an ancestor of c.

For example, let us suppose to have a simple ontology modelling all possible soccer actions performed by soccer agents. Every item represents an action and has a property hasPassesNumber to indicate the number of passes of an action. The ontology has the form of a taxonomy organized according to the kind of action (e.g. Action, GoalAction, Non-GoalAction, CornerAction, etc.). The predicates *prune*(Non-GoalAction) and $prune_{hasPassesNumber>5}$(GoalAction) allow to prune all the "Non-GoalAction" actions and all the "GoalAction" actions with a number of passes greater than 5, from the input transactions. Notice that hasPassesNumber is a property of the upper concept Action, while Non-GoalAction and GoalAction are sub-concepts of Action.

The Abstraction constraints enable to explore different levels of the ontology concepts. The generalization to a predefined level of the hierarchy increases the support of association rules and, consequently, avoids the discovery of a massive quantity of useless rules, especially in case of sparse data. More interesting is the exploitation of rules in which items are extracted at different levels of the hierarchy. The system offers three different predicates:

- The predicate *abstract*(e, c) generalizes a single item e (or all the items belonging to the concept e) to the super-concept c.
- The predicate $abstract_{cond<c1>}(c_1, c_2)$ is similar to the previous one, but it generalizes items according to a condition ($cond<c_1>$) defined on a set of properties of the concept c_1. Only items satisfying this condition are abstracted to the super-concept c_2. As for $prune_{cond}$(c), this condition can also refer to a property of an ancestor of c_1.
- Finally, the $abstract^{l}_{cond<e>}$(e) generalizes every element satisfying the condition $cond<e>$ to the super-concept placed at level l of the hierarchy. In this case, e can be either a

single item or a concept. If the condition is not specified, all the items belonging to e are generalized to the super-concepts at level l.

The symbol "ALL"in Table 1 can be used to select all the items defined in the ontology.

Suppose that the ontology provides also a description of all the soccer field areas, organized as an hierarchy; this permits to specify in which soccer field area each action is performed.

Lets assume that LittlePenaltyArea is a sub-concept of PenaltyArea. The predicate *abstract*(LittlePenaltyArea, PenaltyArea) generalizes all the actions performed in the "Little Penalty Area" to a more generic "Penalty Area". It is equivalent to the predicate $abstract^5$(LittlePenaltyArea), if PenaltyArea is placed at level 5 of the ontology.

The set of constraints V_D is therefore a conjunction of pruning and abstraction constraints.

The **Interpretation Module** in Fig. 4 translates the requests of an user into a set of formal constrains (defined on the DO) so that they can be supplied to the Ontology Query Engine by means of a suitable query language. In our system we used the Sesame RDF Query Language - SeRQL (Broekstra, 2004).

The **Ontology Query Engine** interacts with the DO by performing some queries and retrieving the desired set of items (IS). This module uses a particular ontology query language according to the ontology management system that supports the operations.

The output is used by the **DB Query Engine** for retrieving the DB transactions that contain the items specified in IS.

The **Association Rule Extraction Module** contains an implementation of the *Apriori* algorithm that computes the set of association rules (AR). Such a rules are then stored in the RO. Finally, the **Post-filter Module** evaluates the post filtering constraints for selecting the most relevant association rules w.r.t. the user needs. Since the extracted rules are stored in the OR, this module interacts with the Ontology Query Engine for retrieving the correct items by means of directed queries. The post-filter module interacts with the GUI for receiving the SERQL queries and for presenting the output rules. It is important to notice that the constraints are conceived not only on the structure of the rules, but also on their semantics, since the domain ontology provides a mean to express the semantics of single items belonging to the rules. For example, let us consider the case in which the ontology models all the possible soccer actions performed by soccer agents. Association rules can be mined to highlight relations among the actions performed by all players. Given a set of output rules, we are able to focus only on those in which the antecedent contains at least one kick action made by a player having a particular role. In this case, the semantics of the item "kick" and "player role" are encoded in the domain ontology.

4. Discovering strategic behaviors by association rules mining

In this section we describe how the association rule mining system presented above can be instantiated in the context of behavior discovery in multi-agents system (in Section 4.1), and we describe (in Section 4.2) an application example in the field of simulated robotic soccer.

4.1 Multi-agent planning with ontology-based association rules

Earlier in the chapter, we have discussed the increasing trend of behavior discovery in multi-agent systems. Here, we focus on the role of association rule mining as a means for extracting such strategic action patterns: in particular, we point out the capital importance of an ontology based approach for extracting high-level planning knowledge that can understood both by the human users as well as by the software agents. For this reason, we present an ontology-based planning scheme for intelligent computational entities.

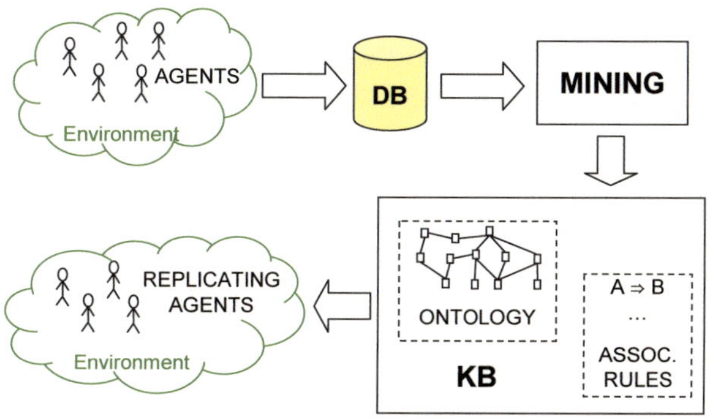

Fig. 5. Behavior mining with association rules in a multi-agent system

Consider, for instance the scenario in Fig. 5: a set of reference agents, either human or artificial, is interacting within an environment, each of them pursuing a particular goal based on its internally defined policy. The *configurations* of the system, including the courses of actions of the agents as well as characteristic information (e.g. spatial location, battery level and signal strength in radio agents) are monitored and logged to a database (DB in Fig. 5). After a sufficient amount of time, the database will contain massive amounts of *flat* information referring to the development of the agents' behavior in time, as well as the strategies used to pursue their goals. Our work aims at analyzing such large data repository to extract a knowledge base (KB in Fig. 5) holding higher level information content that can be used to refine the planning strategies of the *replicating* intelligent agents in Fig. 5.

Consider, for instance, the elements in Fig. 5 as being players from two soccer teams: by monitoring their actions during the whole match we can log interesting information, such as player location, speed, stamina, as well as events of interest, e.g. kick, pass, header, slide, dash, etc. Then, by applying association analysis, we can extract a set of association rules that describes the most likely decisions to be taken for the most frequent configurations of the game. For instance, the knowledge base can end up containing rules such as

> ***IF*** *LeftPlayer10.location IS RightQuarterMidfield AND LeftPlayer10.pass AND LeftPlayer11.location IS RightPenaltyArea*
> ***THEN*** *LeftPlayer11.dash;*

which describes a typical soccer attack strategy, that is "when player 10 is on the attack and passes the ball, then player 11 should run for the pass in the penalty area". The extracted rules are, in turn, used to populate our knowledge repository KB, which includes a description of our reference *world* given by the *domain ontology*.

For the particular problem at hand, i.e. soccer matches, the ontology can include, for instance, a taxonomy of the areas of the field, a description of the players (e.g. in terms of physical/physiological parameters such as height, weight, stamina, speed, etc.) as well as a multi-layered description of the actions and events (e.g. the pass action that can be further differentiated in long and short pass, heal pass, etc.).

The novelty of our approach lies in the full integration between the ontology and the association analysis: the ontological description can be used to navigate the results of the association analysis at a given level of abstraction by imposing ontology-based constraints to filter the association rules in the KB. More formally, this is obtained by applying the *rule ontology* population scheme in Fig. 4: by exploiting the knowledge representation in the *domain ontology* we perform queries that filter the association rules at particular level of abstraction. The resulting rules serve to populate the rule ontology that is in turn used for the decision making process.
Association rules filtering is of capital importance when approaching behavior mining in a multi-agent system, due to the explosive rule base complexity that may arise as a result of the association analysis. The interaction of multiple agents, in fact, generates a massive amount of data that populates the DB; as a consequence the traditional association analysis can generate a large number of rules addressing the behavior of several agents. Thus, a pruning technique is needed to reduce the cardinality of the association rules by filtering the results of the association analysis to obtain personalized views of the discovered procedural knowledge. In our approach this mechanism is implemented by means of ontology-based filtering. For instance, if we are interested in knowledge regarding heal pass actions performed in opponent's team area, we apply the ontological definitions of heal pass and opponent's area to the association rules in the knowledge repository. As a result only those rules satisfying the constraints are retained and used to plan the next decision. This approach offers a two-fold advantage: on the one hand, it allows human users to selectively tune their view on the extracted knowledge so that their analysis of the data can be more tightly focused. On the other hand, such an approach is fully exploitable by intelligent software agents to obtain behavioral patterns that are tailored to the particular decision scenario they are addressing. The ontological description of the world can, in fact, be easily coded into the software agent. The agent itself can then use it to gather knowledge from the mare-magnum of the association rule base at the required descriptive level (i.e. abstraction constraints) or for a particular ontological relation (i.e. property constraints). By this means, we can define intelligent software agents that exploit their world description, i.e. the ontology, to learn useful behavioral patterns, i.e. the ontology-filtered rules, from a vast amount of flat data.
Going back to our soccer example, we can interpret the replicating agents in Fig. 5 as being players in a virtual soccer match: the rule base extracted from the *actual* soccer players is accessed by each replicating agent that applies its ontology-based filter to extract those behavioral patterns that best fit the current game configuration. Then, the planning module of the replicating agent determines the next action (or sequence of actions) based on the retrieved decision rules (see Fig. 6). In a sense, the replicating agent is learning a behavioral pattern by analyzing the most frequent configurations generated by the *actual* agents and selecting only those that fits best its role and its personal view of the world. Notice that, in general, nothing prevents the real agents to be the same as the replicating agents; hence, by devising *smart* association rule filters, an agent can be made to learn from the history of its past actions.
Fig. 6 depicts a prototypical planning module for a replicating agent that is based on the proposed ontology-driven behavior mining model. The agent builds the current configuration descriptor by gathering information from the environment, the proprioceptive

sensors as well as from its internal state. This information is used to determine the preconditions for the decision rules and is exploited by the planner to elaborate the next course of actions. In order to do so, the agent planner queries its knowledge base at the required level of abstraction and obtains a set of decision rules that match the current system configuration. Based on these rules, the planner determines the next actions that are forwarded to the actuator sub-system.

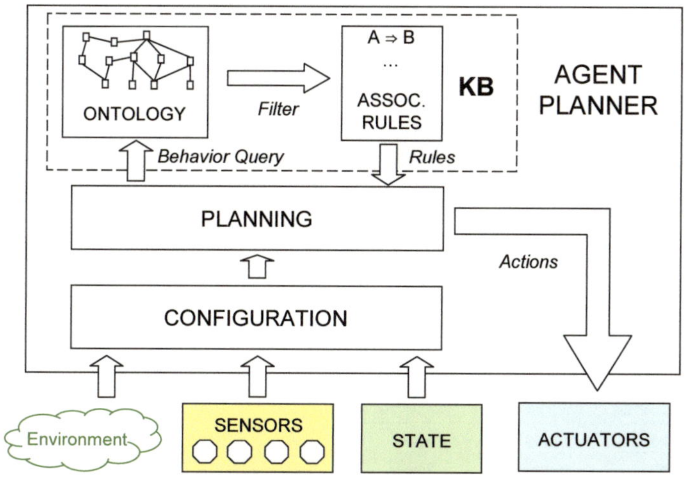

Fig. 6. Agent planner based on the ontology-driven behavior mining model

The simple soccer scenario that we have adopted to introduce the behavior mining scheme is not far from a real-world application of the model. Several soccer teams, in fact, have started monitoring their players with sensors both during training sessions and during games (e.g. see the MilanLab[3] structure of AC Milan). Such information, if accessible, can in principle be used to extract strategic action patterns that can be exploited, for instance, to devise intelligent software agents for virtual soccer matches, such as in the Robocup league. Unfortunately, such data has not been disclosed, yet: hence, in the next section, we present a behavior extraction example based on the freely available logs from the Simulated Soccer League.

4.2 Extracting strategic behaviors from robot soccer data: an example

In this paragraph we provide a schematic example describing the steps of the analysis process. The example is based on data collected from the RoboCup simulated league. RoboCup is an international initiative to foster artificial intelligence and intelligent robotics by providing a standard test bed where several technologies can be integrated and examined. The RoboCup simulated league consists of teams of autonomous software agents competing against one another in soccer games. Logs were taken from each simulation game consisting of about 6000 simulation cycles. Each cycle contains information about player location and orientation, ball location, ball possessor, game

[3] http://www.acmilan.com/InfoPage.aspx?id=565

score and play modes such as throw-ins, goal-kicks and so on. Since our aim is to "mine" the behaviour of a team in the soccer RoboCup league in the 2004 edition[4], we have collected all the log events related to the match played by the winner team, focusing on position and action information.

Step1: The ontology definition.

We defined an ontology for describing the entities of the soccer domain, focusing on the variables actually contained in the collected data. We built the ontology by creating a taxonomy of concepts, adding interesting data as well as object properties according to the target of our analysis.

In Fig. 7 is reported the ontology of the main concepts (superClasses) and their data and object properties. For our purposes it was necessary to model the concepts *Player*, *Kick*, *Body*, *Ball*, *PlayerRole* and *FootballFieldArea*. Concerning the kick action, it is interesting to know the actor and the receiver (2 players), where it is directed (the field area) as well as the body part that has produced it. The corresponding object properties are *madeBy*, *towards*, *hasDestinationArea*, *madeByUsing*.

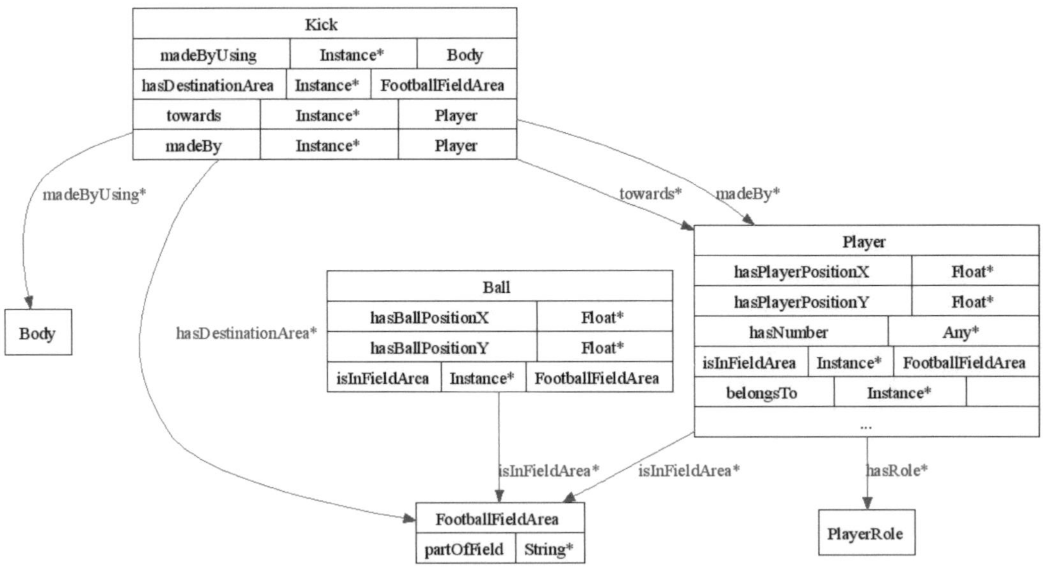

Fig. 7. The domain ontology for the soccer domain: the super-concepts.

For a player is interesting to know its role and its position. The latter information is of interest also for the ball concept.

For the concepts *Body*, *FootballFieldArea* and *PlayerRole* we realized three taxonomies of sub-concepts (Fig. 8, Fig. 9 and Fig. 10 respectively), that in the following will be used to express AR constraints.

[4] http://staff.science.uva.nl/~jellekok/robocup/rc04/

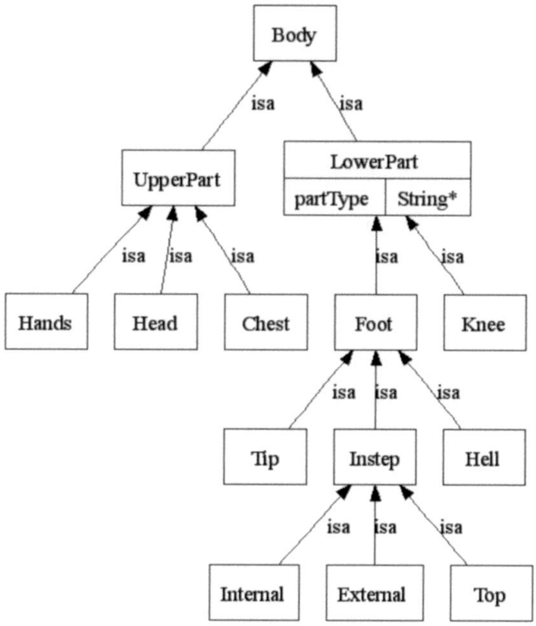

Fig. 8. The *Body* taxonomy.

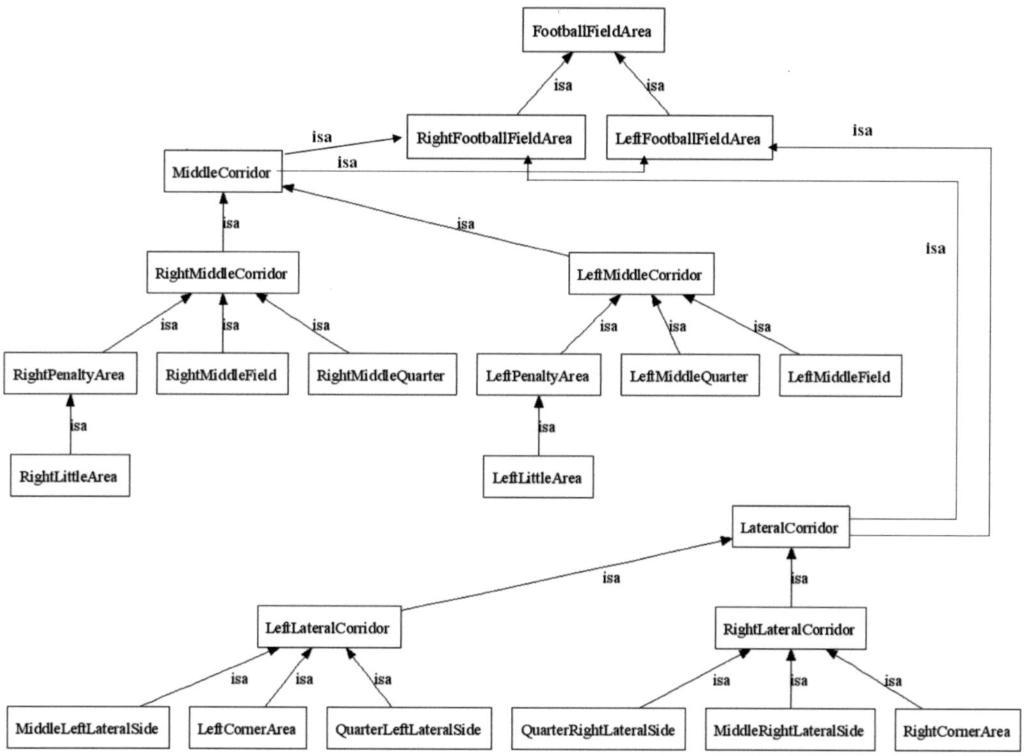

Fig. 9. The *FootballFieldArea* taxonomy.

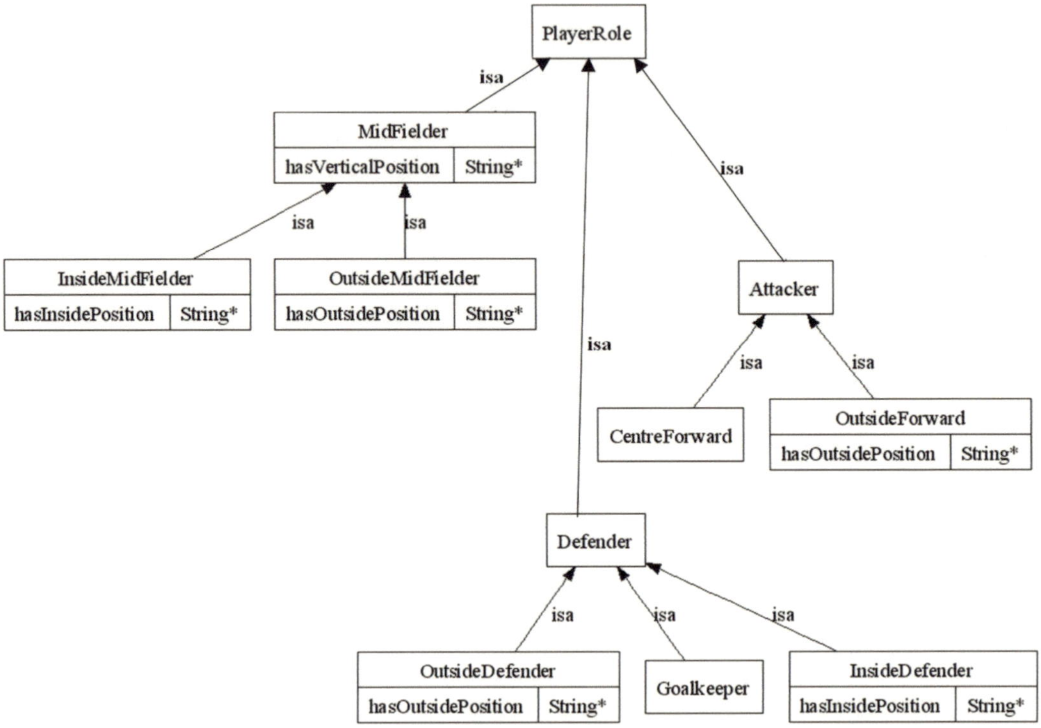

Fig. 10. The *PlayerRole* taxonomy.

The information related to all the previous ontological entities, was not explicitly represented in the DB. Starting from the information contained in the DB, we inferred the new information according to our ontology entities. For instance, in the *FootBallFieldArea* taxonomy, we split the soccer field into 24 disjointed parts, that are used to characterize players and ball positions, as showed in figure Fig. 11.

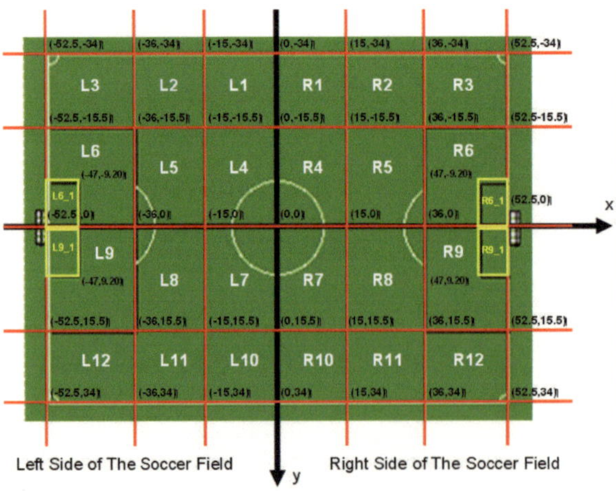

Right Side Field	Left Side Field
R1 = Middle Right Lateral Side	L1 = Middle Right Lateral Side
R2 = Quarter Right Lateral Side	L2 = Quarter Right Lateral Side
R3 = Right Corner Area	L3 = Right Corner Area
R4 = Right Middle Field	L4 = Right Middle Field
R5 = Right Middle Quarter	L5 = Right Middle Quarter
R6 = Right Penalty Area	L6 = Right Penalty Area
R6_1 = Right Little Area	L6_1 = Right Little Area
R7 = Left Middle Field	L7 = Left Middle Field
R8 = Left Middle Quarter	L8 = Left Middle Quarter
R9 = Left Penalty Area	L9 = Left Penalty Area
R10 = Middle Left Lateral Side	L10 = Middle Left Lateral Side
R9_1 = Left Little Area	L9_1 = Left Little Area
R11 = Quarter Left Lateral Side	L11 = Quarter Left Lateral Side
R12 = Left Corner Area	L12 = Left Corner Area

Fig. 11. Parts of the soccer field.

Starting from the coordinates of each position, inference rules are used for identifying each part of the soccer field. The SWRL rules have the following form:

$$\forall x,y,z\ \ Player(?z) \wedge hasPlayerPositionX(?z, ?x) \wedge hasPlayerPositionY(?z, ?y) \wedge \text{GreaterOrEqualThan}(?x, -52.5) \wedge \text{LessOrEqualThan}(?x, -47) \wedge \text{GreaterThan}(?y, -9.2) \wedge \text{LessThan}(?y, 0) \rightarrow isInFieldArea(?z, LLA\_L).$$

This rule states that a certain player (identified by the variable *z*), belonging to the *Left* team, is in the left part of the little area (*LLA_L*), where *LLA_L* is an individual of the *LeftLittleArea* ontology class. Similar rules have been used for deducting all the other ontology entities. According to the architectural description in Section 3, we introduce an ontology for storing the ARs, that is shown below in Fig. 12.

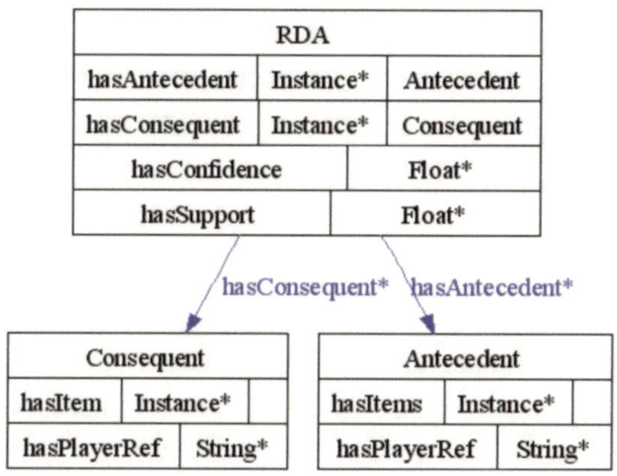

Fig. 12. The Rule Ontology.

This ontology models the structure of an association rule. The concept *RDA* identifies a rule: it has the reference to the confidence and the support, as well as the link (1:n relation) to concepts *Antecedent* and *Consequent*, that are, in turn, used to model the left and right hand side of a rule.

Step2: Data collection and preparation.

Raw data consists of about 65.000 transactions and 108 attributes. The structure of each one of the 22 players is reported in Table 2. Moreover, we have some actual attributes related to

Attribute	Type	Description
pos_x	real	X Coordinate of a player position in the football field
pos_y	real	Y Coordinate of a player position in the football field
actions_kick	{0, 1}	1 if actual action is equal to kick, 0 otherwise
actions_dash	{0, 1}	1 if actual action is equal to dash, 0 otherwise
actions_turn	{0, 1}	1 if actual action is equal to turn, 0 otherwise
actions_catch	{0, 1}	1 if actual action is equal to catch, 0 otherwise
actions_move	{0, 1}	1 if actual action is equal to move, 0 otherwise
calculated_possessor	{0, 1}	1 if player has ball, 0 otherwise

Table 2. Player Attributes.

the ball state, such as the coordinate of the ball in the football field (*ball_position*), the distance of the ball from a goal (*ball_goal_distance*) and the possessor team of the ball (*ball_possessor*).

The rule extraction algorithm is not applied directly to the collected data described above, whereas such information is abstracted or filtered in agreement with the user specified constraints. For example a player position, expressed by x-y coordinates, can be abstracted to whatever field zone defined by the ontology.

We selected all the actions where the team under analysis is in "offensive state". To determine an "offensive state", we abstract the *ball_position* attribute to an ontology high level concept considering only those actions in which the ball is into a specific part of the football field, for example *PenaltyArea*, or simply in the opponent midfield. Then, we filtered the data considering only the actions where at least one player executes a kick. Finally, we observed the behavior of the forward players (e.g. with numbers from 9 to 11) related to the opposite defenders (e.g. with numbers from 2 to 5).

It is important to notice that all the data transformations introduced above are expressed by means of constraints defined on the basis of a well-defined domain ontology, and automatically executed by the software interpreting the results provided by the ontology.

Step3: The constraints definition.

Our simple experimentation concerns two queries. In the first case, the objective is to find frequent itemsets involving "kicking players" and actions performed in the right middle soccer field. In the second case, we are interested in these specific kicking players: PlayerL9, PlayerL10, PlayerL11, PlayerR2, PlayerR3 and PlayerR4. By using our constraint language, an user can formulate the first query in the following way:

$$\begin{aligned} query1 = \; & (prune_{(belongsTo\,=\,'R')}(\text{Player}) \\ & \wedge\, prune_{(hasNumber\,>\,5\ or\ hasNumber\,=\,1)}(\text{Player}) \\ & \wedge\, prune_{(belongsTo\,=\,'L')}(\text{Player}) \\ & \wedge\, prune_{(hasNumber\,<\,9)}(\text{Player}) \\ & \wedge\, prune_{(isInField\,=\,RightPart)}(\text{Player}) \\ & \wedge\, prune_{(madeBy\text{-}1\,=\,'null')}(\text{Player}) \\ & \wedge\, abstract^{3}(\text{FootBallFieldArea})) \end{aligned}$$

In *query1*, the first four clauses are used for pruning all the players *PlayerR1, PlayerR5, PlayerR6, PlayerR7, PlayerR8, PlayerR9, PlayerR10, PlayerR11, PlayerL1, PlayerL2, PlayerL3, PlayerL4, PlayerL5, PlayerL6, PlayerL7* and *PlayerL8*. The 5th clause permits to prune all the players performing actions in the left part of the soccer field, while the 6th one permits to prune all the players not kicking (by using the inverse property of madeBy). The last clause abstracts all the positions in the filed to level 3 of the FootBallFielArea taxonomy.

The second example is defined by the following query:

$$\begin{aligned} query2 = \; & (prune_{(belongsTo\,=\,'R'}(\text{Player}) \\ & \wedge\, prune_{(hasNumber\,>\,5\ or\ hasNumber\,=\,1)}(\text{Player}) \\ & \wedge\, prune_{(belongsTo\,=\,'L')}(\text{Player}) \\ & \wedge\, prune_{(hasNumber\,<\,9)}(\text{Player}) \\ & \wedge\, prune_{(isInField\,=\,RightPart)}(\text{Player}) \\ & \wedge\, prune_{(madeBy\text{-}1\,=\,L1\ or\ madeBy\text{-}1\,=\,L2\ or\ madeBy\text{-}1\,=\,L3\ or\ madeBy\text{-}1\,=\,L4\ or\ madeBy\text{-}1\,=\,L5\ or\ madeBy\text{-}1\,=\,L6\ or\ madeBy\text{-}1\,=\,L7\ or\ madeBy\text{-}1\,=\,L8)}(\text{Player}) \end{aligned}$$

$\wedge$ *prune*(madeBy-1 = R1 or madeBy-1 =R5 or madeBy-1 = R6 or madeBy-1 = R7 or madeBy-1 =R8 or madeBy-1 = R9 or madeBy-1 = R10 or madeBy-1 = R11)(Player)

$\wedge$ *abstract*^{3}(FootBallFieldArea)).

Query2 is very similar to *query1*, but instead of selecting all the kicking players, it retrieves only those items in which either *PlayerL9, PlayerL10, PlayerL11, PlayerR2, PlayerR3* or *PlayerR4* performs the kick.

Step4: Extracting the Behavior Patterns.

Once that the ontology-based query has gone through all the steps discussed above, we obtain a populated rule ontology specifying a set of association rules that describes the behavioral patterns of interest.

As an example, let us focus on defensive strategies by centering our analysis on the behavior of the defenders *D = {PlayerR1, PlayerR2, PlayerR3, PlayerR4}* in the right-side team with respect to the forwards *F = {PlayerL9, PlayerL10, PlayerL11}* of the left team. To study the positioning of the players at a sufficient level of detail, we abstract to the bottommost layer of the taxonomy in Fig. 9 (see Fig. 11 for a visual representation of the field areas). Moreover, to quantize actions to *relevant* time instants we filter out those transactions where the ball is not kicked by any player. If we are interested in discovering knowledge regulating the in-field positioning of the defenders, we can query for relevant association rules regarding the players position on the field. An example of high-confidence ontological rules produced by this association analysis is:

a. *PlayerL10_RMQ_R, PlayerL9_LMQ_R, PlayerR2_QLLS_R, PlayerR4_RMQ_R --> PlayerR3_LMQ_R;*
b. *PlayerL10_RMQ_R, PlayerL11_LMQ_R, PlayerR2_LMQ_R --> PlayerR4_RMQ_R;*
c. *PlayerL11_LPA_R, PlayerL9_LPA_R, PlayerR3_LPA_R PlayerR5_RPA_R, --> PlayerR2_LPA_R;*
d. *PlayerL10_RMF_R, PlayerR5_MRLS_R --> PlayerR4_RMF_R.*

Rule (a), for instance, describes the typical defense scheme where a player, i.e. *PlayerL10,* is controlled by the defender *PlayerR4,* while a second attacker, i.e. *PlayerL9,* has to be defended by *PlayerR3* which, indeed, has to position in the same area of the field of *PlayerL9.* Notice that the rule states that a third defender, i.e. *PlayerR2,* is already holding a position and should not defend on *PlayerL9.* Similarly, rule (b) requires *PlayerR4* to defend on *PlayerL10* while *PlayerR2* is on *PlayerL11.* These kinds of rules can be used by those agents implementing the defender players to determine their position on the field in standard defense schemes. Rule (c) describes a dangerous situation in which two forwards, i.e. *PlayerL11* and *PlayerL9,* are in the left side of the penalty area: *PlayerR3* is defending on both players while *PlayerR4* is on the right side of the penalty. Hence, a smart defense strategy requires *PlayerR2* to help *PlayerR3* on the left side. On the other hand, rule (d) describes a typical pattern where *PlayerR5* should keep its position, not defending on *PlayerL10,* while *PlayerR4* should localize in the same field position as *PlayerL10.*

If we are interested in discovering action patterns correlated with the defense schemes, we can let action concepts into the itemset together with the players' positions. An example of rules that we obtained from the simulated soccer data is:

e. *PlayerR2_LMQ_R_dash, PlayerR3_LMQ_R_dash --> PlayerR4_RMQ_R_dash;*
f. *PlayerL9_RMQ_R_dash --> PlayerR4_RMQ_R_dash.*

The former rule describes a typical joint defensive movement, that is: when two defenders located in the same field area dash, then neighbor players, i.e. *PlayerR4*, should follow them.

The latter rule, on the other hand, states probably the most typical defensive movement: if a forward *PlayerL9* dashes in the field area of a defender *PlayerR4*, then *PlayerR4* should follow him.

The same approach that has lead us to discover defensive patterns can be used to extract knowledge concerning attack strategies. For instance, let us study first the positioning on the field for the forwards *F = {PlayerL9, PlayerL10, PlayerL11}* of the left team. By querying the rule ontology we obtain rules such as:

g. *PlayerL11_QLLS_R, PlayerL9_LMQ_R, PlayerR4_LMQ_R --> PlayerL10_RMQ_R;*
h. *PlayerL10_hasBall, PlayerL9_RMQ_R --> PlayerL11_LMQ_R;*
i. *PlayerL10_RPA_R, PlayerL10_hasBall, PlayerL9_RPA_R, PlayerR4_RPA_R --> PlayerL11_LPA_R.*

Rule (g) describes a basic attack strategy where the forward *PlayerL11* is off-center on the left lateral side while *PlayerL9* is controlled by the defender *PlayerR4* on the center-left; therefore, the third forward *PlayerL10* should chase the space on the center-right area. Rule (h) identifies another fundamental, yet very simple, attack strategy that is: if *PlayerL10* has the ball and is about to kick (notice that we have filtered kick-only timeframes) then the two other forwards, i.e. *PlayerL9* and *PlayerL11*, should position outside the penalty area, respectively on the right and on the left, waiting for a pass or a rebound. Rule (i), on the other hand, depicts a clear scoring chance where *PlayerL10* is holding the ball in the right penalty area and *PlayerL9* is defended by *PlayerR4* on the same position, hence *PlayerL11* should chase the space, and a possible rebound or pass, on the left of the penalty.

Likewise with defensive strategies we can query for attack action patterns by including action concepts in the itemset. Notice that there are two approaches for achieving this: the former, that was already shown in rules (e)-(f) considers items as a concatenation of actions and positions; the latter considers actions and positions separately. An example of attack strategies is:

j. *PlayerL9_LMQ_R_dash --> PlayerL10_RMQ_R_dash;*
k. *PlayerL10_hasBall, PlayerL9_RMQ_R --> PlayerL10_kick, PlayerL11_LMQ_R;*
l. *PlayerL11_hasBall, PlayerL9_LMQ_R, PlayerR3_LMQ_R --> PlayerL10_RMQ_R, PlayerL11_kick.*

Rule (j) is, again, an example of the first approach: it simply states that *PlayerL10*, positioned on the right of the penalty area, should follow *PlayerL9* when he dashes outside the left of the penalty area (i.e. joint attack). Rules (k) and (l), on the other hand, are two examples of the second approach. The first rule, for instance, states that, if *PlayerL10* has the ball and *PlayerL9* is positioned outside the right of the penalty area, then *PlayerL11* has to chase the space on the left and *PlayerL10* should shoot or pass. Taking a closer look at the database transactions that falls in the support of rule (k) confirms that, actually, *PlayerL10* tries to pass or shoot the ball whenever this rule is enabled. Similarly, rule (l) states that *PlayerL10* should occupy the area on the right of the penalty and *PlayerL11* should pass or shoot if *PlayerL9* is defended by *PlayerR3* on the left.

The rules described above are just a small sample of those obtained by the proposed ontology-based association analysis on the simulated soccer data. In general, each agent implementing a particular role should filter the rules of its interest, that is those containing interesting information on the consequent. By exploiting its internally defined ontology, each agent can obtain personalized views of the knowledge concealed in the flat data.

Typically, an agent could be interested only in association rules whose consequent refers to a specific player, e.g. *PlayerL10*. By using our constraint language, the agent can express such a post-processing constraint in the following way:

$$\text{query\_post} = \textit{prune}_{(\text{hasPlayerRef} <> \text{PlayerL10})}(\text{Consequent}).$$

Indeed, the value of the retrieved expert knowledge will depend strictly on the quality on the world description, i.e. the domain ontology, as well as on the available data, that is to say, if we rely solely on the strategies of loser teams, we will most probably obtain also-ran agents.

5. Conclusion

We have introduced an ontology-based approach for association analysis in the context of behavior mining in multi-agent systems. Our proposal is based on the idea of the ontological description of the domain as an essential via-point for accessing the expert knowledge concealed underneath massive amounts of "flat" data. The introduction of a multi-layered and multi-relational representation of the domain allows approaching the information content from several, diverse, viewpoints. Within the multi-agent area, this approach offers considerable advantages since it allows the agents to gather personalized views of the extracted knowledge, represented by means of rule ontologies. By exploiting this "relativistic" representation, an agent can dynamically generate and selectively access, at the desired level of abstraction, the knowledge that is of higher relevance for its current decision-making activity. Besides presenting the clear advantage of offering personalized views of the world, which is totally consistent with the multi-agent model, this approach relieves the agents from the burden of acquiring, maintaining and mastering a monolithic, and encyclopedic, representation of their knowledge.

The model presented in this chapter tackles the association analysis task with the standard static approach where each transaction is considered in isolation and not as part of a spatiotemporal trajectory. Sequence mining, on the other hand, studies how to approach the problem of finding frequent patterns for trajectory data. An interesting future development for our model would be to extend it to sequence mining (Agrawal & Srikant, 1995): in particular, this would be of great interest for behavioral pattern mining, since it can naturally tackle the problem of planning medium to long term strategies comprising lengthy sequences of inter-dependent actions. However, we would like to point out that our model already offers a means for processing multiple transactions in a sort of time trajectory. Through the domain ontology is possible, in fact, to describe concepts whose definition transcends the single transaction. Consider, for instance, the pass concept in the soccer ontology: since it requires to specify source and destination of the action, its instantiation needs to process multiple transactions to find all the requested information, e.g. the receiver of the pass.

6. References

Agrawal, R.; Imielinski, T. & Swami, A. (1993). Mining Association Rules Between Sets of Items in Large Databases. *Proceedings of the SIGMOD Conference* 1993, pp. 207-216, Washington, D.C., May 1993

Agrawal, R.; Methta M.; Shafer, J. & Srikant, R (1994). Fast algorithms for mining association rules in large databases. *Proceeding of the 20th International Conference on Very Large Databases (VLDB'94)*, pp. 478–499, ISBN 1-55860-153-8, Santiago de Chile, Chile, September 1994, Morgan Kaufmann, San Francisco, CA

Agrawal, R. & Srikant, R. (1995). Mining Generalized Association Rules. *Proceedings of the 21st International Conference on Very Large Data Bases (VLDB'95)*, pp. 407-419, ISBN 1-55860-379-4, Zurich, Switzerland, September 1995, Morgan Kaufmann, San Francisco, CA

Agrawal, R. & Srikant, R. (1995). Mining Sequential Patterns. *Proceedings of the Eleventh International Conference on Data Engineering*, pp. 3-14, ISBN 0-8186-6910-1, Taipei, Taiwan, May 1995

Aristotle. (350 B.C.) Categories, The Internet Classic Archive

Baader, F.; Calvanese, D.; McGuinness, D.; Nardi, D. & Patel-Schneider, P. (2003) The description logic handbook. Cambridge University Press (2003)

Bellandi, A.; Furletti, B.; Grossi, V. & Romei, A. (2008) Ontological Support for Association Rule Mining, In: *Proceedings of IASTED International conference on Artificial Intelligence and Applications – AIA 2008*

Bezek, A.; Gams, M. & Bratko, I. (2006) Multi-agent strategic modeling in a robotic soccer domain, In: *Proceedings of the Fifth international Joint Conference on Autonomous Agents and Multiagent Systems*, Hakodate, Japan, May 2006, 457-464, ACM, New York, NY

Boekstra, J. & Kampman, A. (2004) SeRQL: An RDF Query and Transformation Language

Cao, L.; Luo, C. & Zhang, C. (2007) Autonomous Intelligent Systems: Multi-Agents and Data Mining, LNCS, Vol. 4476, 60-73, Springer, Berlin

Bonchi, F. & Lucchese, C. (2005). Pushing tougher constraints in frequent pattern mining. *Proceedings of the Ninth Pacific-Asia Conference on Knowledge Discovery and Data Mining (PAKDD '05)*,pp. 114-124, ISBN 3-540-26076-5, Hanoi, Vietnam, May 2005, Springer

Bonchi, F. & Lucchese C. (2007). Extending the state-of-the-art of constraint-based pattern discovery, In: *Data and Knowledge Engineering*, Vol. 60, No. 2, (February 2007) pp. 377–399, ISSN 0169-023X

Bucila, C.; Gehrke, J.; Kifer, D. & White, W. (2002). Dualminer: a dual-pruning algorithm for itemsets with constraints. *Proceedings of the eighth ACM SIGKDD international conference on Knowledge Discovery and Data mining (KDD 2002)*, pp. 42–51, Edmonton, Alberta, Canada , July 2002

Dudek, D. (2007) Using Data Mining Algorithms for Statistical Learning of a Software Agent, In: Agent and Multi-Agent Systems: Technologies and Applications, LNCS, Vol. 4496, 111-120, Springer, Berlin

Esposito, R.; Meo, R. & Botta, M. (2006). Answering constraint-based mining queries on itemsets using previous materialized results. In: *Journal of Intelligent Information Systems*, Vol 26, No. 1, (January 2006) pp. 95-111, ISSN 0925-9902

Fu, Y. and Han, J. (1995). Discovery of multiple-level association rules from large databases, *Proceedings of the 21st International Conference on Very Large Data Bases (VLDB '95)*, pp. 420-431, ISBN 1-55860-379-4, Zurich, Switzerland, September 1995, Morgan Kaufmann, San Francisco, CA

Gallo A., Esposito R., Meo, R. & Botta, M. (2005). Optimization of Association Rules Extraction Through Exploitation of Context Dependent Constraints. *Proceedings of the 9th Congress of the Italian Association for Artificial Intelligence* (AI*IA 2005), pp. 258-269, Milan, Italy, September 2005

Göckelin, R. (1980) Lexicon philosophicum. Reprinted by Georg Olms 2 edition 1980.

Goethals, B. & Van den Bussche, J. (2000). Interactive Constrained Association Rule Mining. *Proceedings of the Second International Conference on Knowledge Discovery and Data Mining (DaWaK 2000),* pp. 307-316, ISBN 3-540-67980-4, London, UK, September 2000, Springer-Verlag, London, UK

Han, J.; Lakshmanan, L. V. S.; Ng, R. T. & Pang, A. (1999). Optimization of constrained frequent set queries with 2-variable constraints. *Proceedings of the 1999 ACM SIGMOD international conference on Management of data (SIGMOD '99),* pp. 157–168, ISBN 1-58113-084-8, Philadelphia, PA, ACM, New York, NY

Han, J.; Lakshmanan, L. V. S. & Ng, R. T. (1999). Constraint-Based Multidimensional Data Mining. In: *IEEE Computer,* Vol. 32, No. 8, (August 1999) pp. 46-50, ISSN 0018-9162

Hellström, T. (2003). Learning robotic behaviors with association rules. *WSEAS Transactions on Circuits and Systems,* Vol. 2, No. 3, July 2003, 534-546, ISSN 1109-2734

Horrocks, I. (2004). SWRL: A Semantic Web Rule language Combining OWL and RuleML. http://www.w3.org/Submission/2004/SUBM-SWRL-20040521/

Kaya, M. & Alhajj, R. (2005) "Fuzzy OLAP association rules mining-based modular reinforcement learning approach for multiagent systems," *IEEE Transactions on Systems, Man, and Cybernetics, Part B,* vol. 35, no. 2, April 2005, 326-338

Lorhard J. (1613) Theatrum philosophicum, Basilia

Makio, K.; Tanaka, Y. & Uehara, K. (2007) Discovery of Skills from Motion Data, In: *New Frontiers in Artificial Intelligence,* LNCS, Vol. 3609, 266-282, Springer, Berlin

Mealy G. H. (1967) Another look at data, In *Proceedings of the Fall Joint Computer Conference,* Vol 31, 525-534. Thompson Books, London: Accademic Press

Mitkas, P.A.; Kehagias, D.; Symeonidis, A.L. & Athanasiadis I.N. (2004) A Framework for Constructing Multi-agent Applications and Training Intelligent Agents, In: *Agent-Oriented Software Engineering IV,* LNCS, Vol. 2935, 255-290, Springer, Berlin

Mori, T.; Takada, A.; Noguchi, H.; Harada, T. & Sato, T. (2005) Behavior prediction based on daily-life record database in distributed sensing space, *Proceedings of the 2005 IEEE/RSJ International Conference on Intelligent Robots and Systems (IROS 2005),* 1703-1709, 2-6 Aug. 2005, IEEE

Pei, J. & Han, J. (2000). Can we push more constraints into frequent pattern mining? *Proceedings of the sixth ACM SIGKDD international conference on Knowledge Discovery and Data mining,* pp. 350–354, ISBN 1-58113-233-6, Boston, MA, August 2000, ACM, New York, NY

Srikant R.; Vu, Q. & Agrawal, R. (1997). Mining Association Rules with Item Constraints. *Proceedings of the 3rd International Conference of Knowledge Discovery and Data Mining (KDD '97),* pp. 67-73, ISBN 1-58113-233-6, Newport Beach, CA, August 1997, ACM, New York, NY

Tan, P. N.; Steinbach M. & Kumar V. (2005). *Introduction to Data Mining (First Edition),* Addison-Wesley Longman Publishing Co., ISBN 0321321367, Inc. Boston, MA, USA

Viamonte, M.J.; Ramos, C.; Rodrigues, F. & Cardoso, J.C. (2006) ISEM: a multiagent Simulator for testing agent market strategies, *IEEE Transactions on Systems, Man, and Cybernetics, Part C,* vol. 36, no. 1, Jan. 2006, 107-113

Wang, K.; He, Y. & Han, J. (2003). Pushing Support Constraints Into Association Rules Mining. In: *IEEE Transactions on Knowledge and Data Engineering,* Vol. 15, No. 3, (May-June 2003) pp. 642-658, ISSN: 1041-4347

Wojciechowski, M., Zakrzewicz, M., Dataset Filtering Techniques in Constraint-Based Frequent Pattern Mining, In: *Lecture Notes in Computer Science,* Vol. 2447, pp. 77-91, ISBN 3-540-44148-4, Springer-Verlag, London, UK

Xu W.L., Kuhnert L., Foster K., Bronlund J., Potgieter J. & Diegel O. (2007) Object-oriented knowledge representation and discovery of human chewing behaviours, *Engineering Applications of Artificial Intelligence,* Vol. 20, No. 7, October 2007, 1000-1012

Zou, X.; Lu, J.; Luo, L.; Luo, X. & Zhou, Y. (2006) Extracting Behavior Knowledge and Modeling Based on Virtual Agricultural Mobile Robot, In: *Advances in Artificial Reality and Tele-Existence,* LNCS, Vol. 4282, 29-37, Springer, Berlin

11

A New Algorithm for Initialization and Training of Beta Multi-Library Wavelets Neural Network

Wajdi Bellil, Mohamed Othmani, Chokri Ben Amar and Mohamed Adel Alimi
University of Sfax (National school of engineers of Sfax)
Tunisia

1. Introduction

The resolutions of neurons networks training problems by gradient are characterized by their noticed inability to escape of local optima [Mich93], [Fabr94] and in a least measure by their slowness [Wess92], [Zhan92]. The evolutionist algorithms bring in some domains a big number of solutions: practice of networks to variable architecture [With90], automatic generation of Booleans neurons networks for the resolution of a class of optimization problems [Grua93]. However the effort of research was especially carried on the generation and the discreet network training.

In this chapter, we propose a new algorithm of wavelets networks training, based on gradient that requires:

- A set of training examples: the wavelets networks are parametrables functions, used to achieve statistical models from examples (in the case of classification) or of measures (in the case of modeling); their parameters are calculated from these examples or couples {input, output}.
- The definition of a cost function that measures the gap between the input of the wavelets network and the desired output (in the case of classification) or the measured values (in case of modeling) present on the set of training.
- A minimization algorithm of the cost function.
- An algorithm of selection of basic function to initialize the network parameters.

We try then to show the importance of initialization of the network parameters. Since the output is non linear in relation to these parameters, the cost function can present local minima, and the training algorithms don't give any guarantee to find the global minimum.

We note that if we have a good initialization, the local minimum problem can be avoided, it is sufficient to select the best regressions (the best based on the training data) from a finished set of regressors. If the number of regressors is insufficient, not only some local minima appear, but also, the global minimum of the cost function doesn't necessarily correspond to the values of the searched parameters, it is useless then in this case to put an expensive algorithm to look for the global minimum.

With a good initialization of the network parameters the efficiency of training increases. A very important factor that it is necessary to underline is: whatever the chosen algorithm, the quality of training wavelets networks is as much better than we have an optimal initialization.

2. New wavelets networks architecture

2.1 Presentation

From a given wavelets network architecture, it is possible to generate a family of parametrables functions by the values of the network coefficients (weight, translations, dilations).

The objective of the wavelets networks training phase is to find, among all these functions, the one that approaches the most possible regression (Beta function for example). This one is unknown (otherwise it would not be necessary to use an approximation by wavelets networks); we only know the observed values (values of the regression to which are added noise) for several values valued by the input (points of the training set).

We consider wavelets networks as follows [Belli07]:

$$\hat{y} = \sum_{i=1}^{N_1} \omega_i^1 \psi_i^1(x) + \sum_{i=1}^{N_2} \omega_i^2 \psi_i^2(x) + ... + \sum_{i=1}^{N_M} \omega_i^M \psi_i^M(x) + \sum_{k=0}^{N_i} a_k \; x_k \tag{1}$$

$$= \sum_{j=1}^{M} \sum_{i=1}^{N_{Mw}} \omega_i^j \psi_i^j(x) + \sum_{k=0}^{N_i} a_k \; x_k \tag{2}$$

$$= \sum_{l(i,j)=1}^{N_{Mw}} \omega_l \psi_l(x) + \sum_{k=0}^{N_i} a_k \; x_k \tag{3}$$

$$with \quad N_{M_W} = \sum_{l=1}^{M} N_l \,, \quad i = [1,....,N], j = [1,....,M] \,, \; x_0 = 1 \tag{4}$$

Where $\hat{y}$ is the network output and x = {x_1,x_2, ..., x_{Ni} } the input vector; it is often useful to consider, in addition to the wavelets decomposition, that the output can have a linear component in relation to the variables: the coefficients ak (k = 0, 1, ... , N_i).

N_l is the number of selected wavelets for the mother wavelet family Ψ_l .

The index l depends on the wavelet family and the choice of the mother wavelet.

The network can be considered as constituted of three layers:

- A first layer with N_i input.
- A hidden layer constituted by N_{Mw} wavelets of M mothers wavelets each to a wavelet family of size N_l.
- A linear output neuron receiving the pondered wavelets outputs and the linear part.

This network is illustrated by the figure 1.

2.2 Description of the procedure of library construction

The first stage of the training procedure consists in the construction of the Beta library. We intend to construct a several mother wavelets families library for the network construction. Every wavelet has different dilations following different inputs.

This choice presents the advantage to enrich the library, and to get a better performance for a given wavelets number. The inconvenience introduces by this choice concerns the size of the library. A wavelet library having several wavelets families is more voluminous than the one that possesses the same wavelet mother. It implies a more elevated calculation cost during the stage of selection.

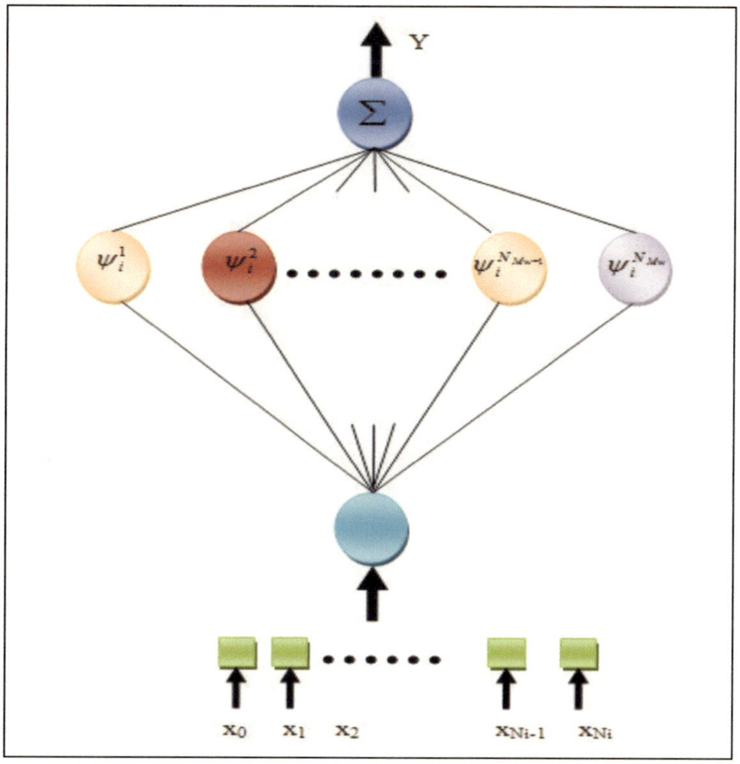

Fig. 1. Graphic representation of the new wavelets network architecture

Nevertheless, using classic algorithms optimization, the selection of wavelets is often shorter than the training of the dilations and translations; the supplementary cost introduced by different dilations can be therefore acceptable.

We have a sequence of training formed of N examples distributed in the interval [a, b]. Let Ψ^j a mother wavelet family, x the variable, ti the translation parameter and di the dilation parameter. The wavelet Ψ_i^j of the family Ψ^j having for parameters ti and di is defined as:

$$\Psi_i^j(x) = \Psi^j\left(d_i\left(x - t_i\right)\right) \tag{5}$$

The wavelet library W, generated from the mother wavelet family, is defined as:

$$W = \left\{ \Psi^j\left(d_i\left(x-t_i\right)\right), d_i \in R_+^*, t_i \in R, j = \left[1....,M\right] \right\} \tag{6}$$

$$= \left\{ W^1, ..., W^N \right\} \tag{7}$$

$$= \left\{ \psi^1\left(d_i\left(x-t_i\right)\right), ...; \psi^M\left(d_i\left(x-t_i\right)\right) \right\} \tag{8}$$

3. New algorithm: MLWNN (Multi-Library Wavelet Neural Network)

In this paragraph, we propose a new selection wavelets algorithm [Bellil07-a], based on the architecture given on the figure 1that permits to make:

- The initialization of weights, translations and dilations of wavelets networks,
- The optimization of the network parameters,
- The construction of the optimal library,
- The construction of the wavelets networks based on discreet transform.

The new architecture of wavelets networks founded on several mother wavelets families having been defined. Consequently, we can ask the question of construction of a model, constituted of wavelets network for a given process.
The parameters to determine for the construction of the network are:

- The values to give to the different parameters of the network: structural parameters of wavelets, and direct terms.
- The necessary number of wavelets to reach a wanted performance.

The essential difficulty resides in the determination of the parameters of the network. Because the parameters take discreet values we can make profit to conceive methods of wavelets selection in a set (library) of discreet wavelets. The conceived performance depends on the initial choice of the wavelets library, as well as a discriminating selection in this library.

3.1 Principle of the algorithm

The idea is to initialize the network parameters (translations, dilations and weights) with values near to the optimal values. Such a task can be achieved by the algorithm "Orthogonal Forward Regression (OFR)" based on the algorithm of orthogonalization of Gram - Schmidt [Chen89], [Ouss98], [Chen06], [Ho01].
Contrary to the OFR algorithm in which the best regressors are first selected [Lin03], [Rao04], [Xiao04], [Angr01], then adjusted to the network, the algorithm presented here integrates in every stage the selection and the adjustment. Before every orthogonalization with a selected regressor, we apply a summary optimization of the parameters of this one in order to bring it closer to the signal.
Once optimized, this new regressor replaces the old in the library and the orthogonalization will be done using the new regressor.
We describe this principle below in detail.

3.2 Description of the algorithm

The proposed algorithm depends on three stages:

3.2.1 Initialization

Let's note by Y the input signal; we have a library that contains N_{Mw} wavelets. To every wavelet Ψ_i^j we associate a vector whose components are the values of this wavelet according to the examples of the training sequence. We constitute a matrix thus constituted Vw of the blocks of the vectors representing the wavelets of every mother wavelet where the expression is:

$$V_w = \left| \begin{matrix} V_1^1(x_1) & \dots & V_N^1(x_1) & \dots & V_1^J(x_1) & \dots & V_N^J(x_1) & \dots & V_1^M(x_1) & \dots & V_N^M(x_1) \\ V_1^1(x_2) & \cdots & V_N^1(x_2) & \dots & V_1^J(x_2) & \cdots & V_N^J(x_2) & \cdots & V_1^M(x_2) & \cdots & V_N^M(x_2) \\ \vdots & \dots & \vdots & \dots & \vdots & \dots & \vdots & \dots & \vdots & \dots & \vdots \\ \vdots & \dots & \vdots & \dots & \vdots & \dots & \vdots & \dots & \vdots & \dots & \vdots \\ \vdots & \dots & \vdots & \dots & \vdots & \dots & \vdots & \dots & \vdots & \dots & \vdots \\ V_1^1(x_N) & \dots & V_N^1(x_{N_i}) & \cdots & V_1^J(x_N) & \dots & V_N^J(x_{N_i}) & \cdots & V_1^M(x_{N_i}) & \cdots & V_N^M(x_{N_i}) \end{matrix} \right| \quad (9)$$

$$V_w(t,d) = \{V_i^j\} i = [1..N],\ j = [1..M] \quad (10)$$

We note by:

- g(x) the constructed network,
- Nw=1the number of wavelet,
- T={ti } i=[1..N] the translations,
- D={di } i=[1..N] the dilations.

3.2.2 Selection

The library being constructed, a selection method is applied in order to determine the most meaningful wavelet for modeling the considered signal. Generally, the wavelets in W are not all meaningful to estimate the signal. Let's suppose that we want to construct a wavelets network g(x) with m wavelets, the problem is to select m wavelets from W.

To the first iteration, the signal is $Y = Y_1$, and the regressors vectors are the $V_w(t,d)$ defined by (10). The selected regressor is the one for which the absolute value of the cosine with the signal Y1 is maximal. The most pertinent vector from the family V_1 carries the index i_{pert1} that can be written as the following manner:

$$i_{pert1}(i,j) = \arg \max_{i,j} \frac{\langle Y | V_i^j \rangle}{\|Y\| . \|V_i^j\|} \text{ with } i=[1..N], j=[1..M] \quad (11)$$

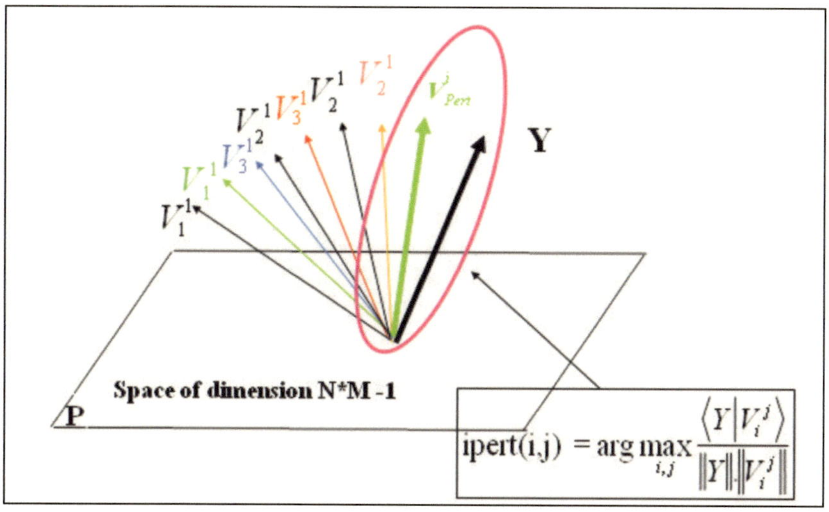

Fig. 2. Selection of the pertinent vector

Once the regressor $V_{i_{pert}}1$ is selected, it can be considered like a parametrable temporal function used for modeling Y. We calculate the weight W_i defined by:

$$w_1 = \frac{Y}{V_{i_{pert}}1} \quad (12)$$

We define the normalized mean square error of training (NMSET) as:

$$NMSET(\omega_{i\,pert}, t_{i\,pert}, d_{i\,pert}) = \frac{1}{N}\sum_{k=1}^{N}(Y(k) - \omega_{i\,pert} * V_{i_{pert}})^2 \quad (13)$$

With Y(k) is the desired output corresponding to the example k, and $\omega_{i\,pert} * V_{i_{pert}}$ is the wavelets network output corresponding to the example k.

3.2.3 Optimization of the regressor

The optimization of the regressor is made by using the gradient method. Let's note by: $e(x) = Y_d(x) - Y(x)$, with Yd : desired output and Y : network output.

$$\frac{\partial NMSET}{\partial w_{ipert}1} = \sum_{i=1}^{Nw} e(x)\Psi\left(\frac{x - t_{ipert}1}{d_{ipert}1}\right) \quad (14)$$

$$\frac{\partial NMSET}{\partial d_{iper}1} = \sum_{i=1}^{Nw} e(x) w_{ipert} 1 \frac{\partial \Psi\left(\frac{x - t_{ipert}1}{d_{ipert}1}\right)}{\partial d_{ipert}1} \quad (15)$$

$$\frac{\partial NMSET}{\partial t_{iper}1} = \sum_{i=1}^{Nw} e(x) w_{ipert} 1 \frac{\partial \Psi\left(\frac{x - t_{ipert}1}{d_{ipert}1}\right)}{\partial t_{ipert}1} \quad (16)$$

This optimization has the advantage to be fast because we only optimize here the three structural parameters of the network.

After optimization, the parameters : $t_{i_{pert}}^{opt}1, d_{i_{pert}}^{opt}1$, $\omega_{i_{pert}}^{opt}1$ of the regressor $V_{i_{pert}}1$ are adjusted, and are solutions of the optimization problem defined by:

$$V_{i_{pert}}^{opt} = V_{i_{pert}}(\omega_{i_{pert}}^{opt}, t_{i_{pert}}^{opt}, d_{i_{pert}}^{opt}) \quad (17)$$

Considering the optimal regressor, we reset the network with this regressor that is going to replace the old in the library and the orthogonalization will be done using the new regressor.

$$V_W = \begin{vmatrix} V_1^1(x_1) & \dots & V_N^1(x_1) & \dots V_{i_{pert}}^{opt}(x_1) \cdots & V_1^M(x_1) & \dots & V_N^M(x_1) \\ V_1^1(x_2) & \cdots & V_N^1(x_2) & \cdots V_{i_{pert}}^{opt}(x_2) \cdots & V_1^M(x_2) & \cdots & V_N^M(x_2) \\ \vdots & & \vdots & \vdots & \vdots & & \vdots \\ \vdots & & \vdots & \vdots & \vdots & & \vdots \\ \vdots & & \vdots & \vdots & \vdots & & \vdots \\ V_1^1(x_N) & & V_N^1(x_{N_i}) & \cdots V_{i_{pert}}^{opt}(x_{N_i}) \dots & V_1^M(x_{N_i}) & \cdots & V_N^M(x_{N_i}) \end{vmatrix} \tag{18}$$

After one iteration we will have:

$$V_{w1}(t,d) = \{V_i^j\}_{i=[1..N],\, j=[1..M]} \cup \left\{V_{i_{pert}}^{opt}\right\} \tag{19}$$

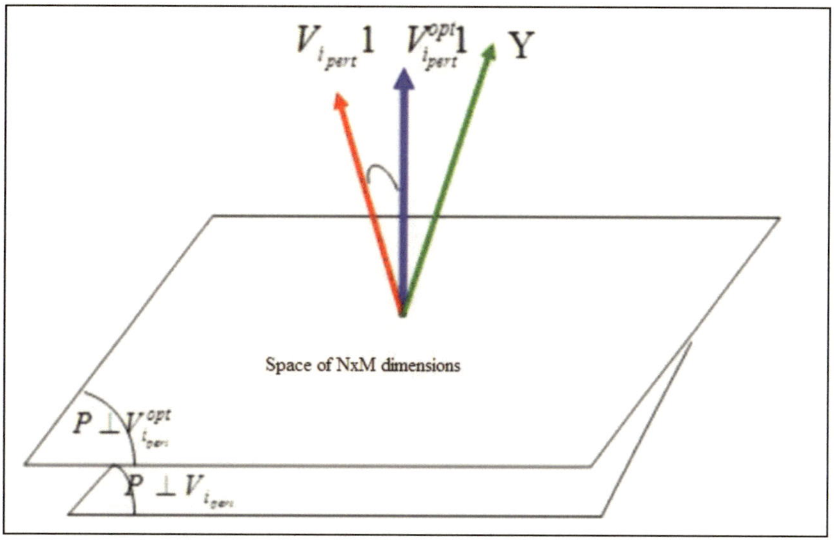

Fig. 3. Regressor optimization

3.2.4 Orthogonalization

The vectors V_i^j are always linearly independent and non orthogonal (because N >> M_W). The vectors V_i^j generate a sub-vector-space of M*N dimensions. We orthogonalize the M*N -1 remaining regressor, and the vector Y1 according to the adjusted regressor $V_{i_{pert}}^{opt}1$:

$$V_i^{j\perp} = V_i^j - \left\langle V_i^j \middle| V_{i_{pert}}^{opt}1 \right\rangle V_{i_{pert}}^{opt}1 \tag{20}$$

$$Y^{\perp}1 = Y - \left\langle Y \middle| V_{i_{pert}}^{opt}1 \right\rangle V_{i_{pert}}^{opt}1 \tag{21}$$

Therefore, we make the library updating:

$$V_W = \begin{vmatrix} V_1^{1\perp}(x_1) & \dots & V_N^{1\perp}(x_1) & \dots V_1^{J\perp}(x_1) & \dots & V_N^{J\perp}(x_1) & \dots & V_1^{M\perp}(x_1) & \dots & V_N^{M\perp}(x_1) \\ V_1^{1\perp}(x_2) & \cdots & V_N^{1\perp}(x_2) & \cdots V_1^{J\perp}(x_2) & \cdots & V_N^{J\perp}(x_2) & \cdots & V_1^{M\perp}(x_2) & \cdots & V_N^{M\perp}(x_2) \\ \vdots & & \vdots & \vdots & & \vdots & & \vdots & & \vdots \\ \vdots & & \vdots & \vdots & & \vdots & & \vdots & & \vdots \\ \vdots & & \vdots & \vdots & & \vdots & & \vdots & & \vdots \\ V_1^{1\perp}(x_N) & & V_N^{1\perp}(x_{N_i}) & \cdots V_1^{J\perp}(x_N) & & V_N^{J\perp}(x_{N_i}) & \cdots & V_1^{M\perp}(x_{N_i}) & \cdots & V_N^{M\perp}(x_{N_i}) \end{vmatrix} \tag{22}$$

We will have:

$$V_w(t,d) = \{V_i^{j\perp}\}_{(i=[1..N],\, j=[1..M]) \setminus \{ i_{pert} \}} \tag{23}$$

$Y^{\perp}1$ and $\left\{V_i^{j\perp}\right\}$ are respectively what remains from the signal and regressors in the orthogonal space to $V_{i_{pert}}^{opt}1$.

The model being at this stage, $g(X) = \omega_i 1 * V_{i_{pert}}^{opt} 1$, can be represented by the figure 4.

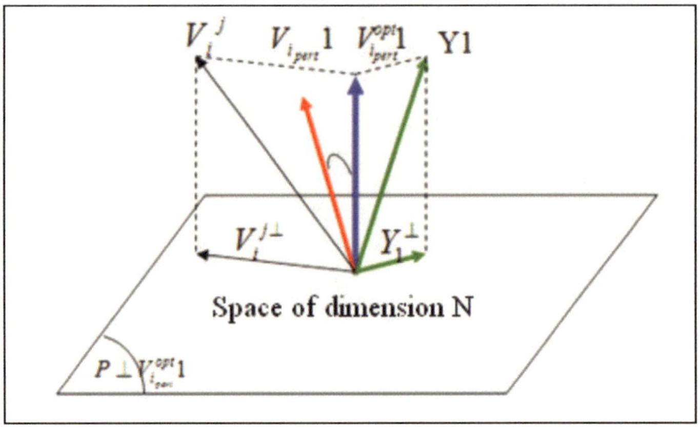

Fig. 4. Orthogonal projection on the optimal regressor

To the following iteration we increment the number of N_w=N_w+1 wavelet. We apply the same stages described above. Let's suppose achieved i-1 iterations: We did i-1 selections, optimizations, and orthogonalizations in order to get the i-1 adjusted regressors ($V_{i_{pert}}^{opt}1$,….., $V_{i_{pert}}^{opt}$ i-1) we reset i-1 parameters of the network.

The network g(x) can be written at the end of the iteration i-1 as:

$$g(x) = \sum_{i=1}^{i-1} \omega_i^{opt} * V_{i_{pert}}^{opt} \tag{24}$$

We have Nw-i+1 regressors to represent the signal Y_i in a space of N*M-i+1 dimensions orthogonal to $(V_{i_{pert}}^{opt}1, \dots, V_{i_{pert}}^{opt} i-1)$

We apply the same principle of selection as previously. The index i_{perti} of the selected regressor can be written as:

$$i_{pert}i(i,j)=\arg\max_{i,j} \frac{\langle Y | V_i^j \rangle}{\|Y\|.\|V_i^j\|} \tag{25}$$

with (i=[1..N], j=[1..M]) \ $\{i_{pert}1, \ldots, i_{pert}i-1\}$

Since the regressor is orthogonal to ($V_{i_{pert}}^{opt}1, \ldots, V_{i_{pert}}^{opt}i-1$), we make the updating of the library, then we optimize the regressor and finally an orthogonalization.

Finally, after N iteration, we construct a wavelets network of N wavelets in the hidden layer that approximates the signal Y.

As a consequence, the parameters of the network are:

$$T^{opt} = \left\{t_{i_{pert}}^{opt}\right\}_{i_{pert}=[1..Nopt]} \tag{26}$$

$$d^{opt} = \left\{d_{i_{pert}}^{opt}\right\}_{i_{pert}=[1..Nopt]} \tag{27}$$

$$\omega^{opt} = \left\{\omega_{i_{pert}}^{opt}\right\}_{i_{pert}=[1..Nopt]} \tag{28}$$

The obtained model g(x) can be written under the shape:

$$g(x) = \sum_{i=1}^{N_{opt}} \omega_i^{opt} * V_i^{opt} \tag{29}$$

4. Interpolation of 1D data

4.1 Mathematical formulation

The mathematical formulation of the interpolation of 1D data can be presented in the following way:

Let the set of the points $E = \{(x_k, y_k) / k = 0,1,....,k\}$. We want to recover N samples of f(x) as $f(x) = y_k \; for \; k = 0,1,...,k$.

The set E represents the constraints of the problem. With this formulation the function f(x) pass inevitably by the set points of E. In practice, the constraints can contain noise. In this case, the signal that we want to rebuild doesn't necessarily pass by the points of the set E, the interpolation becomes then a problem of approximation:

Once we know the function f(x) on the set of the domain x $\in$[0,…,N], the problem is to recover f(x) for x> N. This formulation will be called extrapolation of the signal f(x).

Interpolation is a problem of signal reconstruction from samples is a badly posed problem by the fact that infinity of solutions passing by a set of points (Figure 5). For this reason supplementary constraints that we will see in the presentation of the different methods of interpolation, must be taken in consideration to get a unique solution.

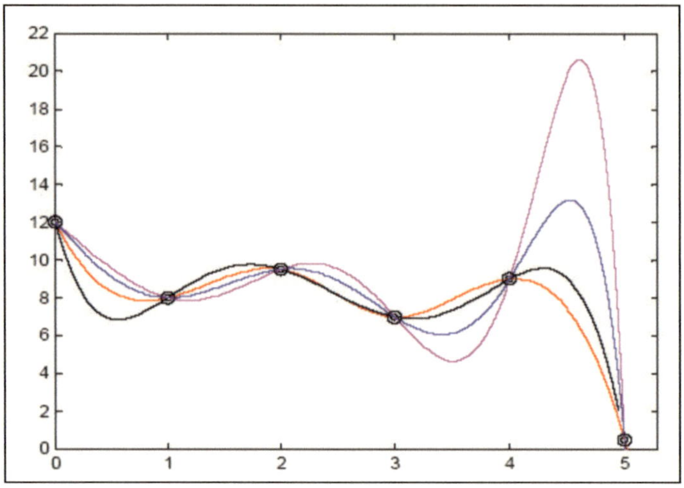

Fig. 5. Infinity of curves passing by a set of points

4.2 Interpolation of 1D data using wavelets networks

4.2.1 Regularization

The variationnel method is often used for the regularization [Buli80], [Boor63], [Terz86]. The principle of this method is to minimize the function ξ(f) defined as the following manner:

$$\xi(f) = \alpha S(f) + \gamma C(f) \text{ with } \alpha = 1 - \gamma \tag{30}$$

The function ξ(f) is the sum of a stabilizing function S(f) and of a cost function C(f). The parameter $\gamma \in [0, 1]$ is a constant of adjustment between these two functions. When γ goes toward zero, the problem of interpolation turns into a problem of approximation. The stabilizing function S(f) fixes the constraint of curve smoothing and it is defined as the following way:

$$S(f) = \int_D \left(\frac{\partial^2 f}{\partial x^2} \right)^2 dx \tag{31}$$

Where, D represents the domain of interest. The cost function C(f) characterizes the anomalies between the rebuilt curve and the initial constraints, this function can be written by:

$$C(f) = \sum_{x_k \in E_p} [\, f(x_k) - y_k \,]^2 \tag{32}$$

Where Ep = {(xk,yk) | k = 1,2,…,K} represents the set of the known points or the signal constraints.

4.2.2 Discretization

With the regularization, it is difficult to get an analytic solution. The discretization is useful and several methods can be used. Grimson [Grim83] uses the finished differences to approximate the differential operators; while Terzopoulos [Terz86] uses finished elements to get and solve an equation system (Grimson and Terzopouloses used the quoted methods while treating the case 2D). In the approach that follows, the function f is written like a linear combination of basic functions:

$$\hat{f}(x) = \sum_{i=0}^{N-1} W_i \, \Psi_i \, (x) \tag{33}$$

Where N is the domain dimension and W_i the coefficients. The basic functions $\Psi_i(x)$ are localized to $x = i\Delta$, $\Psi_i(x) = \Psi(x - i\Delta)$

While substituting (33) in (32) and (31), the function $\xi(f)$ can be rewritten as the following way:

$$\xi = \alpha \sum_{i=0}^{N-1}\sum_{j=0}^{N-1} W_i \, t_{ij} \, W_j \, + \gamma \sum_{k=1}^{K}\left[\sum_{i=0}^{N-1} W_i \, \Psi_i(x_k) - y_k\right]^2 \tag{34}$$

Where tij is a function of basic functions:

$$t_{ij} = \int_D \left(\frac{\partial^2 \Psi_i}{\partial x^2}\right)\left(\frac{\partial^2 \Psi_j}{\partial x^2}\right) dx \tag{35}$$

Several wavelets functions can be used like activation function. The Figure 6 gives the curve of a new wavelets based on Beta function given by the following definition:

4.2.3 Definition

The 1D Beta function as presented in [Alim03] and in [Aoui02] is a parametrable function defined by $\beta(x) = \beta_{x_0, x_1, p, q}(x)$ with x_0, x_1, p and q as real parameters verifying: $x_0 < x_1$, and:

$$x_c = \frac{p x_1 + q x_0}{p + q} \tag{36}$$

Only the case for p>0 and q>0 will be considered. In this case the Beta function is defined as:

$$\beta(x) = \begin{cases} \left(\dfrac{x - x_0}{x_c - x_0}\right)^p \left(\dfrac{x_1 - x}{x_1 - x_c}\right)^q & if \;\; x \in [x_0, x_1] \\ 0 & else \end{cases} \tag{37}$$

4.2.4 Axiom [Ben Amar 2006]

$\forall$ n $\in$ IN, $(p,q) \in \Re_+^2$, p = q, and n < p; the nth derivatives of 1D Beta function are wavelets [Amar06]. Let's note by $Beta^n$ the nth derivative of Beta function.

$$\Psi_n(x) = Beta^n = \frac{d^n \beta(x)}{dx^n} \tag{38}$$

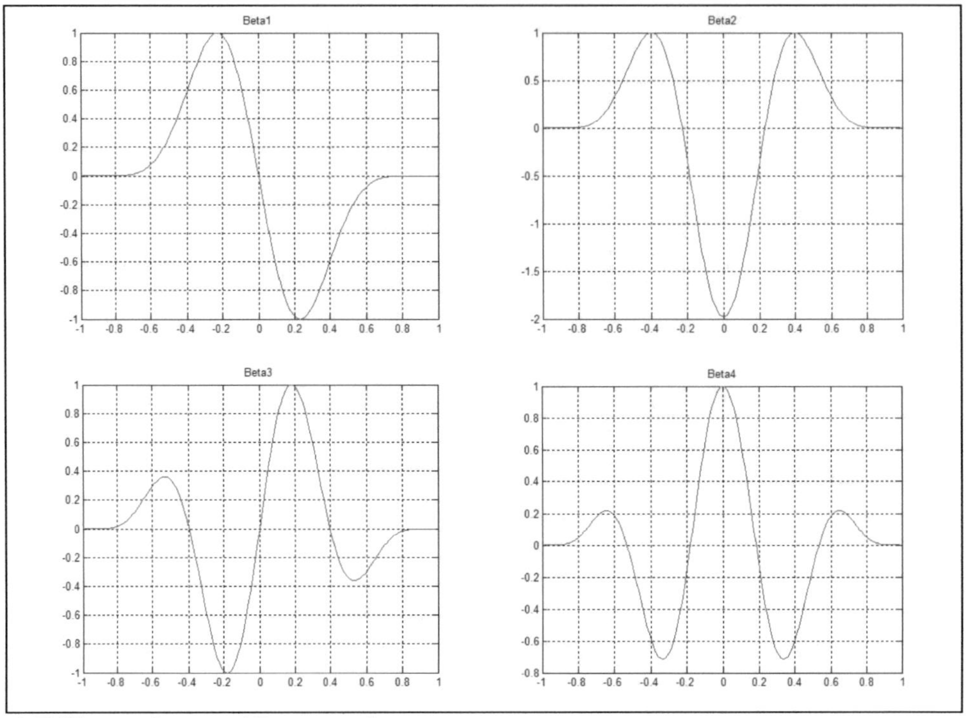

Fig. 6. Different shapes of Beta wavelets

4.2.5 Example: Interpolation of 1D data using classical wavelets network CWNN

In this example, we want to rebuild three signals $F_1(x)$, $F_2(x)$ and $F_3(x)$ defined by equations (39), (40) and (41). We have a uniform distribution with a step of 0.1 known samples. For the reconstruction, we used a CWNN composed of 12 wavelets in hidden layer and 300 trainings iterations. We note that for Beta wavelets we fix the parameter p=q=30.

$$F_1(x) = \begin{cases} -2.186x - 12.864 \ \ for\ x \in [-10,-2[\\ 4.246x \ \ for\ x \in [-2,-0[\\ 10\exp(-0.05x-0.5)\sin(x(0.03x+0.7)) \ \ for\ x \in [0,10[\end{cases} \quad (39)$$

$$F_2(x) = \ 0.5x\sin(x) + \cos^2(x)\, for\ x \in \left[-2.5, 2.5\right] \quad (40)$$

$$F_3(x) = \sin c(1.5x)\, for\ x \in \left[-2.5, 2.5\right] \quad (41)$$

Table 1. gives the final normalized root mean square error (NRMSE) of test given by equation (42) after 300 trainings iterations for the F_1, F_2 and F_3 signals.

We define the NRMSE as:

$$NRMSE = \frac{1}{N}\sqrt{\frac{\sum_{i=1}^{N}\left(\overline{y_i} - y_i\right)^2}{\sum_{i=1}^{N}\left(y_i\right)^2}} \quad (42)$$

Where N is the number of sample and $\overline{y}_i$ the real output.

	M-hat	Polywog¹	Slog¹	Beta¹	Beta²	Beta³
F_1	7.2928e-2	4.118e-2	3.1613e-2	1.7441e-2	1.2119e-2	1.4645e-1
F_2	7.6748e-8	1.0343e-6	2.3954e-6	3.7043e-8	5.3452e-9	1.2576e-9
F_3	3.1165e-7	2.3252e-6	1.2524e-6	5.5152e-7	1.9468e-7	1.9674e-6

Table 1. Normalized root mean square error of test for different activation wavelets function

4.2.5 Example: interpolation of 1D data using MLWNN

We intend to approximate the F_1, F_2, F_3 using MLWNN, composed of a library of 6 mother wavelets (from $Beta^1$ to $Beta^3$, Mexican hat, $polywog^1$ and $Slog^1$), in the same condition as the example of approximation using CWNN.

	M-hat	Polywog¹	Slog¹	Beta¹	Beta²	Beta³	NRMSE
F_1	4	0	3	3	0	2	3.79841e-3
F_2	1	3	3	1	1	3	4.66143e-11
F_3	1	3	1	1	1	5	3.84606e-8

Table 2. Normalized root mean square error of test and selected mother wavelets

To reconstruct the F_1 signal with a NRMSE of 3.79841e-3 using 12 wavelets in hidden layer the best regressors for MLWNN are: 4 wavelets from the Mexican hat mother wavelet, 0 wavelet from the $polywog^1$, 3 wavelets from the $Slog^1$, 3 wavelets from $Beta^1$, 0 wavelet from $Beta^2$ and 2 wavelets from the $Beta^3$ mother wavelets. When using a CWNN the best NRMSE of reconstruction is obtained with $Beta^2$ mother wavelet and it is equal to 1.2119e-2. For F_2 signal the NRMSE is equal to 4.66143e-11 using MLWNN whereas it is of 1.2576e-9 using CWNN with $Beta^3$ mother wavelet. Finally for F_3 signal we have a NRMSE of 3.84606e-8 for a MLWNN over 1.9468e-7 as the best value for a CWNN.

5. 2 Dimensional data interpolation

Previously, for every described method in 1D, the case in two dimensions is analogous, while adding one variable in the equations.

5.1 Mathematical formulation

The mathematical formulation of 2D data interpolation can be presented in an analogous way to the one described in the 1D case [Yaou94] (we will suppose that we want to rebuild an equally-sided surface):

Let E the set of the points $E=\{(x_k, y_k, z_k)/k=0,1,....,k\}$ we wants to recover N×N samples of f(x, y) as $f(x_k, y_k) = z_k$ for $k = 1,..., K$.

The set E represents the constraints of the problem. With this formulation the function f(x, y) passes inevitably by the points of the set E. In practice, the constraints can be noisy. In this case, the signal that we want to rebuild doesn't necessarily pass by the points of the set E. The interpolation becomes then a problem of approximation.

5.2 Method using wavelets networks

The formulations for the 2D case are given by:

$$S(f) = \iint_{\Omega} \left[\left(\frac{\partial^2 f}{\partial x^2} \right)^2 + 2 \left(\frac{\partial^2 f}{\partial x \partial y} \right)^2 + \left(\frac{\partial^2 f}{\partial y^2} \right)^2 \right] d\Omega \tag{43}$$

$$C(f) = \sum_{x_k \in E} \sum_{y_k \in E} \left[f(x_k, y_k) - Z_k \right]^2 \tag{44}$$

$$f(x, y) = \sum_{i=0}^{N-1} \sum_{j=0}^{N-1} w_{ij} \Psi_{ij}(x, y) \tag{45}$$

Since the interpolated basic functions (wavelets) are separable, this will always be the case in this survey:

$$\Psi_{ij}(x, y) = \Psi_i(x) \Psi_j(y) \tag{46}$$

$$t_{ijrs} = \iint_{\Omega} \left[\left(\frac{\partial^2 \Psi_{ij}}{\partial x^2} \right) \left(\frac{\partial^2 \Psi_{rs}}{\partial x^2} \right) + 2 \left(\frac{\partial^2 \Psi_{ij}}{\partial x \partial y} \right) \left(\frac{\partial^2 \Psi_{rs}}{\partial x \partial y} \right) + \left(\frac{\partial^2 \Psi_{ij}}{\partial y^2} \right) \left(\frac{\partial^2 \Psi_{rs}}{\partial y^2} \right) \right] d\Omega \tag{47}$$

5.2.1 Example: approximation of 2D data using CWNN

To compare the performances of the method using wavelets networks and classic wavelets networks, four surfaces have been chosen. These surfaces are used by Franke [Fran79] and are also described in [Renk88] for x and y$\in$ [0, 1]:

$$S_1(x, y) = \frac{e^{-\frac{81}{16}\left[(x-0.5)^2 + (y-0.5)^2\right]}}{3} \tag{48}$$

$$S_2 = \left(\frac{x+2}{2} \right)^5 \left(\frac{2-x}{2} \right)^5 \left(\frac{y+2}{2} \right)^5 \left(\frac{2-y}{2} \right)^5 \tag{49}$$

$$S_3(x, y) = \frac{3.2(1.25 + \cos(5.4y))}{\left(6 + 6(3x-1)^2\right)} \tag{50}$$

$$S_4(x, y) = \left(x^2 - y^2\right) \sin(5x) \tag{51}$$

The Figure 7 represents these four surfaces of 21X21 samples.

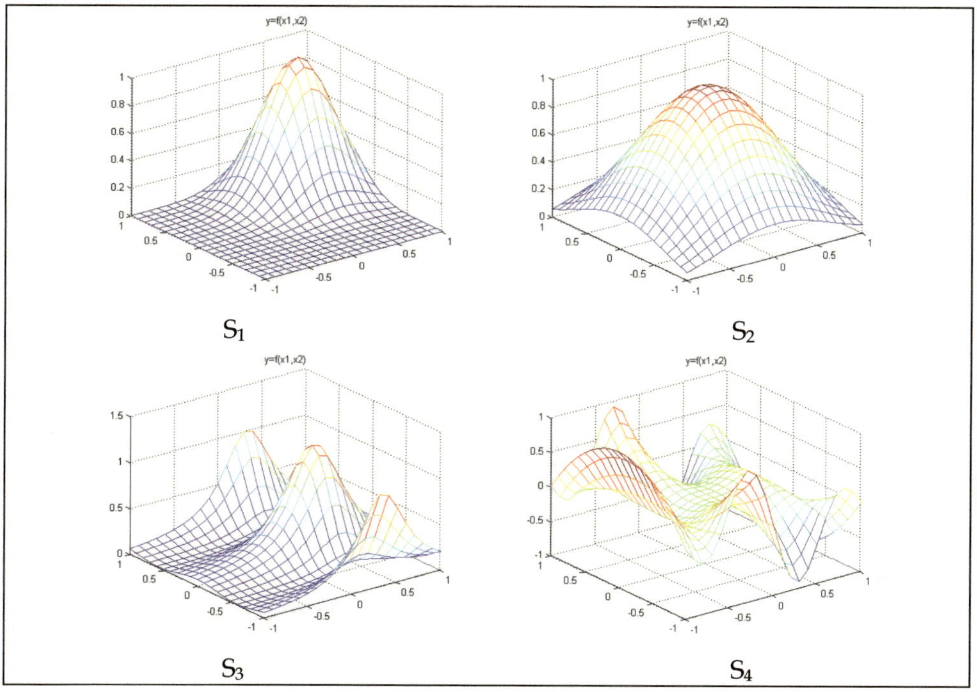

Fig. 7. Surfaces $S_1(x, y)$, $S_2(x, y)$, $S_3(x, y)$ and $S_4(x, y)$ represented by their 21X21 samples

The four following samplings are considered in order to rebuild these surfaces:

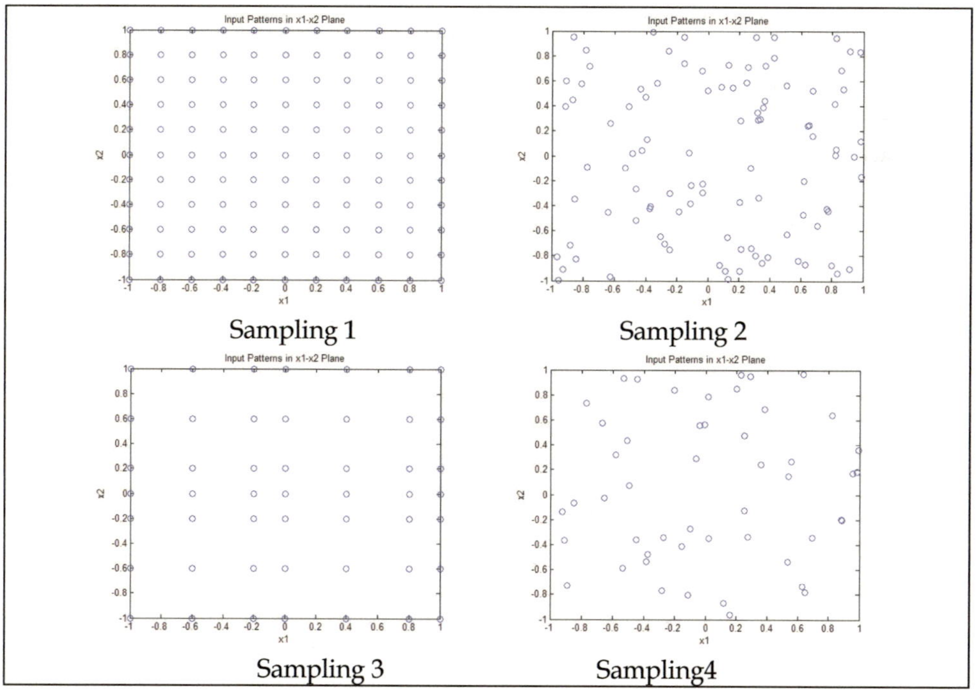

Fig. 8. Samplings considered for reconstruction of the surfaces $S_1(x, y)$, $S_2(x, y)$, $S_3(x, y)$ and $S_4(x, y)$

Surfaces	Sampling	Beta[1]	Beta[2]	Beta[3]	M-hat	Slog[1]	Polywog[1]
S_1	1	4.5472e-6	1.5881e-5	1.2783e-5	3.6561e-5	6.2610e-6	8.0144e-7
S_1	2	3.6612e-6	1.9917e-6	2.6849e-6	4.2004e-5	1.4369e-5	7.6421e-6
S_1	3	1.7730e-6	2.3737e-6	1.5610e-6	3.5810e-6	9.3080e-6	5.2441e-7
S_1	4	2.4055e-7	4.1873e-6	8.6403e-7	7.6317e-6	5.2528e-6	6.7456e-6
Surfaces	Sampling	Beta[1]	Beta[2]	Beta[3]	M-hat	Slog[1]	Polywog[1]
S_2	1	5.2167e-6	7.9647e-6	3.0107e-6	2.2445e-5	8.4106e-5	5.2539e-6
S_2	2	6.2544e-7	1.2262e-6	7.9519e-7	4.3444e-5	4.5899e-5	9.1619e-7
S_2	3	7.2351e-7	2.8619e-6	1.9735e-6	8.7916e-5	3.0003e-5	4.7101e-7
S_2	4	2.1238e-7	8.2108e-7	5.5915e-7	1.7681e-5	1.0731e-5	1.0287e-6
Surfaces	Sampling	Beta[1]	Beta[2]	Beta[3]	M-hat	Slog[1]	Polywog[1]
S_3	1	9.3210e-4	7.4748e-4	6.0998e-4	4.4958e-3	2.7186e-3	9.4392e-4
S_3	2	4.0176e-4	2.6073e-4	1.5930e-4	1.2234e-3	1.2249e-3	7.5690e-4
S_3	3	2.3361e-4	2.7506e-4	2.7690e-4	1.1170e-3	8.5758e-4	1.3850e-4
S_3	4	1.1620e-4	2.6754e-4	1.9940e-4	2.1405e-4	1.1916e-4	2.2120e-4
Surfaces	Sampling	Beta[1]	Beta[2]	Beta[3]	M-hat	Slog[1]	Polywog[1]
S_4	1	3.8187e-3	4.6391e-3	2.2186e-3	3.3708e-2	3.0934e-3	4.8993e-3
S_4	2	1.5365e-3	2.4675e-3	1.3099e-3	9.9976e-3	4.8904e-3	2.5167e-3
S_4	3	3.7501e-3	1.7048e-3	5.4345e-4	1.2712e-2	3.0359e-3	2.2196e-3
S_4	4	3.2641e-4	1.4836e-3	6.2990e-4	1.5668e-3	2.0464e-4	2.2768e-4

Table 3. Normalized Root Mean square error of test for the surfaces $S_1(x, y)$, $S_2(x, y)$, $S_3(x, y)$ and $S_4(x, y)$ using CWNN

Table 3 represents the NRMSE of reconstruction of the four considered surfaces, using classical wavelets network constructed with 12 wavelets in hidden layer and based on Beta, Mexican Hat, Slog[1] and Polywog[1] wavelets [Belli05]. This table informs that the number of samples to consider as well as their disposition for the reconstruction is important.
For a same number of samples, it is preferable to use a uniform sampling than a non uniform one, the more the number of samples is important and the better is the quality of reconstruction.

5.2.1 Example: approximation of 2D data using MLWNN [Bellil07-b]

The same surfaces are used in the same conditions but using MLWNN with a library composed of two mother wavelets (Beta[1] and Beta[3]). Experimental results are given in the following table.
Table 3 and table 4 inform that the number of samples to consider as well as their disposition for the reconstruction is important:
For a same number of samples, it is preferable to use a uniform sampling than a non uniform one, the more the number of samples is important and the better is the quality of reconstruction.

When comparing table 3 and table 4 we can say that the performances obtained in term of NRMSE using the MLWNN algorithm are often very better that the one obtained with the CWNN. This shows that the proposed procedure brings effectively a better capacity of approximation using the parametrable Beta wavelets [Belli07-b].

Surfaces	Sampling	Beta[1] + Beta[3]		
		NMSE	P_1	P_2
S_1	1	6.0509e-7	5	7
S_1	2	3.4558e-7	5	7
S_1	3	5.7987e-7	4	8
S_1	4	4.0519e-7	5	7
Surfaces	Sampling	Beta[1] + Beta[3]		
		NMSE	P_1	P_2
S_2	1	6.1536e-6	3	9
S_2	2	1.1164e-6	2	10
S_2	3	7.0519e-7	1	11
S_2	4	1.1629e-6	4	8
Surfaces	Sampling	Beta[1] + Beta[3]		
		MSET	P_1	P_2
S_3	1	5.1778e-4	6	6
S_3	2	8.8298e-5	7	5
S_3	3	7.8227e-5	6	6
S_3	4	1.1098e-4	6	6
Surfaces	Sampling	Beta[1] + Beta[3]		
		NMSE	P_1	P_2
S_4	1	2.1e-3	8	4
S_4	2	1.4e-3	6	6
S_4	3	1.1e-3	10	2
S_4	4	1.1e-3	6	6

Table 4. Normalized Root Mean square error of test for the surfaces $S_1(x, y)$, $S_2(x, y)$, $S_3(x, y)$ and $S_4(x, y)$ using MLWNN

6. 3 Dimensional data interpolation

The case of 3 dimensions is analogous to 1D or 2D case. The reconstruction of sampled data using wavelets networks is deduced from the 1D or 2D case.

6.1 Mathematical formulation

The mathematical formulation of 3D data interpolation can be presented in an analogous way to the one described for the 1D case (we will suppose that we want to rebuild an equally-sided volume):

Let E the set of the points $E = \{(x_k, y_k, z_k, q_k) / k = 0, 1, ..., k\}$ we wants to recover N×N×N samples of f(x, y, z) as $f(x_k, y_k, z_k) = q_k$ for k = 1,…, K.

The set E represents the constraints of the problem. With this formulation the function f(x, y, z) passes inevitably by the points of the set E. In practice, the constraints can be noisy.

In this case, the signal that we want to rebuild doesn't necessarily pass by the points of the set E. The interpolation becomes then a problem of approximation:
The problem of extrapolation is to recover the values of the function f(x, y, z) for x, y and z not belonging to the domain of interpolation.
The problem of reconstruction of a volume from samples is a badly definite problem because an infinity volume passing by a set of points exists. For this reason, some supplementary constraints must be taken into consideration to get a unique solution.

6.2 Method using wavelets networks

Formulations for the 3D case are given by:

$$S(f)=\iiint_V \left[\left(\frac{\partial^2 f}{\partial x^2}\right)^2+\left(\frac{\partial^2 f}{\partial y^2}\right)^2+\left(\frac{\partial^2 f}{\partial z^2}\right)^2+2\left(\frac{\partial f}{\partial x}\frac{\partial f}{\partial y}\right)^2 2\left(\frac{\partial f}{\partial x}\frac{\partial f}{\partial z}\right)^2 2\left(\frac{\partial f}{\partial y}\frac{\partial f}{\partial z}\right)^2\right]dV \quad (52)$$

$$C(f)=\sum_{x_k\in E}\sum_{y_k\in E}\sum_{z_k\in E}\left[f(x_k,y_k,z_k)-q_k\right]^2 \quad (53)$$

$$f(x,y,z)=\sum_{i=0}^{N-1}\sum_{j=0}^{N-1}\sum_{j=0}^{N-1} w_{ijr}\Psi_{ijr}(x,y,z) \quad (54)$$

Since the interpolated basic functions (wavelets) are separable, this will always be the case in this survey:

$$\Psi_{ijr}(x,y,z)=\Psi_i(x)\Psi_j(y)\Psi_r(z) \quad (55)$$

$$t_{ijrstu}=\iiint_V \left[\begin{array}{l}\left(\frac{\partial^2\Psi_{ijr}}{\partial x^2}\right)\left(\frac{\partial^2\Psi_{stu}}{\partial x^2}\right)+\left(\frac{\partial^2\Psi_{ijr}}{\partial y^2}\right)\left(\frac{\partial^2\Psi_{stu}}{\partial y^2}\right)+\left(\frac{\partial^2\Psi_{ijr}}{\partial z^2}\right)\left(\frac{\partial^2\Psi_{stu}}{\partial z^2}\right)+\\ 2\left(\frac{\partial\Psi_{ijr}}{\partial x}\right)\left(\frac{\partial\Psi_{ijr}}{\partial y}\right)\left(\frac{\partial\Psi_{stu}}{\partial x}\right)\left(\frac{\partial\Psi_{stu}}{\partial x}\right)+\\ 2\left(\frac{\partial\Psi_{ijr}}{\partial x}\right)\left(\frac{\partial\Psi_{ijr}}{\partial z}\right)\left(\frac{\partial\Psi_{stu}}{\partial z}\right)\left(\frac{\partial\Psi_{stu}}{\partial z}\right)+\\ 2\left(\frac{\partial\Psi_{ijr}}{\partial y}\right)\left(\frac{\partial\Psi_{ijr}}{\partial z}\right)\left(\frac{\partial\Psi_{stu}}{\partial y}\right)\left(\frac{\partial\Psi_{stu}}{\partial z}\right)\end{array}\right]dV \quad (56)$$

6.3 Example: approximation of 3D data using MLWNN

We used the GavabDB 3D face database for automatic facial recognition experiments and other possible facial applications like pose correction or register of 3D facial models. The database GavabDB contains 427 images of 3D meshes of the facial surface. These meshes correspond to 61 different individuals (45 male and 16 female), and 9 three dimensional images are provided for each person. The total of the database individuals are Caucasian and their age is between 18 and 40 years old.

Each image is a mesh of connected 3D points of the facial surface without the texture information for the points. The database provides systematic variations in the pose and the facial expressions of the individuals. In particular, there are 2 frontal views and 4 images with small rotations and without facial expressions and 3 frontal images that present different facial expressions.

The following experiment is performed on the GavabDB 3D face database and its purpose is to evaluate the MLWNN that we employ against the CWNN in term of 3D face reconstruction.

For faces reconstruction quality measurement we adopt the common use of NMSE given by:

$$NRMSE = \frac{1}{N * M} \sqrt{\frac{\sum_{i=1}^{N} \sum_{j=1}^{M} \left(\overline{Z_{i,j}} - Z_{i,j}\right)^2}{\sum_{i=1}^{N} \sum_{J=1}^{M} \left(Z_{i,j}\right)^2}} \tag{57}$$

Where $\overline{Z_{i,j}}$ is the depth value of pixels in the reconstructed face; and $Z_{i,j}$ is the depth value of pixels in the original face. To check the utility of the MLWNN, experimental studies are carried out on the GavabDB 3D face database. We used 61 frontal views with two different dimensions: 20*20 and 50*50. The results of comparison are presented in Table 5.

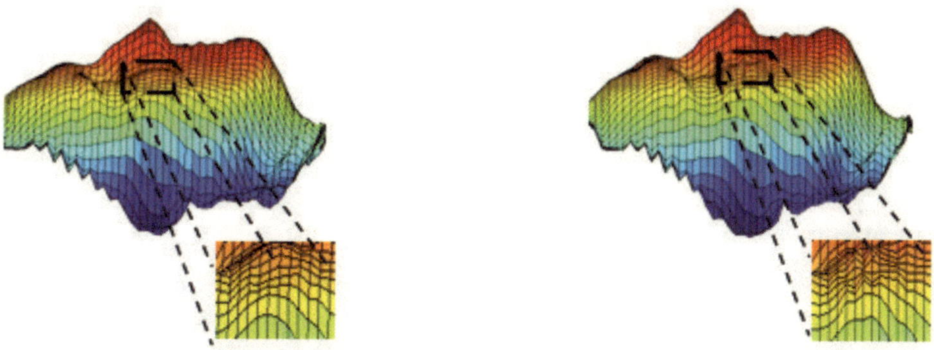

MLWNN CWNN

Fig. 9. 3D face reconstruction using MLWNN and CWNN

	Face 20*20		Face 50*50	
	CWNN	**MLWNN**	**CWNN**	**MLWNN**
5 faces	3.52000*10-5	6.3500*10-6	0.90560*10-5	5.6960*10-6
20 faces	4.55375*10-5	9.0250*10-6	1.13640*10-5	6.8320*10-6
40 faces	4.94625*10-5	9.1562*10-6	1.25040*10-5	7.4600*10-6
61 faces	5.42500*10-5	9.1762*10-6	1.31970*10-5	7.9121*10-6

Table 5. Evaluation in term of NMSE of 3D face reconstruction using MLWNN and CWNN

7. Conclusion

In this chapter, we described a new training algorithm for multi library wavelets network. We needed a selection procedure, a cost function and an algorithm of minimization for the evaluation. To succeed a good training, we showed that it was necessary to unite good ingredients. Indeed, a good algorithm of minimization finds a minimum quickly; but this one is not necessarily satisfactory.
The use of a selection algorithm is fundamental. Indeed, the good choice of regressors guarantees a more regular shape of the cost function; the global minima correspond well to the "true" values of the parameters, and avoid the local minimum multiplication. So the cost function present less local minima and the algorithms of evaluation find the global minimum more easily.
For the validation of this algorithm we have presented a comparison between the CWNN and MLWNN algorithm in the domain of 1D, 2D and 3D function approximation.
Many examples permitted to compare the capacity of approximation of MLWNN and CWNN. We deduce from these examples that:

- The choice of the reconstruction method essentially depends on the type of data that we treat,
- The quality of reconstruction depends a lot on the number of samples used and on their localizations.

Also we have define a new Beta wavelets family that some one can see that they are more superior then the classic one in term of approximation and we demonstrate in [BELLIL07] that they have the capacity of universal approximation.
As future work we propose a hybrid algorithm, based on MLWNN and genetic algorithm and the GCV (Generalised Cross validation) procedure to fix the optimum number of wavelets in hidden layer of the network, in order to model and synthesis PID controller for non linear dynamic systems.

8. References

A. Alimi, (2003). The Bêta-Based Neuro-Fuzzy System: Genesis and Main Properties, *submitted to TASK Quarterly Journal, Special Issue on "Neural Networks"* edited by W. Duch and D. Rutkowska, vol. 7, no. 1, pp. 23-41, 2003.

C. Ben Amar, W. Bellil, M.A. Alimi, (2006). Beta Function and its Derivatives: A New Wavelet Family, *Transactions on Systems, Signals & Devices Volume 1, Number 3*, pp. 275-293, 2005-2006.

C. Aouiti, A.M. Alimi and A. Maalej, (2002). Genetic Designed Bêta Basis Function Neural Networks for Multivariable Functions Approximation, *Systems Analysis, Modeling, and Simulation, Special Issue on "Advances in Control and Computer Engineering", vol. 42, no. 7*, pp. 975-1005, 2002.

C. De Boor, (1978). A practical guide to splines, *Springer-Verlag, New-York*, 1978.

C. J. Lin, C. C. Chin, C. L. Lee, (2003). A wavelet based neuro-fuzzy system and its applications, *Proceedings of the Int. Joint Conference on Neural Networks, vol. 3*, pp. 1921–1926, 2003.

D. W. C. Ho, P. A. Zhang, J. Xu, (2001). Fuzzy wavelet networks for function learning, *IEEE Trans. on Fuzzy Systems, vol. 9, issue 1*, pp. 200–211, 2001.

D. Withleyet al, (1990). Genetic Algorithms and Neural Networks: Optimizing Connections and Connectivity, *Parallel Computing.*

F. Gruau and D.Whitley, "Adding learning to the cellular developmental process: a comparative study", Evolutionary Computation, 1(3), 1993.

Fabrice Rossi and Cédric Gégout, J. (1994). Geometrical Initilization, Parametrization and Control of Multilayer Perceptrons: Application to Function Approximation, *In Proceedings of WCCI ICNN, volume I*, pp. 546-550, June 1994, Orlando (Florida).

L. Angrisani, P. Daponte, M. D'Apuzzo, (2001) Wavelet network-based detection and classification of transients, *IEEE Trans. Instrum. Meas. 50 (5)*, pp. 1425–1435, 2001.

L. Wessels and E. Barnard. (1992). Avoiding False Local Minima by Proper Initialisation of Connections, *IEEE Trans. on Neural Networks*, 3(6), 1992.

Q. Zhang, A. Benveniste, (1992). Wavelet Networks, *IEEE Trans. on Neural Networks* 3 (6) pp. 889-898, 1992.

R. Bulirsch and J. Stoer, (1980). Introduction to numerical analysis, *Springer-Verlag, New-York*, 1980.

R. M. Rao, A. S. Bopardikar, (2004). Wavelet Transform: Introduction to theory and applications, *Pearson education*, 2004.

R.H. Franke, (1979). A critical comparison of some methods for interpolation of scattered data, *Tech. Rep. NPS-53-79-003, Dept. of Mathematics*, Naval Postgraduate School, Monterey, Calif., 1979.

R.J. Renka, (1988). Multivariate interpolation of large sets of scattered data, *ACM Trans. on Math. Software, vol. 14, n° 2*, pp. 139-148, 1988.

S. Xiaoyunl et al. (2004). Neural network with adaptive genetic algorithm for eddy current nondestructive testing, *IEEE Proceedings of the fifth world congress on intelligent control and automation*, pp. 2034–2037, 2004.

S.Chen, C. Cowan, and P. Grant, (1989). Orthogonal least squares learning algorithm for radial basis function networks, *IEEE Trans. On Neuronal Networks, vol. 2*, pp.302-309, March 1989.

W. Bellil, C. Ben Amar C. andM.A Alimi, (2007-a). Multi Library Wavelet Neural Network for lossless image compression, *International Review on Computers and Software (IRECOS) Vol. 2 N. 5*, ISSN 1828-6003, pp 520-526, September 2007.

W. Bellil, C. Ben Amar C. And M.A Alimi, (2005). Beta wavelets networks for function approximation, *International Conference on Adaptative and Natural Computing Algorithms, ICANNGA05*, Coimbra Portugal, SpringerWien NewYork, pp 18-21, 2005.

W. Bellil, M. Othmani, C. Ben Amar, (2007-b). Initialization by Selection for multi library wavelet neural network training, *Proceeding of the 3rd international workshop on artificial neural networks and intelligent information processing ANNIIP07, in conjunction with 4th international conference on informatics in control, automation and robotics, ICINCO*, pp30-37, 2007.

W.E.L. Grimson, (1983). An implementation of a computational theory of visual surface interpolation, *Comp. Vis. Grap. Ima. Proc., vol. 22*, pp. 39-69, 1983.

Y. Chen, B. Yang, J. Dong, (2006). A Local Linear Adaptive Wavelet Neural Network, *Transaction of the Institute of Electric Engineers of Japan*, v. 69, no. 4-6, pp. 449-465, 2006.

Y. Oussar et al, '(1998). Training wavelet networks for nonlinear dynamic input-output modeling, *Neurocomputing*, 20(1-3):pp. 173-188, 1998.
Z. Michalewicz, (1993). A Hieararchy of Evolution Programs: An Experimental Study, *Evolutionnary Computation*.

12

A Tree-Climbing Robot Platform: Mechanical Concept, Control Software and Electronic Architectures

Reinaldo de Bernardi[1,2] and José Jaime da Cruz[2]
[1] *Genius Institute of Technology*
[2] *Department of Telecommunications and Control, São Paulo University*
Brazil

1. Introduction

Robotic platforms can be used in countless applications and in the most varied branches of activities. Such results, presented over the last two decades, can be verified in Armada et al., 2003 and Virk, 2005.

It deals specifically with robots provided with legs and with the capability, or ability, to climb vertical surfaces, while many other applications/solutions can be found. As an examples: robots to climb lower parts of bridges (Abderrahim et al., 1999), to crawl inside pipes for inspection purposes (Galvez et al., 2001), implemented to perform solder inspection works in nuclear plants (White et al., 1998), to climb metallic structures for inspection purposes (Armada et al., 1990). More examples can be found in marine industry applications, such as walking robots to check internal parts solders in ship hulls, climbing robots for parts solders (Santos et al., 2000) and (Armada et al., 2005), climbing robots for paint cleaning, underwater robots for ballast tank inspection, and underwater robots for hull cleaning (Santos et al., 1997a, 1997b).

Robots were demonstrated to be the ideal option for many such applications due to the fact that the working environment is difficult to access or even hazardous or risky for human beings, such as exposure to hazardous substances or environments and risk conditions. Productivity increase and quality issues are also extremely relevant and are considered.

However, besides the varied applications and areas mentioned above, there still remains a little explored area: environmental research. As in any other area, different applications or problems can be addressed or solved with the help of a robotics platform. As an example, one can mention the tasks:

- **Gathering of Botanical Specimens**: gathering flower and plant specimen is fundamental for biodiversity studies. Several host species are found in high trees, and their collection is actually risky. Thus, this is an application where a robotics platform with tree climbing capability can be used to minimize risks for the researchers involved in this type of activity.
- **Gathering of vegetable material**: collection of vegetable material is a fundamental activity for phytochemistry. Every pharmaceutical preparation that uses as raw material plant parts such as leaves, stems, roots, flowers, and seeds, with known

pharmacological effect is considered phytotherapic (extracts, dyes, ointments and capsules). Thus, just as in botanical collection, this is an activity that can be accomplished by a robot operating in large-sized trees, therefore minimizing risk for the humans involved in the collection.

- **Gathering of Insect Specimens**: usually, for collecting insects in higher levels, nets spread around the tree and a source of smoke below it are used. The great discussion about this technique is that one not only captures the desired insects, but one ends up killing most of the insects that inhabit that tree as well. Thus, one proposes to use a trap, positioned in the robot, containing a pheromone or equivalent substance as bait, specific for the insect species one wants to capture. One can even adapt cameras and sensors in the trap to gather images of the moment of capture. With a trap thus implemented, one reduces the negative environmental impact of the capture. Another relevant issue would be the possibility that the robot moves along the day to capture varied insect specimens at different heights in different hours, to check for variations in insect populations according to the height, and the hour of the day.
- **Climatic studies**: climatic or microclimatic studies refer to works on microenvironments in native vegetation and/or reforested areas. It is important to study the different energy flows: horizontal and vertical. The vertical directly reflects the results of solar radiation, which decisively influences the horizontal energy flows: air masses, hot and cold fronts, action centers. Solar radiation determines the whole system, and may be analyzed according to its elements: temperature, pressure, and humidity, greatly influencing biogeographic characteristics, geomorphologic and hydrologic phenomena etc. Thus, the robot can be equipped with sensors for such measures, to collect data on the desired elements.
- **Studies on biosphere/atmosphere interaction**: biosphere environments are the group of biotic or abiotic factors that interfere in the life conditions on a certain region of the biosphere. In the case of forests the aerial environment is studied, and the most important elements to consider are: light, oxygen, ice formation, winds, humidity and carbon gas. In order to register all this information, the robot can be endowed with specific sensors for each kind of required measure, to collect data regarding the elements at hand, and to provide information on them regarding both height and time variations.
- **Studies on arboreal fauna**: fauna studies are hampered by the existence of many leaves, or very dense treetops, as the lack of existing natural light hinders the observation of the species. Other usually relevant points are the difficulty to obtain a proper angle due to the great heights involved, and the very presence of human beings in that specific environment, easily detected by their movements, noise and odors. For this type of task, the robot can be fitted with cameras to capture both static and dynamic images. These can then be stored locally in some type of memory card, or transmitted via communication interface to a base station.
- **Sensors Network**: robots carrying a group of sensors and fitted with a communication interface, for instance Wi-fi or other similar technology can be dispersed in the forest to capture data regarding the ecological behavior in the area at hand. Measurements such as the ones already mentioned in climatic studies and biosphere/atmosphere interaction can be shared among robots or even retransmitted among robots to reach the base station, without the need for the researcher to "pay visits" to the reading points.

2. Kamanbaré platform

Purporting the main goal to climb trees for environmental research applications, to be applied in tasks such as presented above, a bioinspired robotic platform named Kamanbaré was proposed.
The project's main application is climbing trees for non-invasive search purposes, reaching points (at high altitudes) that may offer risk to humans.
The development was driven mainly to seek for robot stability and efficiency regarding the paws. The adopted line of research is an implementation inspired in nature. One must stress here that the system doesn't mimic nature, by copying an animal or insect to accomplish the desired work, but rather, a robot development project that combines the climbers' best characteristics (insects and lizards) considering the available technologies and the associated cost/benefit ratio, in other words, some parts were inspired in certain solutions found in nature, while other parts were inspired in different elements.
Ensemble stability, contact stability (paws), contact position, and contact force (pressure) were defined and used to implement the strategies and techniques for dynamic and static control of the system.

3. Chameleons: the biological inspiration

Differently from any other reptile or lizard, chameleons have paws that were created, or adapted, to adequately clasp the different types of branches or shrubs existing in its environment.
The chameleon paws evolved to provide them with the best possible maneuvering and grip capabilities on trees. The paws are actually forked, with three fingers on one side and two on the other. Frequently they also have powerful and sharp nails (or claws) that can grip the surface which they hold on to. This unique arrangement allows them to position their paws completely around the branches or shrubs on which they move around, giving them an amazingly strong clasping capability.
Chameleon paws have five fingers each, divided in two groups (internal and external), one side composed of three fingers while the other side has two, as seen in Fig 1. Front paws have two fingers pointing to the external side, and three towards the inner side, while rear paws are configured in an opposite arrangement, i.e., two inside fingers and three outside fingers. This provides the chameleon with the same number of fingers on each side of the branch, considering all the paws, allowing a balanced and stable clasping.

Fig. 1. Details of the chameleon paw, presenting the bifurcation and configuration of the fingers

4. Mechanical model

The mechanical structure of the Kamanbaré platform consists of a central rigid body with four legs, identical and symmetrically distributed. Each leg comprises three links connected by two rotating joints, and it is connected to the central body by a third rotating joint, while each joint has 1 DOF (degrees of freedom). Identical motor and reduction groups make the rotary movements. Fig. 2 shows the kinematic configuration of a leg. Each leg also has a paw, which is forked just as the chameleons, but with only two fingers, one on each side, for simplification purposes. The paw connects to the leg by a rotating joint, and also has another motor and reduction group that provides for its opening and closing movements.
As each leg has four joints, this means 4 DOF. Considering the four legs, the platform will have a total of 16 DOF. Therefore, sixteen motor and reduction groups were necessary to produce the global movements.

Fig. 2. Kinematic configuration of a leg

The main dimensions of the platform are: length of 250 mm, and width varying between 260 and 740 mm, depending on the positioning of the legs. Its height also depends on the posture and positioning of the legs, and can vary between 190 and 310 mm.
This configuration, with all the parts implemented using two different materials, aluminum and polyacetal, comprises an approximate total weight of 1.3 kg, not including the batteries. The geometric model obtained can be seen in Fig. 3. This model was used to generate the basic locomotion behavior.

Fig. 3. Mechanical structure of the Kamanbaré platform

5. Control software architecture

An architecture was implemented for local control of the Kamanbaré platform. This architecture corresponds to the robot's functional organization.

Based on the hardware architecture to be presented in the following section, the development of the following systems was accomplished according to Fig. 4. This model is based on the architecture implemented for the MARIUS robot (Pascoal et al., 1997) and has the following main components described below.

- **Support system:** this system controls energy distribution to the platform's electronic and electromechanic hardware, and monitors energy consumption as well. This system is also responsible for the startup of other subsystems and, during operation, for detecting hardware failures, and for starting and controlling the emergency modes.
- **Actuators control system:** this system is responsible for controlling the motors, and also for controlling the movements of the legs. Information on legs positioning is received from the general control system. Data regarding joint positions, as well as the values of the electric currents involved, are sent to the general control system.
- **General control system:** this system receives trajectory reference information from the mission control system. It controls all the robot's movements, sending the necessary

commands to the actuators control system. Problems occurring in the path, such as obstacles and absence of support points for the paws, are handled in this system.

- **Mission control system:** this system is the main module, the highest hierarchical level of the platform. It is responsible for receiving commands via the communications system, and for distributing them to the systems involved. It also stores information on the general status of the platform (battery voltage, position of the legs, angles etc.) keeping them available in case the Base Station (described in the following topics) requests them. This system gathers information from the Environmental inspection system to be subsequently forwarded to the Base Station.
- **Communication system:** this system is the module responsible for the communication interfaces existing in the platform, managing communications via Wi-fi and Bluetooth, and exchanging data with the Mission control system.
- **Environmental inspection system:** this system is responsible for gathering data from the installed sensors, and for controlling any additional hardware necessary for that purpose as well. Every data acquired are sent to the Mission control system.

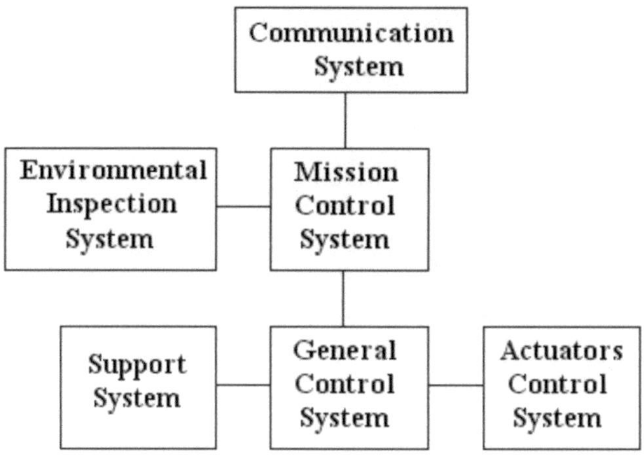

Fig. 4. Kamanbaré's Control Software Architecture

6. Electronic architecture

Considering the implementation of control algorithms, processing information from sensors, the necessary calculations for control and locomotion strategies, interfacing and communication with the base station, added to the need of real time control with position and force feedback, the involvement of high computing complexity was ascertained, thus requiring a processor of advanced architecture.

Eventually, it was selected then a development kit containing a processor based on the ARM9 core, and the deployment of a Linux operating system.

Other motor control boards were also developed using a Texas Instruments microcontroller of the MSP430 family and specific integrated circuits to implement the power actuation section, based on the so-called H-bridge technique.

To implement the control systems for the Kamanbaré platform, an electronic architecture was defined. Initially considering only one joint, it can be represented in Fig. 5, where the main components are seen: a DC motor, a potentiometer and a micro switch.

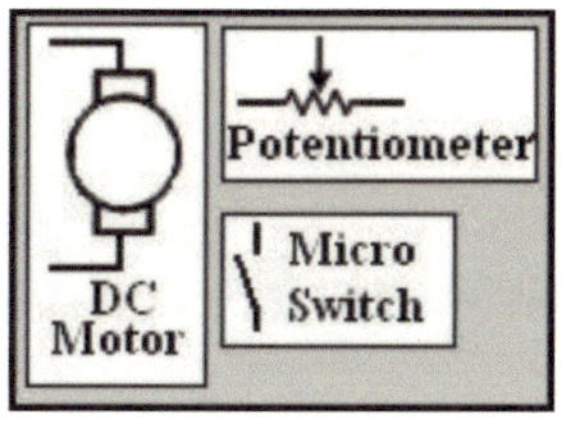

Fig. 5. Representation of a joint

Thus, for control purposes, the need of a PWM output (motor control), an analog input (potentiometer reading, indicating the joint angle), and a digital input (reading the end, or beginning, of the joint course) was ascertained.
As already mentioned, the platform has 16 DOF, corresponding to the need of sixteen copies of the joint control system described above.
As the robot has four legs, one opted for distributing the control individually to each one of them. Thus, each leg control module needs four groups as mentioned, namely, three for the joints, and one for controlling the opening and closing of the claw.
One then developed a motor control board for this specific purpose, Fig. 6, based on the MSP430F149 Texas Instruments microcontroller, and the L298 integrated circuit (H-bridge).

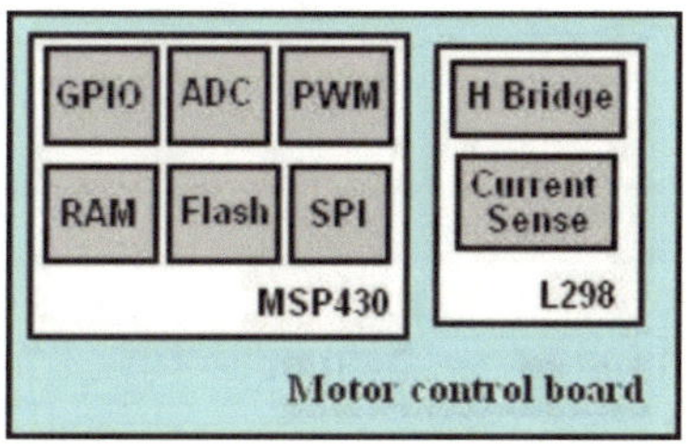

Fig. 6. Motor control board diagram

Due to the control complexity existing in a robotics platform, it was necessary to adopt a main board where the highest hierarchical level control activities were executed.
As a solution for the main board, the model selected was the TS-7250 by Technologic Systems. It was selected because it's compact, contains different standard interfaces, and is based on the EP9302 Cirrus Logic processor, with an ARM9 core, Fig. 7. The EP9302 implements an advanced processor core: 200 MHz ARM920T with support for a memory management unit (MMU). This ensemble allows the use of a high-level operating system, in this case Linux. The ARM920T core has a 32-bit architecture with a 5-stage pipeline, offering high performance and low energy consumption levels. With a 200 MHz clock, the TS-7250 module offers a performance approximately twice as fast as other boards based on 586-core processors.

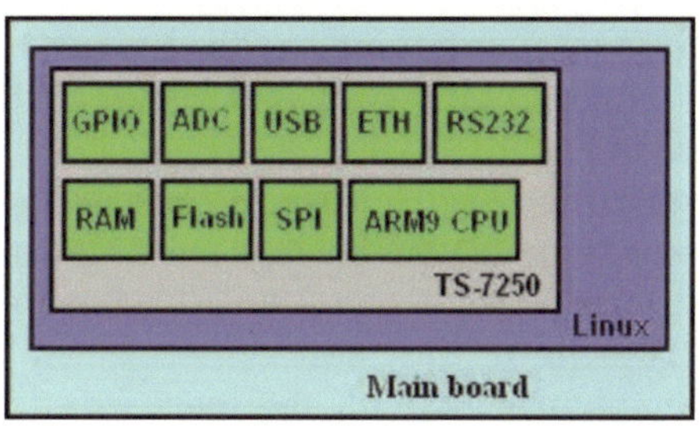

Fig. 7. Main board diagram

Thus, the general electronic architecture for the Kamanbaré platform was deployed according to the diagram in Fig. 8.

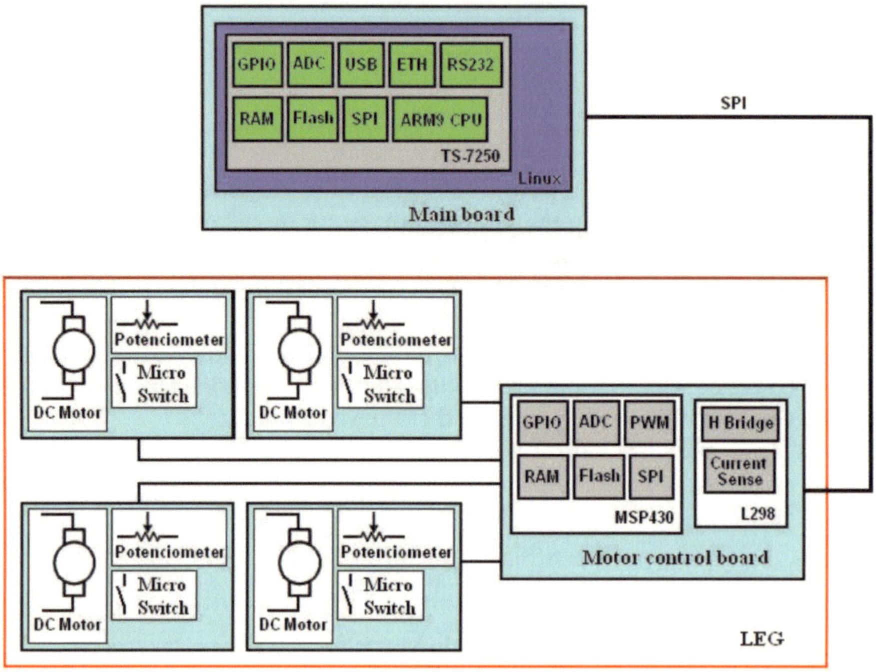

Fig. 8. Electronic architecture of the Kamanbaré platform

7. Simulink models

Simulink® and SimMechanicsTM software use a block diagram approach to model control systems around mechanical devices and simulate their dynamics. With CAD (Computer-aided design) translation, it was possible to combine the power of CAD and SimMechanics software. The translator transforms geometric CAD assemblies into Simulink blocks.
The CAD model of the Kamanbaré robotic platform, described in the section 2 - Mechanical Model, was designed using Soliworks.
Thus a Kamanbaré model was generated using the SolidWorks-to-SimMechanics translator.
This translation process is based on two major steps: to export the CAD assembly into a physical modeling XML format file and to import the generated XML file into a Sim Mechanics model in Simulink.
For the translation procedure, some configurations like the name of the joints and legs were adopted as well the desired movement direction and the gravitational acceleration direction of influence, as demonstrated in Fig. 9.
The resultant model obtained for the Kamanbaré platform is depicted in the next in Fig. 10.
For a better understanding and visualization the legs were represented as model blocks.
Due the fact that all four legs are identical, one detailed leg model block, the Front Left Leg, is described in Fig. 11.
A signal builder was also implemented with the main function of gait generation, here providing the correct angle references for all joints as function of time, Fig. 12.

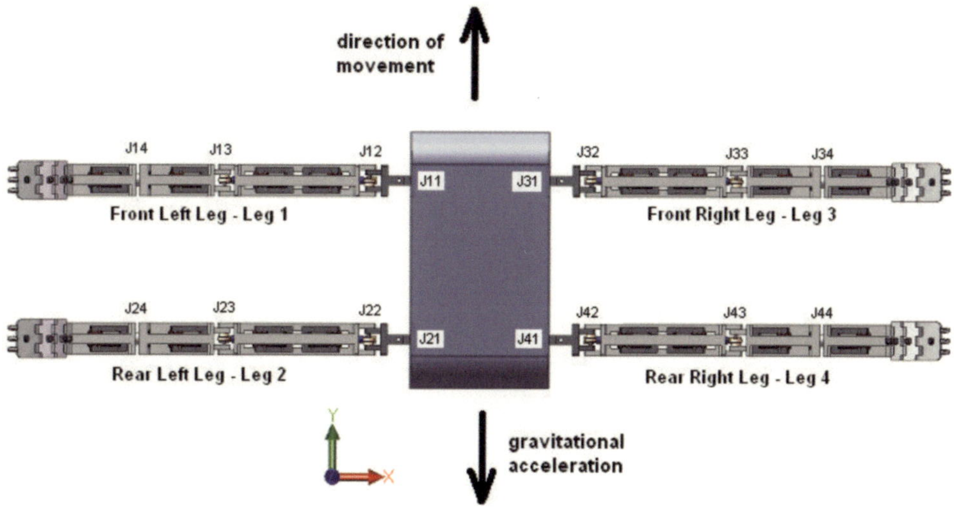

Fig. 9. Kamanbaré's configuration of joints: top view

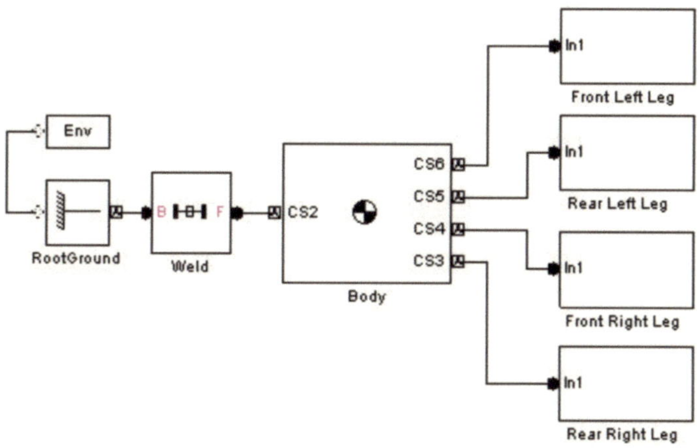

Fig. 10. Kamanbaré's Simulink model

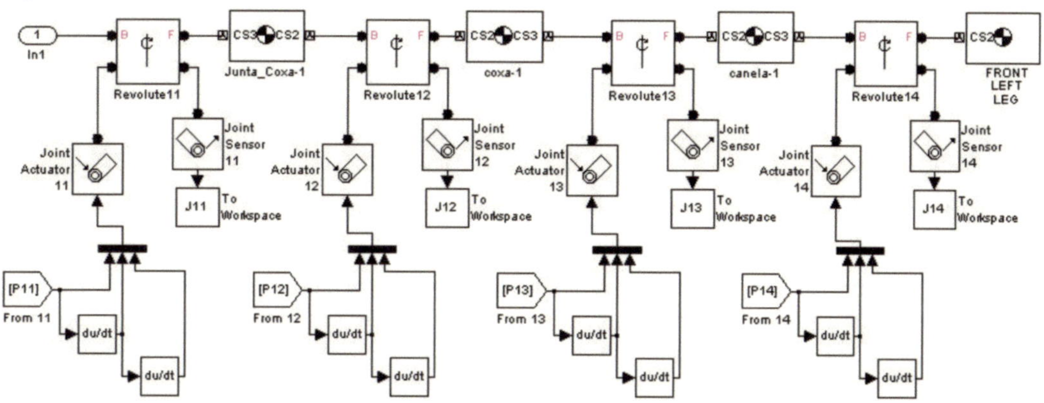

Fig. 11. Front Left Leg model

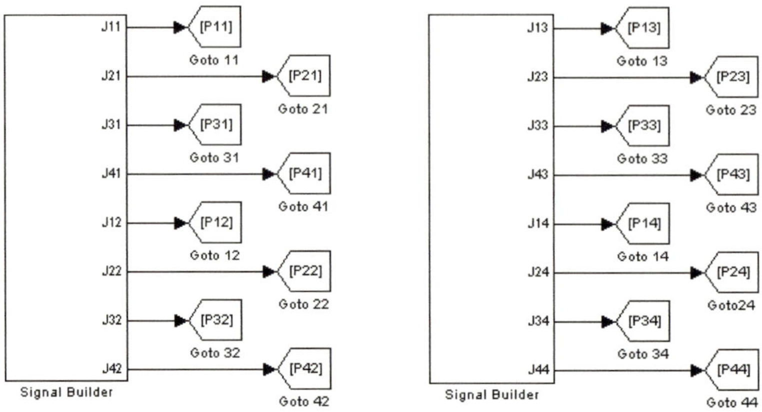

Fig. 12. Gait generator model

8. Gait generation

Gait generation is the formulation and selection of a sequence of coordinated leg and body motions that propel the robot along a desired path.

Free gaits are gaits in which any leg is permitted to move at any time. Fixed, or regular, gaits are those in which a specific pattern of leg movement is imposed. All animals locomote with fixed gaits. A tenet of behavioral systems is that the emulation of animal behavior, hence fixed gaits, is sufficient.

Up to now, just a basic gait was implemented based on the biological inspiration cited before: the chameleon, Fig. 13.

Fig. 13. Biological inspiration for the gait implemented for initial simulations

The gait shown in Fig. 14 represents the variations of the joint angles as function of the time for all the four legs for a movement on a vertical and plane surface.

For a better understanding while analyzing the Fig. 14 a visit on Fig. 9 will help the identification of the names adopted here. Just for clarification, the configuration adopted follows:

- Front Left Leg: Leg 1 with the joints J11, J12, J13 and J14;

- Rear Left Leg: Leg 2 with the joints J21, J22, J23 and J24;
- Front Right Leg: Leg 3 with the joints J31, J32, J33 and J34;
- Rear Right Leg: Leg 4 with the joints J41, J42, J43 and J44.

As explained before, this is a basic gait and it was generated manually observing the chameleon model of movement.

No kind of gait optimization was implemented. The idea here is to have a basic gait to start simulations in order to obtain some results like mechanical interferences between legs while moving and necessary torques for the desired movement.

Gait optimizations in order to obtain a smooth movement as well as the better use of the energy, or energy saving, while moving the joints will be studied and presented on future works.

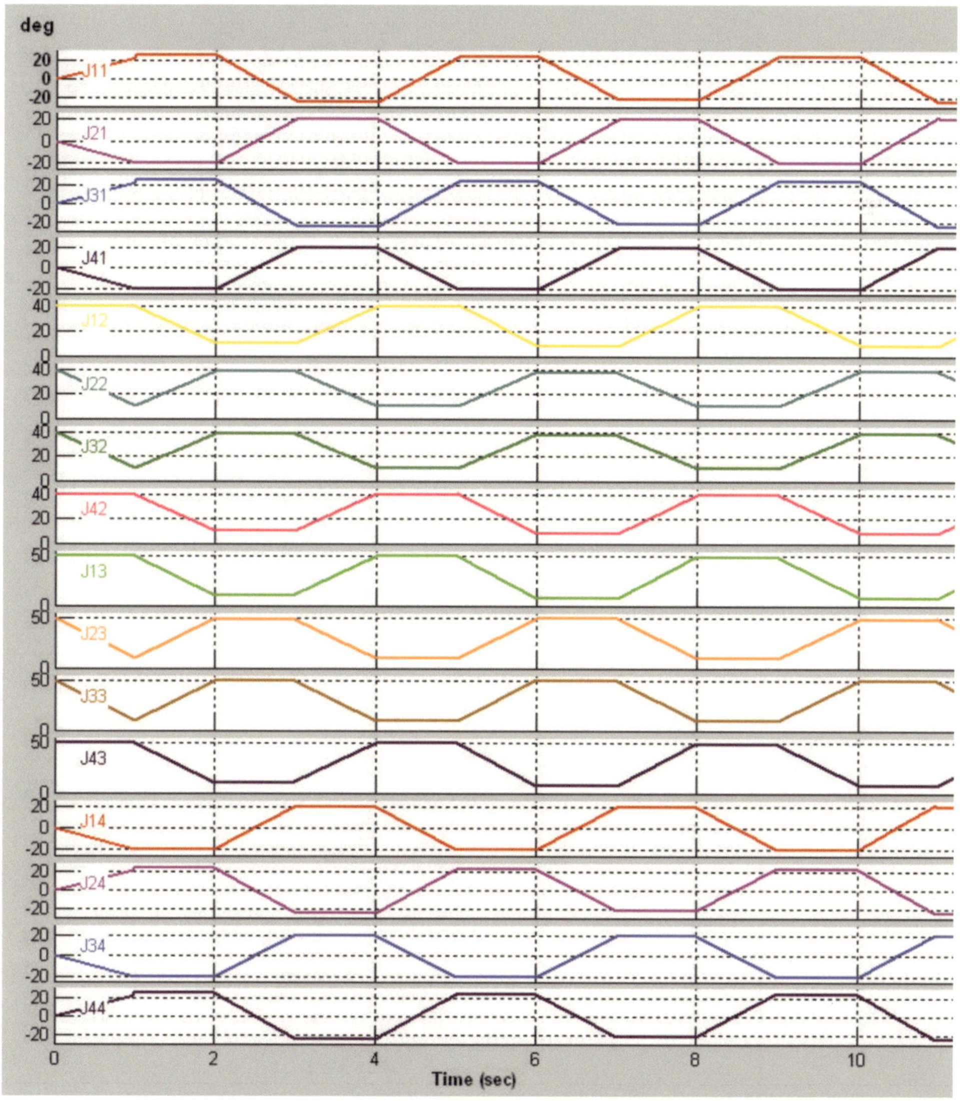

Fig. 14. Gait diagrams corresponding as joint angles positions

9. Simulation results

Regarding the movement generation, or gait, only the simplest way was initially implemented, i.e., the robot follows a straight path at the highest speed allowed by the surface (no waiting time will be introduced, besides those produced by the legs when searching for support points).
The proposed realistic and practical SimMechanics model is further employed to simulate the designed controller which is implemented using Simulink, in this case just a gait generator.
As first result obtained are the models machines presented in Fig. 15 and Fig. 16. These models are quite interesting and useful while running simulations for the reason that all joints and legs movements can be observed.

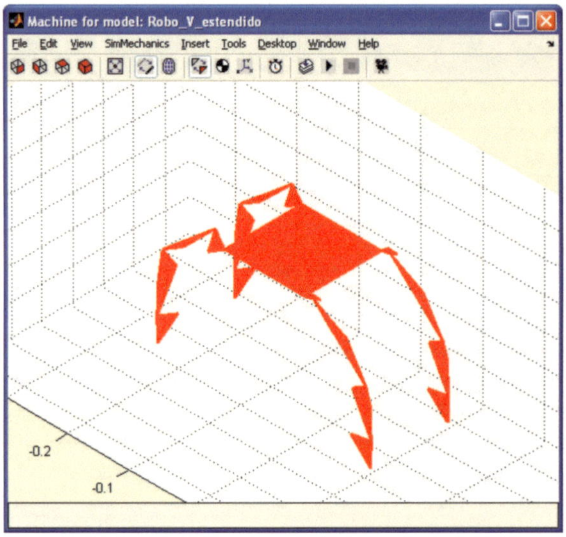

Fig. 15. Kamanbaré's model machine in convex hulls view mode

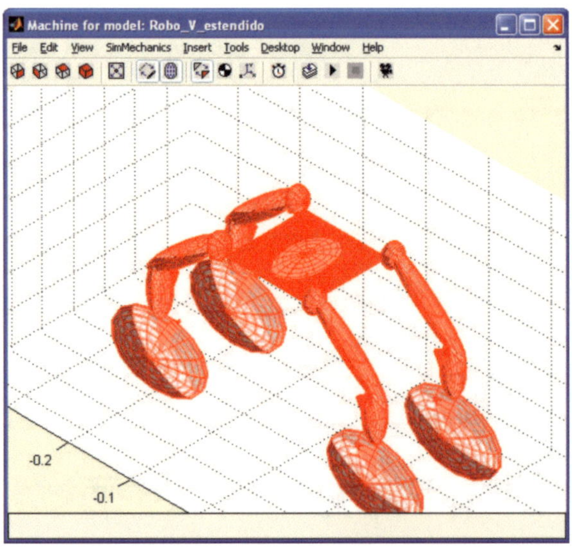

Fig. 16. Kamanbaré's model machine in convex hulls + ellipsoids view mode

As explained before, both the body movement and the foot trajectories are given, as well the joint angles of the legs, as a gait. Using these inputs and running simulations in an inverse dynamics option, all the joint torques can be determined.

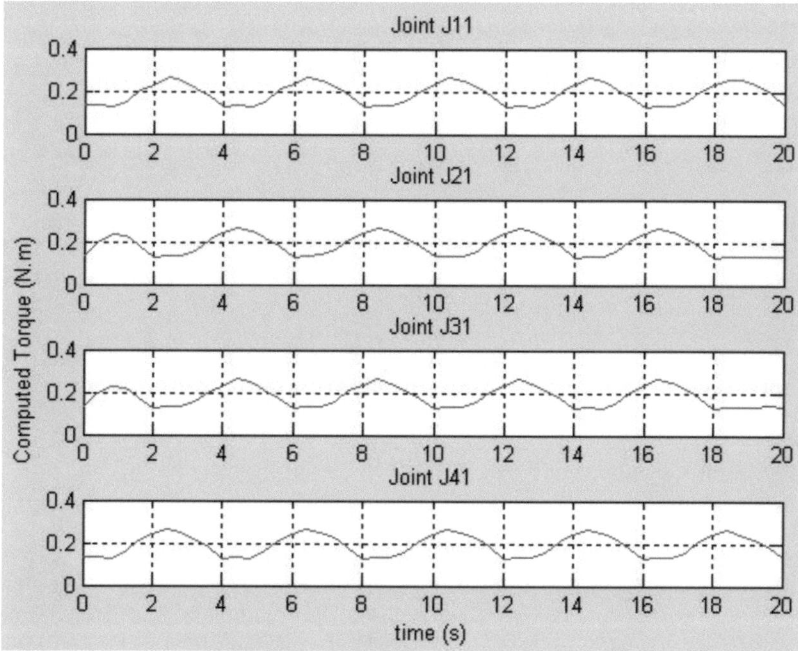

Fig. 17. Computed torques for the group of joints 1

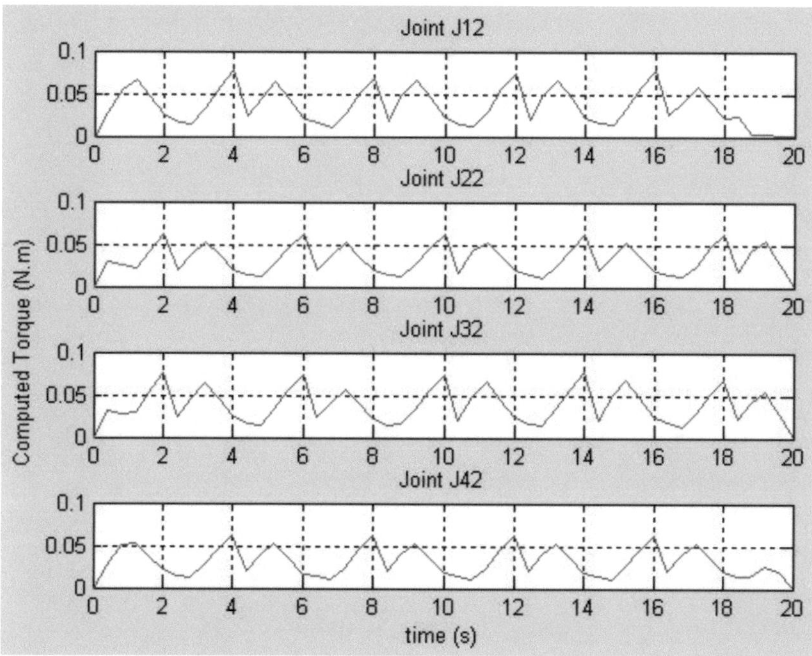

Fig. 18. Computed torques for the group of joints 2

The results presented in the following figures are the necessary torque to produce the desired movement, or gait, presented in Fig. 14. The torques are grouped by the function, i.e, torques for the joints with the same functionality, or the same mechanical position and configuration, for legs 1, 2, 3 and 4.

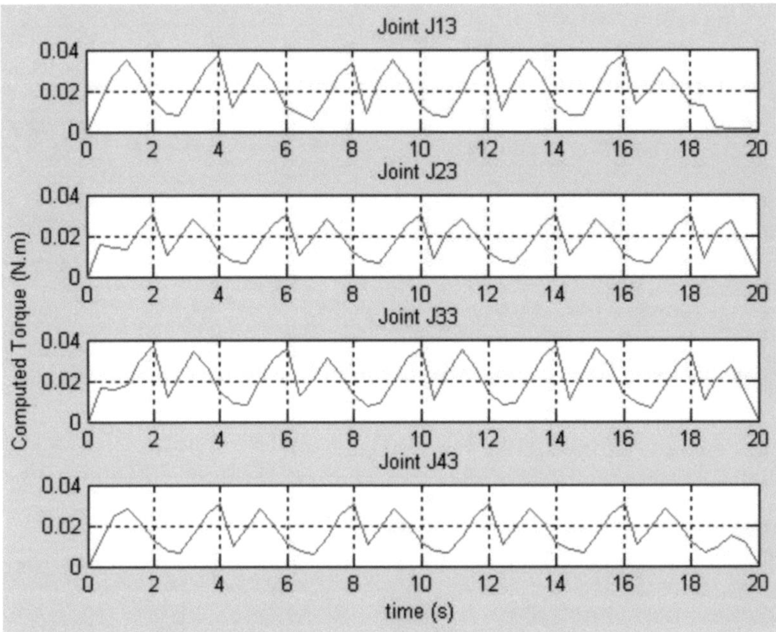

Fig. 19. Computed torques for the group of joints 3

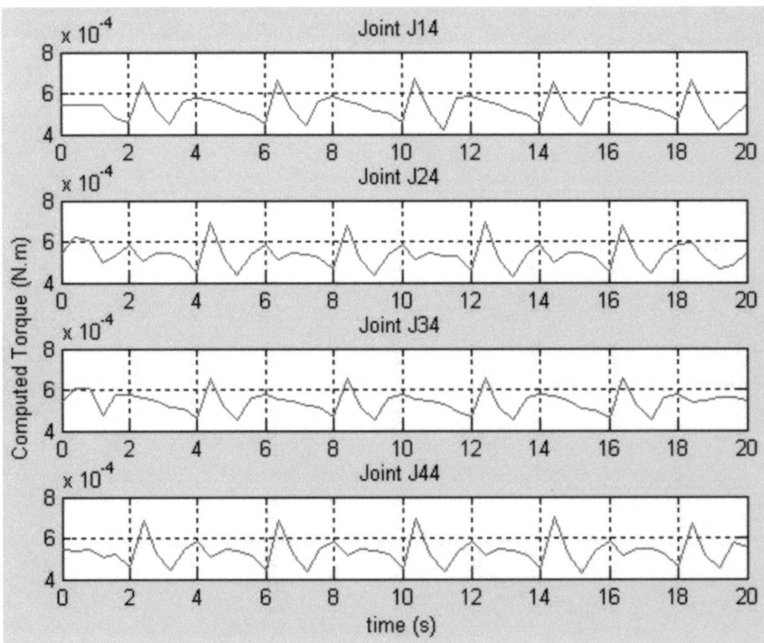

Fig. 20. Computed torques for the group of joints 4

10. Conclusion

This work presented the Kamanbaré robot, which is a four-legged bionspired platform with the main purpose of climbing trees, for environmental research applications. The mechanical and electronics structure, and the software architecture, were presented as well a Simulink mechanical model and simulations results.

Based on its special design, this platform offers the possibility of investigating reptile-like walking and climbing.

As a control approach, just a gait generator based on the chameleon mode of movement was presented, but it was demonstrated that is sufficient to obtain results on legs movement interference and realistic torque estimates.

11. References

Abderrahim, M., Balaguer, C., Giménez, A., Pastor, J. M., Padrón, V. M (1999). ROMA: A Climbing Robot for Inspection Operations. *Proceedings of the IEEE International Conference on Robotics and Automation*, Volume 3, pp. 10-15, 1999, Detroit, USA

Armada, M., Santos, P. G. de, Nieto, J., Araujo, D. (1990). On the design and control of a self-propelling robot for hazardous environments. *In Proc. 21st Int. Symposium on Industrial Robots*, IFS, pp. 159–166.

Armada, M., Santos, P. G. de, Jiménez, M. A., Prieto, M. (2003). Application of CLAWAR Machines. *The International Journal of Robotics Research.* Vol. 22, nº 3-4, pp. 251-264.

Armada, M., Prieto, M., Akinfiev, T., Fernández, R., González, P., Garcia, E., Montes, H., Nabulsi, S., Ponticelli, R., Sarria, J., Estremera, J., Ros, S., Grieco, J., Fernandez, G. (2005). On the Design and Development of Climbing and Walking Robots for the Maritime Industries. *Journal of Maritime Research*, Vol. II. nº 1, pp. 9-32.

Galvez, J. A., Santos, P. G. de, Pfeiffer, F. (2001). Intrinsic Tactile Sensing for the Optimization of Force Distribution in a Pipe Crawling Robot. *IEEE/ASME Transactions on Mechatronics*, Vol. 6, nº 1.

Pascoal, A., Oliveira, P., Silvestre, C., Bjerrum, A., Ishloy, A., Pignon, J. P., Ayela, G., Petzelt, C. (1997). MARIUS: An Autonomous Underwater Vehicle for Coastal Oceanography. *IEEE Robotics & Automation Magazine*, pp. 46-57.

Santos, P. G. de, Armada, M., Jiménez, M. A. (1997a). An industrial walking machine for naval construction. *In IEEE Int. Conf. on Robotics and Automation*, Albuquerque, NM.

Santos, P. G. de, Armada, M., Jiménez, M. A. (1997b). Walking machines. Initial testbeds, first industrial applications and new research. *Computing and Control Engineering Journal*, pp. 233-237.

Santos, P. G. de, Armada, M., Jiménez, M. A. (2000). Ship building with ROWER. *IEEE Robotics and Automation Magazine*, 7(4):35–43.

Virk, G. S. (2005). The CLAWAR Project: Developments in The Oldest Robotics Thematic Network. *IEEE Robotics & Automation Magazine*, pp. 14-20.

White, T. S., Hewer, N., Luk, B. L., Hazel, J. (1998). The Design and Operational Performance of a Climbing Robot Used for Weld Inspection in Hazardous Environments. *Proceedings of IEEE International Conference on Control Applications,* Trieste, Italy.

13

Modeling Virtual Reality Web Application

Berta Buttarazzi and Federico Filippi
University of Roma "Tor Vergata"
Italy

1. Introduction

On the last decade the development of computers and ICT technology has allowed the emergence of a number of new disciplines that have significantly changed daily life and Virtual Reality is certainly one of the most interesting novelty. This technology has focused much of the work researchers. In fact in recent years virtual reality is used in many areas: robotic, simulations, training, entertainment (Video games offers fantastic characters to play and a vast interactive worlds to explore), medicine, etc. Even entire training programs can take place in a virtual reality. In this scenario at first was proposed various languages to introduce the third dimension but not to interact with objects. For against, 3D objects interaction can offer extremely realistic representations that enable users to move and react in a computer simulated environment.

To this purpose few attempts have been made to offer the opportunity to build virtual worlds in what you are completely immersed and where is possible to interact both objects it contains both with people represented by avatar, allowing to hear sounds, see pictures and movies, providing perceptions very similar to real ones.

Though Virtual Reality has reached a mainstream status, actually creating 3D application still requires much effort and only with proper, dedicated tools is it possible to produce convincing result in a timely manner.

Therefore in this chapter we introduce Quest 3D, that is a perfect software package for creating interactive 3D scene including product presentation, architectural visualizations, virtual trainings and computer games.

Quest 3D is a software for creating interactive 3D scene, in a unique style of programming. Instead of having to write thousands of lines of complex code, developers make use of a large set of a powerful building blocks. Both flexible and easy to use, Quest 3D appeals to designers, programmers and artist alike.

Working with Quest 3D means developing in a real-time: you are working directly on the end result. No time will be lost on compiling code or rendering image. While the logic tree scanning goes on, system shows the visualization of scene in the window application.

Quest 3D has a stunning set of graphics features. Large numbers of animated people, vegetation, shadows, fire and smoke effects and realistic water can all be easily added to a scene. Advanced features include physics simulation, path finding routines, database connectivity and networking support. By this way is possible design very simple scene like extremely complex projects: software product always will appear readable and smart. The high number of building blocks can be logically regrouped in folder, characterized by

different name and colour for the various functions: a structure very similar to Java or C class. Blocks take input from the son blocks, which provide services to father. Every scanning of blocks follow a depth first manner from left to right in a single frame, realizing the real time rendering.
Quest3D gives the possibility to manage the logic of scene: the behaviour of objects, their interactions.
Programmer can follow an advanced and sophisticated approach of dynamic loading of the various portions of program, to isolate the different logical parts. This solution provide the possibility to use the all potentialities of graphic card, still limited for household approach of so heavy software programs.

2. User Interface

Quest3D provides an *user-frienly* approach, by a graphic user interface, easy understandable, formed by windows, menu and other direct interaction elements with the developer. Let's show the elements:
This is more productive than worrying about how to achieve consistent results in different hardware and software environments.
Toolbar for working on a scene through the different perspectives

Toolbar for working on a scene through the different perspectives

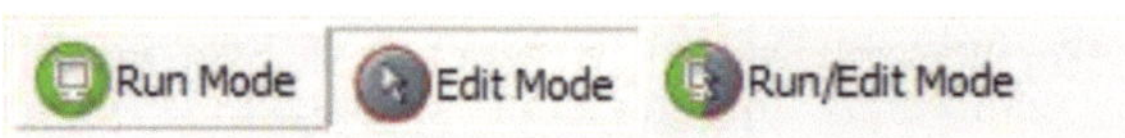

Switch buttons for modifing scenes (*Edit*) and visualize animate scenes (*Run*)
A scene: a set of *channels* connected by functional and hierarchical liabilities

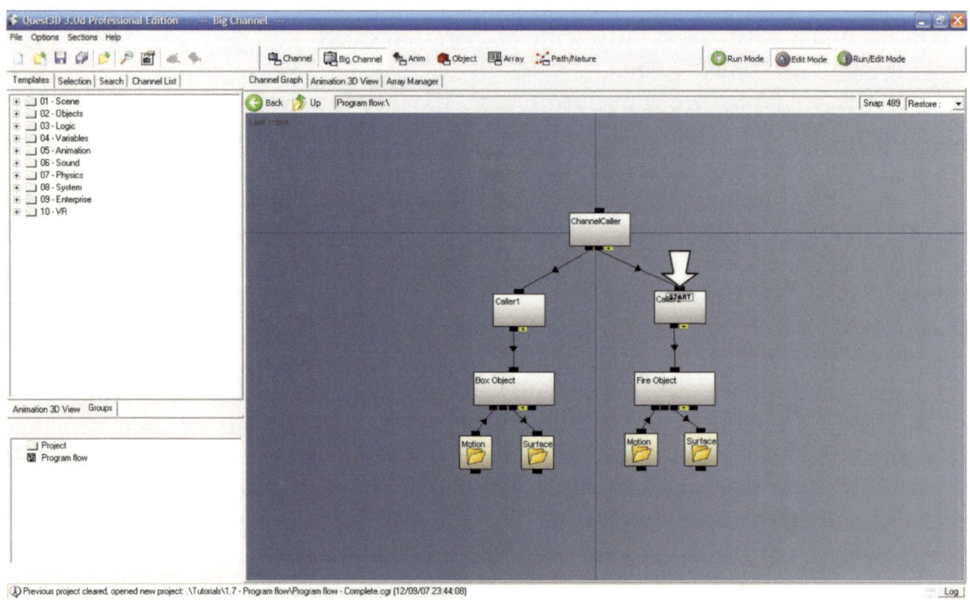

Channel sections show the structure of building blocks of a scene

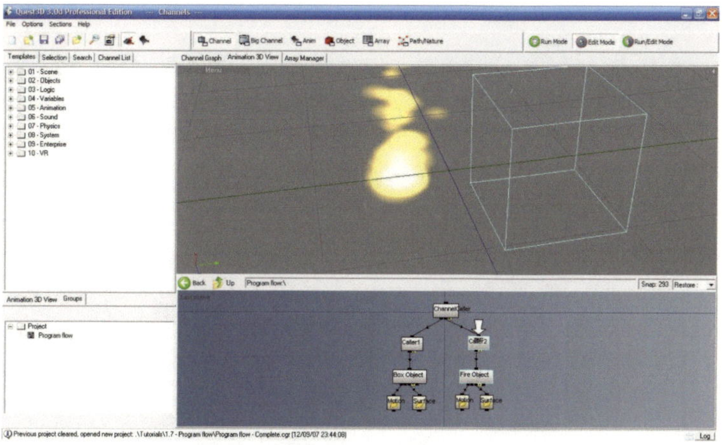

Animation section allow to animate a scene, deciding the behaviour in different temporal instants

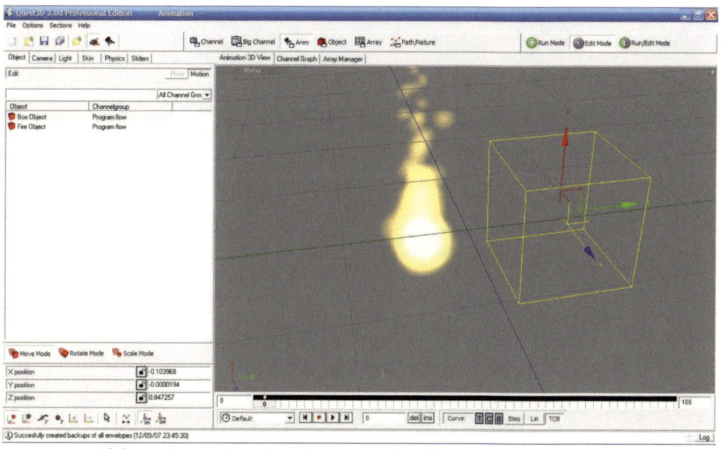

In *Object* section is possible select single objects which form the scene and its surfaces

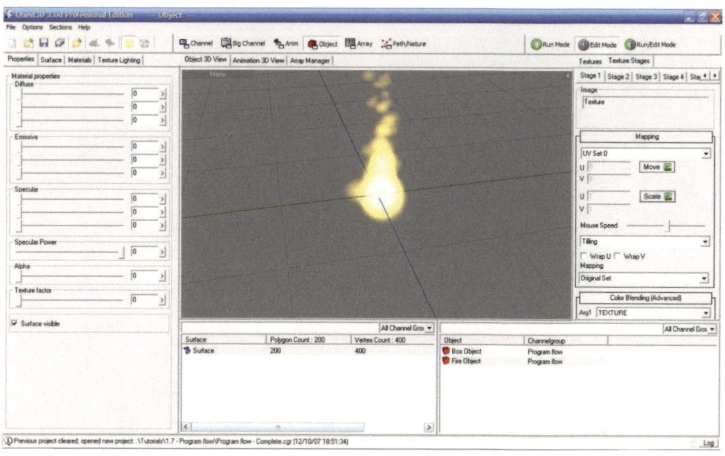

Path/Nature section allow to add naturalistic elements and phenomenon which make more relistic the scene

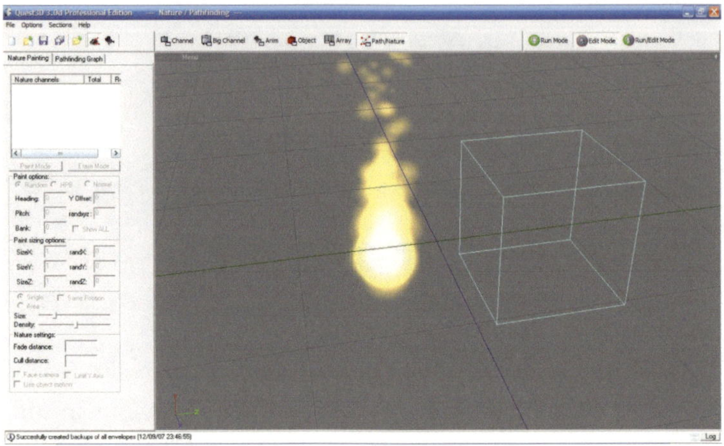

The elements which compose the scene are not created suddenly, but belong to the available channel set

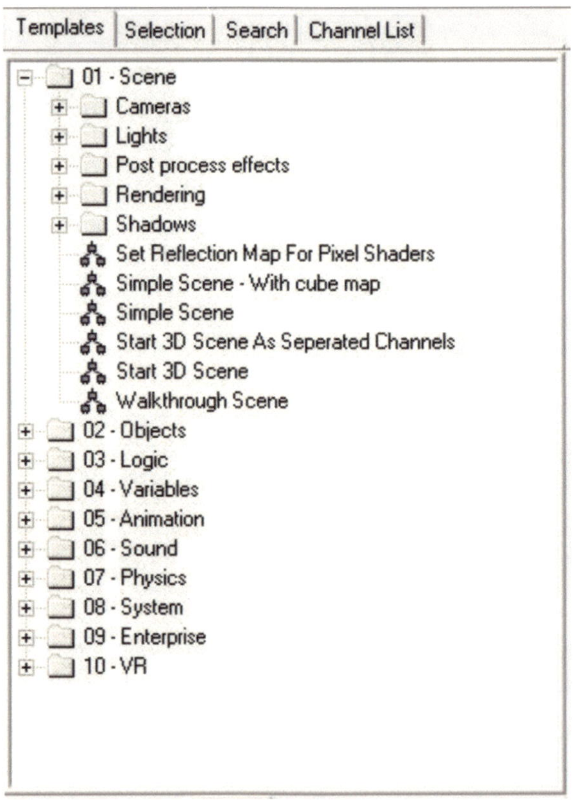

To semplify programing, Quest3D provides every kind of templates: predefined objects, cameras, lights, instruments to add audio file o creating animations, logical-mathematical funtionalities

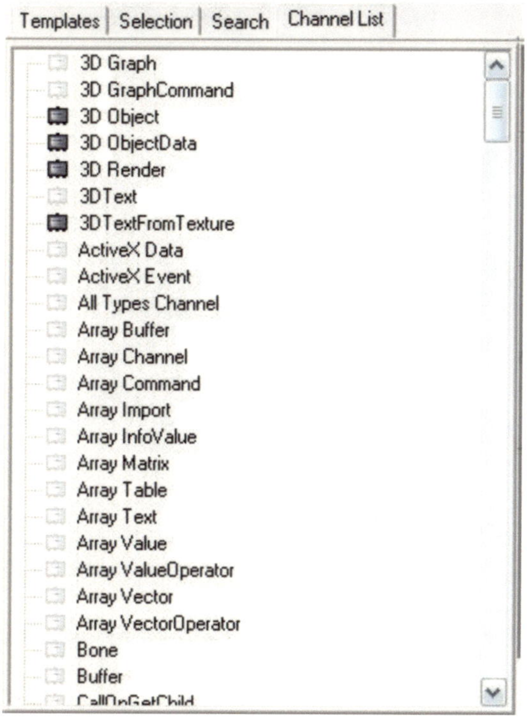

3. Channels

Quest3D programs are made with building blocks. These building blocks are called *channels*. Each channel has a specific function.

The small black squares above and beneath a channel are called *link squares*. Channels can be connected to each other by lines between top and bottom link squares.

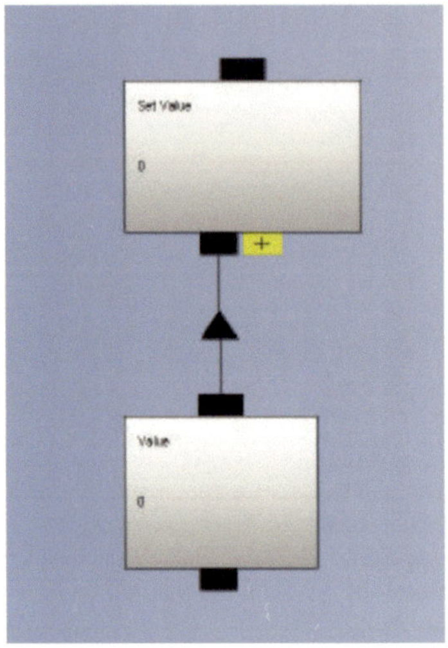

In this image, the upper channel is called the *Parent*. The lower channel is called the *Child*. Children are often used as input or output data for Parents.
The small black squares show which necessarily have to be the *channel* tipology of *childreen*, which *parents* may use like input.
Informations about a building block are visible, throught mouse moving, on specific pop-up windows.

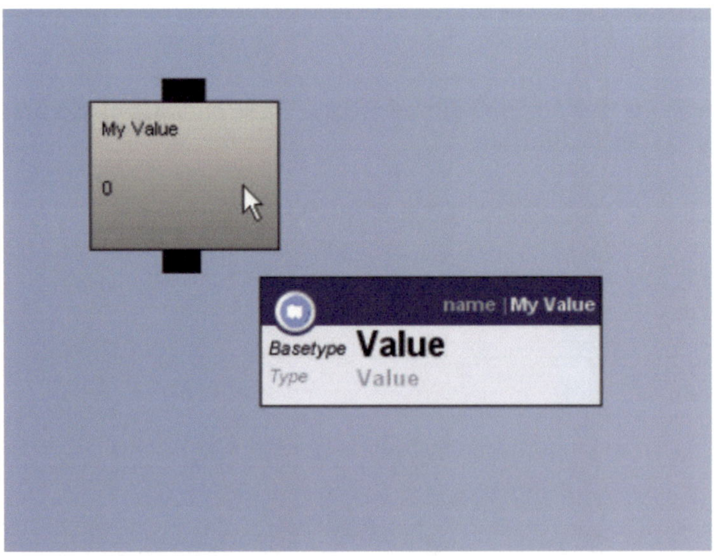

In addition to a name, every channel has a *Basetype* and a *Type*. A *Value* channel is of Basetype *Value* but also of Type *Value*.

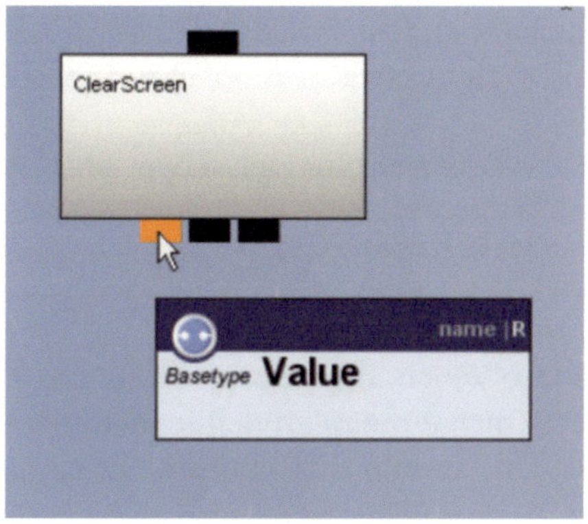

The connected *channels* structures compose a *Channel Group*. A Quest3D project is formed by one or more *Channel Group*.

Double clicking on a channel to visualize *Channel Properties* windows, where are specified the characteristics of a block (name, value, behaviours).

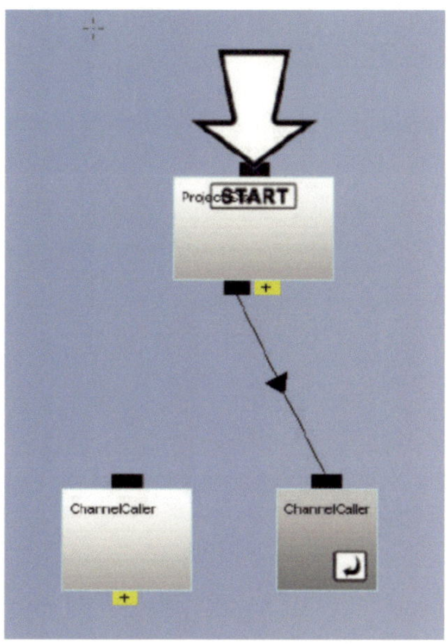

Shortcuts are references to a channel: instead of duplicating a block, is sufficient creating a shortcut (throught the menu by right click of mouse) to inheriting all the characteristics of the block.
Shortcuts are references to the original channel. Their only purpose is to keep Channel Groups more readable.

4. Program flow

A rendering engine is a software which, once download contents (from Internet, like a page HTML or XML, an image, ecc.) format the relative informations and show them on screen.
Quest3D works in real-time. This means that continually executes a complete project and upload the preview.

A complete cycle throught the *channel structure* is called *frame*: the set of all frames produces the rendering.

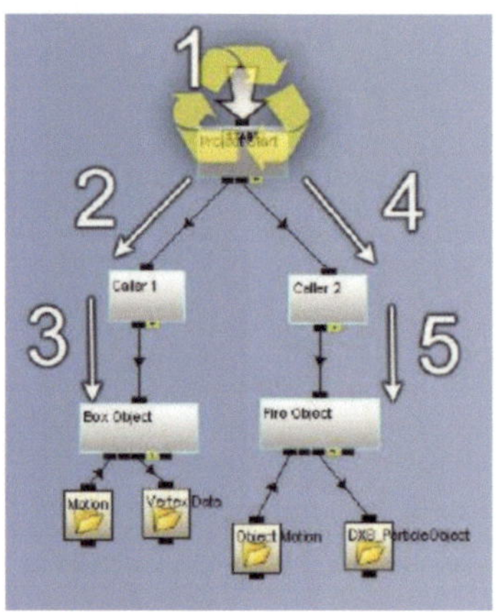

Framerate defines how many times on one second a program is executed, it depends on project complexity and on available hardware.

5. Modeling import

Quest3D is not used to create complex models from beginning. it allow to create interactive 3D scenes architectural presentations, virtual training, computer games, importing entities from available modelling tools.
Models are meanly created by Autocad, 3D Studio Max and Maya, and exported to Quest3D. Exporter of different tools are available on www.quest3d.com.

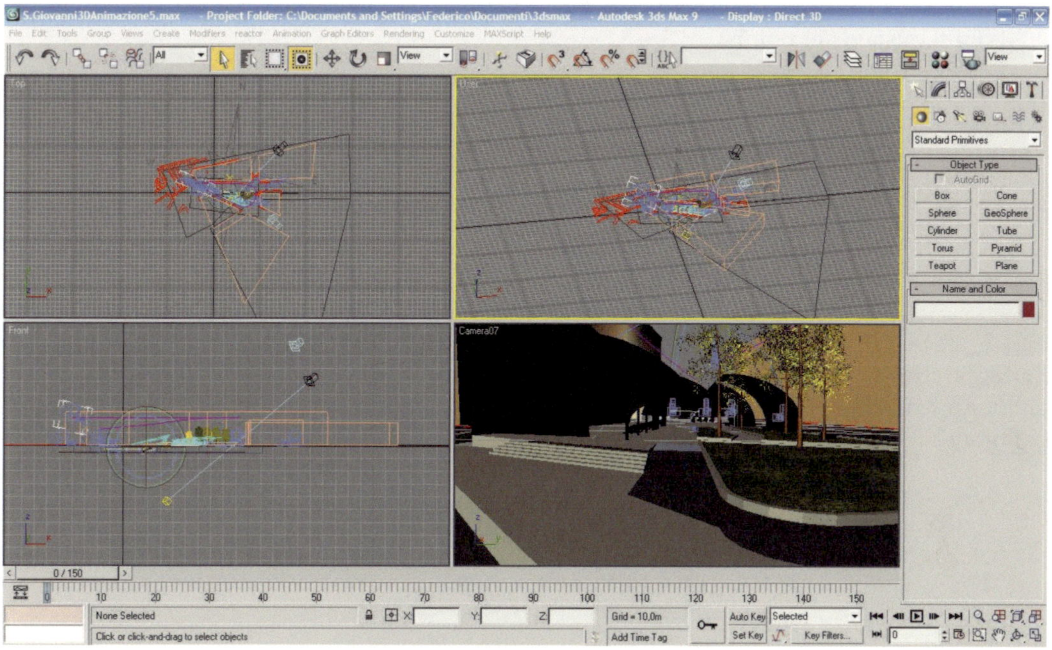

6. 3D Objects

3D scenes consist of virtual objects. Their appearing is defined by a number of elements.

A vertex is a single point in 3D space. Vertexes are defined by X, Y, Z coordinates.

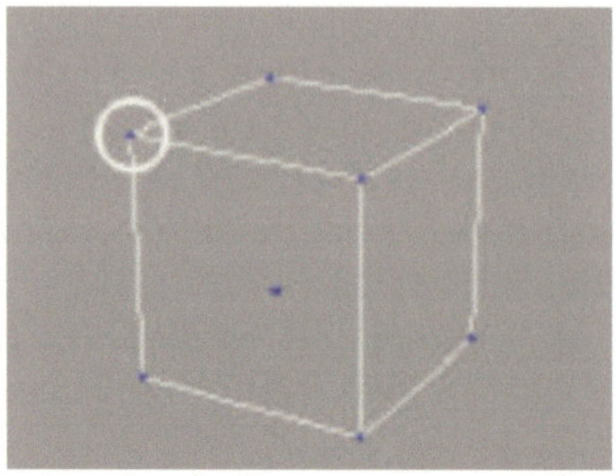

Simple models like a box consist of only few vertexes, while more complex objects can be built out thousands of points!

Surfaces consist of polygons, everyone of them could have an own vertexes number, although real-time graphics usually renders 3 or 4 (3 vertexes for each surface of Quest3D polygons).

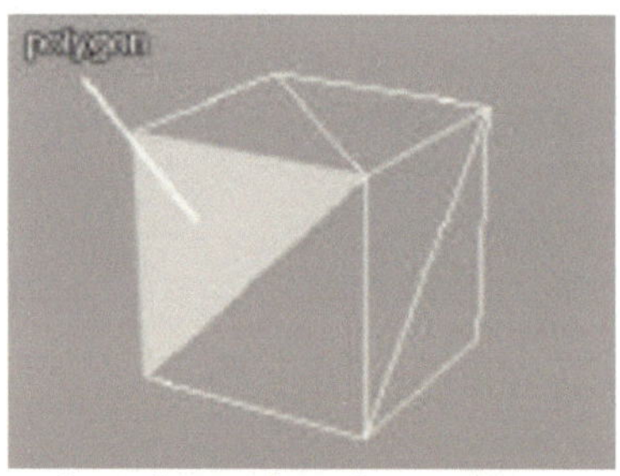

Individual polygons can be assigned a color, but often, another means to add detail is chosen. A 'texture' is an image that can be wrapped around a 3D shape.

Textures are wrapped around an object, to define its color and transparency.

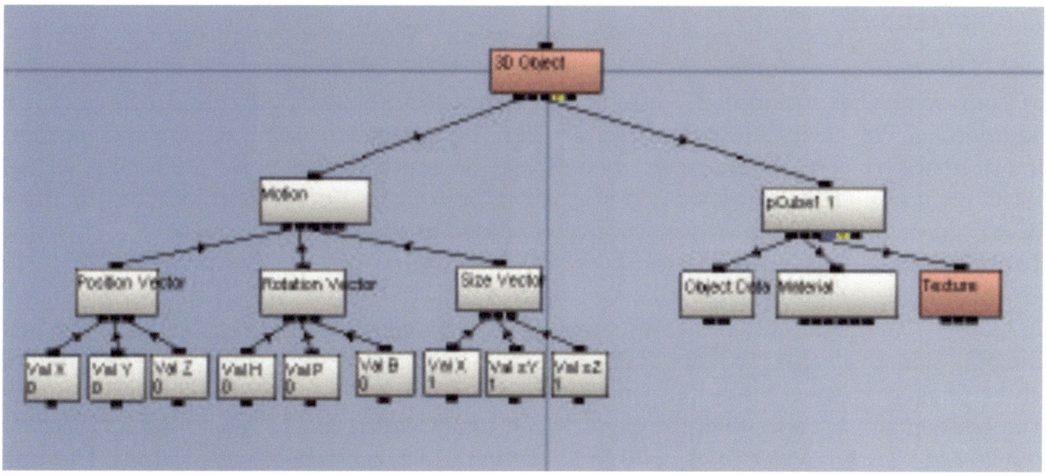

Tipical components of a *3D Object*:

- The *3D ObjectData* channel contains the vertices and polygons of an object. It also contains information on how to wrap a texture around the shape.
- The *Material* channel is subdivided into a number of aspects: Diffuse, Emissive, Specular, Power, Alpha Diffuse and Texture Blend Factor.
- Il*Texture* channel represents the actual image applied on a surface: it contains a diffusive colour image (RGB), and an alpha image.
- Il channel *Motion*, throught its parameters, manages object's position in space.

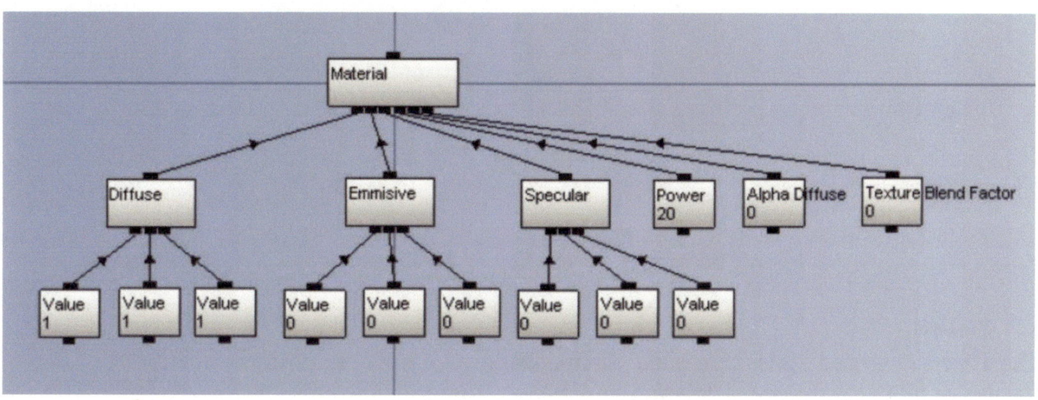

- The *Diffuse vector* contains an object's colour, and consists of three components: red, green and blue, or '*RGB*'.
- The *Emissive vector* describes the amount of self-lighting the model has. Without lights in your scene, an object with an Emissive vector of (0, 0, 0) is black.
- Lo *Specular vector* defines the colour and intensity of the highlight, while the Power value controls the size of the highlight.
- The *Alpha Diffuse* value defines the transparency of an object.
- The *Texture Blend Factor* value defines how much of a texture is blended with the Diffuse colour.

7. Animations

One of the most powerful features of Quest3D is that you can animate virtually every aspect of your scene. For instance, 3D objects can be moved, rotated and scaled over time. In addition, all of their surface properties can be animated, such as their color and transparency values. 3D character models can be brought to life by animating virtual skeletons inside of them. Music and sound can be triggered on certain location- or time-based events. Menu windows can be moved or faded in and out of the screen.
To realize animations Quest3D uses:

- *Envelope*

In Quest3D, animation often means changing values over time. The regular *Value* channel can only store one value. To store animation data, the *Envelope* channel is used. The interface of the *Envelope* channel looks like a mathematical graph. It works in a similar way as well. The horizontal axis represents the input (often time), which can be fed to the *Envelope* by connecting a *(Timer) Value* to its child link. The vertical axis represents the output. Through (X, Y) pairs and interpolation, any input can be converted to an output value.

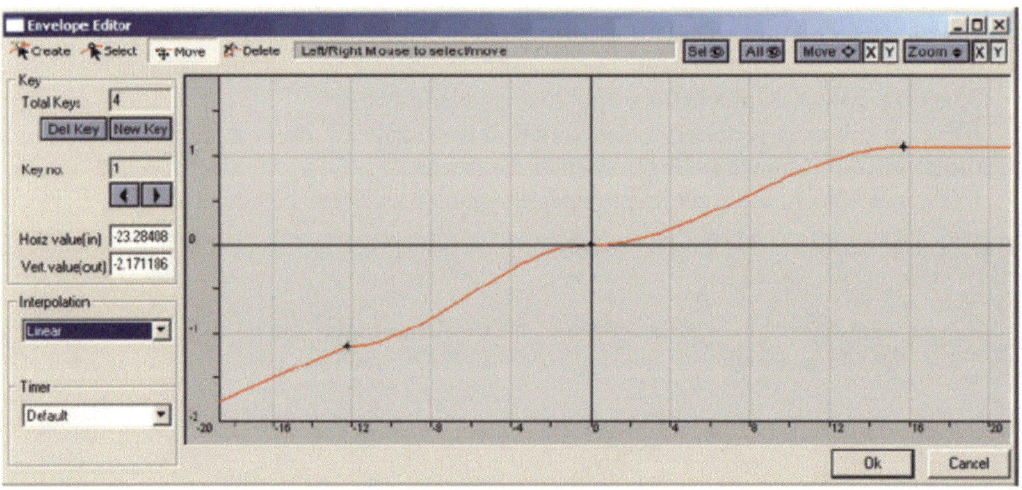

- *Timers*

The *Timer Value* and *Timer Command* channels allow for easy animation control.
Examples of Timer Commands are 'Play', 'Stop', 'Rewind' and 'Play & Loop'.

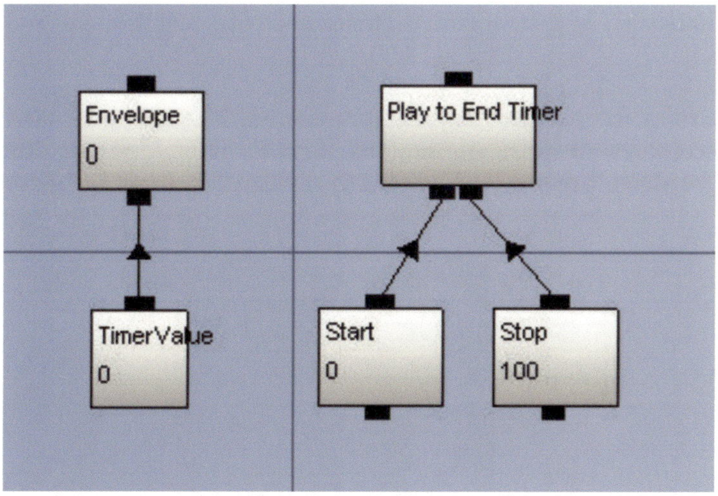

Timer Value channels are automatically adjusted for computer speed. In Quest3D, 25 frames equal exactly 1 second.

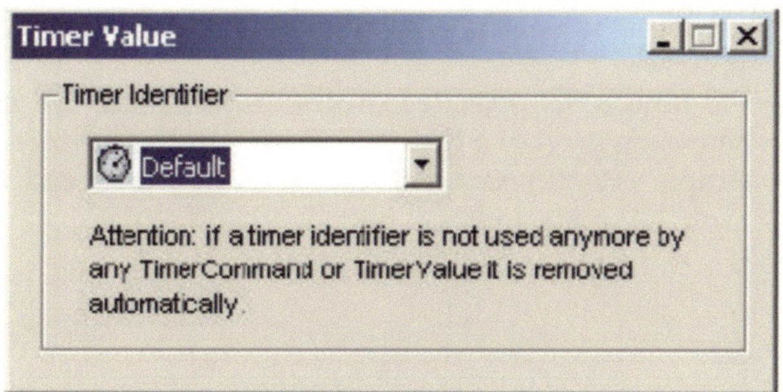

A Timer can be renamed throught the channel property window.
Each element of the scene can need of an own timer.

8. Surface properties

3D object appearing is defined by surfaces properties.

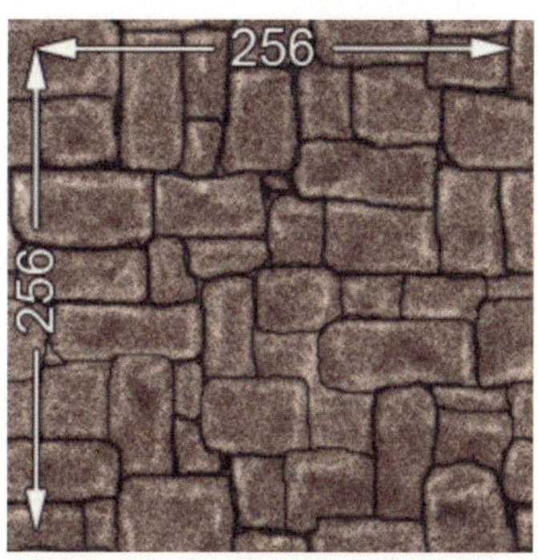

A 'texture' is an image that can be wrapped around a 3D shape. In real-time graphics, for performance reasons, their dimensions in pixels (width and height) must be a power of 2. In

Quest3D, they must also be greater than 8. Dimensions of 32x32, 128x256 and 256x256 pixels are all valid sizes.

Compression:

In addition, textures in Quest3D may be compressed. The image quality might slightly decrease, but compressed textures may use up to four times as little memory as uncompressed versions. Creating a 3D scene is constantly balancing between quality and performance.

Mipmaps:

Larger textures on distant surfaces might produce artifacts. By using various versions of increasingly smaller size, these distortions may be countered. These smaller versions of the same texture are called *mipmaps* and can be generated automatically by Quest3D.

UV Mapping:

Textures are wrapped around a 3D object according to so called 'UV coordinate sets'. UV coordinates specify where each pixel of a texture is drawn onto a surface.

Quest3D can handle simple *UV mapping* methods such as *planar* and *cubic*, shown in the image below. These simple UV mappings can be moved and scaled.

Since UV coordinates are relative, once an object is properly 'UV mapped', textures of any size may be applied to the object.

Wrapping a texture around a 3D object usually requires more control and precision. In practice, UV mapping means placing every vertex of a 3D object onto a 2D image.

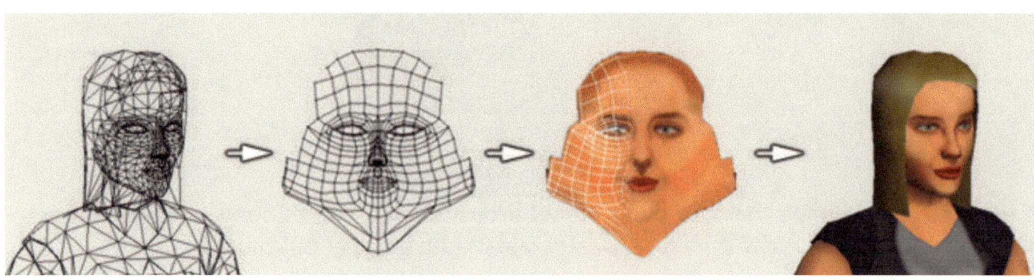

Transparency:

Transparency of surfaces may be handled in a number of ways, managing the characteristics of surfaces in *Object section.*

Texture stages:

More than one texture may be applied on a surface. The maximum number of 'texture stages' that can be used per surface depends on the graphics hardware. Most graphics cards support two stages, and newer graphics cards support three or four.

Quest3D tool allows different methods of textures blending on each stages.

9. Lighting and shadows

Without contrast between light and dark, it is much harder to establish depth in a scene.

A right lightings, the shadow contrast are mean elements to create a realistic 3D environment! A number of lighting techniques are available in Quest3D:

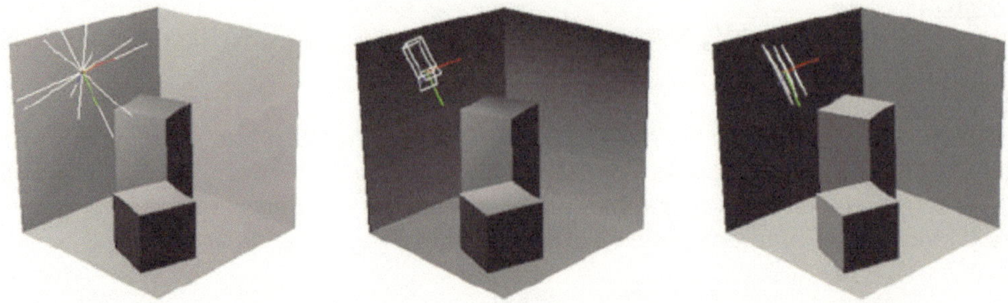

A point light radiates light into every direction. A spotlight is defined by an inner and outer cone angle, resulting in a simulated fall-off effect. A directional light is simply a rotation vector along which the source emits light. All rays of a directional light are parallel to each other.
A Quest3D light can be colored through its 'Diffuse' vector. The 'Ambient' vector serves as a global 'Emissive' vector to all objects affected by the light source.
A 3D object's illumination may be influenced by changing its Diffuse and Emissive vectors. These can also be accessed in the Object Section, on the 'Material' tab. A higher Emissive vector will suggest a certain degree of self-illumination or ambient light.

Shadows for dynamic objects, such as cars or walking characters, can be simulated using texture squares.

Quest3D also supports real-time shadows. The *Stencil Shadow* channel can be seen as a *Render* specialised in calculating shadows. The *Stencil Shadow* channel requires a light source position vector to work. 3D Objects may be linked to the *StencilShadow* by means of a *SoftwareStencilShadowObject*.
Real-time shadows require a lot of processing power.

Light mapping is the process of calculating lighting and shadows once, and storing the information into a texture. The resulting 'light maps' are then applied on an object on a second texture stage, and blended with the diffuse texture.

10. Cameras

A camera defines the point of view from which a scene is presented. Choosing the right camera for a project is important, as it influences the way users experience a scene in a very direct way.
Let's show the camera *templates* available in Quest3D.
Animation Camera:
An *Animation Camera* may be used to guide a user through a scene. The animation camera is similar to a real-life freehand cinematic camera. It may also be the preferred camera for non-interactive scenes such as visualisations, documentaries or screensavers.
Animation Camera with target:
An *Animation Camera with Target* always looks at a dummy object. Both the camera and the helper object can be animated.

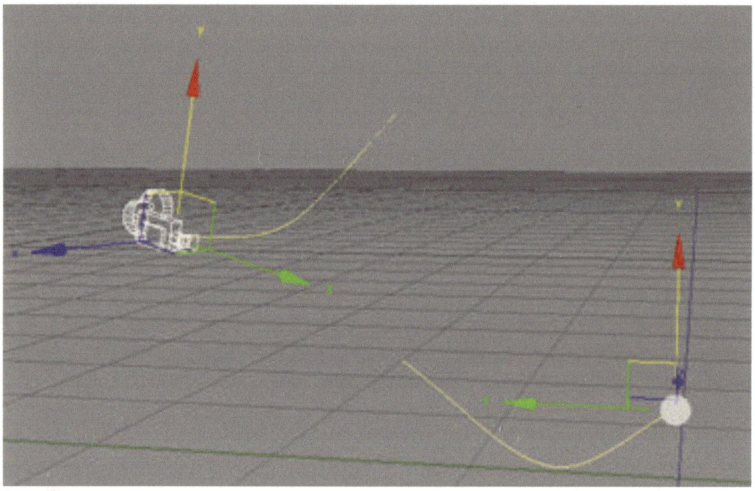

Object Inspection Camera:
Quest3D is the perfect product visualisation package. An *Object Inspection Camera* can orbit around a 3D model, enabling the user to view the object from all angles. Such 'products' might be as small as technological gadgets, or as large as entire cityblocks.
1st Person Walkthrough Camera:
It is often desirable, and rewarding, to experience an architectural visualisation as if you were there. A *1st Person Walkthrough Camera* presents a scene at regular eye level, and allows the user to walk around freely. Since the view is close to real life, this type of camera provides the highest level of immersion in a scene. Many computer games make use of the 1st person perspective.
3rd Person Walkthrough Camera:
In contrast to the 1st person point of view, a 3rd person perspective places the camera just outside of a virtual character. The *3rd Person Walkthrough Camera* is attached to this character, or *avatar*, and follows it wherever it goes.
Every kind of camera is defined by particular characteristics:
zoom factor:
The zoom factor is a quick way to zoom a camera in or out.
The default value of '1.0' results in a normal view. A higher Zoom Factor brings the scene closer, a lower value sets it further away.

<u>*clipping planes:*</u>

The *clipping planes* of a camera are virtual boundaries between which a scene is rendered. Polygon surfaces outside of these boundaries are not calculated.

<u>*fog:*</u>

Fog can be used to blend various elements in a virtual environment together. It can add to the mood of a scene and suggest depth.

Fog can be used to blend various elements in a virtual environment together. It can add to the mood of a scene and suggest depth. A bonus of using fog is that the far clipping plane of the camera may be decreased to just behind the *Fog End* value. If used right, this will result in a smooth transition between the scene and the "nothing" beyond the far clipping plane.

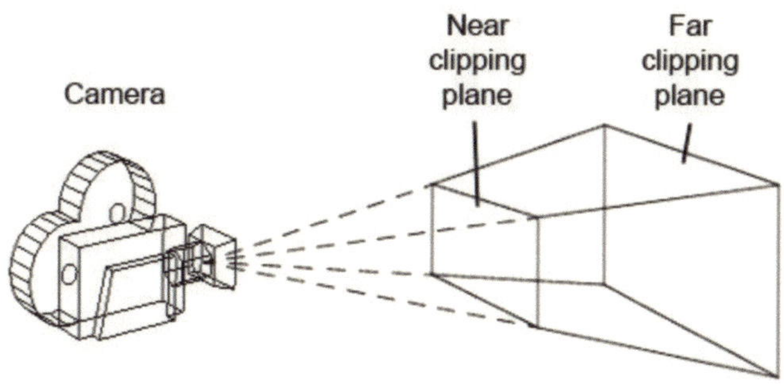

11.Sounds

From simple menu click sounds to full ambient music, *audio* can really add to a scene. It helps to set the mood, just as sharp graphics and lively animation can. The use of spoken voice samples can help users to better understand an application. Turn off the sound while watching television or playing a computer game, and you'll instantly miss the richness of audio.
Sound can be stored digitally on the computer, in many different formats. Quest3D supports both '.wav' and '.mp3' files.
The *Sound File* channel is used to store *.wav* samples within Quest3D. Its properties window contains a number of options including icons to load, play, stop and pause a sample. As the name suggests, the *Repeat* option allows for continuous playback.

Speed ranges from 0% to 200%, and volume ranges from 0% to 100%. Both can be dynamically changed through their respective child links.
Panning controls how the sound is divided between the left and right speaker: 0% means only the left one is used, 100% means the right one. 50% (default value) plays the sound at equal levels through both speakers.
The *Sound Command* channel allows you to control all aspects of a sample, such as playback and volume. It also allows for loading '.wav' files from harddisk. The drop-down list presents all the various options. A *Sound Command* must be called once to be executed, and affects all its children of the *Sound File* channel type.
Besides simply playing a sound, it is also possible to place it in 3D space. 3D Positioned sound requires a *Listener* channel. It is usually connected to the location of the virtual camera, and automatically adjusts volume, panning and frequency by calculating a sound's relative position and distance. Only Mono sound samples can be positioned in 3D space.

The *3D Positioning* options of the *Sound File* channel define the quality of the simulation. Higher quality requires more cpu power.
The '.mp3' format is also supported in Quest3D. The *MP3 File* channel is used to store sound samples of this type.
Through its properties window it is possible to load a file from disk. The *Save File in Channelgroup* option allows you to store the sample data inside of the channel itself. This will of course increase the file size of the channelgroup. The *MP3File*'s only child may also contain a file name in the form of a *Text* channel.
The *MP3 Control* channel can be used to control playback and volume. It can also be used to retrieve information on total length and current playing position, and to skip to a specific position in the sample.

12. Particle sistems

Many special effects in movies and games, such as fire, smoke, explosions and all kinds of magical effects, have one thing in common: they are all created using so called *particle systems*.

A particle system is a collection of objects released from an emitter and usually moved and changed over time. Particles are square objects facing the camera.
For every particle in the system, the position, rotation and size is stored. In addition, other attributes such as speed, color and transparency may be used. Forces such as gravity may be applied to all particles in the system.
Particles are released from an emitter. In Quest3D, this is in the form of an *3D Object Data* channel.
The particle system cycles through all of the vertices in the object and constantly emits a particle from one of these positions.
By randomly placing a number of vertices in 3D space, it appears as if the particles are released randomly as well. This produces convincing results for special effects such as fire and smoke.

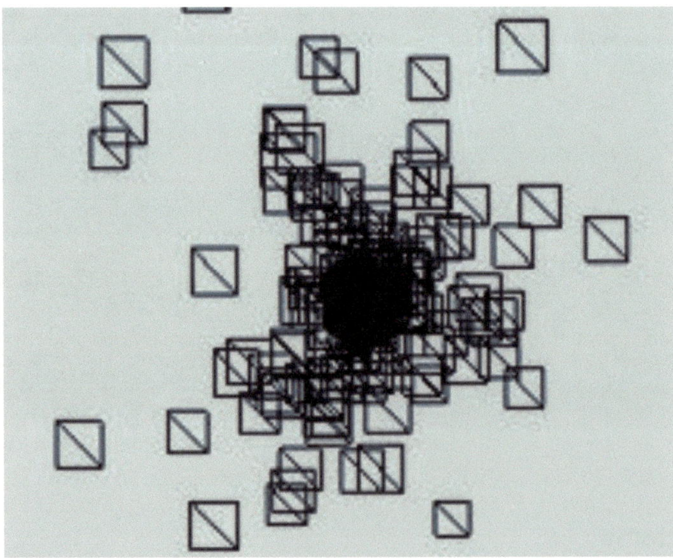

A *Particle Flow Grid* channel can be used to control the flow of the system. It divides a collision object into several parts and calculates forces inside of each of these. Forces can be created by the collision object, external objects and the particles themselves.

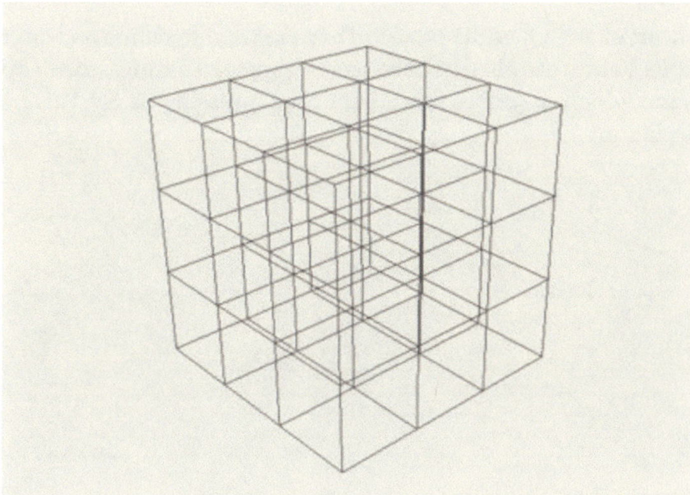

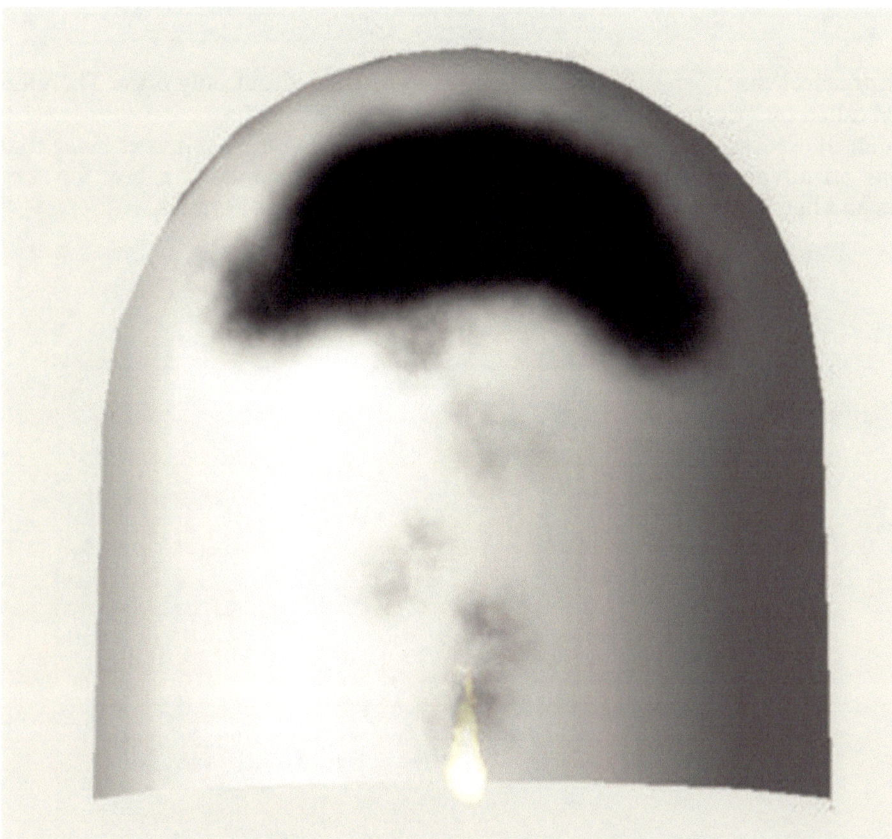

13. Mathematical operators

Values are at the heart of any Quest3D project. They are used to define an object's color, position and size for example. Values can also describe scores, program settings, menu option etc.
A simple instruction which describes mathematical expression is *Set Value*, which realizes a simple value association:

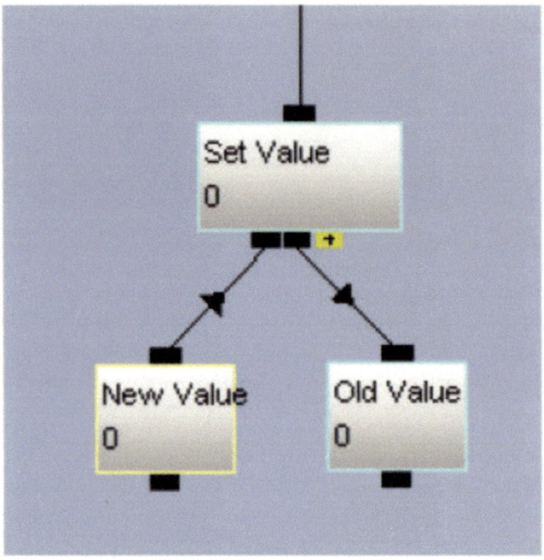

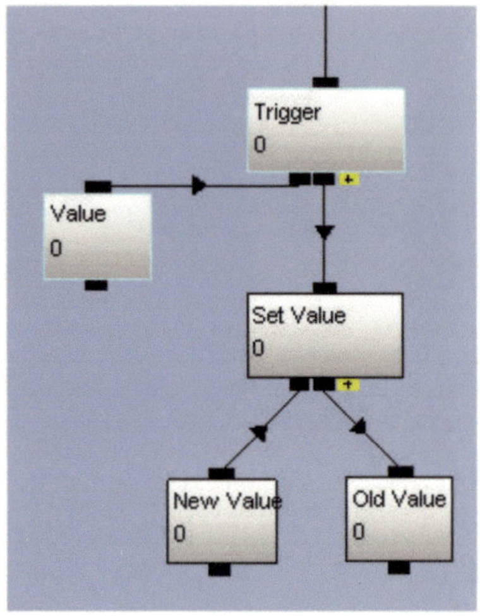

In most practical cases *Set Value* channels need to be carried out only once. Therefore, they are often called by *Trigger* channels.
Mathematical expressions can be described in Quest3D using the *Expression Value* channel. It serves as an advanced calculator. Its properties window contains a box for the actual formula, and help text describing all the usuable variables and operators.

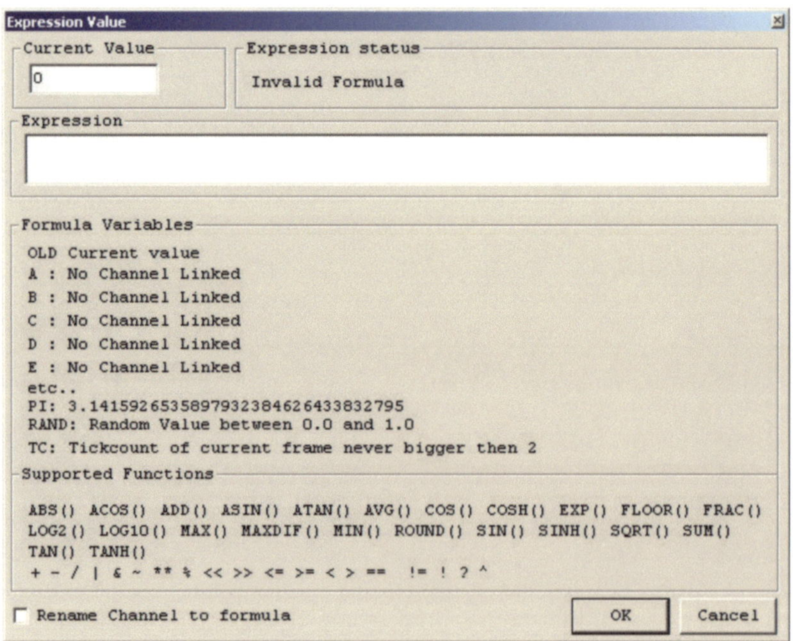

Values connected to the child links of the *Expression Value* can be used as part of the expression within the channel as well. The first child is presented by the letter 'A', the second by the letter 'B', and so forth.

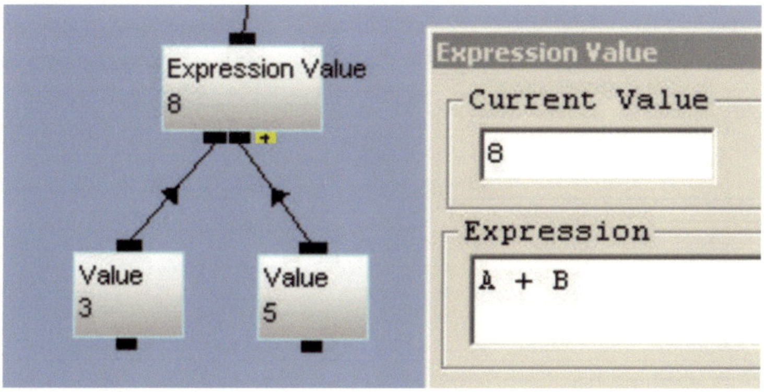

Multiple expressions can be linked to one another within the same channel using the logical operators AND (&&) and OR (| |).
In this example the condition is met only if A==1 && B==2.

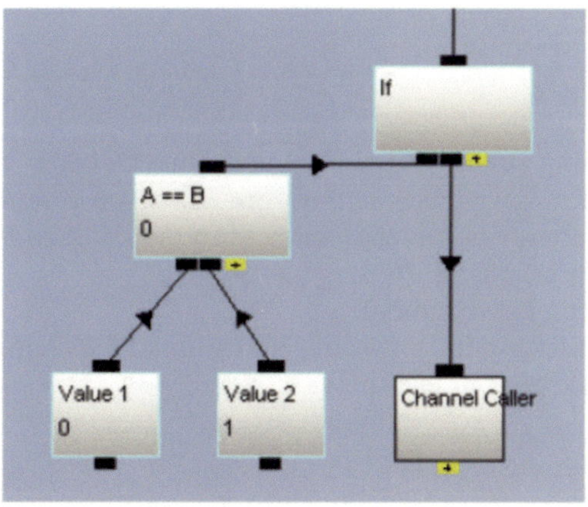

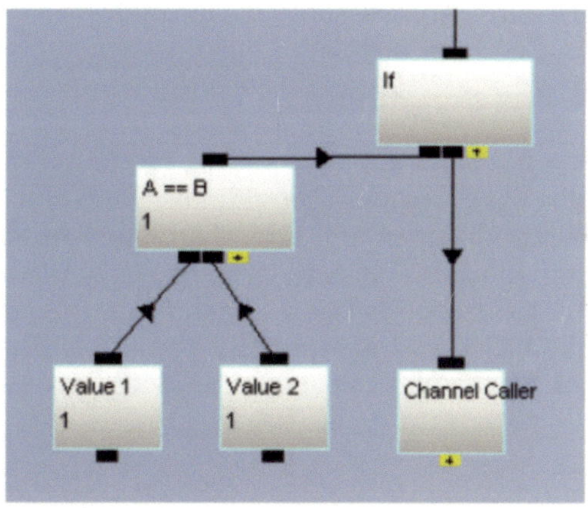

In this other case, the expression A==B ? C:D is based on identity between A and B.

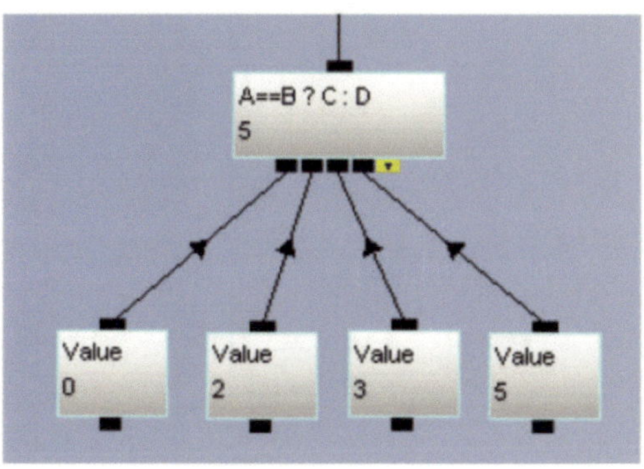

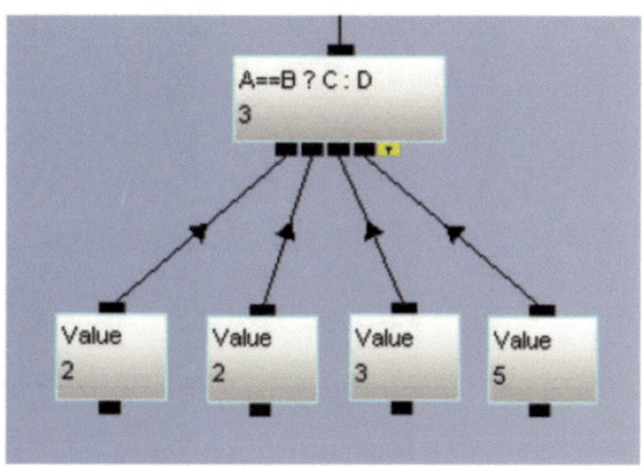

Mathematical functions are realized throught the *Envelopes*, main instruments for managing animations.
For advanced simulations and models, actual formula's may be used by inserting a number of coordinate pairs into the graph. Quest3D will handle the interpolation between these points.

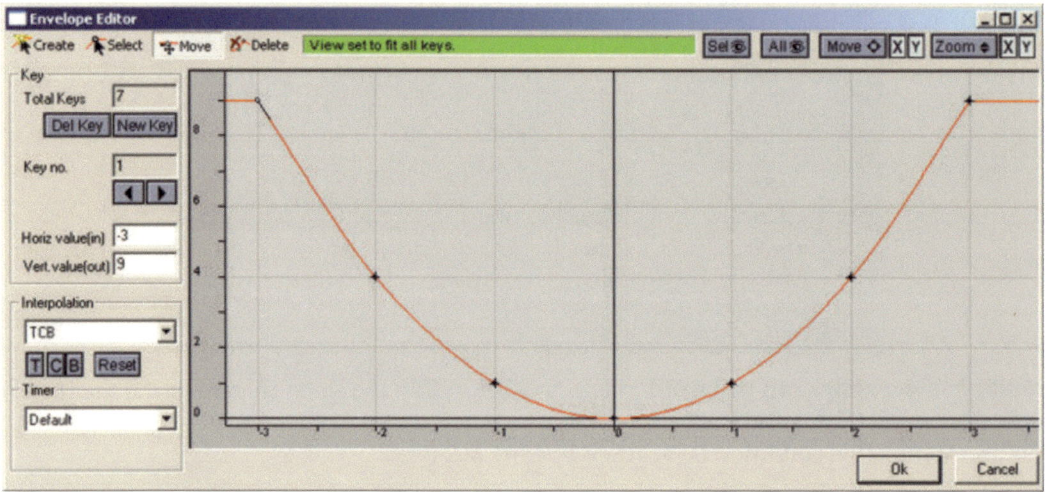

This *Envelope graph* show the formula $y = x^2$

14. Lua Scripting

Quest3D channels are building blocks containing precompiled C++ and DirectX code. The entire set of channels provides functionality for many aspects of real-time 3D development.
Despite the open and logical structure of the channels system, certain functionality can be handled more efficiently or more easily in script language.
Quest3D supports *Lua*, a free third party scripting environment.
Lua in Quest3D is especially applicable for cases such as loading and unloading channel groups, complex calculations and iterative structures (*for loops*).

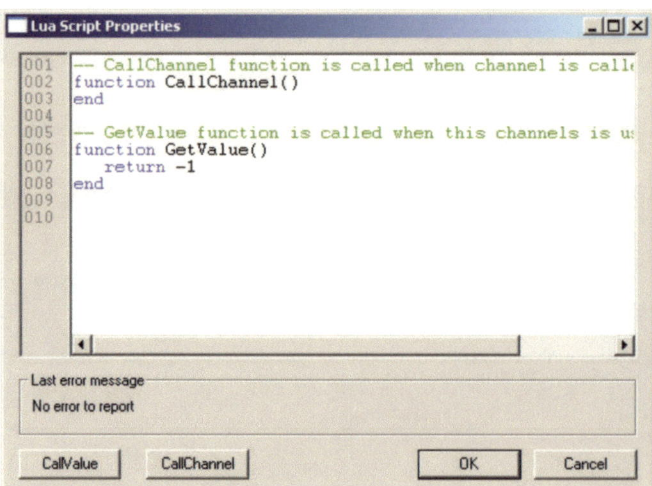

Lua channels can have one of two functions, or both:

- The first is *CallChannel*, and is executed when the channel is simply called;
- The second is *GetValue*, and is executed when the *Lua* channel is used as a value. The actual value of the *Lua* channel itself is equal to the value the script 'returns' at the end of the function.

Just like in other programming and scripting languages, variables used in a *Lua* script must be *declared*. This statement assigns memory to the variable. The word *local* means that the variable is only used in one particular place, in this case the *Lua* channel. As soon as the script is done, the memory is cleared again.

```
local variable
local variable = 1
variable = variable + 1
```

Quest3D supports the following standard Lua group functions: *base*, *string* and *math*.

```
local variable = math.cos(value)
```

The first example calculates the cosine of the value between the brackets.

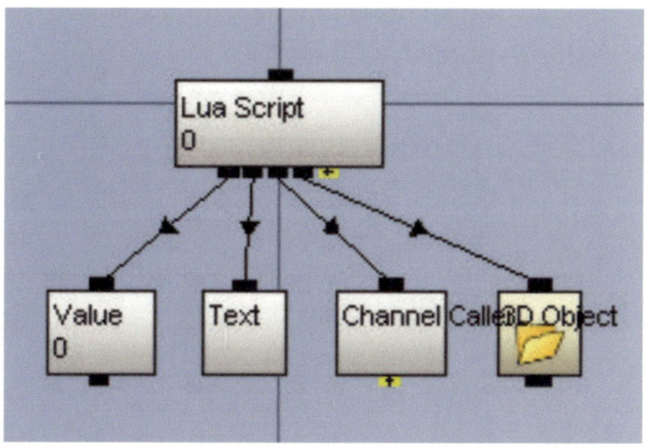

Any number of children of any type may be connected to a *Lua* channel.
Children may be accessed in a *Lua* script using the appropriate function.
As usual, they have to be declared first. The statement below accesses the first child connected to the *Lua* channel (at position '0').

```
local variable = channel.GetChild(0)
```

The actual value of a *Value* channel connected to a *Lua* channel may be accessed using the following commands.

```
local variable = channel.GetChild(0)
local value = variable:GetValue()
```

Text can also be retrieved, using the following statements:

```
local variable = channel.GetChild(0)
local text = variable:GetText()
```

The value or text of a child may be set using the following commands:

```
variable:SetValue(value)
variable:SetText(text)
```

Note the difference between using functions of predefined and local structures. For predefined structures such as 'channel', a function is preceded by a '.' (dot). For local structures such as variables, a function is preceded by a ':' (double colon punctuation mark).
Functions to load or unload *channel group* are very usefull.
For the *LUA group q*, the following instructions realize these statments:

```
q.LoadChannelGroup("group.cgr", "PoolName", instance)
q.RemoveChannelGroup("PoolName", instance)
```

15. Product publications

At a certain point, a Quest3D project may be finished and ready for distribution. Whether it is a modest free screensaver or an extensive commercial training application, the Quest3D program must be exported out of the editor and saved into a format the target audience can run.
Different options are available to publish a Quest3D project:

Can publish to type
.q3d player file
Web application
Screensaver
WinAmp plug-in
Stand-alone self-running .exe
Installer file

Internet ActiveX: .q3d and .htm:
files of type *.q3d* can also be viewed in an internet browser that supports ActiveX. The *.q3d* file must be linked from a regular webpage to show. Web pages containing Quest3D media elements will automatically try to download and install the Quest3D Viewer Package from the Quest3D.com website (www.quest3d.com).
Stand-alone programs: .exe:
a Quest3D project can also be published to a stand-alone self-executing *.exe* file. Besides DirectX, this type of publication requires no additional components.
Installation files: .exe:
installers are *.exe* archive files that copy and expand a program to a specified directory on a harddisk.

16. References

www.quest3d.com

14

Outlier Detection Methods for Industrial Applications

Silvia Cateni, Valentina Colla and Marco Vannucci
Scuola Superiore Sant Anna, Pisa
Italy

1. Introduction

An outlier is an observation (or measurement) that is different with respect to the other values contained in a given dataset. Outliers can be due to several causes. The measurement can be incorrectly observed, recorded or entered into the process computer, the observed datum can come from a different population with respect to the normal situation and thus is correctly measured but represents a rare event. In literature different definitions of outlier exist: the most commonly referred are reported in the following:

- "An outlier is an observation that deviates so much from other observations as to arouse suspicions that is was generated by a different mechanism " (Hawkins, 1980).
- "An outlier is an observation (or subset of observations) which appear to be inconsistent with the remainder of the dataset" (Barnet & Lewis, 1994).
- "An outlier is an observation that lies outside the overall pattern of a distribution" (Moore and McCabe, 1999).
- "Outliers are those data records that do not follow any pattern in an application" (Chen and al., 2002).
- "An outlier in a set of data is an observation or a point that is considerably dissimilar or inconsistent with the remainder of the data" (Ramasmawy at al., 2000).

Many data mining algorithms try to minimize the influence of outliers for instance on a final model to develop, or to eliminate them in the data pre-processing phase. However, a data miner should be careful when automatically detecting and eliminating outliers because, if the data are correct, their elimination can cause the loss of important hidden information (Kantardzic, 2003). Some data mining applications are focused on outlier detection and they are the essential result of a data-analysis (Sane & Ghatol, 2006).

The outlier detection techniques find applications in credit card fraud, network robustness analysis, network intrusion detection, financial applications and marketing (Han & Kamber, 2001). A more exhaustive list of applications that exploit outlier detection is provided below (Hodge, 2004):

- Fraud detection: fraudulent applications for credit cards, state benefits or fraudulent usage of credit cards or mobile phones.
- Loan application processing: fraudulent applications or potentially problematical customers.
- Intrusion detection, such as unauthorized access in computer networks.

- Activity monitoring: for instance the detection of mobile phone fraud by monitoring phone activity or suspicious trades in the equity markets.
- Network performance: monitoring of the performance of computer networks, for example to detect network bottlenecks.
- Fault diagnosis: processes monitoring to detect faults for instance in motors, generators, pipelines.
- Structural defect detection, such as monitoring of manufacturing lines to detect faulty production runs.
- Satellite image analysis: identification of novel features or misclassified features.
- Detecting novelties in images (for robot neotaxis or surveillance systems).
- Motion segmentation: such as detection of the features of moving images independently on the background.
- Time-series monitoring: monitoring of safety critical applications such as drilling or high-speed milling.
- Medical condition monitoring (such as heart rate monitors).
- Pharmaceutical research (identifying novel molecular structures).
- Detecting novelty in text. To detect the onset of news stories, for topic detection and tracking or for traders to pinpoint equity, commodities.
- Detecting unexpected entries in databases (in data mining application, to the aim of detecting errors, frauds or valid but unexpected entries).
- Detecting mislabeled data in a training data set.

How the outlier detection system deals with the outlier depends on the application area.
A system should use a classification algorithm that is robust to outliers to model data with naturally occurring outlier points. In any case the system must detect outlier in real time and alert the system administrator. Once the situation has been handled, the anomalous reading may be separately stored for comparison with any new case but would probably not be stored with the main system data as these techniques tend to model normality and use outliers to detect anomalies (Hodge, 2004).

2. Traditional approaches

The salient traditional approaches to outlier detection can be classified as either distribution-based, depth-based, clustering, distance-based or density-based.

2.1 Distribution-based method

These methods are typically found in statistics textbooks. They deploy some standard distribution model (Normal, Poisson, etc.) and flag as outliers those data which deviate from the model. However, most distribution models typically apply directly to the future space and are univariate i.e. have very few degrees of freedom. Thus, they are unsuitable even for moderately high-dimensional data sets. Furthermore, for arbitrary data sets without any prior knowledge of the distribution of points, expensive tests are required to determine which model best fits the data, if any! (Papadimitriou et al., 2002) Fitting the data with standard distributions is costly and may not produce satisfactory results. The most popular distribution is the Gaussian function. (Kim & Cho, 2006).
A method was proposed by Grubbs which calculates a Z value as the difference between the mean value for the attribute and the query value divided by the standard deviation for the

attribute, where the mean and the standard deviation are calculated from all attribute values including the query value. The Z value for the query is compared with a 1% or 5% significance level. The technique requires no pre-defined parameters as all parameters are directly derived from the data. However, the success of this approach heavily depends on the number of exemplars in the data set. The higher the number of records, the more statistically representative the sample is likely to be (Grubbs, 1969).
A Gaussian mixture model (GMM) and computed outlierness was proposed based on how much a data point deviates from the model (Roberts & Tarassenko, 1995).
The GMM is represented by equation (1):

$$P(t|x)=\sum_{j=1}^{M}\alpha_j(x)j_j(t|x) \tag{1}$$

where M is the number of kernels (ϕ), $\alpha_j(x)$ the mixing coefficients, **x** the input vector and **t** the target vector.
A Gaussian probability density function is defined by equation (2):

$$\phi_j(t \mid x) = \frac{1}{(2\pi)^{\frac{d}{2}} \sigma_j^d(x)} e^{\left\{-\frac{\|t-\mu_j(x)\|^2}{2\sigma_j^2(x)}\right\}} \tag{2}$$

where d is the dimensionality of input space, σ is the smoothing parameter, $\mu_j(x)$ represents the centre of jth kernel and $\sigma_j^2(x)$ is the variance.
Another model was proposed by Laurikkala et al.(Laurikkala et al., 2000) in probably one of the simplest statistical outlier detection techniques. This method uses informal box plots to pinpoint outliers in both univariate and multivariate datasets, produces a graphical representation and allows a human auditor to visually pinpoint the outlying points. This approach can handle real-valued, ordinal and categorical attributes.
Distribution based methods have some advantages such as mathematical justification and fast evaluation once they are built. However, they also have important drawbacks, such as the need for assuming a distribution and a considerable complexity for high dimensional problems.
Indeed statistical models are generally suited to quantitative real-valued data sets at the very least quantitative ordinal data distributions, where the ordinal data can be transformed into suitable numerical values through statistical processing. This fact limits their applicability and increases the processing time if complex data transformations are necessary before processing (Hodge, 2004).

2.2 Depth-based method

This approach is based on computational geometry and computes different layers of k-d convex hulls (Johnson et al., 1998). Based on some definition of depth, data objects are organized in convex hull layers in data space according to peeling depth and outliers are expected to be found from data objects with shallow depth values. In theory, depth-based methods could work in high dimensional data space. However, due to relying on the computational of k-d convex hulls, these techniques have a lower bound complexity of

$\Omega(N^{k/2})$, where N is number of data objects and k is the dimensionality of the dataset. This makes these techniques infeasible for large dataset with high dimensionality (He et al., 2002).

2.3 Clustering

Clustering is a basic method to detect potential outliers. From the viewpoint of a clustering algorithm, potential outliers are data which are not located in any cluster. Furthermore, if a cluster significantly differs from other clusters, the objects in this cluster might be outliers. A clustering algorithm should satisfy three important requirements (Birant & Kut, 2006):

- Discovery of clusters with arbitrary shape;
- Good efficiency on large databases
- Some heuristics to determine the input parameters.

Fuzzy c-means (FCM) is a method of clustering which allows one piece of data to belong to two or more clusters. This method (developed by Dunn in 1973 and improved by Bezdek in 1981) is based on minimization of the following objective function:

$$J_m = \sum_{i=1}^{N}\sum_{j=1}^{C} u_{ij}^m \| x_i - c_j \|^2 \tag{3}$$

where C is the total number of clusters, N is the total number of data, m is any real number greater than 1, u_{ij} is the degree of membership of x_i to the j-th cluster, x_i is the i-th of the d-dimensional measured data, c_j is the d-dimensional center of the cluster, and $\|*\|$ is any norm expressing the similarity between any measured data and the center.

Fuzzy partitioning is carried out through an iterative optimization of the above cited objective function, with the update of membership degree u_{ij} and the cluster centers c_j respectively given by:

$$u_{ij} = \frac{1}{\sum_{k=1}^{C} \left(\frac{\| x_i - c_j \|}{\| x_i - c_k \|} \right)^{\frac{2}{m-1}}} \tag{4}$$

$$c_j = \frac{\sum_{i=1}^{N} u_{ij}^m x_i}{\sum_{i=1}^{N} u_{ij}^m} \tag{5}$$

The iteration stops when the following condition holds:

$$\max_{ij} \left\{ | u_{ij}^{(k+1)} - u_{ij}^{(k)} | \right\} < \varepsilon \tag{6}$$

where ε is a termination criterion lying in the range [0, 1] whereas k is the iteration step. This procedure converges to a local minimum or a saddle point of J_m.

A quite popular technique used for outlier detection is the introduction of a new variable called *credibility* of a vector. This method is proposed by Chintalapudi and Kam (Chintalapudi & Kam, 1998) with regard to the above described fuzzy clustering algorithms. Credibility measures the "how typical" the vector is with respect to the entire data set. An

outlier is expected to have a low value of credibility compared to a non-outlier. The use of the new variable leads to the Credibilistic Fuzzy C Means algorithm. Two formulations are made for the credibility variable, based on the mean nearest statistical distance. Simulations demonstrate that the proposed schemes are robust with respect to the presence of outliers and with respect to the parameter values needed to implement the algorithm.
Most clustering algorithms, especially those developed in the context of Knowledge Discovery in Databases (KDD) (e.g. CLARANS (Ng&Han, 1994), DBSCAN (Ester et al., 1998), BIRCH (Zhang et al., 1996), STING (Wang et al., 1997), Wave Cluster (Sheikholeslami et al., 1998), DenClue (Hinneburg&Keim, 1998), CLIQUE (Aggarwal et al., 1998) , OPTICS (Ankerst et al.,1999), PROCLUS (Aggarwal et al., 1999)) are to some extent capable of handling exceptions.
However, since the main objective of a clustering algorithm is to find clusters, they are developed to optimize the outlier detection. The exceptions (called "noise" in the context of clustering) are typically just tolerated or ignored when producing the clustering result. Even if the outliers are not ignored, the notions of outliers are essentially binary and there are no qualification as to how outlying an object is (Breunig et al., 2000).

2.4 Distance-based method

The notion of distance-based (DB) outlier is been defined by Knorr and Ng (1988):
An object O in a dataset T is a DB(p,D)-outlier if at least fraction p of the objects in T lie greater than distance D from O.
The concept of DB-outlier is well defined for any dimensional dataset. The parameter p is the minimum fraction of objects in a data space that must be outside an outlier D-neighborhood (Li & Hiroyuki, 2007).
This notion generalizes many concepts from distribution-based approach and better faces computational complexity. It is further extended based on the distance of a point from its k-th nearest neighbor (Ramasmamy et al., 2000).
After ranking points by the distance to its k-th nearest neighbor, the top k points are identified as outliers. Alternatively, in the algorithm proposed by Angiulli and Pizzuti (Angiulli & Pizzuti,2000), the outlier factor of each data point is computed as the sum of distances from its k nearest neighbors.
Both the method proposed in (Matsumoto et al., 2007) and the Mahalanobis outlier analysis (MOA) (Marquez et al., 2002) are distance-based approaches which exploit Mahalanobis distance as outlying degree of each data point.
In 1936 P.C. Mahalanobis introduced a distance measure (Mahalanobis, 1936) which is based on correlations between variables by which different patterns can be identified and analyzed and provides a useful way of determining similarity of an unknown sample set to a known one. It differs from Euclidean distance in that it takes into account the correlations of the data set and is scale-invariant, i.e. not dependent on the scale of measurements.
Formally, the Mahalanobis distance from a group of values with mean $\mu=(\mu_1, \mu_2, \mu_3, \ldots, \mu_p)^T$ and covariance matrix Σ for a multivariate vector $x=(x_1, x_2, x_3, \ldots, x_p)^T$ is defined as:

$$D_M = \sqrt{(x-\mu)^T \Sigma^{-1} (x-\mu)} \quad (7)$$

Mahalanobis distance can also be defined as dissimilarity measure between two random vectors **x** and **y** of the same distribution with the covariance matrix P:

$$d(x,y) = \sqrt{(x-y)^T P^{-1}(x-y)} \tag{8}$$

If the covariance matrix is the identity matrix, the Mahalanobis distance reduces to the Euclidean distance. Finally if the covariance matrix is diagonal, then the resulting distance measure is called the normalized Euclidean distance:

$$d(x,y) = \sqrt{\sum_{i=1}^{P} \frac{(x_i - y_i)^2}{\sigma_i^2}} \tag{9}$$

where σ_i is the standard deviation of the x_i over the sample set.

Mahalanobis distance is computed on the basis of the variance of data points. It describes the distance between each data point and the center of mass.

When one data point is on the center of mass, its Mahalanobis distance is zero, and when one data point is distant from the center of mass, its Mahalanobis distance is more than zero. Therefore, datapoints that are located far away from the center of mass are considered outliers (Matsumoto et al., 2007).

2.5 Density-based method

This method assigns to each object a degree to be an outlier. This degree is called the local outlier factor (LOF) of an object. It is "local" in the sense that the degree depends on how isolated the object is with respect to the surrounding neighborhood.

In LOF algorithm, outliers are data objects with high LOF values whereas data objects with low LOF values are likely to be normal with respect to their neighborhood. High LOF is an indication of low-density neighborhood and hence high potential of being outlier (Mansur & Noor, 2005). In order to understand LOF method, it is necessary to define some auxiliary notions. (Breunig et al., 2000).

Definition 1: (k-distance of an object p).

For any positive integer k, the k-distance of object p, denoted as k-distance (p), is defined as the distance d(p,o) between p and an object o∈D such that:

i. For at least k objects o'∈ D\ {p} it holds that d(p,o')≤d(p,o) and
ii. For at most k-1 objects o'∈ D\ {p} it holds that d(p,o')<d(p,o).

Definition 2: (k-distance neighborhood of an object p).

Given the k-distance of p, the k-distance neighborhood of p contains every object whose distance from p is not greater than the k-distance, i.e.

$$N_{k-distance(p)}(p) = N_k(p) = \{q \in D \setminus \{p\} \mid d(p,q) \le k-distance(p)\} \tag{10}$$

These objects q are called the k-nearest neighbors of p.

Definition 3: (reachability distance of an object p w.r.t. object o).

Let k be a natural number. The reachability distance of object p with respect to object o is defined as

$$reach-dist_k = \max\{k-distance(o), d(p,o)\} \tag{11}$$

The higher the value of k, the more similar the reachability distances for objects within the same neighborhood.

In a typical density-based clustering algorithm, there are two parameters that define the notion of density: a parameter *MinPts* specifying a minimum number of objects and a parameter specifying a volume. These two parameters determine a density threshold for the clustering algorithms to operate. That is, objects or regions are connected if their neighborhood densities exceed the given density threshold. To detect density-based outliers, however, it is necessary to compare the densities of different sets of objects, which means that the density of sets of objects must be dynamically determined. An idea is to keep MinPts as the only parameter and use the values reach-dist$_{MinPts}$(p,o), for o $\in N_{MinPts}$(p).

Definition 4: (local reachability density of an object p).

The local reachability density of p is defined as:

$$lrd_{MinPts}(p) = \frac{1}{\dfrac{\sum_{o \in N_{MinPts}(p)} reach-dist_{MinPts}(p,o)}{|N_{MinPts}(p)|}} \tag{12}$$

Intuitively, the local reachability density of an object *p* is the inverse of the average reachability distance based on the MinPts-nearest neighbors of *p*.

Definition 5: (local outlier factor of an object p)

The local outlier factor of p is defined as:

$$LOF_{MinPts}(p) = \frac{\sum_{o \in N_{MinPts}(p)} \dfrac{lrd_{MinPts}(o)}{lrd_{MinPts}(p)}}{|N_{MinPts}(p)|} \tag{13}$$

The outlier factor of object p captures the degree to which we call p an outlier. It is the average of the ratio of the local reachability density of p and those of p's MinPts-nearest neighbors.

It is possible to know the range where LOF lies. By defining ε as

$$\varepsilon = \frac{\max\{reach-dist(p,q)\}}{\min\{reach-dist(p,q)\}-1} \tag{14}$$

it can be demonstrated that (Breunig et al., 2000):

$$\frac{1}{1+\varepsilon} \leq LOF(p) \leq 1+\varepsilon \tag{15}$$

3. Wavelets-based outlier detection approach

The *wavelet transform* or *wavelet analysis* is probably one of the most recent solutions to overcome the shortcomings of the Fourier transform. In wavelet analysis the use of a fully scalable modulated window solves the signal-cutting problem. The window is shifted along the signal and for every position the spectrum is calculated. This process is repeated many times with a slightly shorter (or longer) window for every new cycle. The processing result is a collection of time-frequency representations of the signal corresponding to different resolutions.

The wavelet transform is an operation that transforms a function by integrating it with modified versions of some kernel function. (Combes et al., 1989). The kernel function is called the *mother wavelet*, and the modifications are translations and compressions of the mother wavelet. A function *g(t)* can be a mother wavelet if it is *admissible*, i.e. if the following condition holds (Grossmannn & Morlet, 1984):

$$c_g \equiv \int_{-\infty}^{+\infty} \frac{|G(\omega)|^2}{|\omega|} d\omega < \infty \tag{16}$$

where $G(\omega)$ is the Fourier transform of $g(t)$. The constant c_g represent the admissibility of the function that must be finite for inversion of the wavelet transform. Any admissible function can be a mother wavelet (Weiss, 1994).

Wavelet transforms are classified into discrete wavelet transforms and continuous wavelet transforms on the basis of the applications and data.

Outlier detection by means of the wavelet transform is a recent study area. (Bruce et al., 1994) (Wang, 1995) (Dorst, 1999) and can be an alternative to the previously discussed methods because the data do not need to belong to a known distribution and the discontinuities in the processed data, that can often correspond to outliers, are easily detected. (Kern et al., 2005).

For instance Yu et al. (Yu et al., 2002) introduced a new outlier detection approach, called FindOut (which stands for *Find Outliers),* based on wavelet transform. FindOut identifies outliers by removing clusters from the original data. The main idea is to apply signal processing techniques to transform the space and find the dense regions in the transformed space. The remaining objects in the non-dense regions are labeled as outliers. The primary motivation for applying signal-processing techniques to spatial datasets comes from the observation that the multidimensional data points can be represented as a d-dimensional signal. Wavelet transform is a signal processing technique that decomposes a signal into different frequency sub-bands; this method use two filters: a high-pass filter and a low-pass filter. One important step of wavelet transform is to convolve its input with a low-pass filter that has the main property of removing noise (outlier). The low-pass filter removes the outliers and smoothes the input data. However in FindOut the objective is to find outliers and not remove them so the idea is to remove the clusters from the original data and thus identify the outliers.

4. Outlier detection using artificial intelligence techniques

When dealing with industrial automation, where data coming from the production field are collected with different and heterogeneous means, the occurrence of outliers is more the rule than the exception. Standard outlier detection methods fail to detect outliers in industrial data because of the high dimensionality of the data. In these cases, the use of artificial intelligence techniques has received increasing attention in the scientific and industrial community, as the application of these techniques shows the advantage of requiring poor or no *a priori* theoretical assumption on the considered data. Moreover their implementation is relatively simple and with no apparent limitation on the dimensionality of the data.

4.1 Neural networks

In 1943, McCulloch and Pitts introduced the idea of an artificial neuron to process data. In the 50ies this work was advanced by arranging neurons in layers. Although learning rules to cope with multiple layers of perceptrons were not developed until later, this work formed the basis of the Multi-Layer Perceptron (MLP) that is used today.
Another kind of neural network that is frequently used is the Radial Basis Function (RBF), which exploits gaussian activation functions in the first (or sometimes called hidden) layer.
Once the inputs and the outputs have been defined, it is useful to see if the data set contains any points that violate this limits. If there are many similar examples for a given input pattern, an outlier can be classified as the one which is furthest from the median value.
Other methods that can be applied to detect outliers are the Principle Component Analysis (PCA) and Partial Least Squares (PLS). Outliers can be found by investigating points at the edges of the previously created clusters. (Tenner et al., 1999).
Liu and Gader indicated that including outlier samples in training data and using more hidden nodes than required for classification for MLP and BRF networks and proceeding an RBF with principal Component decomposition can achieve outlier rejection. The further addition of a regularization term to the PCA-RBF can achieve an outlier rejection performance equivalent or better than that of other networks without training on outliers (Liu & Gader, 2000).
Williams et al. propose to apply the so-called Replicator Neural Network (RNN) for outlier detection. RNN are multi-layer perceptron neural networks with three hidden layers and the same number of output neurons and input neurons to model the data. The input variables are also the output variables so that the RNN forms compressed model of data during training. A measure of outlyingness of individuals is developed as the reconstruction error of individual data points. (Hawkins et al., 2002). This method is often compared with other methods, in particle with Hadi94 (Hadi, 1994) and Donoho-Stahel (Knorr et al., 2001).
The Hadi94 is a parametric bulk outlier detection method for multivariate data and Donoho-Stahel uses the outlyingness measure compute by the Donoho-Stahel (Rousseeuw & Leroy, 1997) estimator of location and scatter. These two methods perform well on large and complex data sets. However these are parametric methods and lack of the flexibility of non-parametric methods, such as Minimum Message Length (MLL) (Oliver et al., 1996) and RNN. MLL clustering works well for scattered outlier while RNN degrades with datasets containing radial outliers. However RNN performs satisfactory for small and large datasets (Neural Network methods often have difficulty with such smaller datasets) (Williams et al., 2002).
The self-organizing map (SOM) is an artificial neural networks, which is trained by using unsupervised learning in order to produce a low dimensional representation of the training samples while preserving the topological properties of the input space. (Munoz & Muruzabal, 1997).
Nag et al. (Nag et al., 2005) proposed a SOM based method for outlier detection, which identifies the multidimensional outliers and provides information about the entire outlier neighborhood. The SOM based outlier detection method is non-parametric and can be used to detect outliers from large multidimensional datasets. This method has the advantages that does not require any a priori assumption on the variable, is easy to implement and does not have problems with dimensionality of data.

4.2 Support Vector Machine (SVM)

Support Vector Machine, introduced by V. Vapnik (Vapnik, 1982), is a method for creating functions from a set of labelled training data. The function can be a classification function or a general regression function. In classification tasks, SVMs operate by finding a hypersurface in the space of the possible inputs, which separates the samples belonging to different classes. Such division is chosen so as to have the largest distance from the hypersurface to the nearest of the positive and negative examples (Vapnik, 1998).

Jordaan and Smits (Jordaan & Smits, 2004) propose a robust model-based outlier detection approach that exploits the characteristics of the support vectors extracted by the SVM method (Cortes & Vapnik, 1995). This method makes use of several models of varying complexity to detect outliers based on the characteristics of the support vectors obtained from SVM-models. This approach has the advantage that the decision does not depend on the quality of a single model, which adds to the robustness of the approach. Furthermore, since it is an iterative approach, the most severe outliers are firstly removed. This allows the models in the next iteration to learn from "cleaner" data and thus reveal outliers that were masked in the initial model. The need for several iterations as well as the use of models, however, makes the on-line application of this method difficult for the not negligible computational burden. Moreover, if the data to be on-line processed come from dynamic systems, which tend to change (more or less rapidly) their conditions through time, the models update is required and this can furtherly slow the outlier detection.

4.3 Fuzzy logic

Fuzzy Logic (FL) is linked with the theory of fuzzy sets, a theory which relates to classes of objects with un-sharp boundaries in which membership is a matter of degree.

Fuzzy theory is essential and is applicable to many systems — from consumer products like washing machines or refrigerators to big systems like trains or subways. Recently, fuzzy theory has been a strong tool for combining new theories (called soft computing) such as genetic algorithms or neural networks to get knowledge from real data (Melin & Castillo, 2008).

Fuzzy logic is conceptually easy to understand, tolerant of imprecise data and flexible. Moreover this method can model non-linear functions of arbitrary complexity and it is based on natural language. Natural language has been shaped by thousands of years of human history to be convenient and efficient. Since fuzzy logic is built atop the structures of qualitative description used in everyday language, fuzzy logic is easy to use (Baldwin, 1978).

Fuzzy inference system (FIS) (Ross, 2004) is the process of formulating the mapping from a given input to an output using fuzzy logic. The mapping then provides a basis from which decisions can be made, or pattern discerned. The process of fuzzy inference involves: membership functions (MF), a curve that defines how each point in the input space is mapped to a membership value or degree of membership between 0 and 1; fuzzy logic operators (and, or, not); if-then rules. Since decisions are based on the testing of all of the rules in an FIS, the rules must be combined in some manner in order to make decision. Aggregation is the process by which the fuzzy sets that represents the outputs of each rule are combined into a single fuzzy set. Aggregation only occurs once for each output variable, just prior to the final step, defuzzification (Patyra & Mlynek, 1996).

Due to the linguistic formulation of its rule basis, the FIS provides an optimal tool to combine more criteria among those that were above illustrated according to a reasoning that is very similar to the human one. So doing, in practical application, the knowledge of the technical expert personnel can easily be exploited by the system designer.

In the following, an exemplar method will be proposed which exploits a fuzzy inference system (FIS) in order to combine several outliers detection strategies by jointly evaluating four features of a particular datum within a series of measurements (Cateni et al., 2007).

For each pattern the following four features are extracted and fed as input of the FIS:

- Distance between each element and the centroid of the overall distribution normalized with respect to the average value. (dist)
- Fraction of the total number of elements that are near to the pattern itself. (n-points)
- Mean distance between the considered pattern and the remaining patterns normalized with respect to the maximum value. (memb-deg)
- Degree of membership of the patterns to the cluster to which it has been assigned by the preliminary fuzzy c-means clustering stage. (mean-dist)

In geometry the *centroid* of an object X in n-dimensional space is the intersection of all hyperplanes that divide X into two parts of equal moment about the hyperplane. Roughly speaking, the centroid is a sort of "average" of all points of X.

The FIS is of Mandami type (Mandami & Assilian, 1975) and the FIS output variable, named *outlier-index* (outindx), is defined in the range [0;1]. The output function provides an indication on the risk that the considered pattern is an outlier.

The inference rules relating the output variable to the four inputs is formulated through a set of 6 fuzzy rules, that are listed below:

1. IF (dist is very high) AND (n-points is very small) AND (memb-deg is low) AND (mean–dist is big) THEN (outindx is very high).
2. IF (dist is medium) AND (n-points is small) AND (memb-deg is quite low) AND (mean–dist is small) THEN (outindx is quite high).
3. IF (dist is low) AND (n-points is medium) AND (memb-deg is quite low) AND (mean-dist is very small) THEN (outindx is low).
4. IF (dist is medium) AND (n-points is very small) AND (memb-deg is quite low) AND (mean–dist is small) THEN (outindx is quite high).
5. IF (dist is low) AND (n-points is small) AND (memb-deg is high) AND (mean–dist is quite big) THEN (outindx is low).
6. IF (dist is low) AND (n-points is medium) AND (memb-deg is high) AND (mean–dist is small) THEN (outindx is low).

Figure 1 depicts a scheme of the proposed method. An outlier is detected when its outlier index overcome a prefixed threshold.

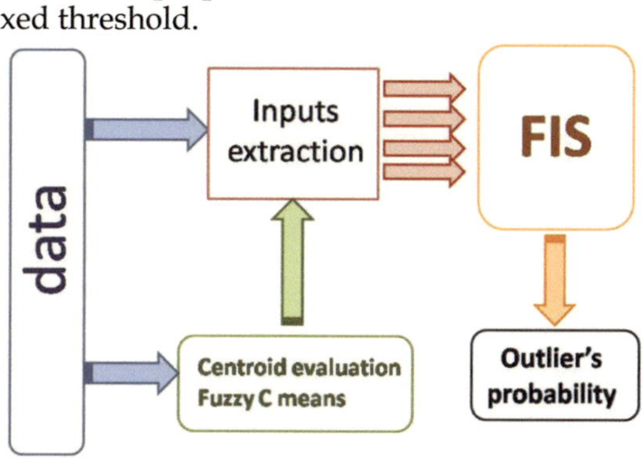

Fig. 1. Block diagram depicting the overall system for outliers detection.

5. Results

The proposed outlier detection method, which does not require a priori assumption on the data, has been tested in the preprocessing of data provided by a steelmaking industry, where outliers can provide indications on malfunctionings or anomalous process conditions. The method has been tested by considering that the technical personnel provided indications of those values that should be considered as outliers.
In this work two applications are proposed that use two different variables that are important to determine the quality of steel and the final destination. The variables are composed by 100 samples normalized respect their mean value.
The performance of the fuzzy logic-based method has been compared with Grubbs test and Local Outlier Factor (LOF) techniques, that are considered among the most important and widely adopted traditional outlier detection methods. The results show that the fuzzy logic-based method outperforms the other approaches, but, on the other hand, the required computational time is approximately ten times greater than the time required by traditional methods, due to the increased complexity of the FIS-based evaluation.
In particular, in the first exemplar application, the considered variable represents the concentration of a chemical element extracted from the analysis made on molten steel. In this piece of database there are five outliers. The samples that are considered outliers are 3, 8, 16, 35 and 92 referring respectively to following values:

0.9349 0.9385 1.0455 0.9626 0.9541.

Figure 2 shows the result of Grubbs test. It clearly appears that only two anomalous samples have been recognized as outliers (the 3rd and 8th samples).
Figure 3 shows the result of LOF technique: only three samples are exactly classified as outliers but there are two samples that this method does not recognize as outliers.
Finally, in figure 4, is shown the result using fuzzy logic method. It clearly appears that all outliers have been correctly recognized.

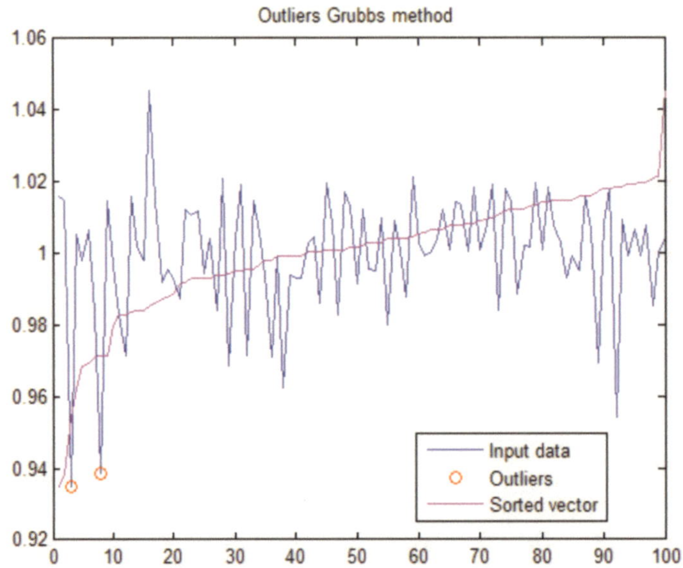

Fig. 2. First example using Grubbs method.

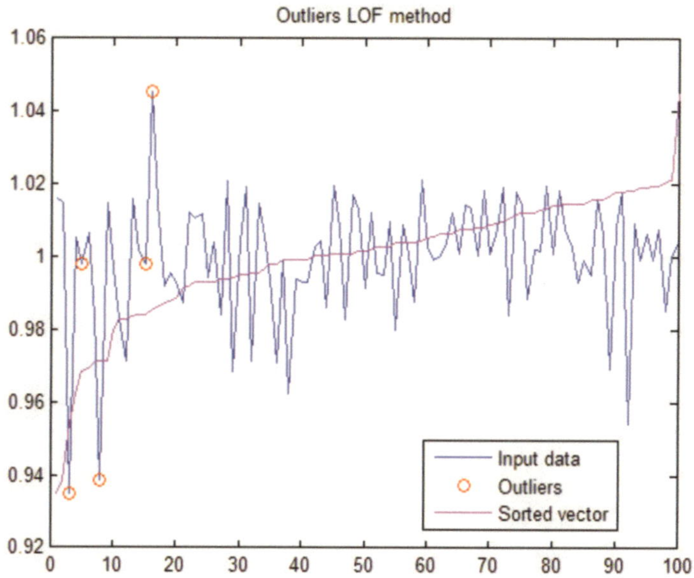

Fig. 3. First example using Local Outlier Factor method.

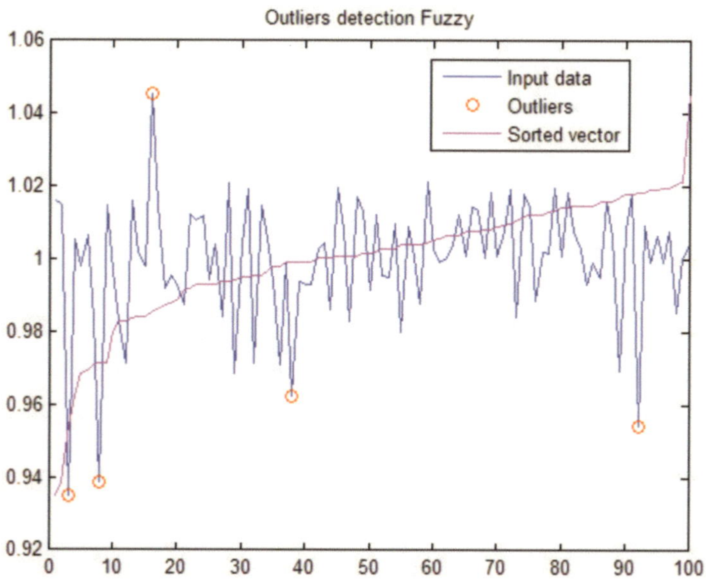

Fig. 4. First example using Fuzzy Logic method.

Similar results are obtained in the second application.

The second examined variable is referred to a sensor detection made within the process. In this case, the samples considered outliers are 3,5,7,40 and 60 referring respectively to following values:

0.4730 0.4779 0.4779 0.4680 1.9217.

The Grubbs test does not detect any outliers. The result is shown clearly in figure 5.

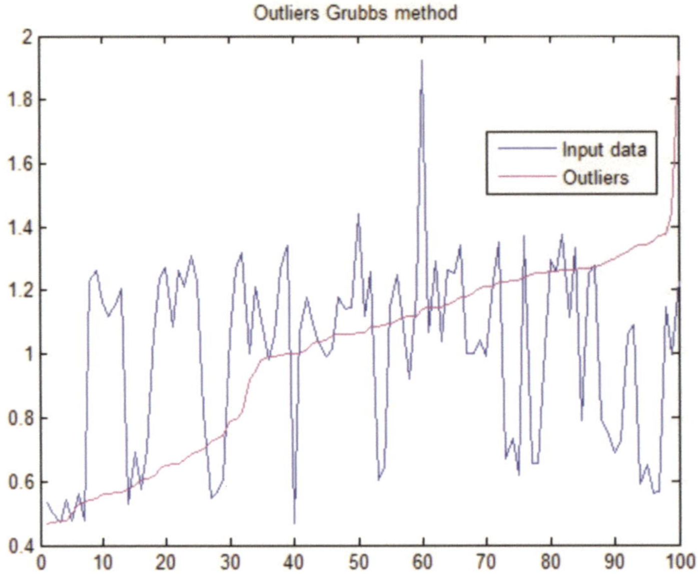

Fig. 5. Second example using Grubbs method.

The Local Outlier Factor technique detects four outliers but only one is really an outlier. The result is shown in figure 6.

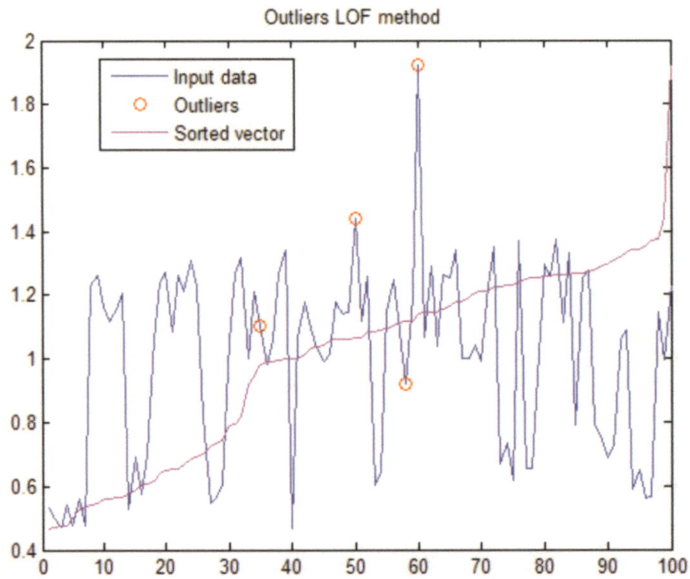

Fig. 6. Second example using Local Outlier Factor method.

The proposed method, as show figure 7, recognize all the outliers which are present in data.

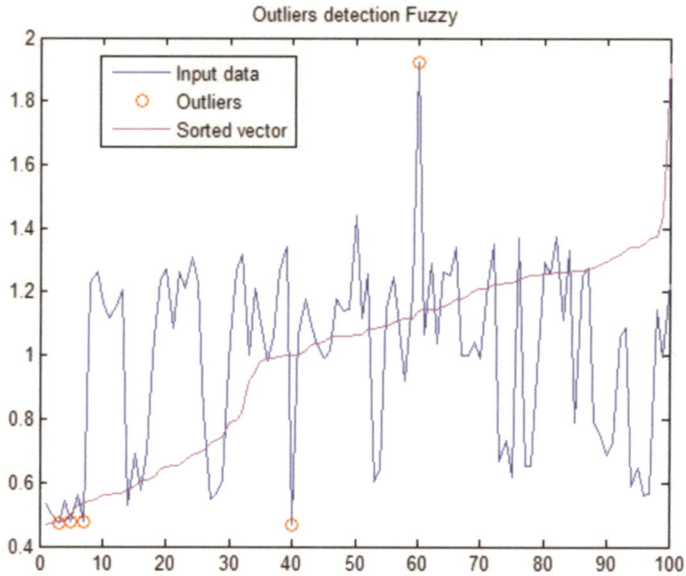

Fig. 7. Second example using Fuzzy Logic method.

6. Conclusions and future work

A description of traditional approaches and of the most widely used methods within each category has been provided. As standard outlier detection methods fail to detect outliers in industrial data, the use of artificial intelligence techniques has also been proposed, because it presents the advantage of requiring poor or no a priori assumption on the considered data.
A procedure for outlier detection in a database has been proposed which exploits a Fuzzy Inference System in order to evaluate four features for a pattern that characterize its location within the database. The system has been tested on a real industrial application, where outliers can provide indications on malfunctionings or anomalous process conditions. The presented results clearly demonstrate that the Fuzzy Logic-based method outperforms the most widely adopted the traditional methods.
Future work on the FIS-based outliers detection strategy will concern the algorithm optimization in order to improve its efficiency and its on-line implementation. Moreover further tests will be performed on different applications.

7. References

Aggarwal, C.C.; Procopiuc, C.; Wolf, J.L.; Yu, P.S. & Park, J.S. (1999). Fast algorithms for projected clustering, *Proceedings of ACM SIGMOD International Conference on Management of Data*, pp. 61–72, Philadephia, Pennsylvania, U.S.A.

Aggarwal, R.; Gehrke, J.; Gunopulos, D. & Raghavan, P. (1998). Automatic Subspace Clustering of High Dimensional Data for Data Mining Applications, *Proceedings of*

ACM SIGMOD International Conference on Management of Data, pp. 94-105, Seattle, WA.

Ankerst,M.; Breunig, M.M.; Kriegel, H.P. & Sander, J. (1999). OPTICS: Ordering points to identify the clustering structure, *Proceedings of the 1999 ACM SIGMOD International Conference on Management of Data*, pp. 49–60, June 1999, Philadelphia, Pennsylvania, U.S.A.

Baldwin, J.F. (1978). Fuzzy Logic and Fuzzy Reasoning. *International Journal of Man-Machine Studies*, Vol. 11, pp. 465-480.

Barnet, V. & Lewis, T. (1994), *Outliers in statistical data,* John Wiley, ISBN 0-471-93094-6, Chichester.

Birant, D.& Kut, A. (2006). Spatio-Temporal Detection in Large Databases, *Proceedings of the 28th International Conference Information Technology Interfaces ITI 2006* June 19-22, Croatia.

Breunig, M.M; Kriegel, H.P.; Ng, R.T. & Sander, J. (2000) LOF: Identifying density-based local outliers, *Proceedings of the 2000 ACM SIGMOD International Conference on Management of Data*, pp. 93-104, May 2000, Dallas.

Bruce, A.G.; Donoho L.G.; Gao, H.Y. & Martin R.D. (2004). Denoising and robust nonlinear wavelet analysis, *SPIE Proceedings Wavelet Applications,* Vol. 2242, pp. 335-336, Harald H.San (ed), The International Society for Optical Engineering (SPIE), Orlando, FL.

Cateni, S.; Colla, V. & Vannucci, M. (2007). A fuzzy logic-based method for outlier detection, *Proceedings of the 25th conference on Proceedings of the 25th IASTED International Multi-Conference: artificial intelligence and applications,* pp.561-566, Innsbruck, Austria.

Chen, Z.; Fu, A. & Tang, J., (2002). *Detection of outliered Patterns*, Dept. of CSE, Chinese University of Hong Kong.

Chintalapudi, K. & Kam,M. (1998), The credibilistic fuzzy c means clustering algorithm, *IEEE Fuzzy Systems Proceedings*, Vol. 2, pp. 2034-2039.

Combes, J.M.; Grossman A. & Tchamitchian P. (1989) *Wavelets: Time-Frequency Methods and Phase Space,* Second Edition, Springer-Verlag, New York.

Cortes, C. & Vapnik, V. (1995). Support vector networks, *Machine Learning*, Vol. 20, pp. 273–297.

Dorst, L. (1999) Current and addional procedures for superconducting gravimer data at the main tidal frequency. *Graduation Report,* Delft University of Technology, Delft.

Ester, M.; Kriegel, H.P., Sander, J. & Xu, X. (1996). A Density-Based Algorithm for Discovering Clusters in Large Spatial Databases with Noise, *Proceedings of the 2nd conference on Proceedings of the 25th IASTED International Conference on Knowledge Discovery and Data Mining*, pp. 226-231, Portland.

Grossmann, A. & Morlet, J. (1984) *Decomposition of Hardy Functions into Square Integrable Wavelets of Constant Shape,* SIAM J.Math. Anal. Vol. 15, N° 4, pp. 723-736.

Grubbs, F.E. (1969), Procedures for detecting outlying observations in samples, Technometrics 11, pp.1-21.

Hadi, A. (1994) A modification of a method for the detection of outliers in multivariate samples, *Journal of Royal Statistical Society B,* Vol. 56, N°2.

Han, J. & Kamber M. (2001) *Data Minings Concepts and Techniques,* Morgan Kauffman Publisdhers.

Hawkins, D. (1980), *Identification of Outliers,* Chapman and Hall, London.

Hawkins, S.; He, X.; Williams, G.J. & Baxter, R.A. (2002). Outlier detection using replicator neural networks. *Proceedings of the 5th international conference on Knowledge Discovery and Data Warehousing.*

He, Z.; Xu, X. & Deng, S. (2002). Discovering cluster-based local outliers. *Pattern Recognition Letters*, Vol. 24, N. 9-10, pp. 1641-1650, ISSN 0167-8655.

Hinneburg, A. & Keim D. A. (1998). An Efficient Approach to Clustering in Large Multimedia Databases with Noise, *Proceedings of the 4th international conference on Knowledge Discovery and Data Mining*, pp. 58-65, New York City, NY.

Hodge, V.J. (2004), *A survey of outlier detection methodologies*, Kluver Academic Publishers, Netherlands, January 2004.

Jiemenez-Marquez, S.A.; Lacroix, L. & Thibault, J. (2002) *Statistical data validation methods for large cheese plant database*. J.Dayry Sci., Vol.85, N°9, pp.2081-2097, Sep 2002.

Johnson, T.; Knok, I.; Ng, R. (1998). Fast computation of 2-dimensional depth contours, *Proceedings of 4thInternational Conference on Knowledge Discovery &Data Mining*, pp.224-228, New York, August 1998.

Jordaan, E.M. & Smits, G.F.(2004) Robust Outlier Detection using SVM Regression, *Proceeding. of the IEEE, International Joint Conference on Neural Networks*, Vol.3, pp. 2017-2022, July 2004.

Kantardzic, M. (2003). Data mining Concepts, Models, Methods and Algorithms. *IEEE Transactions on neural networks*, Vol.14, N. 2, March 2003.

Kern, M.; Preimesberger, T; Allesch, M.; Pail, R.; Bouman, J. & Koop, R. (2005). Outlier detection algorithms and their performance in GOCE gravity field processing, *Journal of Geodesy*, Vol. 78, pp. 509-519, January 2005.

Kim, S. & Cho, S. (2006). Prototype based outlier detection. *Proceedings of International JoinConference on Neural Networks*, ISBN 0-7803-9490-9, Vancouver, BC, Canada, 16-21 July 2006.

Knorr, E.M..; Ng, R. (1988). Algorithms for Mining Distance-Based Outliers in Large Datasets., *Proceedings of VLDB*, pp.392-403.

Knorr, E.M.; Ng, R.T. & Zamar, R.H. (2001) Robust Space Transformation for Distance-based Operations, *Proceeding of the 7th International Conference on Knowledge Discovery and Data Mining KD001*, pp. 126-135.

Laurikkala, J.; Juhola, M. & Kentala, E. (2000) Informal Identification of Outliers in Medical Data. *Proceeding in the Fifth International Workshop on Intelligent Data Analysis n Medicine and Pharmacology IDAMAP-2000 Berlin*, 22 August. *Organized as a workshop of the 14th European Conference on Artificial Intelligence* ECAI-2000.

Li, Y. & Hiroyuki, K. (2007). Example-Based DB-Outlier Detection from high Dimensional Datasets. *Proceedings of DEWS*.

Liu, J. & Gader, P. (2000). Outlier rejection with MLPs and variants of RBF Networks, *Proceedings of the 15th IEEE International Conference on Pattern Recognition*, Vol.2, pp. 680-683, 3-7 September, 2000, Missouri.

Mahalanobis, P.C. (1936). On the generalized distance in statistics. *Proceedings of the National Institute of Science of India*, pp. 49-55.

Mamdani, E.H. & Assilian, S. (1975) An experiment in linguistic synthesis with a fuzzy logic controller, *International Journal of Man-Machine Studies*, Vol. 7, No. 1, pp. 1-13.

Mansur M.O. & Mohd. Noor Md. Sap (2005), Outlier detection technique in data mining : a research perspective, *Proceedings of the postgraduate annual research seminar*.

Matsumoto, S.; Kamei, Y; Monden, A. & Matsumoto K. (2007) Comparison of Outlier Detection Methods in Fault-proneness Models. *Proceedings of the 1st International Symposium on Empirical Software Engineering and Measurement* (ESEM2007), pp.461-463, September 2007.

Melin, P. & Castillo, O. (2008). *Fuzzy logic: theory and applications*, Springer.

Moore, D.S. & McCabe G.P. (1999), *Introduction to the Practice of Statistics.* , Freeman & Company.

Munoz, A. & Muruzabal, J. (1997) *Self organizing maps for outliers detection*, Elsevier, Neurocomputing, N°18, pp.33-60, August 1997.

Nag, A.K.; Mitra, A. & Mitra, S.(2005), Multiple outlier Detection in Multivariate Data Using Self-Organizing Maps Title, Computational Statistical, N.20, pp.245-264.

Ng R. T. & Han J. (1994) Efficient and Effective Clustering Methods for Spatial Data Mining, *Proceeding of the 20th International Conference on Very Large Data Bases*, Santiago, Chile, Morgan Kaufmann Publishers, pp. 144-155, San Francisco, CA.

Oliver, J.J.; Baxter, A. & Wallace, C.S. (1996), Unsupervised Learning using MML, *Proceedings of the 13th International Conference (ICML)*, pp. 364-372, Morgan Kaufmann Publishers, San Francisco, CA.

Papadimitriou, S.; Kitawaga, H.; Gibbons, P.B. & Faloutsos C. (2002), LOCI: Fast Outlier Detection Using the Local Correlation Integral, *Proceedings of the 19th International Conference on Data Engineering (ICDE)*, pp. 315-326.

Patyra, M.J. & Mlynek D.J. (1996). *Fuzzy logic: Implementation and applications.*Wiley & Teubner, ISBN 047195099.

Ramasmawy R.; Rastogi R. & Kyuseok S. (2000). Efficient algorithms for mining outliers from large data sets. *Proceedings of the ACM SIGMOD International Conference on Management of Data*, pp.427-438, ISBN 1-58113-217-4, Dallas, Texas, United States.

Roberts, S. & Tarassenko, L. (1995). A Probabilistic Resource Allocating Network for Novelty Detection. *Neural Computation*, Vol. 6, N. 2, , 1995, pp. 270-284.

Ross, Timothy J. (2004). *Fuzzy logic with engineering appplications*, John Wiley & sons ltd, England.

Rousseeuw, P. & Leroy, A. (1997). *Robust Regression and Outlier detection*, John Wiley & Sons.

Sane, S. & Ghatol, A. (2006), Use of Instance Tipicality for Efficient Detection of Outliers with neural network Classifiers. *Proceedings of 9thInternational Conference on Information Technology*, ISBN 0-7695-2635-7.

Sheikholeslami, G.; Chatterjee, S. & Zhang A.. (1998). WaveCluster: A Multi-Resolution Clustering Approach for Very Large Spatial Databases, *Proceedings International Conference on Very Large Data Bases*, pp. 428-439, New York, NY.

Tenner, J.; Linkens, D.A. & Bailey, T.J. (1999). Preprocessing of Industrial Process Data with Outlier detection and correction, *Proceedings of the 2nd International Conference on Intelligent Processing and Manufacturing of Materials, IEEE*, , pp. 921–926, Vol. 2, 10-15 July 1999.

Vapnik, N. (1982) *Estimation of dependence based on empirical data*, Springer-Verlag, New York.

Vapnik, N. (1998) *Statistical learning theory*. John Wiley & Sons, ISBN 0-471-03003-1, New York.

Yen, J. & Langari, R. (1998) *Intelligence control and information*. Prentice Hall, 1998.

Yu, D.; Sheikholeslami G. & Zhang, A. (2002) FindOut: Finding Outliers in Very Large Datasets. *Knowledge and Informations Systems*, vol.4, pp. 387-412, Springer-Verlag, London.

Wang, Y. (1995) *Jump and sharp cusp detection by wavelets*. Biometrika, Vol. 82, pp. 385-397.

Wang, W.; Yang J. & Muntz, R. (1997) *STING:* A Statistical Information Grid Approach to Spatial Data Mining, *Proceedings of the 23th International Conference on Very Large Data Bases*, Athens, Greece, Morgan Kaufmann Publishers, pp. 186-195, 1997, San Francisco, CA.

Weiss, L.G. (1994). Wavelets and Wideband Correlation Processing, *Signal Processing magazine* IEEE, Vol. 11, pp.13-32, ISSN 1053-5888, January 1994.

Williams, G.; Baxter, R.; He, H. & Hawkison,S. (2002). A comparative study of RNN for outlier detection in data mining, *Proceedings of the IEEE International Conference on Data Mining*, pp. 709–712, 9-12 December 2002, Australia.

Zhang, T.; Ramakrishnan, R. & Linvy M. BIRCH (1996). An Efficient Data Clustering Method for Very Large Databases, *Proceedings of ACM SIGMOD International Conference on Management of Data*, pp. 103-114, ACM Press, New York.

15

Multi-Model Approaches for Bilinear Predictive Control

Anderson Cavalcanti[1], André Maitelli[2] and Adhemar Fontes[3]
[1]*Federal Center of Technological Education of Rio Grande do Norte*
Academic Department of Informatics and Industry
[2]*Federal University of Rio Grande do Norte*
Department of Computer and Automation Engineering
[3]*Federal University of Bahia*
Department of Electrical Engineering
Brazil

1. Introduction

It is well known that linear controllers can exhibit serious performance limitations when applied to nonlinear systems since nominal linear models used during design cannot represent the nonlinear plant in its whole operating range (Arslan et al., 2004). For this reason, several researches has been proposed new techniques in order to supply a solution for this problem. The main alternative technique, proposed by academy, to resolve the referred problem is known as multi-model approach. The basic idea of multi-model approach consists in decompose the system's operating range into a number of operating regimes that completely cover the chosen trajectory as showed in (Foss et al., 1995). There are, basically, two approaches for multi-model. The first one consists of to design a set of suitable controllers (one for each operating regime) and to calculate weighting factors to them as showed in (Arslan et al., 2004) and (Cavalcanti et al., 2007a). The global control signal is a weighting sum of the contributions of each controller. The second one consists of to build a global model as a weighting sum of each local model as showed in (Foss et al., 1995) and (Cavalcanti et al., 2007b). In both cases, a way to measure distances between models is defined. Multivariable Model Predictive Control (MMPC) has been presented in this chapter. MPC is the an of the most important control technique used in industry. Multivariable Bilinear Generalized Predictive Control (MBGPC) is formulated and, its alternative solution, Multivariable Bilinear Generalized Predictive Control with Iterative Compensation (MBGPCIC) is presented. This chapter shows either proposed metrics in order to build multi-model based controllers (based in MBGPC and MBGPCIC) and presents simulation results applied in distillation columns.

2. Multivariable Bilinear Generalized Predictive Control (MBGPC)

MBGPC is a MPC technique based in the minimization of a objective function. This objective function considers the predicted output of a system. The prediction is obtained by a mathematical model of this system.

Considering a multi-input, multi-output (MIMO) system, as showed in Fig.1, with *p*-inputs and *q*-outputs, being *y* the system's output and *u* the system's input.

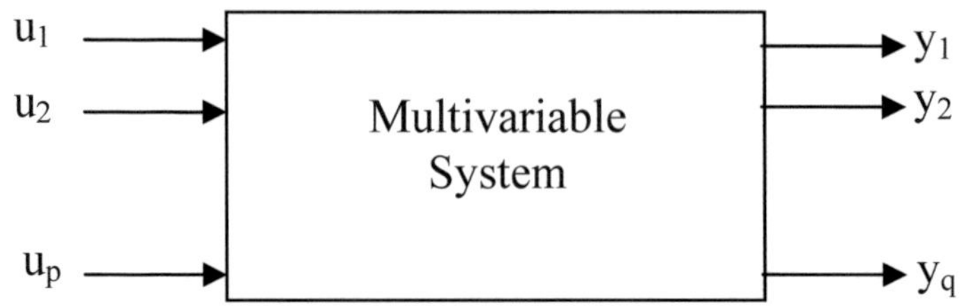

Fig. 1. Block diagram, Multivariable system

Considering that this system is described by the following matrix polynomial expression, showed in (Fontes, 2002):

$$\begin{aligned} &A(q^{-1})\Delta_q(q^{-1})y(k) = B(q^{-1})\Delta_p(q^{-1})u(k-1) + \\ &D_e(q^{-1})D[u(k-1)]D_d(q^{-1})\Delta_q(q^{-1})y(k-1) + \\ &C(q^{-1})e(k) \end{aligned} \tag{1}$$

where $y(k) \in R^q$ is the process output vector, $u(k) \in R^p$ is the process input vector and $e(k) \in R^q$ is the Gaussian white noise with zero mean and covariance $diag(\sigma^2)$. The matrices $A(q^{-1})$, $B(q^{-1})$ and $C(q^{-1})$ are polynomials matrices in shift operator q^{-1} and are defined by:

$$A(q^{-1}) = I_{q\times q} + A_1 q^{-1} + \cdots + A_{na} q^{-na} \tag{2}$$

$$B(q^{-1}) = B_0 + B_1 q^{-1} + \cdots + B_{nb} q^{-nb} \tag{3}$$

$$C(q^{-1}) = I_{p\times p} + C_1 q^{-1} + \cdots + C_{nc} q^{-nc} \tag{4}$$

$$D_d(q^{-1}) = D_{d,0} + D_{d,1} q^{-1} + \cdots + D_{d,nd_d} q^{-nd_d} \tag{5}$$

$$D_e(q^{-1}) = D_{e,0} + D_{e,1} q^{-1} + \cdots + D_{e,nd_e} q^{-nd_e} \tag{6}$$

where $A(q^{-1}) \in R^{q\times q}$, $B(q^{-1}) \in R^{q\times p}$, $C(q^{-1}) \in R^{q\times q}$, $D_e(q^{-1}) \in R^{q\times p}$ and $D_d(q^{-1}) \in R^{p\times q}$. The matrix $D[u(k-1)]$ is defined as:

$$D[u(k-1)] = diag\left| u_1(k-1) \quad u_2(k-1) \quad \cdots \quad u_p(k-1) \right| \tag{7}$$

The nonlinear model presented in (1) is quasi-linearized to be used in MBGPC. The multivariable quasilinear multi-model must be obtained by rewriting the expression (1) of the following form:

$$A(q^{-1},u)\Delta_q(q^{-1})y(k) = B(q^{-1})\Delta_p(q^{-1})u(k-1) + C(q^{-1})e(k) \tag{8}$$

where

$$A(q^{-1},u) = A(q^{-1}) - q^{-1}D_e(q^{-1})D[u(k-1)]D_d(q^{-1}) \tag{9}$$

The polynomial matrix $A(q^{-1},u)$ is calculated considering its parameters as constant in prediction horizon. The polynomial matrix $A(q^{-1},u)$ is considered diagonal in this work. The output prediction *i-step* ahead may be obtained of the expression (8), such that:

$$\widetilde{A}(q^{-1},u)y(k+i) = B(q^{-1})\Delta_p(q^{-1})u(k+i-1) + C(q^{-1})e(k+i) \tag{10}$$

where $\widetilde{A}(q^{-1},u) = A(q^{-1},u)\Delta_q(q^{-1})$. In this case, the polynomial matrix $C(q^{-1})$ is equal to I_{pxp} due the fact that the noise be supposed white. Considering the following Diophantine equation:

$$I_{p\times p} = E_i(q^{-1},u)\widetilde{A}(q^{-1},u) + q^{-i}F_i(q^{-1},u) \tag{11}$$

where

$$E_i(q^{-1},u) = E_{i,o}(u) + \cdots + E_{i,i-1}(u)q^{-(i-1)} \tag{12}$$

$$F_i(q^{-1},u) = F_{i,o}(u) + \cdots + F_{i,na}(u)q^{-na} \tag{13}$$

Pre-multiplying (10) for $E_i(q^{-1},u)$ we obtain:

$$E_i(q^{-1},u)\widetilde{A}(q^{-1},u)y(k+i) = E_i(q^{-1},u)B(q^{-1})\Delta_p(q^{-1})u(k+i-1) + E_i(q^{-1},u)e(k+i) \tag{14}$$

Rewriting (9) of the following form:

$$E_i(q^{-1},u)\widetilde{A}(q^{-1},u) = I_{p\times p} - q^{-i}F_i(q^{-1},u) \tag{15}$$

Substituting (15) in (14) we obtain:

$$y(k+i) = F_i(q^{-1},u)y(k) + E_i(q^{-1},u)e(k+i) + E_i(q^{-1},u)B(q^{-1})\Delta_p(q^{-1})u(k+i-1) \tag{16}$$

As the degree of $E_i(q^{-1},u)$ is $i-1$, then the sub-optimal prediction of $y(k+i)$ is:

$$\hat{y}(k+i) = F_i(q^{-1},u)y(k) + E_i(q^{-1},u)B(q^{-1})\Delta_p(q^{-1})u(k+i-1) \tag{17}$$

In order to separate past and future values, we make:

$$E_i(q^{-1},u)B(q^{-1}) = H_i(q^{-1},u) + q^{-i}H_{ipa}(q^{-1},u) \tag{18}$$

As the degree of $H_i(q^{-1},u)$ is less than $i-1$, the predictor may be written as:

$$\hat{y}(k+i) = F_i(q^{-1},u)y(k) + H_{ipa}(q^{-1},u)\Delta_p(q^{-1})u(k-1) + H_i(q^{-1},u)\Delta_p(q^{-1})u(k+i-1) \tag{19}$$

The last term of (19) considers the future inputs (forced response) and the two first terms consider only past inputs (free response). So, we define:

$$\hat{y}(k+i) = H_i(q^{-1},u)\Delta_p(q^{-1})u(k+i-1) + Y_{li} \tag{20}$$

where

$$Y_{li} = F_i(q^{-1},u)y(k) + H_{ipa}(q^{-1},u)\Delta_p(q^{-1})u(k-1) \tag{21}$$

The expression (21) is the free response. The objective function is given by:

$$J = \sum_{i=N_1}^{NY} \left\| r(k+i) - \hat{y}(k+i) \right\|_R^2 + \sum_{i=1}^{NU} \left\| \Delta u(k+i-1) \right\|_Q^2 \tag{22}$$

where N_1 is minimum prediction horizon, NY is prediction horizon, NU is the control horizon, R and Q are weighting matrices of error signal and control effort in instant k in the chosen trajectory, respectively, $\hat{y}(k+i)$ is the sub-optimum i-step ahead predicted output, $r(k+i)$ is the future reference trajectory. The control effort is obtained, without constraints, by the minimization of the objective function (22).
Consider the predictions set:

$$y_{N_{1y}} = H_{N_{1yu}} \Delta_p U_{NU} + Y_{lN_1y} \tag{23}$$

where

$$y_{N_{1y}} = \left[\hat{y}(k+N_1) \quad \hat{y}(k+N_1+1) \quad \cdots \quad \hat{y}(k+NY) \right]^T \tag{24}$$

$$H_{N_{1yu}} = \begin{bmatrix} H_{N_1-1} & H_{N_1-2} & \cdots & H_{N_1-NU} \\ H_{N_1} & H_{N_1-1} & \cdots & H_{N_1+1-NU} \\ \vdots & \vdots & \ddots & \vdots \\ H_{NY-1} & H_{NY-2} & \cdots & H_{NY-NU} \end{bmatrix} \tag{25}$$

$$\Delta_p U_{NU} = \begin{bmatrix} \Delta_p(q^{-1})u(k) \\ \Delta_p(q^{-1})u(k+1) \\ \vdots \\ \Delta_p(q^{-1})u(k+NU-1) \end{bmatrix} \tag{26}$$

$$y_{lN_{1y}} = \begin{bmatrix} Y_{lN_1} \\ Y_{lN_1+1} \\ \vdots \\ Y_{lNY} \end{bmatrix} \tag{27}$$

The objective function (22) may be rewritten of the following form:

$$J=(H_{N_{1yu}}\Delta_p U_{NU}+y_{lN_{1y}})^T\overline{R}(H_{N_{1yu}}\Delta_p U_{NU}+y_{lN_{1y}})+\Delta_p U^T{}_{NU}\overline{Q}\Delta_p U_{NU} \quad (28)$$

where

$$\overline{R}=diag[R_1,\cdots,R_{q\times NY}] \quad (29)$$

$$\overline{Q}=diag[Q_1,\cdots,Q_{p\times NU}] \quad (30)$$

This minimization is obtained by the calculation of its gradient (making it equals zero), of the following form:

$$\frac{\partial J}{\partial \Delta_p U_{NU}}=0 \quad (31)$$

The minimization of (28) produces the following control law:

$$\Delta_p U_{NU}=(H^T_{N_{1yu}}H_{N_{1yu}}+\overline{Q})^{-1}H^T_{N_{1yu}}\overline{R}(r-y_{lN_{1y}}) \quad (32)$$

Because of the receding control horizon, only the first p rows of (32) are computed.

3. Multivariable Bilinear Generalized Predictive Control With Iterative Compensation (MBGPCIC)

3.1 Motivation

The quasi-linearization presented in (8) produces a prediction error that degrades the controller performance. This prediction error increases with the prediction horizon. In order to solve this problem, several algorithms has been proposed (Fontes et al., 2002), (Fontes et al. 2004), (Fontes and Ângelo, 2006) and (Fontes and Laurandi, 2006). This section presents the multivariable case showed in (Fontes and Laurandi, 2006).

The basic quasi-linear algorithm, presented by (Goodhart, 1994), calculates the output prediction *i*-step ahead considering the terms $A_j(q^{-1},u)$ with $j=1,\cdots,na$ depending only of known values of the input (until *k-1* step). The approximation of this approach generates a prediction error that increases with the prediction horizon and degrades the controller performance.

3.2 The basic idea of iterative compensation algorithm

The idea of the iterative compensation algorithm consists of consider the effort control sequence (obtained by the classic quasi-linear algorithm) to correct the parameters of $A(q^{-1},u)$. Considering the following sequence of effort control, calculated in k step:

$$\Delta_p U_{NU}=[\Delta u_1(k) \ \cdots \ \Delta u_p(k) \ \cdots \ \Delta u_1(k+NU) \ \cdots \ \Delta u_p(k+NU)]^T \quad (33)$$

The classic quasi-linear algorithm, considering (30), calculates the sequence of future control efforts, that is given by:

$$U_{NU} = \left[u_1(k) \quad \cdots \quad u_p(k) \quad \cdots \quad u_1(k+NU) \quad \cdots \quad u_p(k+NU)\right]^T \tag{34}$$

where:

$$u_j(k+i) = u_j(k-1) + \sum_{t=0}^{i} \Delta u_j(k+t),\ j = 1,\cdots,p \tag{35}$$

The iterative compensation algorithm uses the values of (34) to correct the polynomial matrix (9). This correction will produces a new prediction and a new sequence of control efforts. This process is repeated until a stop criterion be achieved. It is important to remember that this algorithm considers the receding horizon too. When the algorithm converges, only the first p rows of (34) are sent to the process.

3.3 Convergence and stop criterion

The stop criterion of the algorithm is based in the norm variation of the vector $\Delta_p U_{NU}$. The procedure of correction of $A(q^{-1},u)$ is repeated until that the norm of the variation calculated in the iteration r be less than a tolerance value ε established:

$$\sqrt{(\Delta_p U_{NU,r} - \Delta_p U_{NU,r-1})^T (\Delta_p U_{NU,r} - \Delta_p U_{NU,r-1})} < \varepsilon \tag{36}$$

It is important to remember that this algorithm may not to converge. In this case, another stop criterion must be established. The second stop criterion is based in the maximum number of iterations of the algorithm:

$$r < N_{\max} \tag{37}$$

where $N_{\max}$ is the maximum number of iterations of the algorithm. If the algorithm stops for the $N_{\max}$ criterion, the control effort sent to the process is the control effort calculated by the classic quasi-linear algorithm.

4. Multi-models approaches for MPC

Considering a bilinear multivariable model showed in (1) that describes the system's behavior in a small region. This structure is valid around its operating regime and more or less invalid outside this regime. Considering yet that the process has been decomposed into *NOR* operating regimes, the first step to develop a multi-model structure is to identify *NOR* local models (in this case bilinear), of the following form:

$$\begin{aligned} &A_{(j)}(q^{-1})\Delta_q(q^{-1})y(k) = B_{(j)}(q^{-1})\Delta_p(q^{-1})u(k-1) + \\ &D_{e(j)}(q^{-1})D[u(k-1)]D_{d(j)}(q^{-1})\Delta_q(q^{-1})y(k-1) + \\ &C_{(j)}(q^{-1})e(k) \end{aligned} \tag{38}$$

where $j = 1,\cdots,NOR$.

4.1 Building a global model

The first multi-model approach consists in to obtain a global model from the local bilinear models showed in (38). To build the global model, we must consider that there is a validity function $\delta_j(k)$ that is designed such that its value is close to one for operating points where the local model structure is a good description of the system and close to zero otherwise, in instant *k*. In this case, each polynomial matrix $P^{(k)}(q^{-1})$ of (1) for each instant *k* would be calculated of the following form:

$$P^{(k)}(q^{-1}) = \sum_{j=1}^{NOR} P_{(j)}(q^{-1}) w_{j,k} \tag{38}$$

where $P^{(k)}(q^{-1})$ the global built polynomial matrix of the bilinear model, $P_{(j)}(q^{-1})$ is the polynomial matrix of the jth bilinear model, and:

$$w_{j,k} = \frac{\delta_j(k)}{\sum_{t=1}^{NOR} \delta_t(k)}; \quad j = 1, \cdots, NOR \tag{39}$$

where $w_{j,k}$ is a weighting factor to the jth bilinear model in instant *k*. Theses approaches has the following defined property:

$$\sum_{j=1}^{NOR} w_{j,k} = 1 \tag{40}$$

4.2 Building a global controller

The second multi-model approach consists of to build a global controller from a set of controllers. In this case, one controller is designed to each operation point. A validity function $\delta_j(k)$ is designed too, in order to evaluate what controller must have a greater weighting factor. Considering the control effort $u_{i,j}(k)$ with $j = 1, \cdots, NOR$ and $i = 1, \cdots, p$ of each controller (one for each operation point), the control effort sent to the process is given by:

$$u_i(k) = \sum_{j=1}^{NOR} w_{j,k} u_{i,j}(k) \tag{41}$$

where $w_{j,k}$ is calculated as showed in (39).

4.3 Metric based in norms

The validity function is usually called by *metric*. In multivariable case, in a process with p-inputs and q-outputs, the output is $y(k) \in R^q$ and the input is $u(k) \in R^p$. In a known trajectory of process output, the distance from the first operation point to the last operation point is given by:

$$d_{1,NOR} = \left\| y_{NOR} - y_1 \right\|_2 \tag{42}$$

To measure the distance from the current operation point to the operation point of *j^th^* designed controller, we can use the expression:

$$\delta_j = \frac{d_{1,NOR}}{\left\| y(k) - y_i \right\|_2}; \quad i = 1, \cdots, NOR \tag{43}$$

To this metric, only monotonic (increasing or decreasing) trajectories must being considered.

4.4 Metric based in phase margin

Phase margin has been chosen as measurement parameter in order to quantitatively estimate the distance between two different models.

One of the most important techniques for measure the robustness of a dynamic system is the margin phase technique. In general terms, in a linear time-invariant system, phase margin is maximum phase angle that can be added, such as this system not becomes instable. In this chapter, a multivariable bilinear model is obtained for each operating regime, and this model is quasi-linearized. Each obtained model has a phase margin value (or multivariable equivalent). An interpolated model is calculated to obtain the current valid model (that depends of the current operation point). The difference between the interpolated model and models of the chosen operating regimes is calculated. This difference is considered to the computation of a set of weighting factors to the controllers. An equivalent method to the multivariable margin phase calculus is showed.

The model showed in (8) is time-step quasi-linear (linear at each time instant). We obtain a transfer function in z^{-1} operator, pre-multiplying (8) for $A^{-1^{(i)}}(z^{-1},u)$ of the following form:

$$y(z^{-1}) = A_{(k)}^{-1}(z^{-1},u) B_{(k)}(z^{-1}) z^{-1} u(z^{-1}) \tag{44}$$

where $A_{(k)}^{-1}(z^{-1},u)$ and $B_{(k)}(z^{-1})$ are interpolated polynomial matrices in instant *k*. From (44), the matrix transfer function for the interpolated model in instant *k* is given by:

$$G_{(k)}(z^{-1},u) = A^{-1}{}_{(k)}(z^{-1},u) B_{(k)}(z^{-1}) z^{-1} \tag{45}$$

In this approach, we consider the $p \times q$ transfer functions from (45), and calculate individually the phase margin from each one transfer function.

For each operating regime, the minor margin phase is chosen. Considering that $MPM^{(j)}$ represents the minor phase margin of the matrix transfer function in the jth operating regime and that $\max(MPM^{(j)})$ is the maximum of phase margin for $j = 1, \cdots, NOR$ and $\min(MPM^{(j)})$ is the minimum phase margin for $j = 1, \cdots, NOR$.

Considering yet that $G_{(k)}(z^{-1},u)$ is an interpolated transfer function given in *k* instant, and $MPM^{(k)}$ is its minor phase margin, the distance factor to the designed controller in the jth operating regime, in instant *k* is given by:

$$\delta_j = 1 - \frac{\left| MPM^{(k)} - MPM^{(j)} \right|}{\max(MPM^{(j)}) - \min(MPM^{(j)})} \tag{46}$$

It is important to observe that the interpolated model's parameter (in this case, obtained by cubic spline interpolation method) is always in the bounds of the estimated model's parameter (it is guaranteed by the algorithm). The algorithm guarantees also that, if a calculated phase margin is out of bounds $\min(MPM^{(i)})$ and $\max(MPM^{(i)})$, the designed controllers of these bounds have weighting factor equals 1 (maximum).

5. Applications of multi-model controllers based in proposed metrics

This section shows some applications of the combinations between:

- approaches showed in sections 4.1 and 4.2;
- controllers showed in sections 2 and 3, and;
- metrics showed in sections 4.3 and 4.4.

The application showed in this chapter is based in a simulated debutanizer distillation column. Debutanizer distillation column is usually used to remove the light components from the gasoline stream to produce Liquefied Petroleum Gas (LPG). The most common control strategy is to manipulate the reflux flow rate and the temperature in column's bottom and, to control the concentrations of any product in *butanes* stream and in C5+ stream as showed in (Almeida, *et al.*, 2000). The chosen process variables are: concentration of i-pentane in butanes stream (y_1) and concentration of i-butene in C5+ stream (y_2). The studied column is simulated in Hysys software and is showed in Figure 2.

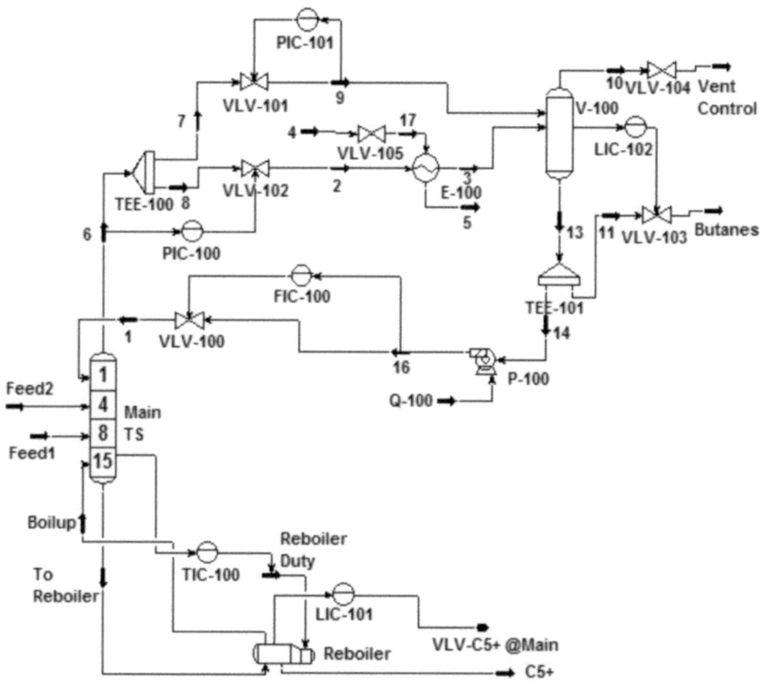

Fig. 2. Distillation Column simulated in Hysys Software

The reflux flow rate (u_1) is manipulated through the FIC-100 controller and the temperature of column's bottom (u_2) is manipulated through the TIC-100 controller. The reflux flow rate is measured in m³/h and the temperature of column's bottom is measured in °C.
In the applications showed in this chapter, three operation points were chosen, as showed in Table 1. The identified bilinear models were obtained using the multivariable recursive least squares algorithm and the model's structure has been chosen by using the Akaike criterion. In all points, the chosen sample rate is 4 minutes. Only monotonic trajectories are being considered. The trajectory of y_1 is monotonically increasing and the trajectory of y_2 is monotonically decreasing.

Operation Point	Input	Output (Mass Fractions)
1	u_1 = 40 m³/h	Y_1 = 0.014413
	u_2 = 147 °C	Y_2 = 0.001339
2	u_1 = 37 m³/h	Y_1 = 0.017581
	u_2 = 147.5 °C	Y_2 = 0.001161
3	u_1 = 34 m³/h	Y_1 = 0.021994
	u_2 = 148 °C	Y_2 = 0.001004

Table 1. Three operation points chosen in distillation column.

In order to quantitatively asses the performance of multi-model quasi-linear GPC, some indices like showed in (Goodhart, et al., 1994) are calculated. Theses indices may be extended to multivariable case, of the following form:

$$\varepsilon_{1,i} = \sum |u_i(k)| / N \tag{47}$$

where $i = 1, \cdots, p$ and N is the amount of control effort applied in the process to achieve the desired response. The index showed in (47) is the account of total control effort to achieve a given response. The variance of controlled actuators is:

$$\varepsilon_{2,i} = \sum (u_i(k) - \varepsilon_{1,i})^2 / N \tag{48}$$

The deviation of the process of integral of absolute error (IAE) is:

$$\varepsilon_{3,j} = \sum |r_j(k) - y_j| / N \tag{49}$$

where $j = 1, \cdots, q$. The overall measure of effectiveness is defined as:

$$\varepsilon_j = \sum_{i=1}^{p} (\alpha_i \varepsilon_{1,i} + \beta_i \varepsilon_{2,i}) + \rho_j \varepsilon_{3,j} \tag{50}$$

The factors α_i, β_i and ρ_j are weightings chosen to reflect the actual financial cost of energy usage, actuator wear and product quality, respectively. In this case, we consider $\alpha_i = 0.1$, $\beta_i = 0.15$ and $\rho_j = 0.5$ because we have established as priority the product quality.

5.1 Application 1 - Controller based in global model, norm-2 metric and MBGPC

In this simulation, the process is in the 3rd operating regime and a deviation in reference is applied in the proposed controller. With this reference deviation, the process will come to close to the 1st operating regime. The proposed quasi-linear multi-model is compared with quasi-linear single-model (using the model of the 3rd operating regime). Figures 3 and 4 show the process's output and Figures 4 and 6 show the control effort.

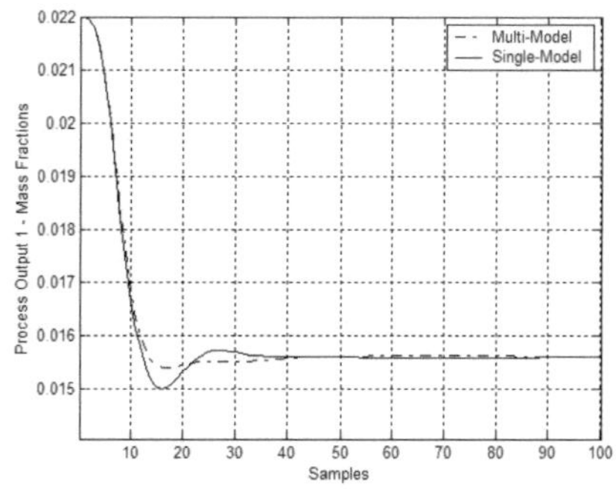

Fig. 3. Process Output 1. Comparison between single-model and multi-model approach (application 1).

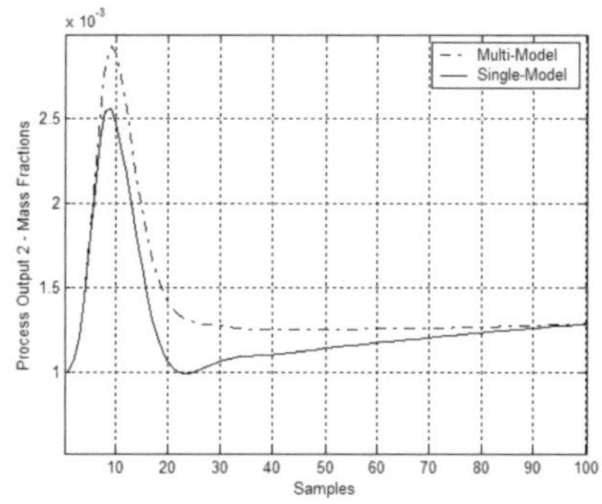

Fig. 4. Process Output 2. Comparison between single-model and multi-model approach (application 1).

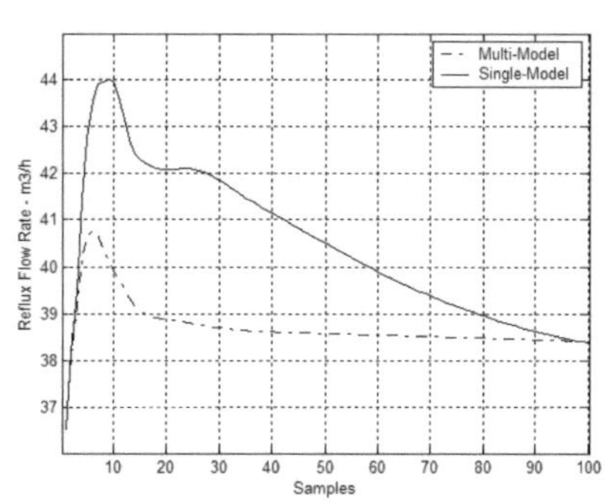

Fig. 5. Reflux Flow rate. Comparison between single-model and multi-model approach (application 1).

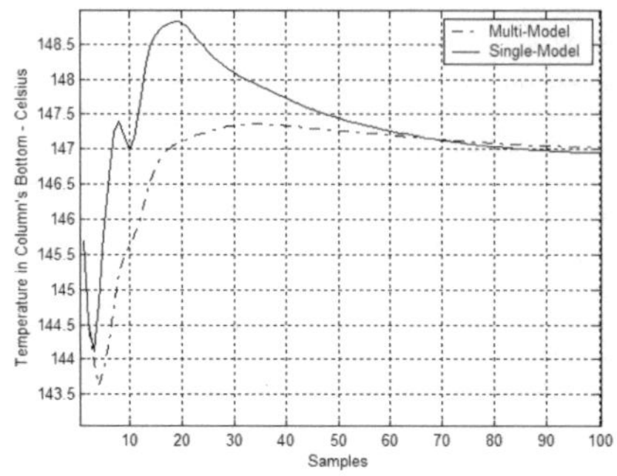

Fig. 6. Temperature in column's bottom. Comparison between single-model and multi-model approach (application 1).

Figures 3, 4, 5 and 6 shows the better performance of multi-model approach, so much of point view of the process response as of the control effort.

5.2 Application 2 - Controller based in global controller, norm-2 metric and MBGPC

In this simulation, the same reference deviation as in section 5.1 is applied. The proposed quasi-linear multi-model is compared with quasi-linear single-model (using the model of the 3rd operating regime). Figures 7 and 8 show the process's output and Figures 9 and 10 show the control effort.

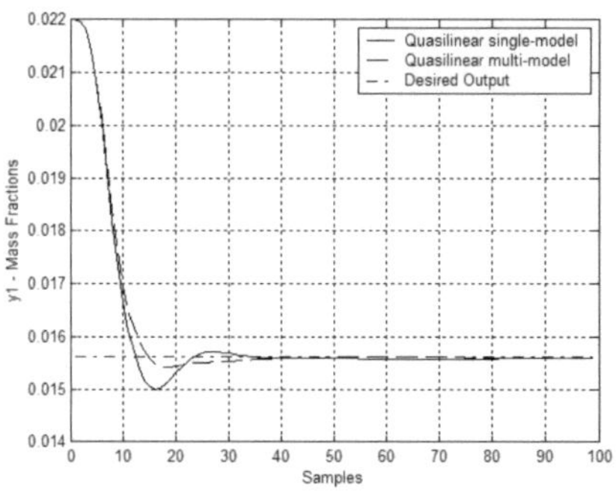

Fig. 7. Process Output 1. Comparison between single-model and multi-model approach (application 2).

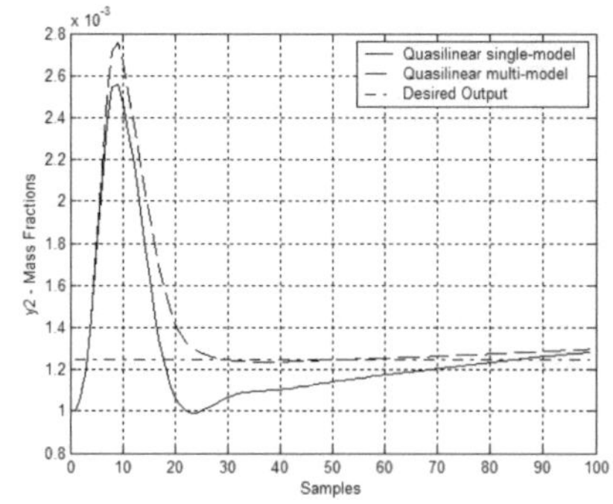

Fig. 8. Process Output 2. Comparison between single-model and multi-model approach (application 1).

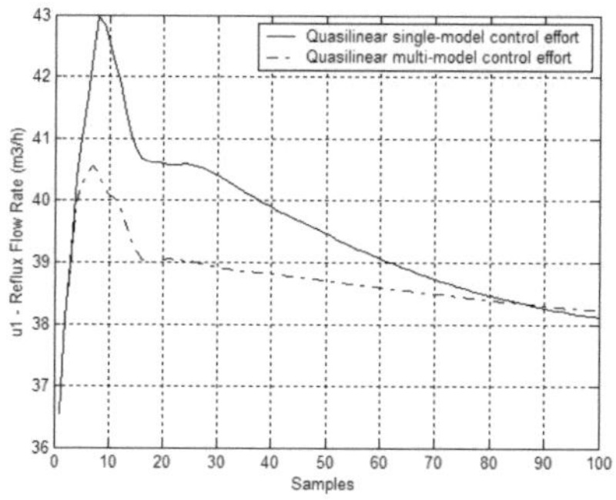

Fig. 9. Reflux Flow rate. Comparison between single-model and multi-model approach (application 2).

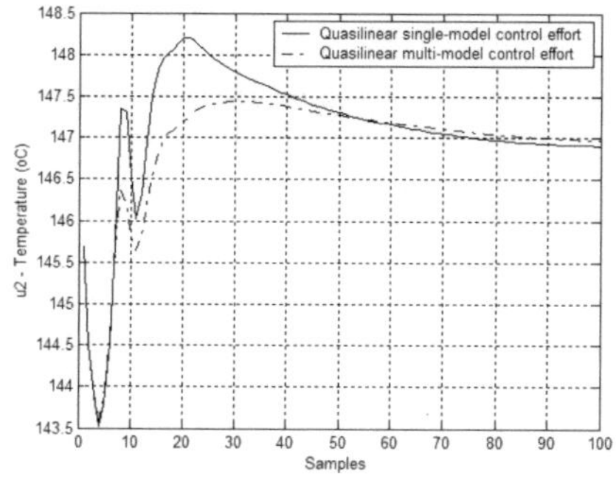

Fig. 10. Temperature in column's bottom. Comparison between single-model and multi-model approach (application 2).

5.3 Application 3 - controller based in global controller, phase margin metric and MBGPC

In this simulation, the same reference deviation as in section 5.1 and 5.2 is applied. The proposed quasi-linear multi-model is compared with quasi-linear single-model (using the model of the 3rd operating regime). Figures 11 and 12 show the process's output and Figures 13 and 14 show the control effort.

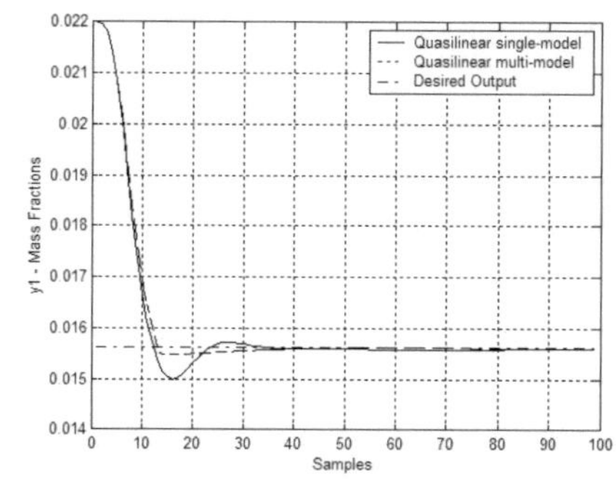

Fig. 11. Process Output 1. Comparison between single-model and multi-model approach (application 3).

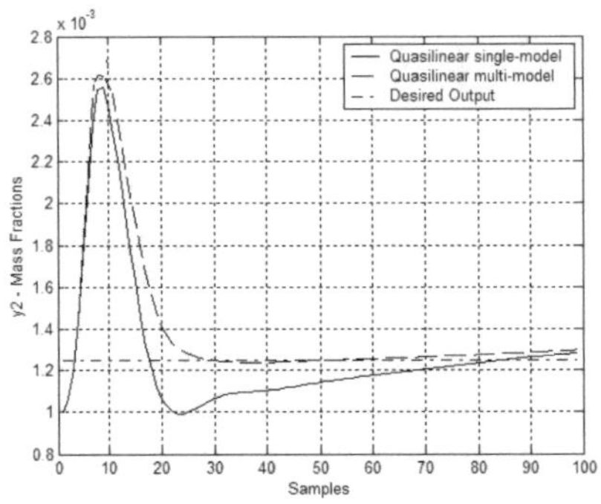

Fig. 12. Process Output 2. Comparison between single-model and multi-model approach (application 3).

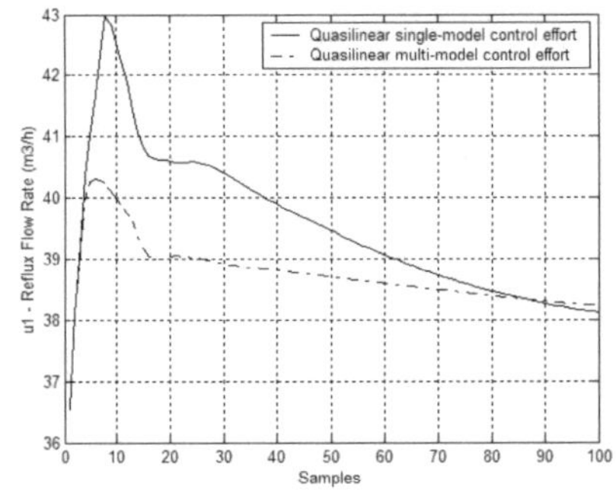

Fig. 13. Reflux Flow rate. Comparison between single-model and multi-model approach (application 3).

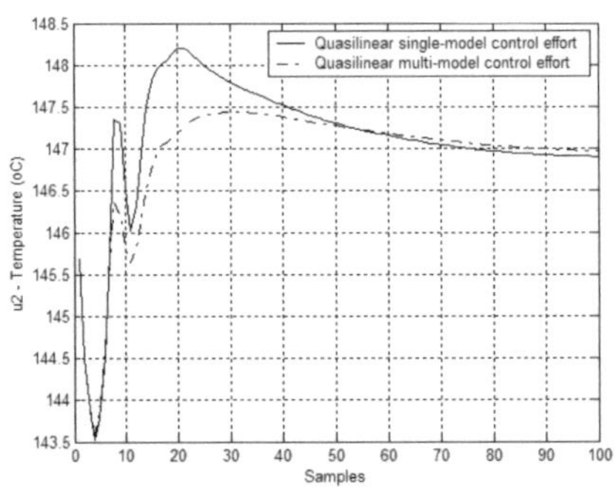

Fig. 14. Temperature in column's bottom. Comparison between single-model and multi-model approach (application 2).

5.4 Qualitative comparison between applications 1,2 and 3

This section shows the qualitative comparison between application 1,2 and 3. The comparison is based in the indices showed in (47), (48), (49) and (50) with $N=100$.

The table 2 shows, in all the cases, that multi-model approach has better performance in relation of single-model approach in terms of less energy usage, less actuator wear and better product quality. Application 2 has presented better performance in relation of the other multi-model approaches.

I/O	Approach	ε_1	ε_2	ε_3	ε
1	Single	40.47	2.61	287.46	163.00
2	Single	147.38	0.63	142.40	90.47
1	Application 1	38.72	0.31	255.20	146.26
2	Application 1	146.88	0.36	117.71	77.52
1	Application 2	38.38	0.32	248.41	142.83
2	Application 2	146.94	0.29	103.48	70.36
1	Application 3	38.55	0.33	253.41	145.36
2	Application 3	147.01	0.31	113.71	75.51

Table 2. Qualitative comparison between applications 1,2 and 3.

5.5 Application 4 - controller based in global controller, norm-2 metric and MBGPCIC

In this simulation, the process is the following operation point: $u_1 = 31\ m^3/h$, $u_2 = 148.5\ ^oC$, $y1=0.028125$ *e* $y2=0.000874$. The reference deviation 0.01371 and 0.000465 are used in this application. This approach has been compared with application 2. Figures 15 and 16 show the process's output and Figures 17 and 18 show the control effort.

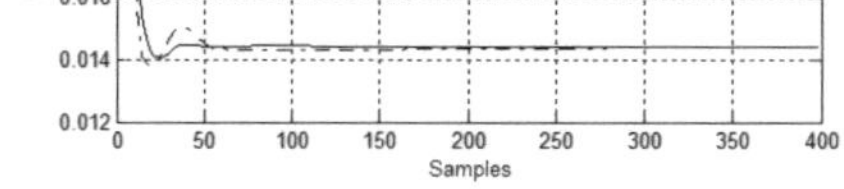

Fig. 15. Process Output 1. Comparison between MBGPC and multi-model approach and MBGPCIC multi-model approach (application 4).

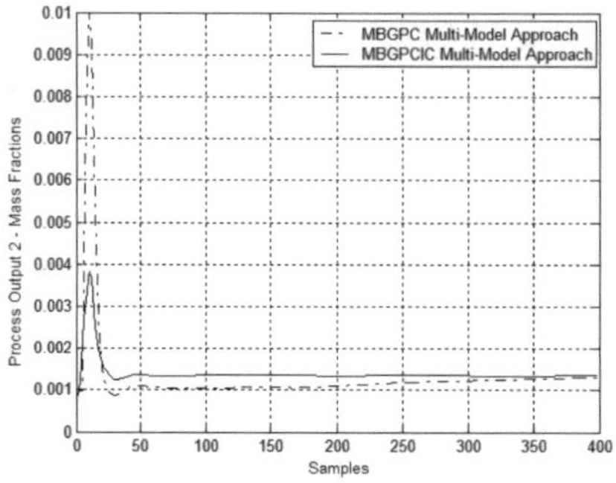

Fig. 16. Process Output 2. Comparison between MBGPC and multi-model approach and MBGPCIC multi-model approach (application 4).

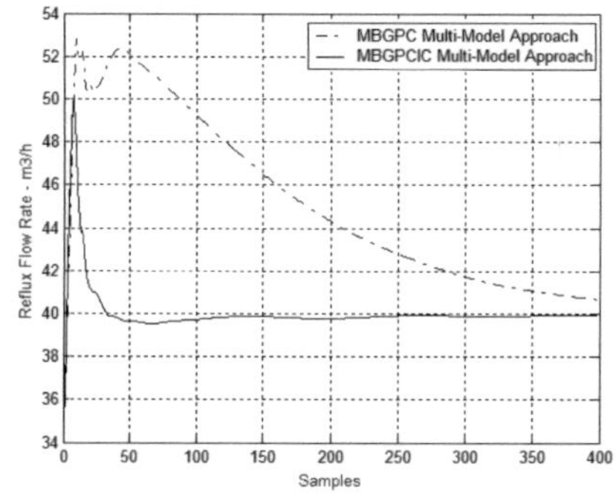

Fig. 17. Reflux Flow rate. Comparison between MBGPC and multi-model approach and MBGPCIC multi-model approach (application 4).

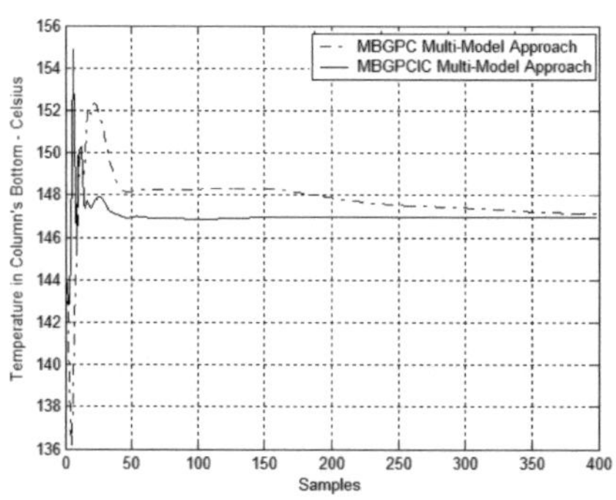

Fig. 18. Temperature in column's bottom. Comparison between MBGPC and multi-model approach and MBGPCIC multi-model approach (application 4).

Figures 15, 16, 17 and 18 show the better performance of MBGPCIC multi-model approach, so much of point view of the process response as of the control effort. The justification to this significant improvement of MBGPCIC multi-model approach in relation to MBGPC

multi-model approach is the wide variation in control effort. So, the iterative compensation procedure founds more space to minimize the prediction error.
Due too the large operating range, this approach has not been compared with single-model approach, because its performance was very poor.
Table 3 shows the performance indices of this comparison for *N=400*.

I/O	Approach	ε_1	ε_2	ε_3	ε
1	MBGPCIC Approach	50.47	6.61	1219.90	632.02
2	MBGPCIC Approach	40.71	5.78	1151.12	595.48
1	MBGPC Approach	148.27	8.00	952.00	498.07
2	MBGPC Approach	147.21	1.73	260.98	150.41

Table 3. Qualitative comparison between MBGPC and multi-model approach and MBGPCIC multi-model approach (application 4).

6. Conclusion

This chapter showed the importance and the relevance of multi-model approaches. Several researches has been proposed in order to solve design problems in process that operates in a large range (like batch processes). Some proposals of multi-model have been presented in this chapter and its comparison with classic approaches. All multi-model approaches presented better performance when compared to single-model approach. The next step of this research is to adjust these approaches to a robust and stable algorithm of multi-model.

7. References

Almeida, E.; Rodrigues, M.A. & Odloak, D. (2000). Robust Predictive Control of a Gasoline Debutanizer Column. *Brazilian Journal of Chemical Engineering*, vol. 17, pp. 11, São Paulo.

Arslan, E.; Çamurdan, M. C.; Palazoglu, A. & Arkun, Y (2004). Multi-Model Control of Nonlinear Systems Using Closed-Loop Gap Metric. *Proceedings of the 2004 American Control Conference*, Vol.3, pp. 2374-2378, Boston.

Cavalcanti, A. L. O; Fontes, A. B. & Maitelli, A. L. (2007a). Generalized Predictive Control Based in Multivariable Bilinear Multimodel. *Proceedings of 8th International IFAC Symposium on Dynamics and Control of Process Systems*, pp. 91-96, Cancún.

Cavalcanti, A. L. O; Fontes, A. B. & Maitelli, A. L. (2007b). A Multi-Model Approach For Bilinear Generalized Predictive Control. *Proceedings of 4th International Conference on Informatics in Control, Automation and Robotics*, pp. 289-295, Angers.

Cavalcanti, A. L. O; Fontes, A. B. & Maitelli, A. L. (2004). Bilinear generalized predictive control: an adaptive approach. *Proceedings of 5th Asian Control Conference*, Vol. 3, pp. 1781-1785, Melbourne.

Fontes, A.; Maitelli, A. L. & Salasar, A. O (2002). A New Bilinear Generalized Predictive Control Approach: Algorithm and Results. *Proceedings of 15th Triennial IFAC World Congress*, pp. 235-239, Barcelona.

Fontes A. B. (2002). Desenvolvimento e Avaliação de Controladores Preditivos Baseados em Modelos Bilineares, *PhD Thesis,* UFRN, Natal.

Fontes, A. & Angelo, E. (2006). Controle Preditivo Generalizado Bilinear Aplicado A Uma Coluna de Butadieno 1,3: A Compensação Iterativa, uma nova abordagem. *Proceedings of 16th Brazilian Congress of Automatic*, pp. 863-868, Salvador.

Fontes, A. & Laurandi, S. (2006). Controlador Preditivo Generalizado Bilinear Multivariável com Compensação Iterativa, Uma Nova Abordagem. *Proceedings of 12th Latin-American Congress on Automatic Control*, pp. 133-138, Salvador.

Foss, B.A.; Johansen, T.A. & Sorensen, A.V (1995). Nonlinear Predictive Control Using Local Models - Applied to a Batch Fermentation Process. *Control Eng. Practice,* pp. 389-396.

Goodhart, S. G.; Burnham, K. J. & James, D.J.G. (1994). Bilinear Self-tuning Control of a high temperature Heat Treatment Plant. *IEEE Control Theory Applications,* Vol. 141, n° 1, pp. 779-783.

16

The Holonic Production Unit: an Approach for an Architecture of Embedded Production Process

Edgar Chacón, Isabel Besembel, Dulce M. Rivero and Juan Cardillo
Universidad de Los Andes, Facultad de Ingeniería Escuela de Ingeniería de Sistemas, Av. Alberto Carnevalli, Complejo la Hechicera, Edf. B. piso 2, Mérida-Venezuela

1. Introduction

The holonic system paradigm derives from the word "Holon" that was coined by Arthur Koestler (Kloester, 1967), and it comes from the composition of the Greek word "holos" that means "the whole" and the suffix "-on" that indicates part. According to Koestler, every holon has the same structure, that is stable, coherent and its components has the same structure; the organization is invariant, the holon can be part of a bigger holon and it has parts that also are holons. This result was obtained by Koestler in the search of auto-organizational models for biological systems, where he identified structural patterns that form hierarchical embedded structures that are copies of their. This embedded hierarchy was named holarchy; this term reflects the holon trend to act as autonomous entities that cooperate in order to achieve an hierarchical system auto-organized similar to cells/weave/organ hierarchical systems in biology (Christense, 1994). In the holarchy, holons, independently of its aggregation level, have a behaviour as a whole and as parts that cooperates to reach an established goal. Inside the holarchy, holons can simultaneously belong to different groups, having precise rules that describe its behaviour. Internally, its own rules define the holon as an individual entity, with a structure and a functional model and its properties. In a holarchy, the autonomy duality, as a principal contradictory opposition, it is balanced by the knowledge model, which defines the functionality of the system as a composition of semiautonomous holons.

The Holonic Enterprise (HE) (McHugh et al, 1995) emerges based on the concepts given in the upper paragraph in order to satisfy the enterprise paradigm of having reconfigurable, open, and flexible production models, which are able to follows the dynamic of the market in a global economic-network. The HE has strategies and relationships that evolve with the time according to the business environment dynamics.

Nowadays, formal enterprises look for the accomplishment of the openness paradigm, able to reconfiguration, to be flexible by the creation of autonomous production units able to cooperate or dynamic enterprise networks build to accomplish a specific goal. In this way, an ideal organization or a virtual enterprise is a set of different production units that are geographically distributed simulating to be a unique organization with a specific localization. For a virtual enterprise, it is necessary the existence of a technological

communication infrastructure that communicate all of the units that belongs to the virtual organization (Ulieru et al, 2002).
Actually, there are several models based on the holonic approach for the manufacturing systems and the architectures that support them (Kloester, 1967), (Chistense, 1994), (Brusel et al, 1998), (Hannover, 2000), (Brussel et al, 1998), (Wyns, 1999). In (Chacon et al, 2007), it is shown a schema for a holonic production processes that incorporate complex - precise models for the whole system.
In this work, we propose a reference model for the production processes in virtual or networked enterprises, which is seen as the cooperating composition of Holonic Production Units. The relationship among the different plants of the production process is derived from the business model, the value chain, and the production flows (Cardillo et al, 2005). These Holonic Production Units (HPU) are described in terms of their components, where each one has the holon characteristics. The invariant composition procedure, without lost generality, is given by the recursive composition of the holons as was mentioned in previous holon definition.
The basic holon is composed of non-autonomous equipment (resources) that performs direct product transformation, storage and transportation tasks, direct control and regulation systems, the supervision system, and the management system for the holon. Resources can be controlled by an exclusive controller for each equipment and the set of resources and controllers shapes the controlled basic holon. This basic holon must have the capability to supervise and coordinate its own set of controllers; in this case we call an autonomous supervised basic holon. If the basic holon must be able to negotiate production goals and to synchronize the production scheduling with other basic holon to ensure a production goal; in this case we have the expected behaviour for the holon. In figure 1 it is shown the functional structure and decision elements for a basic holon.

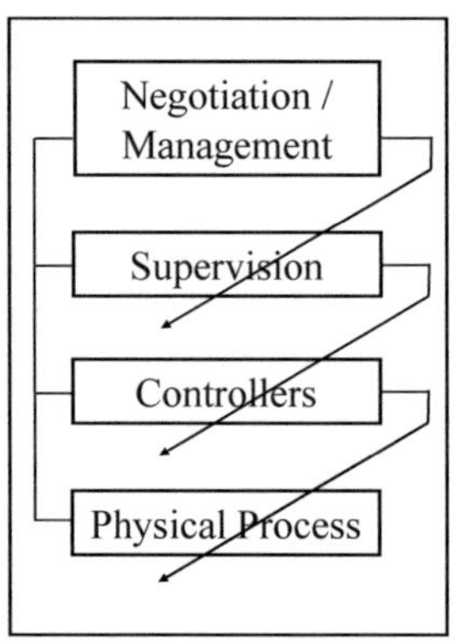

Fig 1. The basic holon

A set of basic holons or a basic holon that attain a goal is defined as the Production Unit Holon (PUH). This PUH keeps the holon structure and the control and regulation functionalities. Then, it must have control mechanisms, a supervision mechanism and negotiation / coordination mechanisms with other PUH at the same level.
In section 2 it is presented how to model formal enterprises from a holonic point of view. Section 3 describes the Holonic Production Unit that is oriented to the achievement of a goal that is derived from the value chain. In section 4, we present the holonic enterprise like a composition of Holonic Production Units. Conclusions and future works are given in section 5.

2. Modelling formal enterprises.

As has been presented in the before section, a production process must be modelled in the same way, independently if the kind of enterprise (formal or virtual) that will be organized to accomplish a production goal. The production process depends only of the value chain and the production flows. The production process necessary to obtain the production goal follows a model where each element of the value chain represents a stage of the production process. Each stage adds value to a product by transformation, transport or storage of that product across the value chain (Chacon et al, 2002) as it is shown in figure 2.
The production flow can be defined as the set of transformation stages determined by the production method using a set of resources specific for each stage (recipe) until to obtain the final product. Each stage must ensure the quality of the process and the products that are obtained.

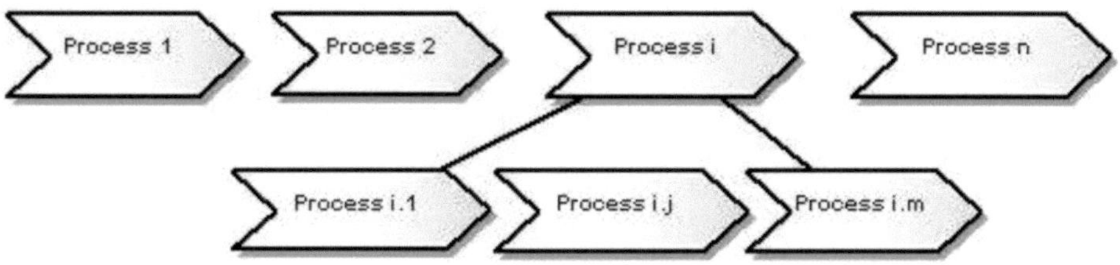

Fig. 2. The Value Chain

2.1 Description of a production unit in production process

Each stage for the value chain (raw material input / processing or transformation / storage) of the production flow is seen as performed by a Production Unit. Then, a production process is an aggregation of cooperating production units. The characterization of each production unit depends on how the resources evolve on the production process at the production unit. The resources dynamic can be: Continuous, Discrete or Event Dynamic (manufacturing), or Hybrid (for example in batch processes). Additionally, each PU performs operations that follow a recipe for the equipment that belongs to the production unit.
It is possible to find common elements that allow describing the PU in a generic way. Then a Production Unit must have:

- A process to obtain raw material
- A transformation process
- A storage process for final or intermediate products

At begin, production resources (equipment, HR, and raw material) are managed internally or requested to other PU by means of UP coordinator. Then resources are localized and translated from a UP to the UP that requires resources and intermediate/final products are translated from a UP to the client that can be extern to the enterprise or other UP. Thus, the resources localization and translation operation is responsible to guarantee the resources for a production process corresponding to a production method. The selection of the production method depends on the quality and quantity of raw material, capability and available capacity of the equipment. When the production process ends, resources are liberated and the obtained products are temporally stored, until other PU or final client request for the product.

In figure 3, we show the structural model of a PU with all its elements. This model includes the control, supervisory, and coordination/management system for each stage of the value chain. A PU drives the production methods and the internal configuration of the internal resources in order to obtain the production goal.

Using the above UP definition, we establish the basis to obtain the information that allow to fix a production schedule and to determine the state across the production process; Key Production Indicators: production quality, production capacity, expected production, etc., all of them necessaries to negotiate with other PU a production goal. A production agreement results from the production capacity and the existence actual or future of raw material.

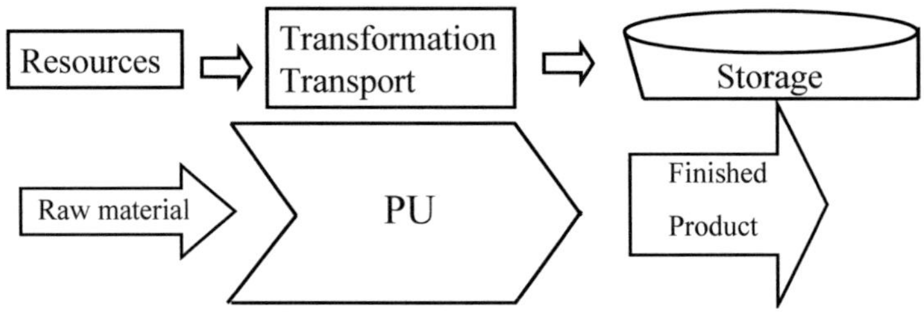

Fig. 3. Production model for a Production Unit

The structural model for the PU is obtained using UML (Cardillo et al, 2005), (Eriksson et al, 2004), (Muller, 1997), (Jacobson, www), and (WBF, www), where boxes represent classes and lines among the boxes describe relations among the classes. Lines describe three kinds of relationships: generalizations / specialization by arrows, associations by plain lines and compositions by lines with a rhombus in the end.

For the model shown in figure 4, we can see different entities that are components for a PU and its relations. Specially, it is shown an association, which record all different resource configurations, the production process, and control and supervision mechanisms for a particular production.

Also, it shows a classification for the resources used in production, that is employed by the planning function belong to the PU. The whole model has a specialization for several classes that they are not shown in the schema; each resource class has a behaviour that is described by means of a Discrete Event Dynamical System. The dynamics describes rules and operations for each class and the composition of dynamics define the behaviour of the PU.

Using the UML model shown in figure 4, a figure 5 represents the embedded procedure of making decision process for each stage in the value chain that allows considering each stage as an autonomous Production Unit.

The embedded model corresponds to the decomposition of the making decision process, which describes control procedures, supervision/coordination schemas. This reference model establishes three levels: the floor level where we can found the physical process (Transformation) and its control loops (observers, controllers, and transducers). All of them installed over the Control, Information, and Communications Technology.

In this way, the physical process is seen as the system to be monitored, supervised, and managed by an upper level that is the supervision level. Without lost of generality, a PU abstraction allows us to see as a physical process that can be part of a new stage in the value chain. This stage receives raw material, handle it, perform transformations by means of its own resources and generates new products for the next stage on the value chain or for the final client.

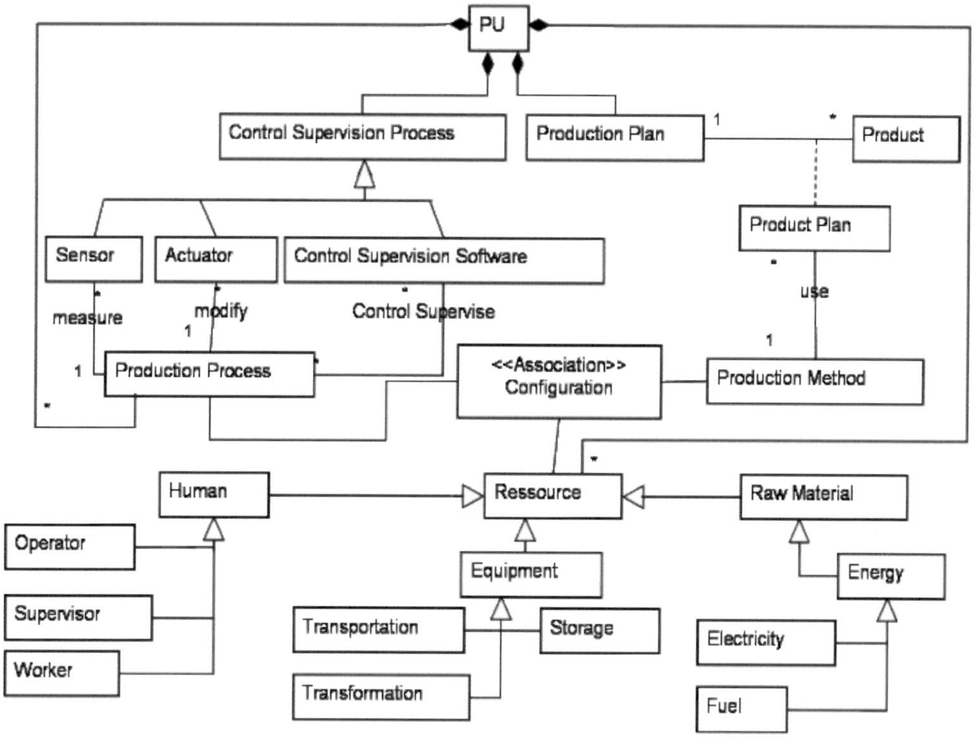

Fig. 4. The UML model for a PU

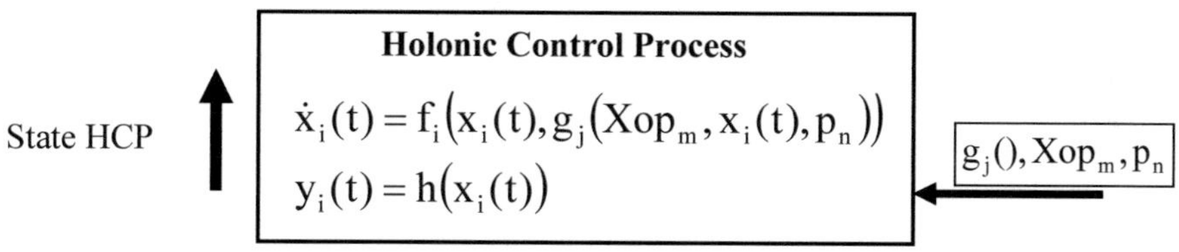

Fig. 5. Embedded Model of the Production Process

3. The holonic production unit

A holon for manufacturing enterprises is defined as a cooperative and autonomous construct able to transform, transport, store and/or validate information and/or physical products, [8, 12, 17, 18, 21]. A holon has the autonomy to create and control the execution of its own plans; it can cooperate with other holons to develop a viable plan in order to reach the global goal for the system. The cooperation among the holons is accomplished by means of a holarchy that evolves from the same organization. We have a holonic system.

For a holonic production system, the goal is to obtain all the facets for the decision and control procedures from the establishment of a production plan for the whole PU until the control/regulation at the floor level. The integrated automation is based upon a global vision of the production process, where each element that make part of the system must be take into account in order to control, supervise, and manage the production; see figure 6.

The automation schema is given by the construction of models that represents the PU in its structural and behavioural aspects. Control schemas ensure that the PU behaviour can be guaranteed inside a set of restrictions. The knowledge of the state of the physical system allows evaluating which will be the control actions that are viable in order to attain the desired state. At the system, the control is restraint by the accepted goal for the PU. The accepted goal is evaluated and negotiated taking into account the capabilities and availability of the PU.

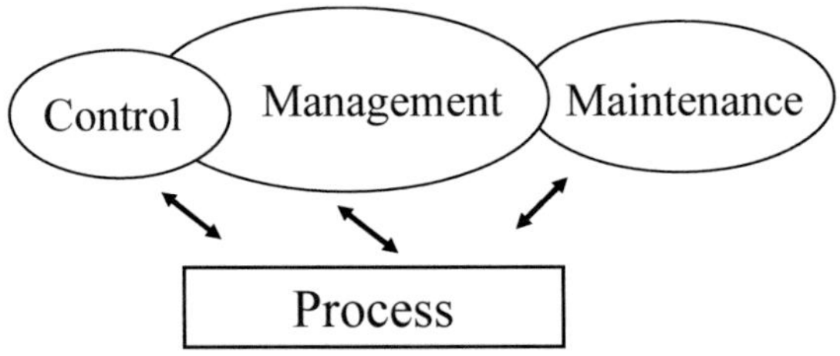

Fig. 6. Integral Automation Model of Production Systems

All the knowledge of the PU is internal to the PU, which has a behaviour that follows the physical/chemical laws for an operation mode. The composition of the laws for each component gives the behaviour of the whole PU. This description is the base to establish the control and supervision methods that can be performed by humans or by the technological system. Each production resource has a proper dynamic; the changes across the time and the information recovered from the process determine the accomplishment of the production goal or the impossibility to attain the goal due to the failure of one or more resources. Those changes must be managed by the proper PU, in order to solve internally with its own resources by means of a new internal configuration, or renegotiate in case to be necessary and establish a new internal goal.

To build a PU, it is necessary to fix implementation architecture for the PU. We propose to fix a structure composed by a "BODY" where the physical process (transformation /storage/transport) are instantiated, and they are performed using the physical assets of the enterprise, such as: reactors, warehouses, pipelines, etc. A "HEAD" that instantiate all the decision process, which are based on the knowledge of the plant, and is constituted by human and machines. A "NECK" that is the interface between the BODY and the HEAD that supports de decision making system and store, transport information, data, and knowledge of the PU; the NECK is spread on the Information & Communication Technology.

In figure 8, it is shown the correspondence between the HPU structure and the common technology found in production systems. Then, the body is the production process itself, the neck is composed by the networks, sensors, actuators, that allow obtain, transport data and by computers, industrial computers that process and store information/knowledge. Head is constituted by the humans and algorithms that take decisions.

In this way, we propose a recursive holonic structure that describes the desired composition for the Holonic Production Unit (HPU). The HPU is composed by one PU or by an aggregation of PU, which are holons. The Production Resource Holon, which be named Basic Holon results from the aggregation of basic process that are at this time controlled, supervised, coordinated, and managed.

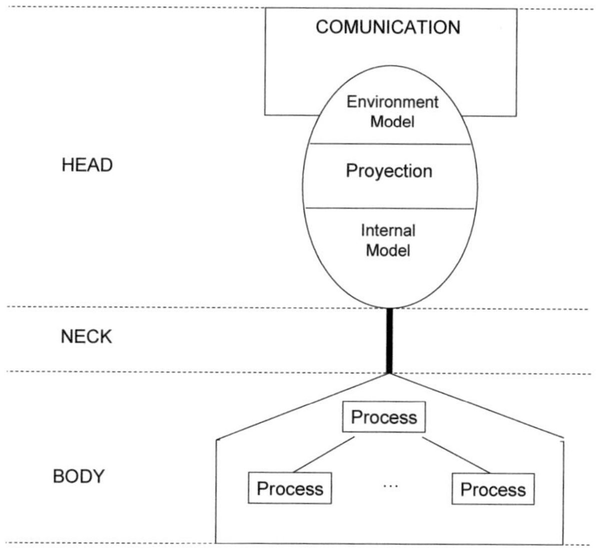

Fig. 7. Composition of a Holon in Production Process

Each basic holon has as body a controlled process, that it is formed by all the control loops. Each holon has the structure shown in figure 9, where its intelligence allows attaining the desired state by the Controller.

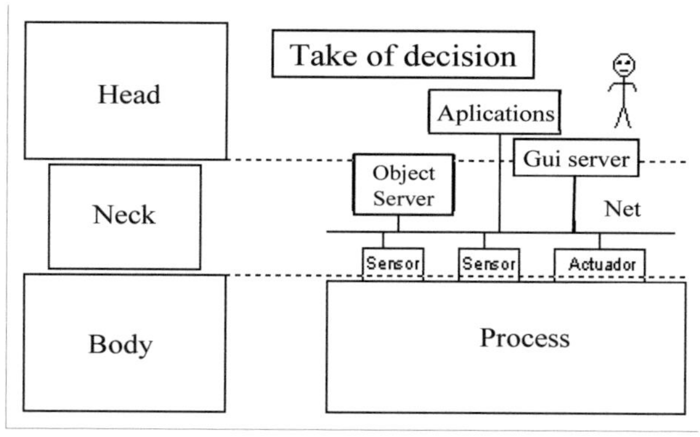

Fig. 8. Relationship between holon and the TIC infra-structure

3.1. Holonic control loop

In classical control, the basic functional unit is the control loop. If to the control loop we add as part of it, the fault detection capability and the capability of selection of new control strategies according to the operational condition, we talk about a Holonic Control Loop, the set of dynamical equations for different operational conditions must be part of the knowledge of the holon, also the controllers associated to each conditional operation. Neck for this holon is the set of elements that measure, adequate, transmit, and store physical signals from/to the process. Head has the mechanisms that allows determine the operational conditions and perform the control algorithms. Those algorithms are derived from the physical models of the plant and the control mechanisms that were selected for each operational condition.

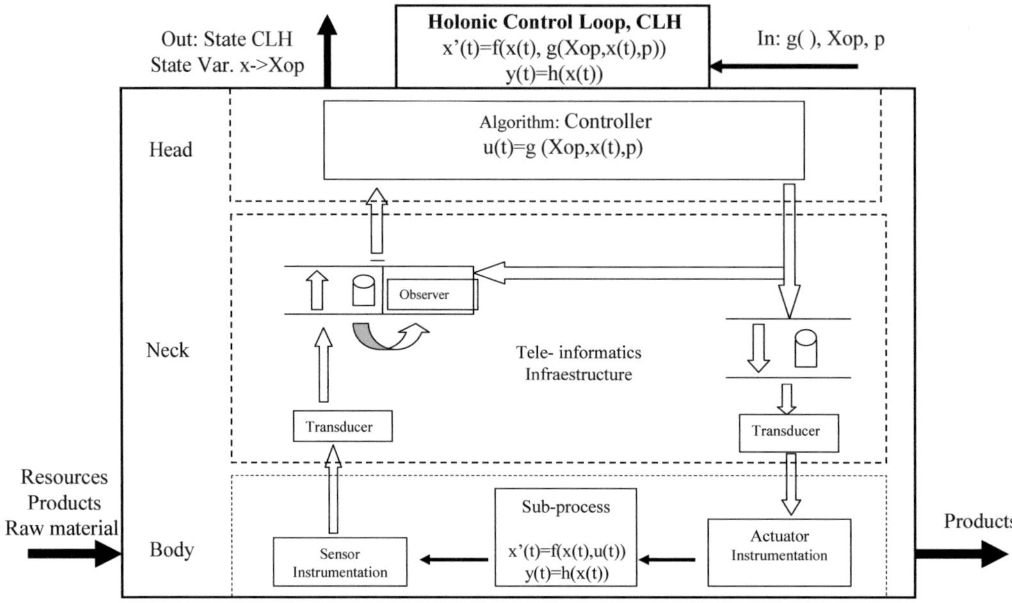

Fig. 9. Holonic Control Loop

The self-diagnostic mechanism allows to measure efficiency, conditions of the equipment; to establish requirements for raw material and equipment in order to perform the negotiation and to manage the maintenance activities. Then the holon with control loop is an autonomous system where raw material is received and final products are generated according a negotiated goal; it fix itself set-points select control strategies and control parameters. See figure 9.

3.2. Holon with process supervision

As we know, a process can have several control loops. Each control loop corresponds to an element that must be coordinated with others in order to have a valid operation. The set of control loops corresponds to a controlled holon. The process management of the set of control loops corresponds to a supervisor/coordinator. This making decision mechanism in classical supervision establishes set points, controller parameters or a control selection for a production method. If we add mechanisms to detect abnormal situations due to failures in control loops and incoherent states on the individual control loops for the global system, we have a supervised holon or holons with process supervision.

Without losing generality, the Holonic Controlled Process is a process that has *i* models (maybe more than one for a controlled loop), where each model has a combination of *m* of the nominal values of operation that can be reached with a combination of j of the controllers' types which are adjusted under an criteria/judgement determined by the loop with n parameters as it can seen in figure 10.

The Supervised Process Holon is conformed by a body that contains to the Holonic Controlled Process whose output is the state of the equipment and deviations of the consign, and whose input is the consign, type, and parameters of the controller to use, due to a set of tasks to complete that they have been evaluated and negotiated previously.

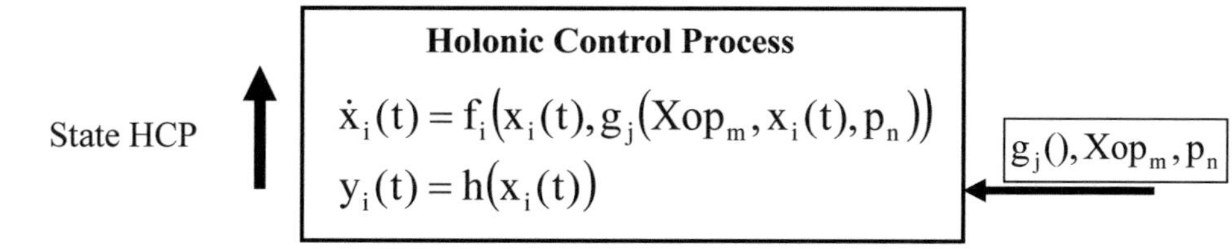

Fig. 10. Holonic Controlled Process

Figure 11 presents the Holonic Supervised Process, the neck conformed by all of the tele-informatics architecture and applications, which allow to detect and to send events (they are able to capture, to try, to store, to adapt, continuous information in events and vice versa). The neck sends consign to the Holonic Controlled Process. From the method of production instanced and it detects the states generated by the Holonic Controlled Process. The head or fault tolerant supervisor, this conformed by the mechanism of taking of decisions capable of to establish by agreement/negotiation the set of tasks to carry out for Holonic Controlled Process, starting from the knowledge of the state of the Holonic Controlled Process and of the group of activities to complete due to the method instantiated. The model of the mechanism of taking of decisions it is seen in a natural way as a model of hybrid system that is generally implanted as a Discrete Event Dynamical System (DEDS) described by the quintuple (X, U, Y, f (.), g (.)), where X is the group it didn't empty finite of states, U it is a group it didn't empty finite of the group or alphabet of controls, Y it is the group it didn't empty of output values.

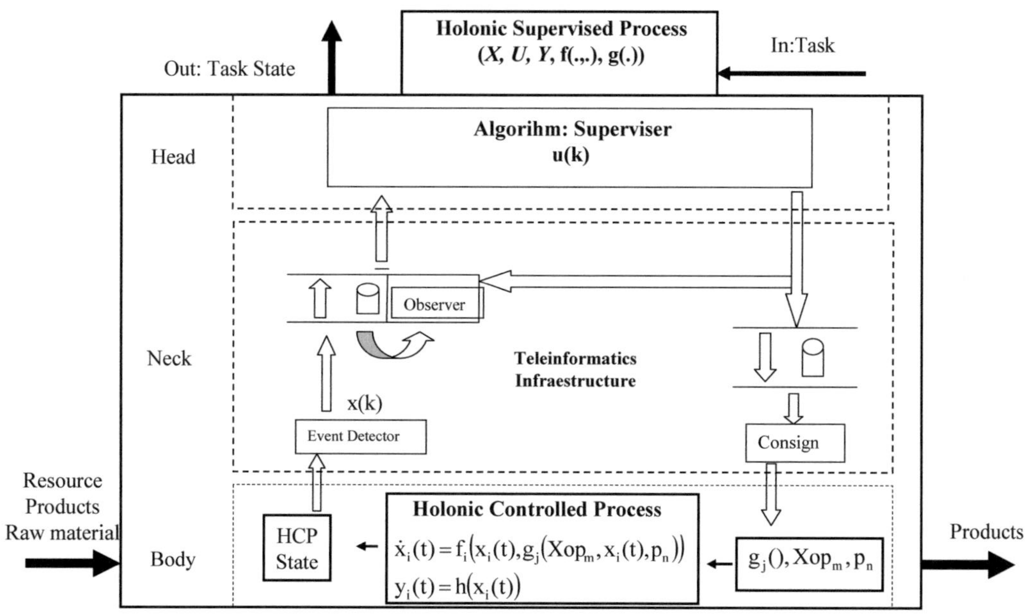

Fig. 11. Holonic Supervised Process

3.3 Production unit holon

In complex industrial processes, it is possible to have more than one supervisor where, conventionally, it is necessary a Coordinator for managing all of the supervisors. Figure 12 presents this to facilitate the coordination tasks to carry out, all of the set of Supervised Process Holon, that we will call Holonic Supervised Production.

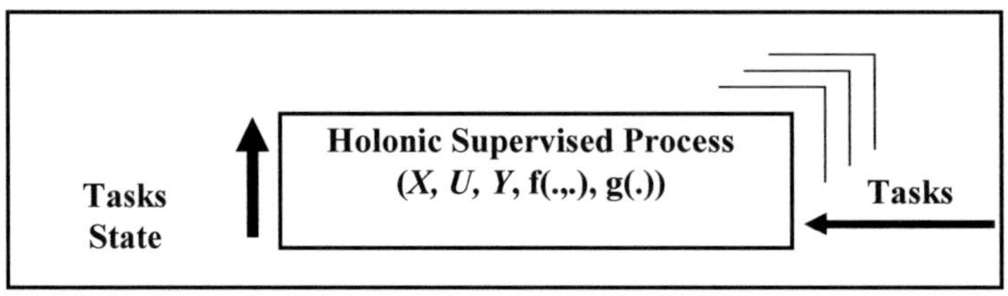

Fig. 12. Holonic Supervised Production

Without losing generality, figure 13 presents the Holonic Supervised Production can have i models, each model has a combined m of tasks nominal instanced of the production method that can be reached with a combined j types of configurations which are adjusted under an approach determined by the supervisor with n of parameters.

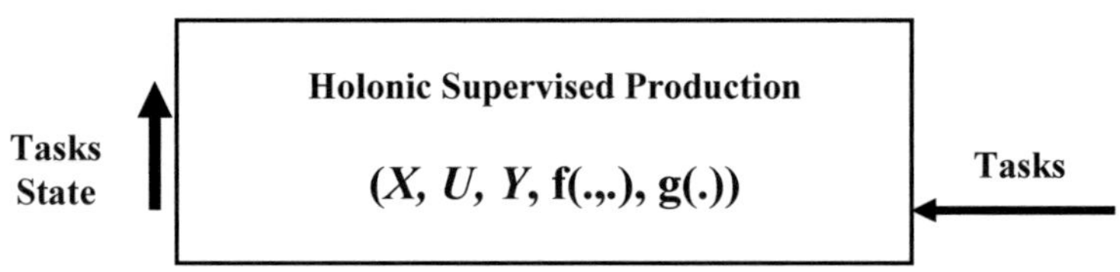

Fig. 13. Holonic Supervised Production

Thus, figure 14 shows the Production Unit Holon is conformed by a body that contains to the Holonic Supervised Production whose output is the state of the set of tasks agreed and the capacity of its resources (equipments, raw material, human resource, etc), the input are the set of planning tasks previously to complete the establish objective that it comes from an evaluated and negotiated recipe previously.

The neck is conformed by the all tele-informatics architecture and applications that allow to detect events (they are able to capture, to try, to store, to adapt, event in events information) like to leave a recipe to project the set of tasks to the Holonic Supervised Production. The head or Fault Tolerance Coordinator, this conformed by the mechanism of taking of decisions capable of, starting from the knowledge of the state of the Holonicly Supervised Production and of the set the commitments with the clients and supply, to indicate that it will be manufactured and with which recipe previously was do it the negotiation. The model of the mechanism of decisions making is seen in a natural way as a model of hybrid system, that is generally implanted as a Discrete Event Dynamical System (DEDS) described by the quintuple one (X, OR, AND, f (.), g (.)), where X is the group it didn't empty finite of states, OR it is a group it didn't empty finite of the group or alphabet of controls, AND it is the group it didn't empty of exit values.

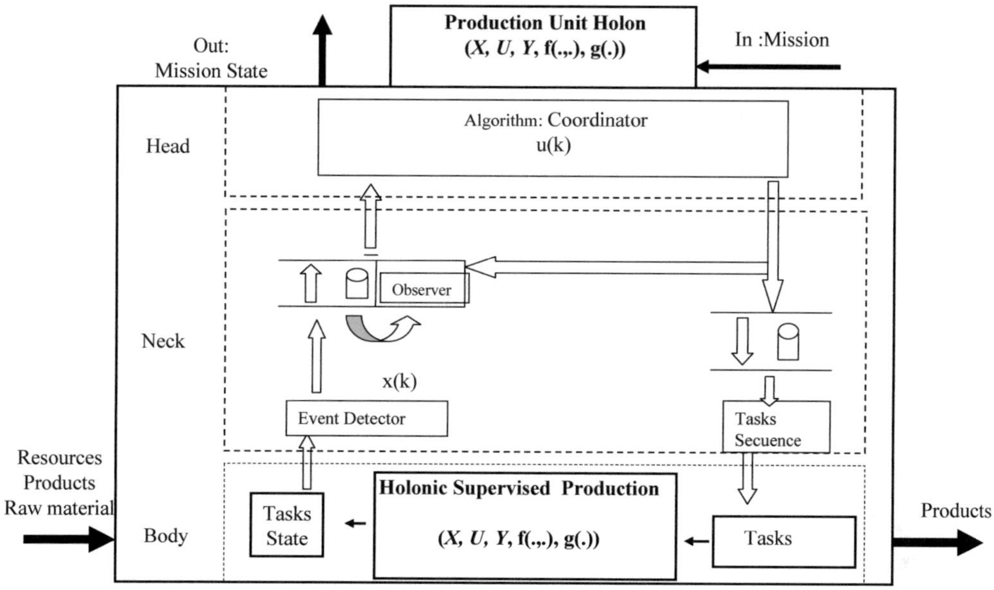

Fig. 14. Production Unit Holon

4. Holonic enterprise

A conventional enterprise is view like the set of production units associated to the chain of value of the productive process in a permanent manner since for this was constituted. A virtual enterprise is view like a set of enterprise associated to a chain of value of the productive process in a temporary manner since for this was constituted. Both use the chain of value of the productive process and for this reason the approach presented in this work, to each link of that chain of value we associate its a Production Unit Holon, see figure 15. Thus, a Holonic Enterprise this constituted for a body conformed by the set of Production Unit Holon associated to the chain of value of the productive process.

Under our focus and in order to facilitate the management to carry out, all of the set of the Production Unit Holon that we will call to Holonic Coordinated Production is shown in figure 16.

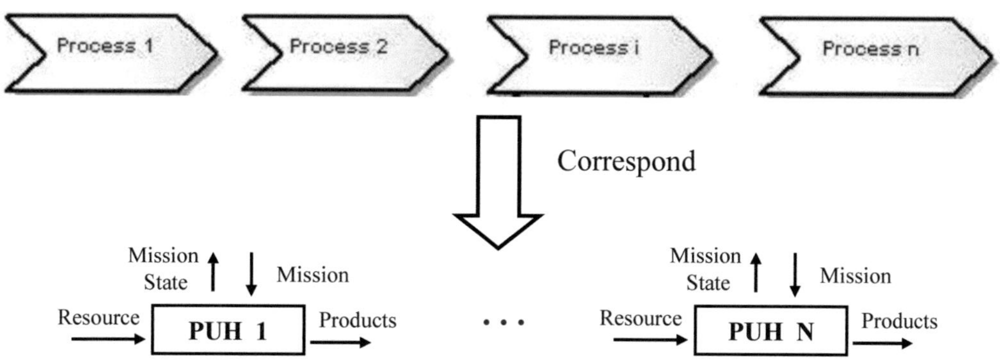

Fig. 15. Value Chain vs. Production Unit Holon

In figure 17, the Holonic Coordinated Production can have i models, each model has a combined m of tasks nominal instanced of the production process that can be reached with a combined j types of configurations which are adjusted under an approach determined by the coordinator with n parameters.

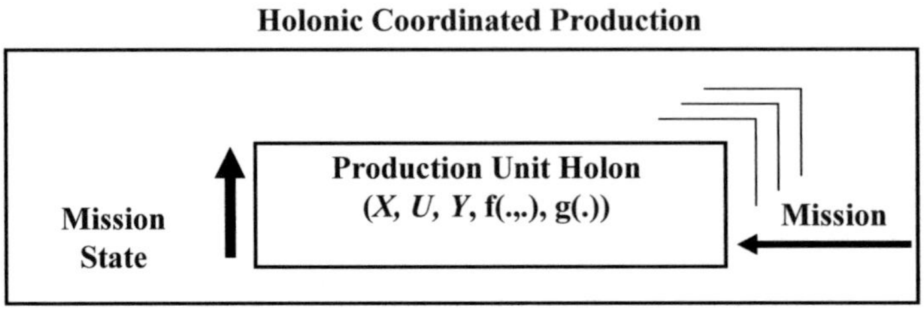

Fig. 16. Holonic Coordinated Production

The neck is constituted by the all of the tele-informatics architecture and applications, that allow to detect events (they are able to capture, to try, to store, to adapt, event in events information) like to leave a recipe to project the set of tasks to the Holonic Coordinated Production.

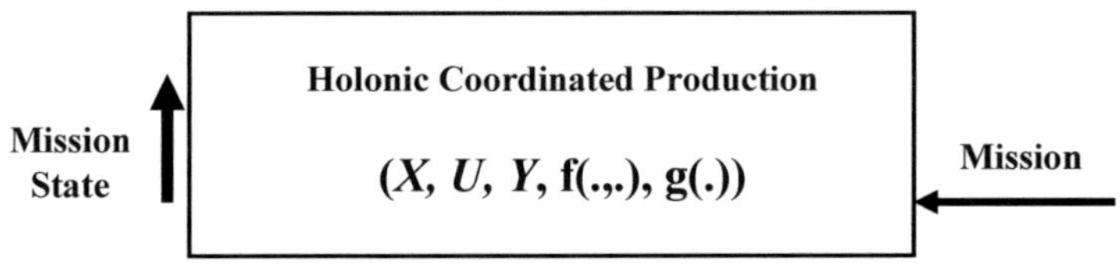

Figure 17. Holonic Coordinated Production

As we can see in figure 18, the head or Fault Tolerance Manager, this conformed by the mechanism of taking of decisions capable of, starting from the knowledge of the state of the Holonic Coordinated Production and of the set the commitments with the clients and supply, to indicate that it will be manufactured and with which process previously negotiation. The model of the mechanism of decisions making it is seen in a natural way as a model of hybrid system that is generally implanted as a Discrete Event Dynamical System (DEDS) described by the quintuple one (X, OR, AND, f (.), g (.)), where X is the group it didn't empty finite of states, OR it is a group it didn't empty finite of the group or alphabet of controls, AND it is the group it didn't empty of exit values.

4. Gratefulness

- Scientific, Humanistic, and Technological Council (CDCHT), University of Los Andes, by mean of the project I-867-05-02 –A, Production Unit Holon.
- To the program France - Venezuela, ECOS NORD: Production Enterprise Nets.

5. Conclusion and future work

In a natural manner the business model, the chain of value, the product flow and the holonic approach allow us to capture a coherent model of the production process in all of their globallity and complexity giving origin to the shown embedded model.

The description invariant embedded presented of the Holon in production processes allows us to catch in a clear and precise way, the holonic recursive definition of the biology starting from the holarchy cell/tissue/organ like holonic functional basic unit of a body.

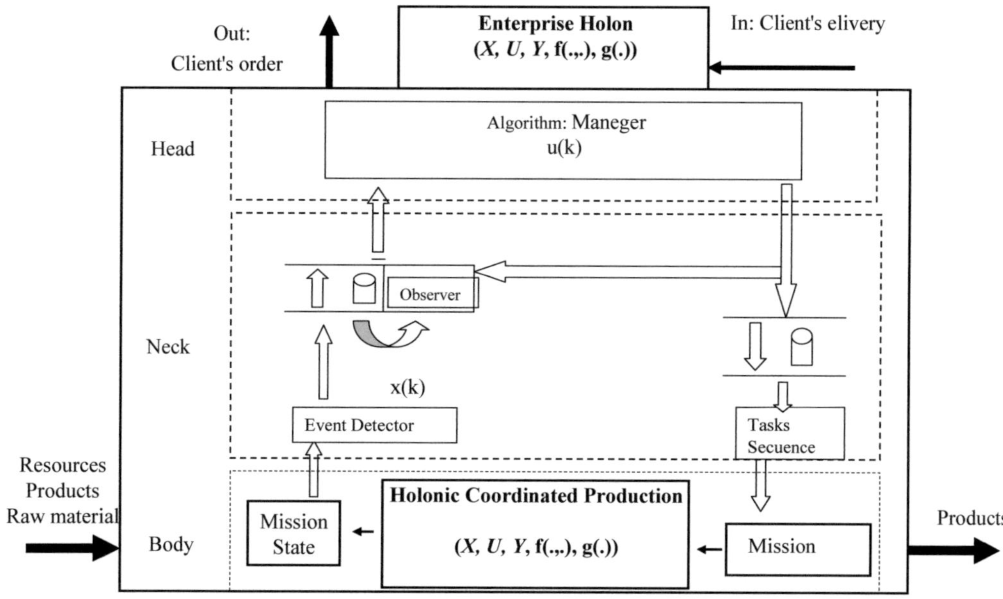

Fig. 18. Enterprise Holon

This proposed recursive structure of the Holon to describe production processes allows us to present one scheme that is topologically the same, which should be configured in each of the instances required to model a company, including the established description of layers of the standard SP95.

Our future work is devoted to complete a set of elaborated cases of study, where we show the implementation of these concepts in several domains.

It is necessary to complete the description of the proposed Production Unit for each case. The negotiation mechanism must be implemented using a protocol that arrives to an agreement, which determines the cooperation among the units making part of the holonic enterprise.

6. References

A. Muller, (1997), Modelado de Objetos con UML. *Eyrolles y Ediciones Gestión 2000, S. A.*

Chritense J. H, (1994), Holonic Manufacturing systems: initial architecture and standards directions, *Proceedings of the First European Conference on Holonic Manufacturing Sistems*, European HMS Consortium, Hanover.

E. Chac0n, I. Besembel, F. Narciso, J. Montilva y E. Colina, (2002), An Integration Architecture for the Automation of Continuous Production Complexes. *ISA Transactions. Journal of the Americna Intitute of Physics*. Vol 41, Nº 1, pag. 95-113.

E. Chacon, I. Besembel, M. Rivero, J. Cardillo, (2007), Holonic Production Process: A model of Complex, Precise, and Global Systems, *ICINCO´07*, Anger-Francia.

Eriksson, Penker, Lyons, Fado. UML 2 Toolkit. *Jhon Wiley*, 2004.

H. Brusel, J. Wyns, P. Valckenaers, L. (1998) Bongaerts y P. Peeters. Reference architecture for holonic manufacturing systems: PROSA. *Computer in Industry*. Vol. 37. pp. 255-274.

I. Jacobson, G. Booch y J. Rumbaugh. Specification of the UML. Rational Software. www.ibm.com/software/rational/uml/.

J. Cardillo, E. Chacón, I. Besembel, M. Rivero, (2005), Unidad de Producción como célula fundamental de los procesos Holónicos de Producción, *V Congreso de Automatización y Control, CAC 2005*, Universidad Simón Bolívar, Caracas-Venezuela.

Koestler A. (1967), The Ghost in the machine, Arkana Paris.

McHugh P., Merli G. and Wheeler W. A., (1995) *Beyond Business Process Reengineering: Towards the Holonic Enterprise*, John Wiley, New York, N.Y.

Ulieru, M, Brennan R., Walker S., (2002), The Holonic enterprise: A model for internet-enabled global manufacturing supply chain and workflow management. *Integrated Manufacturing System*, 8 (13), 538-550.

University of Hannover, (2000). Holonic Manufacturing Systems. Web Page, http://hms.ifw.uni-hannover.de/.

Van Brussel H., J. Wyns, P. Valckenaers, L. Bongaerts, and P. Peeters, (1998) Reference Architecture for Holonic Manufacturing Systems: PROSA; *Computers in Industry*, 37, 255 - 274. Elsevier.

WBF.BPML: Business Process Modelling Language. World Batch Forum. http://www.wbf.org.

Wyns J., (1999) Architecture for Holonic Manufacturing Systems: The Key to Support Evolution and Reconfiguration. *PhD thesis, K.U.Leuven*, PMA Division.

17

Artificial Intelligence Rationale for Autonomous Vehicle Agents Behaviour in Driving Simulation Environment

Vassilis Charissis[1] and Stylianos Papanastasiou[2]
[1]*University of Glasgow, Glasgow School of Art, Digital Design Studio*
[2]*University of Glasgow, Department of Computing Science*
UK

1. Introduction

In recent times, in-vehicle notifications have proliferated with a focus on the exhibition of technological prowess rather than the fulfilment of actual driving needs. In effect, information portrayed by automotive infotainment devices, while useful, is often ignored by the driver due to field of view limitations associated with traditional instrumentation panels. Not surprisingly, under poor visibility conditions and at motorway-level driving speeds, such systems do not effectively present useful information to the user.

Contemporary Head-Up Display (HUD) experiments have focused on adapting the aviation-specific characteristics of HUDs to driver-specific needs, obsolescing functionality and simplifying operations where necessary. The more mature approach of these preceding works has revealed that although in-vehicle HUD technological advances have overcome most implementation issues, the related user-centred interface design is in its infancy prohibiting the HUD's unique features from being successfully exploited. Towards addressing this issue, in previous work, we have designed and implemented a functional prototype of a Human Machine Interface (HMI). Specifically, the proposed HMI system introduces a novel design for an automotive HUD interface, which aims to improve the driver's spatial awareness and response times under low visibility conditions. Particular emphasis has been placed on the prioritisation and effective presentation of information available through vehicular sensors, which would assist, without distracting, the driver in successfully navigating the vehicle. In order to evaluate the effectiveness of the proposed HUD system, we developed a driving simulator based on an open source racing program. Leveraging an open source solution has resulted in manageable levels of incurred expenses whilst allowing us to deliver a flexible simulation application for HUD evaluation.

This chapter discusses the artificial intelligence (AI) as developed for the agent vehicles of our open source driving simulator. The simulator was explicitly designed to measure driver's performance when using the proposed HUD interface and compares its effectiveness to traditional instrumentation techniques. Intuitively, human cognition complexity poses the largest challenge for creating a model of life-like driver's behaviour for any type of traffic flow. Presuming that specific driving characteristics apply to all human drivers, as dictated by common sense, an attempt was made to form a generic reaction

pattern from a list of possible reactions to given situations. The AI embedded in participating vehicles in the simulations embraced certain assumptions, which are outlined later in the chapter. In this way the development time required for the AI routines was reduced by discounting rare occurrences that would indicate a severely misbehaving and irrational human driver. However, it was deemed necessary to develop AI controlled vehicles that could perform potential human miscalculations in order to enhance the realism of the simulation scenarios and their degree of mapping to real-life situations,. Imitation of human error by the robot vehicles, such as failure to brake on time, became a simulation feature that substantially enhanced the driver's immersion in the synthetic environment.
Overall, this chapter elaborates on the modelling process of the agent vehicle's AI and discusses contemporary attempts to develop similar AI simulations for other simulation facilities. Notably, throughout this work, a concerted effort was made to ensure that the simulator scenarios were fair representations of potential real-world accident-prone situations; this has been made possible after attentive examination of the statistical data on driving scenarios provided by the Strathclyde Police Department situated in the City of Glasgow, UK.

2. Proposed HUD simulation

Early attempts to employ Head-Up Displays (HUDs) in automotive environments were greatly influenced by research in aviation and subsequently exhibited limitations and side-effects derived from the misuse of Human Machine Interaction (HMI) principles. Overall, the largely uncritical adoption of aviation HUD technologies held back the potential of automotive HUD use as conveyors of information provided by vehicular sensors. Even under such unfavourable conditions, however, research results have indicated an improvement in the driver's response time in some studies, which in turn hint on the potential of such systems. Recently, contemporary HUD experiments have focused on adapting the aviation specific characteristics of HUDs to driver-specific needs, obsoleting functionality and simplifying operations where necessary. The more mature approach of these preceding works has revealed that although in-vehicle HUD technological advances have overcome most implementation issues, the related user-centred interface design is still in its infancy prohibiting the HUD interface's capabilities from being successfully exploited.
This study introduces a novel design for an automotive HUD interface which aims to improve the driver's spatial awareness and response times under low visibility conditions, offering only the crucial incoming information in a graphical representation manner. As such, a working prototype of a HMI has been designed and implemented to fulfil these requirements (Charissis et al. 2008a). Particular emphasis was placed on the prioritisation and effective presentation of information available through vehicular sensors, which would assist, without distracting, the driver in successfully navigating the vehicle under low visibility conditions (Charissis et al. 2008c). The harmonic collaboration between the human (driver) and machine (vehicle) elements has been supported by utilizing the machine as a collecting and distilling hub of information. Essentially, the human agent has been urged towards improved decision making through careful consideration of user characteristics and needs. That positive effect on the driver has been achieved by conveying the distilled information through the visual cues of the HUD interface reducing driver's response times

in critical situations (Steinfeld & Green, 1995; Charissis et al. 2008b). In order to evaluate the system's effectiveness we used two simulation scenarios in subsequent user trials. These simulations required that substantial attention be given to the AI development in the computer-controlled vehicles. The following sections elaborate on the development of the driving simulator and the programming of the AI vehicles.

3. Driving simulator as an evaluation tool

Driving simulators are an indispensable tool used both in the automotive industry and academic research. Current state of the art simulators include the U.S. National Advanced Driving Simulator (NADS) (Papelis, 1994; Chen et al., 2001), the Swedish National Road and Transport Research Institute (VTI) (Nordmark et al., 2004; VTI, 2007), the Transport Research Laboratories (TRL, 2007) and the Leeds Advanced Driving Simulator (LADS, 2007) in the U.K. These sophisticated simulators can take into consideration numerous driving factors such as deceleration and acceleration forces, weather conditions etc., as well as be able to record, analyse and evaluate the results in real time (Kantowitch, 1999). However, the construction, upgrade and servicing costs of such elaborate hardware and software components can often be prohibitively high. Overall, and in view of conducting academic research, the task of designing and implementing a driving simulator, even one of low fidelity, can involve substantial financial overhead. Hence, academic institutes often rent the facilities of traffic research centres or automotive industry studios which offer specialised driving simulators for testing various automotive systems and devices. Intuitively, the existence of certain financial constraints may require the generation of alternative solutions.

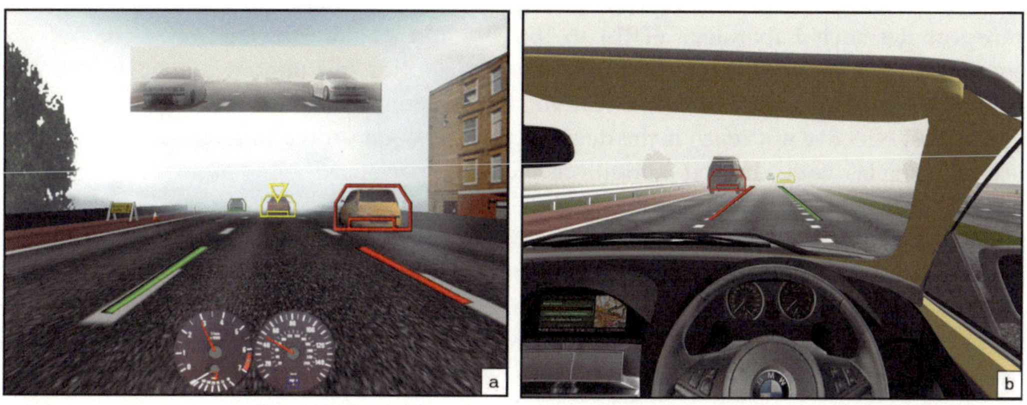

Figure 1. A screenshot of the (a) open source driving simulator and (b) the VR driving simulator developed based on the open source set-up.

As the use of a driving simulator was necessary for the evaluation of the prototype HUD interface, we opted for an alternative low-cost approach. After careful consideration of the available funds and time constraints, the decision was reached to purchase off-the-shelf hardware components and develop the code on the open source TORCS platform (Wymann, 2006; Centelles, 2006). This solution accounts for a manageable level of expenses whilst delivering a flexible simulation application for HUD evaluation purposes.

The reconstruction of accident events in a synthetic driving environment required the deployment of life-like driving virtual participants in the scene, thus a fleet of "intelligent" vehicles which could make maneuvering decisions in real-time and mingle orderly to form mixed traffic environments was deployed in the simulations examined here. The following section presents the challenges encountered during the implementation of realistic AI for the traffic vehicles involved in the events.

4. Open Source Driving Simulator (OSDS)

Recent advancements in processing capacity and graphic representation technologies for personal computers have significantly reduced the cost of realistic computer-generated environments. This vastly increased computational ability allows for vehicle and environment simulation that can reasonably approximate the definition performance of significantly more expensive driving simulator models (Hogue et. al. 2005). Singular PC units are able to support various features of a simulation process. Nevertheless, multitasking computation can potentially overload the system and affect the simulator's performance (screen refresh rate); therefore the simulation process has to be optimised with the view to meet the needs of the experiment (Rosenthal et al., 2003).

A solution to the issue of singular unit performance can be achieved by clustering numerous PC simulators to effectively form a much more capable whole. In this case, PC simulators can divide the tasks and execute them separately; this combinatory approach amplifies the system's capabilities and may even reach the processing power of high-end servers as used for complex simulations. This newly formed breed of driving simulators has been adopted by high-end industrial and academic simulation facilities worldwide such as the Advanced Institute of Science and Technology (AIST) driving simulator at Tsukuba, Japan, the Transport Research Laboratory (TRL) in the UK and so on. However, even though the computational costs have been minimised, these facilities still uphold immense operating costs for upgrading, maintaining and running their systems.

A more cost effective approach is the development on open source software able to simulate various vehicles and driving conditions. These specialised simulators can be easily customised to create a sufficiently realistic environment for testing various driving scenarios. Nonetheless, as the majority of such programs have been developed for gaming purposes, namely racing, they require substantial changes to the core of the programme in order to comply with a real driving scenario as discussed later in this chapter. The combination of off-the-shelf hardware (PC) and open source driving simulation software was, for our purposes, the most cost-efficient and flexible solution. The idea of creating such a custom-driving simulator benefited from input by various researchers from British, European and American universities (Charissis et al. 2006)[1]. A screenshot of the developed simulator in action is shown in Figure 2.

---

[1] This work has benefited from the feedback and contributions of a number of colleagues from academic institutions including the computing science departments of the University of Glasgow, the University of California, Berkeley, Glasgow Caledonian University and Aristotle University, Thessalonica. Where applicable each contribution is clearly marked and identified in this chapter.

Figure 2. Screenshot of the driving simulator with the HUD interface

4.1 Open source driving simulator software development

We conducted extensive market research which indicated that financial costs for software and hardware development would be considerable for building a medium fidelity driving simulator. Renting a simulator facility was also not a viable option as the high daily rates could not be covered by the research funds. It was therefore vital for the study to develop a custom simulating system with minimum expenditure. This necessity lead to collaborative work along with other researchers on the development of such a simulator.

The open source "The Open Racing Car Simulator (TORCS)" provided a suitable solution for the development of a custom driving simulator. Note that TORCS is a multiplatform, 3D car racing simulator based on OpenGL. This software formed the basis for the development of the OSDS. The source code was originally developed on the Linux platform but soon efforts were concentrated on the Windows operating system in order to ensure maximum compatibility with peripherals. The simulator's vehicles, i.e. the "robot" drivers, can be treated as agents with customisable AI. The purpose of programming the AI for each "robot" enables the participant to programme the attribute of their own racing vehicles (agents). As a result racing competitions can be held amongst pre-programmed vehicles by different authors on the Internet, or locally with the user racing against the computer vehicles. The navigation of the vehicles can be achieved through the use of a keyboard, mouse, joystick or steering wheel. Due to the racing nature of the software, the robot vehicles imported into the OSDS required substantial reprogramming in order to adhere to the British Highway Code. The following section will elaborate on the simulation of the HUD interface and the environment visualisation.

A. 2D HUD interface visualisation

The functionality of the HUD interface components have been transferred to the simulated environment they it would purportedly be in a real vehicle. The symbols were programmed to follow the movement of the vehicles and alter their size and colour according to the relevant distance from the user's vehicle. The two dimensional projection of the OSDS did not offer depth of field, which could have further enhanced the realism of the experiment. However, the advanced AI of the robot vehicles and the accurate superposition of the HUD symbols enhanced the sense of presence; the derived results were based on genuine driving efforts to respond to the accident events.

B. Simulation of low visibility conditions

Given that the primary aim of the study was to measure and compare drivers' response times in adverse weather conditions, such environmental conditions had to be replicated accordingly. A survey of weather conditions literature defines that dense fog results in low visibility conditions, which produce significant traffic flow disruptions typically below the 100m viewing distance mark (Cavallo et al. 1999; BBC Accidents, 2007). As such heavy fog, with visibility less than 50m, was simulated in four of the driving scenarios. Note that the visibility percentage could be adjusted via a simulation parameter achieve a satisfactory fog quality. Specific parts of the code were improved in order to simulate the desired fog conditions and realistic depth of field. The area of fog visualisation was revisited in our virtual reality driving simulator (VRDS) which particularly investigated the HUD's focusing distance as discussed in (Charissis et al., 2007b; Charissis & Naef, 2007).

C. 3D visualisation

To achieve a semblance to reality the testing track was designed and modelled based on a section of the M8 motorway in the outskirts of Glasgow, Scotland as depicted in Figure 3 (item b). The simulation model was set to deal only with motorways with three lanes in each direction and one hard-shoulder lane. In order to reduce the amount of 3D geometry, the track was simplified to only one direction, in this case the route from Glasgow to Edinburgh (anticlockwise). A closed circuit was created by seamlessly connecting three main motorways thus forming a triangular track shape. This model did not incorporate ramps on motorways or intersections on rural roads. 3D visualisations of the important intersections and most recognisable landmarks were incorporated in the final model as illustrated in Figure 4. Such visual cues helped the drivers to recognise the environment.

To populate the track, the 20 most popular vehicles, as seen on British roads, were modelled in low polygons geometry (gaming quality). Whereas the original program offered a wide choice of vehicles, the majority of them were racing cars and hence not appropriate for this experiment. The 3D objects and track were modelled using Alias Maya 6.5, and after a distilling process of exports in various formats, they were imported into the simulator using Track Editor and AC3D software. Additional alterations to the XML code were needed to allow compatibility with Windows XP.

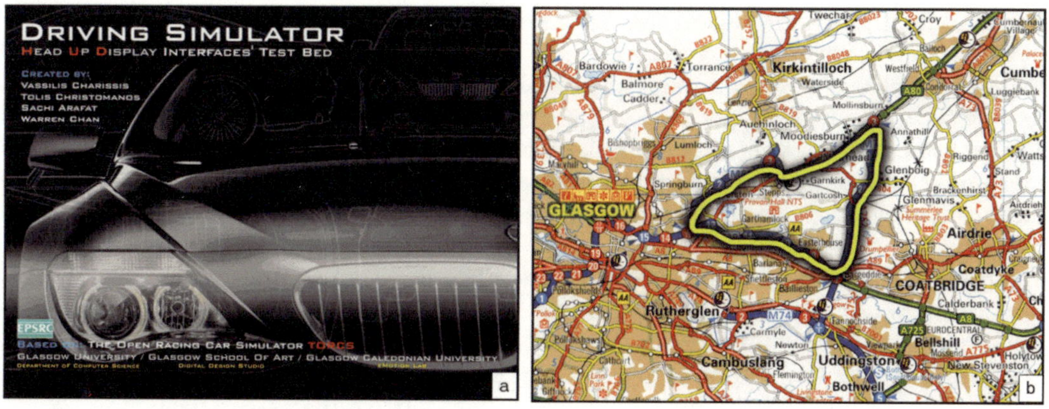

Figure 3. (a) Introductory screen of the OSDS, (b) Simulation track based on real motorway routes.

Figure 4. (a) Landmarks in real environment and (b) in the simulated track

4.2 Open source driving simulator hardware

The development of a cost efficient simulator could be mainly achieved by maintaining the costs of the hardware at minimum levels. However the keyboard or mouse solution did not seem appealing, as it could subconsciously be reminiscent to the user of a computer game. As a full-windshield HUD evaluation required a "close-to-real" driving environment it was deemed essential to purchase an "off-the shelf" steering wheel (Logitech Driving Force Pro GT4). This hardware item is categorised in the range of gaming products, however that particular steering wheel model offered a quite realistic feel (three rotation circles from left to right) and a well-implemented force feedback function (realistically transferring road bumps and the feeling of vehicle's drifting or braking). Note that his component required additional programming as the open source programme failed at first to recognise the device automatically. The steering wheel was also accompanied with a bundle of other equipment which included foot pedals for accelerator and brake (see Figure 6).

The software was run on a custom built PC with dual 64bit processors (AMD Opteron-242) and a Quadro FX1100 graphics card. A frame-rate of 60-90 fps was effortlessly maintained by the system. Although the simulator could perform well on a low specification laptop, this powerful dual processor system was built with the view to accommodate simultaneously more than one driver for future experiments.

Emphasis was also placed on the implementation of a driving cockpit environment; a real driving seat (a Gamepod, "bucket seat") formed in large part the driver's virtual "cockpit". The seat stood on a railing system on the top of a metallic structure which also supported the minimalistic dashboard and the steering wheel. The railing system allowed users to alter the positioning of the seat, the steering wheel and the pedals to match their physical dimensions and driving preferences.

The relevant hardware and software equipment had to be accommodated in a facility that could host two-week long simulation trials and experimentation as driver reactions were recorded. The Glasgow Caledonian University generously offered the E-motion Lab of the Mathematical and Computing Sciences Department which was explicitly equipped with numerous recording and observation devices such as video cameras, eye-fixation recorders, motion detectors and a fully developed observation suite and control room. The academic staff aided us greatly in setting up the connection of the driving simulator devices to the facility's additional equipment.

The observation suite hosted the driving simulator, which was projected on the lab's Panasonic 42 inch plasma screen positioned approximately 2m away from the driver's seat, as shown in Figure 5. The control room was equipped with monitors showing, in real time, the video stream from the cameras in the simulation area. The focus and the position of the cameras could be adjusted also in real time through a joystick on the control console in the observation room. The two videos could be seen either on separate monitors or both on the same window (screen on screen) thus enabling the researcher to observe the driver's driving process (via a screen camera) and his/her facial expression and posture (via a camera focused on the user) The following section presents the challenges encountered during the implementation of realistic AI for the traffic vehicles involved in the events.

5. Accident scenarios simulation

As Boer et al. (Boer & Ward, 2003) define it, an event is a situation requiring a corrective response such as, say, a situation arising when safety margins are violated. In order to evaluate the driver's response to accident events, the AI vehicles employed in the simulation scenarios had to generate the right circumstances, which would eventually lead to such an event occurring.

Note that the agents involved in the simulated scenarios were programmed to react and manoeuvre according to the flow of traffic and external events, whilst demonstrating natural driving behaviour. Every vehicle would obey certain rules which "urged" it to imitate its real life counterpart while its overall behaviour was influenced by several factors. In particular, each vehicle calculated, in real time, information drawn from its surroundings (i.e. other traffic vehicles) as well as information about visible accident hotspots such as enclosing side barriers, bridges, traffic cones and so on. The above information was clustered in the agents' intelligence creating logical links among possible events and responses, thus forming the framework for each individual agent's AI.

The following subsection elaborates on the driving scenario development. Section 5.2 explicitly analyses the macroscopic and microscopic driving scenario approaches, which consequently form the generic AI attributes of the vehicle agents presented in Section 5.3. In turn Section 5.4 presents the drivers' patterns infused in the agents' driving attributes.

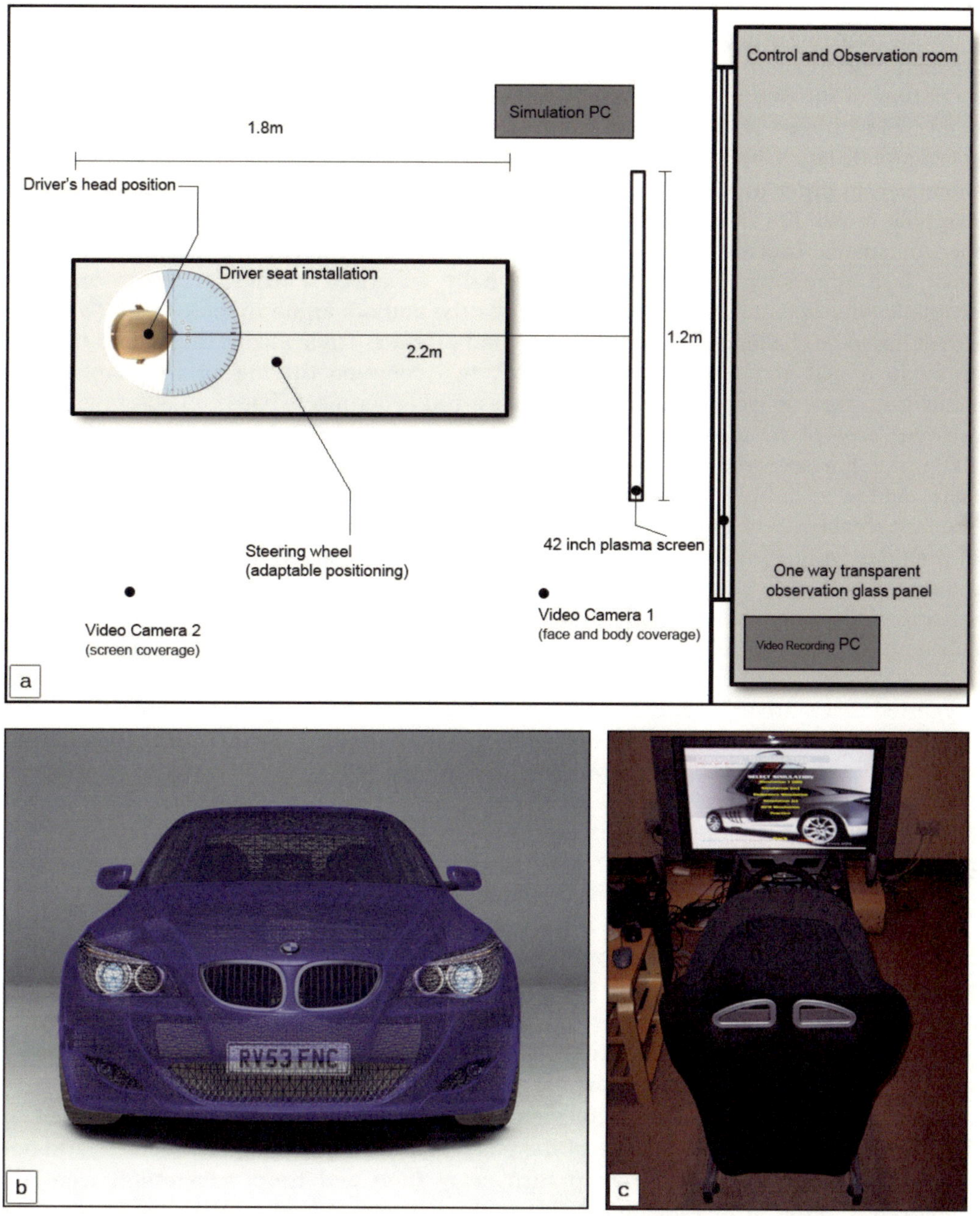

Figure 5. (a) Open Source Driving Simulator (OSDS) schematic top view, (b) example of 3D visualisation of simulated vehicle, (c) driving seat and positioning overview

5.1 Driving scenario development

The development of traffic scenarios was accomplished through careful inspection of data provided by actual traffic police reports. These statistics and planning diagrams aided in predicting drivers' possible reactions (SPD, 2004). Subsequent analysis showed that two particular car-following scenarios occur fairly frequently and reveal a high fatality rate. A detailed description of both such scenarios is given later in this subsection; however, before an outline of the two scenarios may be presented, it is useful to denote certain constraints and considerations evident in both. Firstly, note that for validation purposes, the movement, speed and distances between the vehicles had to adhere to the British Highway Code.

Moreover, in order to enhance the realism of the simulation scenarios and their degree of mapping to real-life situations, the AI controlled vehicles had to perform potential human misjudgements. Imitation of human errors by the robot vehicles (i.e. failure to brake on time) is a simulation feature that substantially enhances a driver's immersion in the synthetic environment (Allen et. al., 1997). As the authors argue in Park et. al. (2004), the driver has to be challenged in order to react and produce driving skills that would normally apply in a real accident situation. Overall, two common driving situations of a "car-following" scenario were developed for the test-bed experiments based on observations and accident prompt strategies produced in previous research (Daganzo, 1999; Smith et al., 2003). All the scenarios were presented in a motorway environment with heavy fog featuring low visibility (clear view being available at under 50m distance).

The first scenario used in this work is a variation of a generic car-following model described in previous work (Brackstone & McDonald, 2003). In that work, the user drives along the motorway and after having travelled a distance of 2km, the lead vehicles brake abruptly, causing following vehicles to decelerate rapidly. Intuitively, this event increases substantially the chances of vehicle collision. A previous study focusing on mapping of driver's possible reactions in similar car following accident scenarios (Smith et. al., 2003) has suggested that a driver's performance is comprised of four driving states: low risk, conflict, near crash and crash imminent, corresponding to four different warning levels respectively. As such, the first scenario was developed along these guidelines in order to evaluate the HUD's ability to effectively convey these four collision states to the driver. Analysing the driver's performance under such discrete time segments, each mapping onto these four pre-collision periods, has provided the study with the ability to identify, for every stage, the impact of the added HUD information over a typical HDD.

The second scenario is a variation of the original car-following situation in which the user drives for 5km following a lead vehicles' group, without any major occurrences taking place. After 5km the road forms a sharp turn (120 degrees) underneath a bridge thus creating an accident-prone situation. The difficulty of the scenario is amplified by the adverse weather conditions, in this case heavy fog, which dramatically decrease the driver's visibility, and by the addition of slow moving traffic congestion positioned at the exit of the turn. As before, the four states of collision warnings observed in other work (Smith et. al., 2003), were infused in the rationale of the potential accident driving scenarios. In addition, collision-prone issues such as sharp turn negotiation and traffic congestion were also employed in order to investigate driver's performance with and without the assistance of the proposed HUD interface.

In both scenarios the user is forced to respond instantly, either by manoeuvring around the accident point or by braking. The robot vehicles involved in the scenarios are programmed to minimise the possibility of accident avoidance, as the experimental focus was to gather detailed measurements of drivers' response times and distance from the lead vehicle with respect to the accident event. A schematic overview of both scenarios is shown in Figure 6.

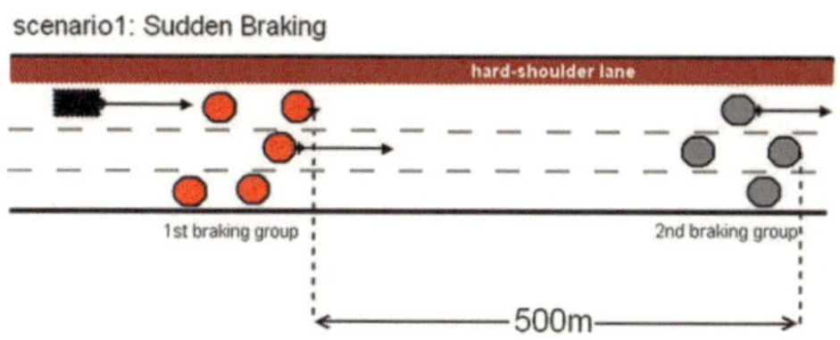

(a) Sudden break scenario

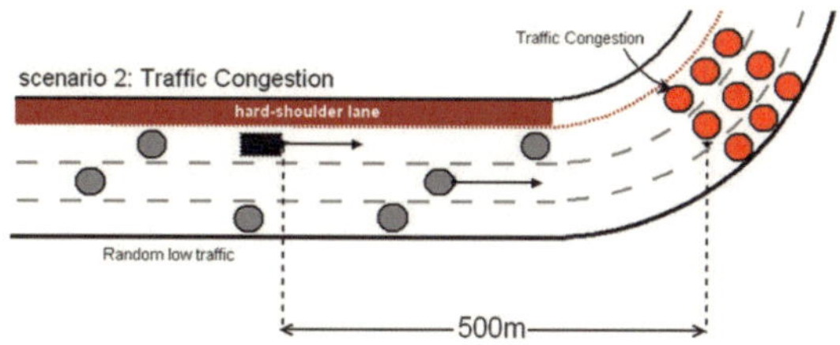

(b) Traffic congestion scenario

Figure 6. Simulation Scenarios

5.2 Macroscopic simulation approach

Investigation of driver behaviour modelling research demonstrated that the majority of high-fidelity traffic simulators utilise the macroscopic simulation method (Kuhl et al., 1995; Cremer, 1995; Papelis & Bahauddin, 1995). Predominantly, this method exploits mathematical models - often originated from fluid dynamics - that treat vehicles as particles in a homogenous flow. A sophisticated version of the macroscopic simulation method can identify three or more groups of vehicles inside the traffic flow which share identical characteristics (typically size and speed) such as cars, motorbikes and buses. This allows for some degree of control over the simulated traffic as these three main groups are characterised by different driving speeds and movements (Ehlert & Rothkrantz, 2001). Yet, these groups do not have any differentiating attributes between individual vehicles in the same group; individual driving characteristics are ignored in favour of simulation of very large numbers of vehicles moving in large urban or motorway complexes. Therefore the utility of the macroscopic approach lies mostly within generic traffic models and estimation of traffic pattern rather than the drivers' individual behaviour.

Evidently, imitation of real life occurrences requires collection of different vehicles, representations of their driving patterns and creation of potentially unpredictable traffic flows. Yet, a number of driving constraints should be considered and a simplification model should be applied in order to ease the simulator's calculating demands. Typically, when considering a macroscopic simulation scenario, variations therein depend on the number of different groups of vehicles. A group is defined as a set of vehicles that have the same driving pattern within a driving session. The scenarios used in this work were initially developed using the macroscopic method to define traffic flow complexity. In this simulation we opted for clustering of the vehicles into two main groups, named "Stop-aheads" and "Jammies", better control over the generation of the accident scenarios was achieved (Charissis et al., 2007a). These two groups had been particularly developed to play different roles in the accident simulation. Although both of them were seamlessly integrated into the generic flow, at a specific moment of the simulation they were scheduled to act according to a master-plan that initiated a potentially accident-causing event. At a predetermined moment the group of lead vehicles (named "Stop-aheads") brakes abruptly challenging the user to respond immediately. The second group (named "Jammies") mainly populated most of the motorway and in some cases developed condensed traffic sections or mini traffic jams. However, the integration of the two groups forming the traffic flow was not identifiable by the driver as an artificial construct because the robot vehicles were developed with additional singular driving characteristics, which provided the impression of autonomous behaviour. These are presented in Figure 7, which shows a diagram of the main robot categories and their functions in the macroscopic and microscopic models respectively.

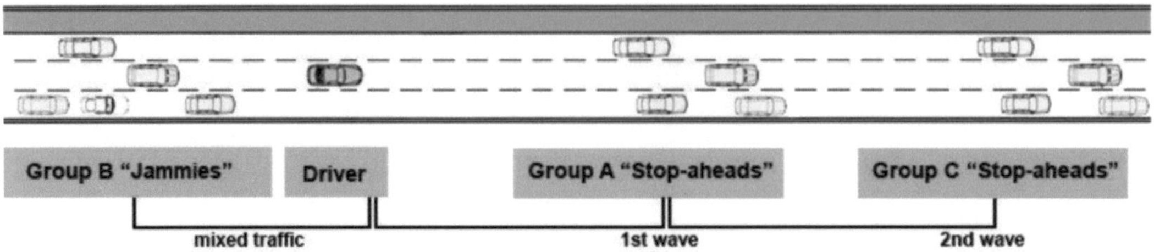

Figure 7. A diagram of the main robot vehicle categories.

5.3 Microscopic simulation approach

In addition the microscopic method was utilised to embed the AI into the robot vehicles and to create different driving profiles (one for each agent). This technique is predominantly used for accident reconstruction where specific agents have to perform particular roles in order to effectively reproduce the circumstances that created the accident. This in-detail approach could not be achieved solely by the macroscopic approach, as it does not address the impact to individual vehicles' problems on the traffic (accidents, vehicle break downs etc). Contrary it solely investigates the impact of such individual incidents on the whole traffic flow (Ehlert & Rothkrantz, 2001) rather the individual behaviour of neighbouring vehicles. Thus the individual behaviour of each agent catered for realistic interactions between the robots regardless of their group identity.

In the simulator the agent's behaviour can be altered by tweaking a set of parameters such as general speed, top speed for session, speed close to turns or linearity to lane (distance from centre of lane). Nevertheless, certain restrictions had to be pre-programmed for all the agents in order to keep autonomous behaviour within acceptable and law-abiding levels.

Summarising, both macroscopic and microscopic traffic simulation systems were embedded in the driving simulator. This combinatory simulation approach has been favourably rated in the literature for its successful realistic depiction of real traffic interactions (Shiraishi et al., 2004; Sahraoui & Jayakrishnan, 2005). As a measure of comparison it should be stated that the simulator used in this work handled approximately twenty vehicles; state of the art simulators which facilitate virtual traffic using real time generated scenarios may incorporate up to 100 participating vehicles created and controlled independently (Suda, 2006). However, a repetitive flow of vehicles inside the looped triangular circuit used in both scenario setups provided the illusion of constant moderate traffic of an endless number of travelling vehicles.

6. Embedding AI into vehicle agents

Intuitively, anticipation of human reactions presents a grandiose challenge for creating a model of life-like driver's behaviour for any type of traffic flow. Hypothesising that specific driving characteristics apply to all human drivers, as dictated by "common sense", an attempt was made to form a generic reaction pattern from a list of possible reactions to given situations (Yamanaka et. al., 2005; Sukthankar et al., 1996). The AI embedded in participating vehicles in the simulations embraced certain assumptions, which are outlined below. In this way the development time required for the AI routines was reduced by discounting rare occurrences that would indicate a severely misbehaving and irrational human driver.

6.1 Simulated driving constraints

Exhaustive investigation into drivers' possible reactions identified a number of constraints and rules that define the unpredictable factors involved in such simulation model which seeks to re-enact specific accident, real-life scenarios. The constrains identified could be divided in two main categories namely; soft and hard constrains, which are introduced in turn.

The *soft-constraints* in a driving scenario are set either by road rules (the motorway code), which are obeyed in normal situations (i.e. avoiding collision with the neighbouring vehicles) or by the agent's behaviour (i.e. speed limited by 'fear of crashing'). Minor deviations in soft constrains are favourable in a simulated environment in order to depict closely the real-life behaviour of various drivers. However it is worthwhile to note that such soft-constraints are exactly the ones violated heavily in crash situations.

On the other hand, there are also certain *hard-constraints* (for instance driving opposite to the direction of the traffic flow), which if violated would render a model void of realism. The simulation parameters can therefore vary and a semblance to reality may be maintained as long as hard-constraints are not violated. If a particular set of simulation parameters stays within the soft-constraints then normal driving conditions are denoted; alternatively, any violation of these signifies a particular accident-prone traffic set-up such as the one in the car-following scenario where a collision might occur due to adverse weather conditions.

6.2 Vehicles' agents categories

Both driving accident events have been simulated by the agent vehicles that populated the track. In order to attain the seamless integration of the robot vehicles into the traffic flow, a large part of the initial open source code had to be re-written as it was originally designed for racing purposes only. The driving behaviour of 20 vehicles was individually reprogrammed to follow the Highway Code (e.g. speed limits, lane keeping, signalling, etc.). In the first scenario (sudden braking of the lead vehicles), the robot vehicles were divided into 3 groups called "traffic waves". The primary purpose of these waves was to intentionally constrain the users from exceeding the speed limit, which eventually implicated them in the car following accident events. The secondary objective of the waves was to augment the sense of realism. In particular, when the simulation started, the test car was positioned amongst the third wave of vehicles ensuring that the driver would be accompanied at all times during the simulation. At a predetermined point, the lead vehicles of the second wave were programmed to brake abruptly, thus instigating the accident scenario. The succeeding vehicles would then respond randomly to that event either by braking on time or by colliding with the front vehicles. Given that limited visibility due to simulated thick fog, the driver would have little time to decide on what action to perform. There were two common reactions: harsh braking or manoeuvring around the stopped vehicles via the hard shoulder lane. In the case where collision with the second wave was avoided, the first wave of vehicles repeated the same scenario after 300m therefore maximising in this way the possibilities of accident involvement.

Similarly, for the purposes of the second scenario, two traffic wave groups were formed: the accompanying group and the congestion group. Whilst the accompanying vehicles had the same driving behaviour as in the first scenario, the congestion group had been allocated virtually motionless behind a sharp left turn, under a complex of bridges. This accident scenario can be equally hard to avoid when visibility is low as it is difficult to identify the sharp turn and traffic congestion in advance. Achieving a high level of realism for the scenarios was the most important element for the evaluation of the proposed HUD. The sense of immersion can be initially achieved by the presence of a sufficient number of vehicles on the track with a distinguishable variation of behaviours comparable to the variation found in reality.

A particularly interesting re-inaction of that driving pattern variation was achieved in the second scenario. As the user is driving along in the light traffic she encounters a sharp bend under a bridge and traffic congestion at the exit of the curve. It is worth noting that this scenario is actually a depiction of a real problem, so there were no other pre-programmed accident events taking place (like abrupt braking of the lead vehicles in the first scenario). The robot vehicles forming the traffic around the driver were expected to improvise and react in an appropriate manner. A variation of AI reactions was recorded when robot vehicles attempted to stop behind the static traffic: some tried to manoeuvre around and utilise the hard-shoulder lane while others crashed into the rear of the static traffic. The interaction amongst the different agents and groups considerably enhanced the feeling of realism as the driver could witness a realistic conclusion to the event.

As described in the microscopic method of the simulation, different categories of agents were deployed depending on their role in the forthcoming events. The major agents strongly instigated the particular key event (e.g. an imminent collision) and the observer agents indirectly influenced the final event. The ambiguity in any model is that in real life every driver can alter her category status with regards to a hypothetical accident. However, on hindsight of a particular key event the majority of the drivers behave in a similar fashion so as to be grouped appropriately.

6.3 Driving pattern investigation

In order for the simulation to become more realistic, it is often useful to add a degree of randomness to the distances between cars at the initial simulation setup as well as provide some variation to the simulation time.

Evidently the simulated groups have three specified soft-constraints, nevertheless, there are some hidden constraints generated from the vehicle models and road model. As more groups of vehicles with specific purpose are introduces, the number of constraints on driving behaviour increases and some control is introduced in the environment, leading to a 'test scenario' from which conclusive judgements can be drawn. When unpredictable events occur in a scenario it becomes harder to deduce the reason for the key event happening – as the user might get distracted by 'non-key events' taking place due to vehicles' variation; these affect the driver's reaction to the key event.

However, such variations do exist in reality due to the numerous differences in driving behaviours of real drivers. We discovered that to achieve both realistic variation and sufficient control necessary for the key event to take place, the driving time has to be sufficiently long. In this way, each group in a simulation can exhibit "personality" whilst their influence gets slightly diluted over time, so that their behaviour will not appear more peculiar than the key-event.

While developing the simulator and its associated AI we also took into consideration the possible design of the networking and systems backend which would be necessary for the implementation of our HUD design in actual vehicles. We have argued, in related work, that the detection of the headway and positioning of the neighbouring vehicles can be accomplished by a collaborative Mobile Ad-hoc Networking System (MANETS) between vehicles and with the aid of GPS and sensors (Charissis & Papanastasiou, 2006).

7. AI simulation challenges and solutions

During the development of the driving simulation environment, the agent vehicles' autonomous reactions had been extensively tested in order to identify potential issues. Initial presumptions with regard to different driving patterns resolved most anticipated issues, yet some participants' unexpected behaviour reactions required more creative solutions (Green, 2005). In order to address the problems that occurred during the preliminary trials, we employed fundamental techniques used in similar cases in AI such as the Cognitive Simulation introduced by Newell, Shaw and Simon at the RAND Corporation and at Carnegie Institute of Technology (Dreyfuss, 1992). Originally this approach involved the collection of protocols from human subjects which were subsequently analysed to discover the heuristics that these subjects employed (Dreyfuss, 1992). More specifically, the following two solutions were applied to retain the driver amongst at least one of the traffic waves.

7.1 "Bandit-Driver" phenomenon

This phenomenon typically occurs in a driving simulation environment when the user reacts as if driving a "racing computer game". In order to avoid such behaviour which alters considerably the real results, the users had been instructed to follow the Motorway Code for the entire duration of the experiment. Yet a number of users exceeded the speed limit substantially. This, however, resulted in an event-free journey as they had over-taken all the traffic waves. After experimenting with different speeds and chase techniques for the driving agents, we decided that one of the vehicles should play the role of the "super-car". Therefore one of the vehicles had been given the ability to chase and dangerously over-take the "bandit-driver" forcing him/her to slow down until the rest of the traffic would merge into the picture. That ability was triggered by a chosen agent as soon as the "bandit-driver" was out of range of any possible event (1km).

7.2 "Timid-Driver" phenomenon

The opposite of the previous case was the exceptionally slow driver, also referred to as a "timid-driver" (Daganzo 1999). Even when the robot vehicles were decelerating significantly, such drivers were travelling with a speed of less than 30 km/h thus maintaining a considerable distance from the preceding main traffic. This resulted once more in the participant missing out the accident events. The remedy to this was the incorporation of the "fear factor". A model of a black truck participated in that role. The truck was travelling at a conveniently slow speed, following the slow vehicle in close proximity (not visible due to the fog-effect). So, when the user was permitting a greater distance than 1km from the last leading vehicle, the agent of that robot vehicle had been programmed to initiate a close-vicinity pursuit. If the driver persisted with the same driving pattern the truck would alarmingly sound its horn.

We considered that it might seem reasonable to exclude the results of both "bandit" and "timid" users from the final analysis. However, driving behaviours similar to the ones described above exist in real life; these extra agents were developed in order to embrace a wider breadth of driving patterns.

7.3 "Follower" phenomenon

This category of drivers typically followed closely the movements of the lead agent vehicles. As such they increased their level of anticipation significantly by simply imitating the driving behaviour of the simulated vehicles. Although such behaviour appears in real-life as well and in some respect may even be desirable, during simulation it does not accurately reflect the driver's real intentions. This occurs as drivers feel that the simulation experiment will evaluate their own driving abilities instead of the effectiveness of the interface. In order to avoid such issues we enhanced the ability of the "jammies" to overtake and partially merge with the lead vehicles. That position exchanging and multiple unexpected manoeuvres blur significantly the roles between the agent vehicles for the participant.

8. Conclusions

This chapter presented the rationale and development process of an open source driving simulator which was exclusively created for the evaluation process of a prototype HUD interface. Financial limitations led to the exploration of new avenues for evaluating the prototype HUD. The simulator hardware was built using off-the-shelf components and the software was developed through customisation of an open source code game engine.
For evaluation purposes, equal emphasis was placed on was the development of an appropriate testing environment and suitable driving scenarios. To achieve a realistic mix of agent behaviour, the AI infused in the vehicles' agents was thoroughly customised and designed to facilitate the accident events created to gauge driver reactions.

9. References

Allen. R. W., Magdeleno, R. and Serafin, C., (1997), Driver Car Following Behaviour Under Test Track and Open Road Driving Conditions, SAE Paper 970170.

BBC, (2007), One dead after 40-vehicle pile-up, Available from: http://news.bbc.co.uk/1/hi/england/kent/6257456.stm, [accessed on 01 July 2007].

Boer, E.R., and Ward, N.J., (2003), Event-Based Driver Performance Assessment, in Proceedings of the 2nd International Symposium on Driving Assessment, Training, and Vehicle Design, Park City, Utah, USA.

Brackstone M. & McDonald M., (2003), Driver Behaviour and Traffic Modelling. Are We Looking at the Right Issues?, in Proceedings of the IEEE Intelligent Vehicles Symposium 2003, Columbus, Ohio, USA, June.

Cavallo V., Dore J., Colomb M., and Leqoueix, G., (1999), Distance Over-Estimation of Vehicles Rear Lights in Fog, in A. G. Gale et al. (Eds.), Vision in Vehicles-VII, pp. 373-379, Marseilles, France: North-Holland-Elsevier.

Centelles, V. M., (2006), Formula 1 TORCS: Build your Track for TORCS in 20 minutes, Available from: usuarios.lycos.es/f1torcs/build_your_trocs_track_in_20_minutes.pdf, last accessed, [accessed on 12 July 2007].

Chen, L.D. et al., (2001), NADS At the University of Iowa: A Tool for Driving Safety Research, in Proceedings of the 1st Human-Centered Transportation Simulation Conference.

Charissis V. and Papanastasiou S., (2006), Exploring the ad hoc network requirements of an automotive Head-Up Display Interface, 5th International Conference on Communication Systems, Networks and Digital Signal Processing, CSNDSP'06 (IEEE), Patras, Greece.

Charissis, V., Arafat, S., Chan, W., and Christomanos, C., (2006), Driving Simulator for Head-Up Display Evaluation: Driver's Response Time on Accident Simulation Cases, in Proceedings of the Driving Simulation Conference DSC'06, Asia/Pacific, Tsukuba/Tokyo, Japan, 2006.

Charissis, V., Arafat, S., Patera, M., Christomanos, C. and Chan, W., (2007a), AI for Driving Simulator for Driving Vehicles and Accident Events, in Proceedings of: International Conference in Artificial Intelligence and Applications, (AIA '07), as part of the 25th International Multi-Conference on Applied Infomatics, Innsbruck, Austria.

Charissis, V., Naef, M. and Patera, M., (2007b), Calibration Requirements of an Automotive HUD Interface Using a Virtual Environment: Methodology and Implementation, in Proceedings of: International Conference in Graphics and Visualisation in Engineering, (GVE '07), Clearwater, Florida, USA.

Charissis V., and Naef M., (2007), Evaluation of Prototype Automotive Head-Up Display Interface: Testing Driver's Focusing Ability through a VR Simulation, in Proceedings of the IEEE Intelligent Vehicle symposium, (IV '07), Istanbul, Turkey.

Charissis V., Naef M., Arafat S., and Vlachos G., (2008a), On the Impact of User's Computer Knowledge on Driving Simulation Test Results - HUD Simulation Case Study, in Proceedings of the European Annual Conference on Human Decision-Making and Manual Control, EAM'08, TU Delft, Netherlands.

Charissis V., Papanastasiou S., and Vlachos G., (2008b), Comparative Study of Prototype Automotive HUD vs. HDD: Collision Avoidance Simulation and Results, in Proceedings of the Society of Automotive Engineers World Congress 2008, 14-17 April, Detroit, Michigan, USA.

Charissis V., Papanastasiou S., and Vlachos G., (2008c), Investigation of Users' Age and Driving Performance With the Use of Prototype Automotive HUD System, in Proceedings of the 15th World Congress in Intelligent Transportation Systems, IEEE, 16-20 November, New York City, NY, USA.

Cremer, J., Kearney, J., and Papelis, Y., (1995), Driving Simulation: Challenges for VR Technology, in Journal of IEEE Computer Graphics and Applications, Volume 16, Issue 5, Sept. 1996, pp 16-20.

Daganzo, C.F., (1999), A Behavioural Theory of Multi-Lane Traffic Flow. Part 1 Long Homogenous Freeway Sections, Research Report UCB-ITS-RR-99-5, Institute of Transportation Studies, University of California, Berkley.

Dreyfus, H., (1992), What Computers Still Can't Do: A Critique of Artificial Reason, Cambridge, Massachusetts, MIT press.

Ehlert, P.A.M., and Rothkrantz, L.J.M, (2001), Microscopic Traffic Simulation With Reactive Driving Agents, in Proceedings of the 4th International IEEE Conference on Intelligent Transportation Systems, ITSC 2001, August, Oakland, CA, USA, pp 860-865.

Green, P., (2005), How Driving Simulator Data Quality Can Be Improved, in Proceedings of the Driving Simulator Conference (DSC 05) North America, Orlando, USA.

Kantowitz, B. H., (1999), Using Simulators to Study Driver Response to Advanced In-Vehicle Systems, 78th Annual Meeting of transportation Research Board.

Kenny T., Anthony D., Charissis V., Darawish Y. & Keir P., (2004), Integrated Vehicle Instrument Simulation: i-ViS Initial Design Philosophy, in Proc. of the 3rd International Conference on Total Vehicle Technology, Brighton, UK, pp 93-102.

LADS, (2007), Leeds Advanced Driving Simulator-Investigating Driver's Behaviour in a Safe and Controllable Research Environment, Available from: http://www.its.leeds.ac.uk/facilities/lads/ [accessed 01 July 2007].

Nordmark, S. et al., (2004), The New VTI Driving Simulator. Multipurpose Moving Base with High Performance Linear Motion, in Proceedings of the Driving Simulation Conference DSC Europe2004, pp.45-55.

Papelis, Y., (1994), Terrain Modeling on High Fidelity Ground Vehicle Simulators, in Proceedings of the Fifth Annual Conference on AI Simulation and Planning in High-Autonomy System, Gainesville, Florida, USA, pp. 48–54.

Papelis, Y. and Bahauddin, S., (1995), Logical Modeling of Roadway Environment to Support Real-Time Simulation of Autonomous Traffic, in Proceedings of the SIVE95: The First Workshop on Simulation and Interaction in Virtual Environments, pages 62--71, Iowa, USA.

Park, G., Rosenthal, T.J., and Aponso, B.L., (2004), Developing driving scenarios for research, training and clinical applications, in International Journal of Advances in Transportation Studies, Special Issue, 19-28

Kuhl, J., Evans, D., Papelis, Y., Romano, R., and Watson, G., (1995), The Iowa Driving Simulator: an immersive research environment, Computer, Volume 28, Issue 7, July 1995 pp 35 - 41.

Rosenthal, T. J., Allen, R. W., and Aponso, B. L., (2003), Configuration Management and User's Interface for a PC Based Driving Simulator, presented at Image Conference, Scottsdale, AZ, USA July 14-18 2003. (STI-P-613).

TRL, Transport Research Laboratory (2007), Driving Simulation, Available from: http://www.trl.co.uk/content/main.asp?pid=70, [accessed 1 July 2007].

VTI,(2007), VTI Driving simulators, Available from: http://www.vti.se/templates /Page____3257.aspx [accessed 1 July 2007].

Wymann, B., (2006), T.O.R.C.S. Manual Installation and Robot Tutorial, Manuscript, Available from: http://torcs.sourceforge.net/index.php?name=News&file =print&sid=57, [accessed 12 July 2007].

SPD, Strathclyde Police Department, (2004), Accident Statistics 2001-2004, Glasgow, UK.

Sahraoui Abd-El-Kader and Jayakrishnan, R., (2005), Microscopic-Macroscopic Models Systems Integration: A Simulation Case Study for ATMIS, in SAGE Simulation Journal, Simulations councils Inc. Vol. 81, No. 5, 353-363

Shiraishi, T. et al., (2004), Development of a Microscopic Traffic Simulation Model for Interactive Traffic Environment, in Proceedings of the Intelligent Transportation Systems World Congress '04, 3446.

Smith, D. L., Najm, W. G., Lam, A. H., (2003), Analysis of Braking and Steering Performance in Car-Following Scenarios, Society of Automotive Engineers.

Steinfeld A. & Green P., (1995), Driver Response Times to Full-Windshield, Head-Up Displays for Navigation and Vision Enhancement, Tech. Rep. UMTRI9529, Transportation Research Institute, University of Michigan, USA.

Suda, Y., Shladover, S. E., Takahashi Y., Onuki, M., Matsushita, K., Kanoh, M., Honda, K., Tomoyoshi, Shiraishi, (2006), Validation of the Universal Driving Simulator with Interactive Traffic Simulation, in Proceedings of the Intelligent Transportation Systems World Congress 2006, London, UK.

Sukthankar, R., Hancock J., Pomerleau, D. and Thorpe C., (1996), A Simulation and Design System for Tactical Driving Algorithms, in Proceedings of the Sixth Annual Conference on AI, Simulation and Planning in High Autonomy Systems, La Jolla, California, USA.

Yamanaka, K., Maeda, K., Miyoshi, T., Hayashi, H. and Nakayasu, H., (2005), Evaluation of Response Properties on Automobile Driver Based on Design of Experiment, in International Journal of Advances in Transportation Studies, vol. 6, section B, pp 37-56.

18

Humanoid Robot Balancing

Youngjin Choi[1] and Doik Kim[2]
[1]*Division of Electrical Engineering and Computer Science*
Hanyang University, Ansan, 426-791,
[2] *Korea Institute of Science and Technology (KIST), Seoul, 136-791,*
Rep. of Korea

1. Introduction

There have been many researches about humanoid robot motion control, for example, walking pattern generation (Huang et al, 2001) (Kajita et al, 2002, 2003), walking control (Choi et al, 2006) (Grizzle et al, 2003) (Hirai et al, 1998) (Kajita et al, 2001) (Kim & Oh, 2004), (Lohmeier et al, 2004) (Park, 2001) (Takanishi et al, 1990) (Westervelt et al, 2003), running control (Nagasaki et al, 2004, 2005), balancing control (Kajita et al, 2001) and whole body coordination (Choi et al, 2007) (Kajita et al, 2003) (Sentis & Khatib, 2005) (Goswami & Kallem, 2004) (Harada et al, 2003) (Sugihara & Nakamura, 2002). Especially, the whole body coordination algorithm with good performance becomes a core part in the development of humanoid robot because it is able to offer the enhanced stability and flexibility to the humanoid motion planning. In this chapter, we explain the kinematic resolution method of CoM Jacobian with embedded motion which was suggested in (Choi et al, 2007), actually, which offers the ability of balancing to humanoid robot. For example, if humanoid robot stretches two arms forward, then the position of CoM(center of mass) of humanoid robot moves forward and its ZMP(zero moment point) swings back and forth. In this case, the proposed kinematic resolution method of CoM Jacobian with embedded motion offers the joint configurations of supporting limb(s) calculated automatically to maintain the position of CoM fixed at one point.

Also, a design of balancing controller with good performance becomes another important part in development of humanoid robot. In balancing control, the ZMP control is the most important factor in implementing stable bipedal robot motions. If the ZMP is located in the region of supporting sole, then the robot will not fall down during motions. In order to compensate the error between the desired and actual ZMP, various ZMP control methods have been suggested; for example, direct/indirect ZMP control methods (Choi et al, 2004) (Kajita et al, 2003) and the impedance control (Park, 2001). Despite many references to bipedal balancing control methods, research on the stability of bipedal balancing controllers is still lacking. The exponential stability of periodic walking motion was partially proved for a planar bipedal robot in (Grizzle et al, 2003) (Westervelt et al, 2003). Also, the ISS (disturbance input-to-state stability) of the indirect ZMP controller was proved for the simplified bipedal robot model in (Choi et al, 2004, 2007). In this chapter, we will explain the balancing control method and its ISS proof which were suggested in (Choi et al, 2007). The

suggested method offers the stable balancing control as well as the function of whole body coordination to humanoid robot. Also, the suggested control method includes both the CoM/ZMP trajectory tracking control and CoM/ZMP set-point regulation control based on the kinematic resolution of CoM Jacobian with embedded task motions. Due to the modeling uncertainties and the complexity of the full dynamics of a bipedal walking robot, we will represent the dynamic walking robot as a simple rolling sphere model on a constraint surface. And then the ISS is proved for the simplified model of bipedal robot.
This chapter is organized as follows: section 2 introduces a simplified model for bipedal walking robot, section 3 explains the kinematic resolution method of CoM Jacobian with an embedded task motion, section 4 suggests the balancing controller for humanoid robot and proves its ISS for the simplified bipedal robot model, section 5 shows the experimental results about the stable balancing functions obtained by using the kinematic resolution method of CoM Jacobian with embedded motion, and section 6 concludes the chapter.

2. Rolling sphere model

The bipedal walking mechanism is an essential part of humanoid robot. Since humanoid legs have high degrees of freedom for human-like motion, it is difficult to use their full dynamics to design controller and to analyze stability. Therefore, we will simplify the walking related dynamics of bipedal robot as the equation of motion of a point mass concentrated on the position of CoM as shown in Fig. 1.

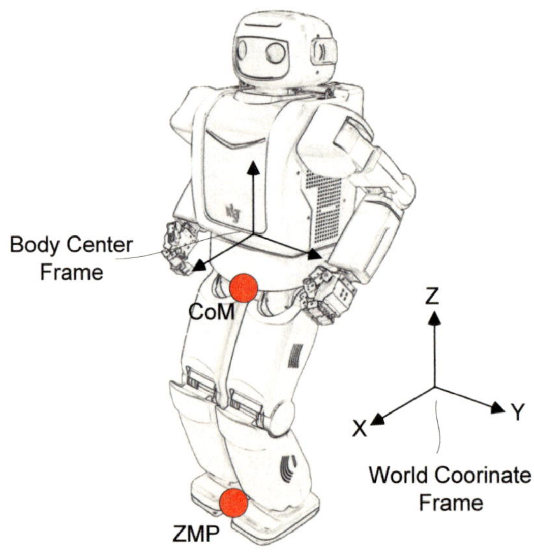

Fig. 1. Humanoid Robot

First, let us assume that the motion of CoM is constrained on the surface, $z = c_z$, then the rolling sphere model with the concentrated point mass m can be obtained as the simplified model for bipedal robot as shown in Fig. 2. In Fig. 2, the motion of the rolling sphere on a massless plate is described by the position of CoM, $c = [c_x, c_y, c_z]^T$, and the ZMP is described by the position on the ground, $p = [p_x, p_y, 0]^T$.

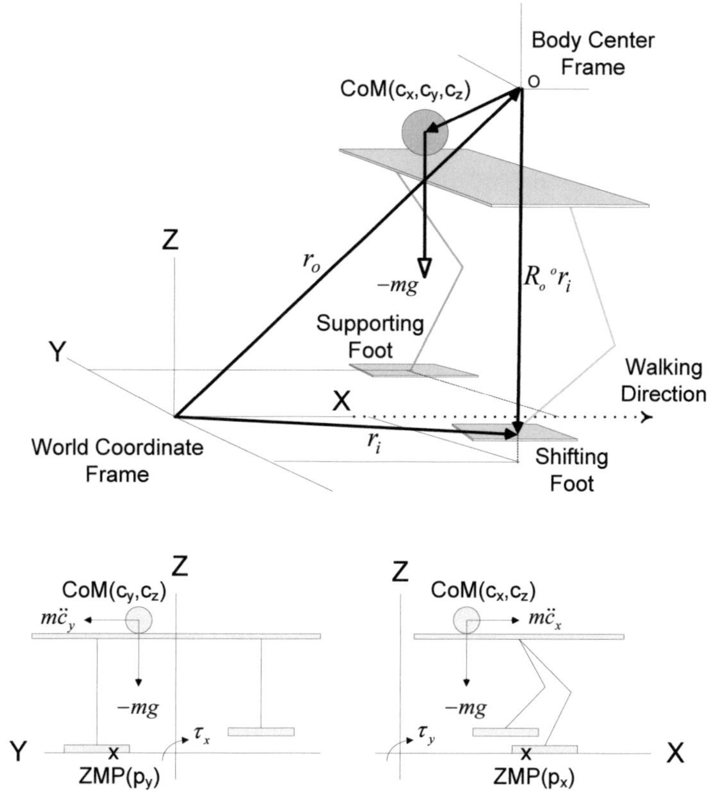

Fig. 2. Rolling Sphere Model for Humanoid Robot

Second, the equations of motion of the rolling sphere (mass =m) in Fig. 2 are expressed on the plane $z = c_z$ by:

$$\tau_x = mgc_y - m\ddot{c}_y c_z \tag{1}$$

$$\tau_y = -mgc_x + m\ddot{c}_x c_z \tag{2}$$

$$\tau_z = -m\ddot{c}_x c_y + m\ddot{c}_y c_x$$

where g is the acceleration of gravity, c_z is a height constant of constraint plane and τ_i is the moment about i-th coordinate axis, for $i = x, y, z$. Now, if we introduce the conventional definition of ZMP as following forms to Eq. (1) and (2):

$$p_x = -\frac{\tau_y}{mg}$$

$$p_y = \frac{\tau_x}{mg}$$

then ZMP equations can be obtained as two differential equations:

$$p_x = c_x - \frac{c_z}{g}\ddot{c}_x \tag{3}$$

$$p_y = c_y - \frac{c_z}{g}\ddot{c}_y. \tag{4}$$

Actually, above equations can be easily obtained by finding the position of zero moment on the ground in Fig. 2 like these:

$$\sum M_{p_y} = -mg(p_y - c_y) - (m\ddot{c}_y)c_z = 0$$

$$\sum M_{p_x} = mg(p_x - c_x) + (m\ddot{c}_x)c_z = 0.$$

Also, the state space realization of ZMP equations (3) and (4) can be written as:

$$\frac{d}{dt}\begin{bmatrix} c_i \\ \dot{c}_i \end{bmatrix} = \begin{bmatrix} 0 & 1 \\ \omega_n^2 & 0 \end{bmatrix}\begin{bmatrix} c_i \\ \dot{c}_i \end{bmatrix} + \begin{bmatrix} 0 \\ -\omega_n^2 \end{bmatrix} p_i,$$

for $i = x, y$, where $\omega_n = \sqrt{g/c_z}$ means the natural radian frequency of the simplified biped walking robot system. These state space equations describe the relation between the dynamics of CoM and the ZMP. Above equations will be used to prove the stability of the balancing controller in the following sections.

3. Kinematic resolution method for balancing

In this section, we will explain the kinematic resolution method of CoM Jacobian with embedded motion, ultimately for humanoid balancing. Let a robot has *n* limbs and the first limb be the base limb, for example, *n*=4 for humanoid robot except the neck. The base limb can be any limb but it should be on the ground to support the body. Each limb of a robot is hereafter considered as an independent limb. In general, *i*-th limb has the following relation:

$${}^o\dot{x}_i = {}^oJ_i\dot{q}_i$$

for $i = 1, 2, \cdots, n$, where ${}^o\dot{x}_i \in \Re^6$ is the velocity of the end point of *i*-th limb, $\dot{q}_i \in \Re^{n_i}$ is the joint velocity of *i*-th limb, ${}^oJ_i \in \Re^{6\times n_i}$ is the usual Jacobian matrix of *i*-th limb, and n_i means the number of active links of *i*-th limb, at least, $n_i \geq 6$ for the limb related with leg. The leading superscript o implies that the elements are represented on the body center coordinate system shown in Fig. 1 and Fig. 2, which is fixed on a humanoid robot.

3.1 Compatibility condition

In our case, the body center is floating, and thus the end point motion of *i*-th limb about the world coordinate system is written as follows:

$$\dot{x}_i = X_i^{-1}\dot{x}_o + X_o\,{}^oJ_i\dot{q}_i \tag{5}$$

where $\dot{x}_o = [\dot{r}_o^T; \omega_o^T]^T \in \Re^6$ is the velocity of the body center represented on the world coordinate system, and

$$X_i = \begin{bmatrix} I_3 & [R_o {}^o r_i \times] \\ 0_3 & I_3 \end{bmatrix} \in \Re^{6\times 6}$$

is a (6×6) matrix which relates the body center velocity and the *i*-th limb velocity. I_3 and 0_3 are an (3×3) identity and zero matrix, respectively. $R_o {}^o r_i$ is the position vector from the body center to the end point of the *i*-th limb represented on the world coordinate frame. $[(\cdot)\times]$ is a skew-symmetric matrix for the cross product. The transformation matrix X_o is

$$X_o = \begin{bmatrix} R_o & 0_3 \\ 0_3 & R_o \end{bmatrix} \in \Re^{6\times 6}$$

where $R_o \in \Re^{3\times 3}$ is the orientation of the body center represented on the world coordinate frame, and hereafter, we will use the relation $J_i = X_o {}^o J_i$. Also, the concrete proof of Eq. (5) is in appendix 9.1.

All the limbs in a robot should have the same body center velocity, in other words, from Eq. (5), we can see that all the limbs should satisfy the compatibility condition that the body center velocity is the same, and thus, *i*-th limb and *j*-th limb should satisfy the following relation:

$$X_i(\dot{x}_i - J_i\dot{q}_i) = X_j(\dot{x}_j - J_j\dot{q}_j). \tag{6}$$

From Eq. (6), the joint velocity of any limb can be represented by the joint velocity of the base limb and Cartesian motions of limbs. Actually, the base limb should be chosen to be the support leg in single support phase or one of both legs in double support phase. Let us express the base limb with the subscript 1, then the joint velocity of *i*-th limb is expressed as:

$$\dot{q}_i = J_i^+ \dot{x}_i - J_i^+ X_{i1}(\dot{x}_1 - J_1\dot{q}_1), \tag{7}$$

for $i = 2,\cdots,n$, where J_i^+ means the Moore-Penrose pseudoinverse of J_i and

$$X_{i1} = X_i^{-1}X_1 = \begin{bmatrix} I_3 & [R_o({}^o r_1 - {}^o r_i)\times] \\ 0_3 & I_3 \end{bmatrix}.$$

Note that if a limb is a redundant system, any null space optimization scheme can be added in Eq. (7). With the compatibility condition of Eq. (6), the inverse kinematics of humanoid robot can be solved by using the information of base limb like Eq. (7), not by using the information of body center like Eq. (5).

3.2 CoM Jacobian with fully specified embedded motions

Now, let us rewrite the conventional CoM Jacobian suggested in (Sugihara & Nakamura, 2002) as follows:

$$\dot{c} = \dot{r}_o + \omega_o \times (c - r_o) + \sum_{i=1}^{n} R_o {}^oJ_{c_i} \dot{q}_i. \tag{8}$$

where n is the number of limbs, c is the position vector of CoM represented on the world coordinate system, namely, $c = [c_x, c_y, c_z]^T$, ${}^oJ_{c_i} \in \Re^{3\times n_i}$ means CoM Jacobian matrix of i-th limb represented on the body center coordinate frame, and hereafter, we will use the relation $J_{c_i} = R_o {}^oJ_{c_i}$. Also, the concrete proof of Eq. (8) is in appendix 9.2.

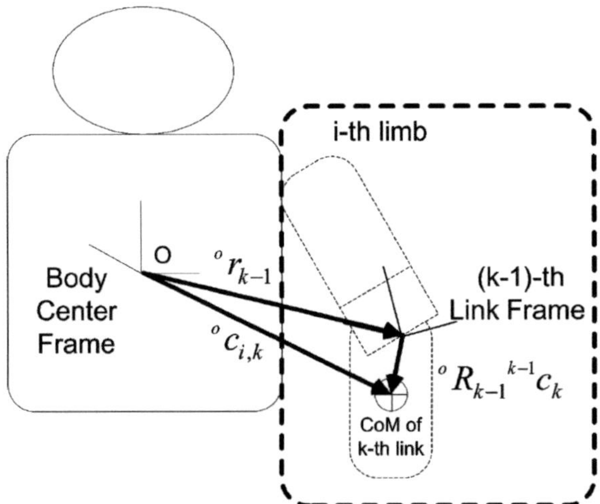

Fig. 3. Position of CoM of the k-th link in i-th limb

Remark 1 *The CoM Jacobian matrix of i-th limb represented on the body center frame is expressed by*

$${}^oJ_{c_i} = \sum_{k=1}^{n_i} \mu_{i,k} \frac{\partial^o c_{i,k}}{\partial q_i}, \tag{9}$$

where ${}^oc_{i,k} \in \Re^3$ *means the position vector of center of mass of k-th link in i-th limb represented on the body center frame as shown in Fig. 3 and the mass influence coefficient of k-th link in i-th limb is defined as follow:*

$$\mu_{i,k} = \frac{m_{i,k}}{\sum_{i=1}^{n}\sum_{k=1}^{n_i} m_{i,k}}, \tag{10}$$

where $m_{i,k}$ *is the mass of k-th link in i-th limb. Also, the systematic derivation of CoM Jacobian matrix of Eq. (9) is in appendix 9.3.*

The motion of body center frame can be obtained by using Eq. (5) for the base limb as follows:

$$\dot{x}_o = X_1 \{\dot{x}_1 - J_1 \dot{q}_1\}$$

$$\begin{bmatrix} \dot{r}_o \\ \omega_o \end{bmatrix} = \begin{bmatrix} I_3 & [R_o{}^o r_1 \times] \\ 0_3 & I_3 \end{bmatrix} \left\{ \begin{bmatrix} \dot{r}_1 \\ \omega_1 \end{bmatrix} - \begin{bmatrix} J_{v_1} \\ J_{\omega_1} \end{bmatrix} \dot{q}_1 \right\}, \tag{11}$$

where J_{v_1} and J_{ω_1} are the linear and angular velocity part of the base limb Jacobian J_1 expressed on the world coordinate frame, respectively. Now, if Eq. (7) is applied to Eq. (8) for all limbs except the base limb with subscript 1, the CoM motion is rearranged as follows:

$$\dot{c} = \dot{r}_o + \omega_o \times (c - r_o) + J_{c_1}\dot{q}_1 + \sum_{i=2}^{n} J_{c_i} J_i^+ (\dot{x}_i - X_{i1}\dot{x}_1) + \sum_{i=2}^{n} J_{c_i} J_i^+ X_{i1} J_1 \dot{q}_1. \tag{12}$$

Here, if Eq. (11) is applied to Eq. (12), then the CoM motion is only related with the motion of base limb:

$$\dot{c} = \dot{r}_1 + \omega_1 \times r_{c1} - J_{v_1}\dot{q}_1 + r_{c1} \times J_{\omega_1}\dot{q}_1 + J_{c_1}\dot{q}_1 + \sum_{i=2}^{n} J_{c_i} J_i^+ (\dot{x}_i - X_{i1}\dot{x}_1) + \sum_{i=2}^{n} J_{c_i} J_i^+ X_{i1} J_1 \dot{q}_1 \tag{13}$$

where $r_{c1} = c - r_1$. Also, if the base limb has the face contact with the ground (the end-point of base limb represented on world coordinate frame is fixed, $\dot{x}_1 = 0$, namely, $\dot{r}_1 = 0$, $\omega_1 = 0$), then Eq. (13) is simplified as follows:

$$\dot{c} - \sum_{i=2}^{n} J_{c_i} J_i^+ \dot{x}_i = -J_{v_1}\dot{q}_1 + r_{c1} \times J_{\omega_1}\dot{q}_1 + J_{c_1}\dot{q}_1 + \sum_{i=2}^{n} J_{c_i} J_i^+ X_{i1} J_1 \dot{q}_1.$$

Finally, $3 \times n_1$ CoM Jacobian matrix with embedded motions can be rewritten like usual kinematic Jacobian of base limb:

$$\dot{c}_{\text{fsem}} = J_{\text{fsem}}\, \dot{q}_1, \tag{14}$$

where

$$\dot{c}_{\text{fsem}} = \dot{c} - \sum_{i=2}^{n} J_{c_i} J_i^+ \dot{x}_i, \tag{15}$$

$$J_{\text{fsem}} = -J_{v_1} + r_{c1} \times J_{\omega_1} + J_{c_1} + \sum_{i=2}^{n} J_{c_i} J_i^+ X_{i1} J_1.$$

Here, if the CoM Jacobian is augmented with the orientation Jacobian of body center ($\omega_o = -J_{\omega_1} q_1$) and all desired Cartesian motions are embedded in Eq. (15), then the desired joint configurations of base limb (support limb) are resolved as follows:

$$\dot{q}_{1,d} = \begin{bmatrix} J_{\text{fsem}} \\ -J_{\omega_1} \end{bmatrix}^+ \begin{bmatrix} \dot{c}_{\text{fsem},d} \\ \omega_{o,d} \end{bmatrix}, \tag{16}$$

where the subscript d means the desired motion and

$$\dot{c}_{\mathrm{fsem},d} = \dot{c}_d - \sum_{i=2}^{n} J_{c_i} J_i^{+} \dot{x}_{i,d}. \tag{17}$$

All given desired limb motions, $\dot{x}_{i,d}$ are embedded in the relation of CoM Jacobian, thus the effect of the CoM movement generated by the given limb motion is compensated by the base limb. The CoM motion with fully specified embedded motions, $\dot{c}_{\mathrm{fsem},d}$, consists of two relations: a given desired CoM motion(the first term) and the relative effect of other limbs(the second term). The CoM Jacobian with fully specified embedded motions, J_{fsem} also consists of three relations: the effect of the body center(the first and the second term), the effect of the base limb(the third term), and the effect of other limbs(the last term).
The CoM Jacobian with fully specified embedded motions J_{fsem} is a $(3 \times n_1)$ matrix where n_1 is the dimension of the base limb, which is smaller than that of the original CoM Jacobian, thus the calculation time can be reduced. After solving Eq. (16), the desired joint motion of the base limb is obtained. The resulting base limb motion makes a humanoid robot balanced automatically during the movement of the all other limbs. With the desired joint motion of base limb, the desired joint motions of all other limbs can be obtained by Eq. (7) as follow:

$$\dot{q}_{i,d} = J_i^{+}(\dot{x}_{i,d} + X_{i1} J_1 \dot{q}_{1,d}),$$

for $i = 2, \cdots, n$. The resulting motion follows the given desired motions, regardless of balancing motion by base limb. In other words, the suggested kinematic resolution method of CoM Jacobian with embedded motion offers the WBC(whole body coordination) function to the humanoid robot automatically.

3.3 CoM Jacobian with partially specified embedded motion

In some cases, the desired motion of any limb is specified in the joint configuration space. For example, let us consider that the walking motions (for the leg limbs of i=1,2) are partially specified in Cartesian space and the other limb motions (for i=3,4) in joint space, then the kinematic resolution method of Eq. (14) in the previous section should be slightly modified as follows:

$$\dot{q}_{1,d} = \begin{bmatrix} J_{\mathrm{psem}} \\ -J_{\omega_1} \end{bmatrix}^{+} \begin{bmatrix} \dot{c}_{\mathrm{psem},d} \\ \omega_{o,d} \end{bmatrix}, \tag{18}$$

$$\dot{q}_{2,d} = J_2^{+}(\dot{x}_{2,d} + X_{21} J_1 \dot{q}_{1,d}) \tag{19}$$

where the CoM desired motion and CoM Jacobian with partially specified embedded motion are expressed by, respectively,

$$\dot{c}_{\mathrm{psem},d} = \dot{c}_d - J_{c_2} J_2^{+} \dot{x}_{2,d} - \sum_{i=3}^{n} J_{c_i} \dot{q}_{i,d}, \tag{20}$$

$$J_{\mathrm{psem}} = -J_{v_1} + r_{c1} \times J_{\omega_1} + J_{c_1} + J_{c_2} J_2^{+} X_{21} J_1.$$

The walking motion is generally specified as the Cartesian desired motions for dual legs but the motions of other limbs can be specified as either joint or Cartesian desired motion. Also, if we are to implement the robot dancing expressed by desired joint motions, then the desired dancing arm motions as well as desired CoM motion are embedded in the suggested resolution method and then the motion of base limb (support leg) are automatically generated with the function of auto-balancing. This is main advantage of the proposed method.

4. Design of balancing controller

Since a humanoid robot is an electro-mechanical system including many electric motors, gears and link mechanisms, there are many disturbances in implementing the desired motions of CoM and ZMP for a real bipedal robot system. To show the robustness of the controller to be suggested against disturbances, we apply the following stability theory to a bipedal robot control system. The control system is said to be disturbance input-to-state stable (ISS) (Choi & Chung, 2004), if there exists a smooth positive definite radially unbounded function $V(e,t)$, a class K_∞ function γ_1 and a class K function γ_2 such that the following dissipativity inequality is satisfied:

$$\dot{V} \le -\gamma_1(|e|) + \gamma_2(|\varepsilon|), \tag{21}$$

where $\dot{V}$ represents the total derivative for Lyapunov function, e is the error state vector and ε is disturbance input vector. In this section, we propose the balancing (posture/walking) controller for bipedal robot systems as shown in Fig. 4.

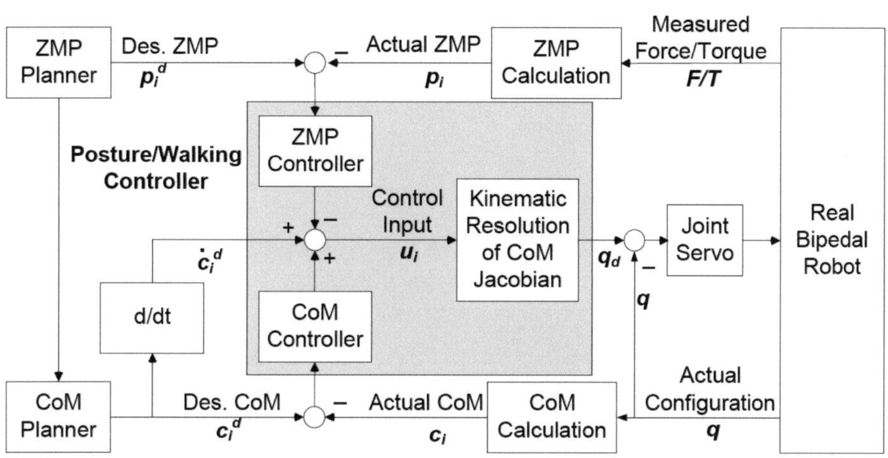

Fig. 4. Balancing Controller for Humanoid Robot

In this figure, first, the ZMP Planer and CoM Planer generate the desired trajectories satisfying the following differential equation:

$$p_{i,d} = c_{i,d} - 1/\omega_n^2 \ddot{c}_{i,d} \quad for \quad i = x, y. \tag{22}$$

Second, the simplified model for the real bipedal walking robot has the following dynamics:

$$\dot{c}_i = u_i + \varepsilon_i \tag{23}$$

$$p_i = c_i - 1/\omega_n^2 \ddot{c}_i \quad for \quad i = x, y$$

where ε_i is the disturbance input produced by actual control error, u_i is the control input, c_i and p_i are the actual positions of CoM and ZMP measured from the real bipedal robot, respectively. Actually, the real bipedal robot offers the ZMP information from force/torque sensors attached to the ankles of humanoid and the CoM information from the encoder data attached to the motor driving axes, respectively, as shown in Fig. 4. Here, we assume that the disturbance produced by control error is bounded and its differentiation is also bounded, namely, $|\varepsilon_i| < a$ and $|\dot{\varepsilon}_i| < b$ with any positive constants *a* and *b*. Also, we should notice that the control error always exists in real robot systems and its magnitude depends on the performance of embedded local joint servos. The following theorem proves the stability of the balancing controller to be suggested for the simplified bipedal robot model.

Theorem 1 *Let us define the ZMP and CoM error for the simplified bipedal robot control system (23) as follows:*

$$e_{p,i} = p_{i,d} - p_i$$

$$e_{c,i} = c_{i,d} - c_i \quad for \quad i = x, y.$$

If the balancing control input u_i *in Fig. 4 has the following form:*

$$u_i = \dot{c}_i^d - k_{p,i} e_{p,i} + k_{c,i} e_{c,i} \tag{24}$$

under the following gain conditions:

$$k_{c,i} > \omega_n \quad and \quad 0 < k_{p,i} < \left(\frac{\omega_n^2 - \beta^2}{\omega_n} - \gamma^2 \right) \tag{25}$$

with any positive constants satisfying the following conditions:

$$\beta < \omega_n \quad and \quad \gamma < \sqrt{\frac{\omega_n^2 - \beta^2}{\omega_n}},$$

then the balancing controller gives the disturbance input($\varepsilon_i, \dot{\varepsilon}_i$ *)-to-state(* $e_{p,i}, e_{c,i}$ *) stability (ISS) to a simplified bipedal robot, where, the* $k_{p,i}$ *is the proportional gain of ZMP controller and* $k_{c,i}$ *is that of CoM controller in Fig. 4.*

Proof. First, we get the error dynamics from Eq. (22) and (23) as follows:

$$\ddot{e}_{c,i} = \omega_n^2 (e_{c,i} - e_{p,i}). \tag{26}$$

Second, another error dynamics is obtained by using Eq. (23) and (24) as follows:

$$k_{p,i} e_{p,i} = \dot{e}_{c,i} + k_{c,i} e_{c,i} + \varepsilon_i, \tag{27}$$

also, this equation can be rearranged for $\dot{e}_c$:

$$\dot{e}_{c,i} = k_{p,i}e_{p,i} - k_{c,i}e_{c,i} - \varepsilon_i. \tag{28}$$

Third, by differentiating the equation (27) and by using equations (26) and (28), we get the following:

$$\begin{aligned}\dot{e}_{p,i} &= 1/k_{p,i}\left(\ddot{e}_{c,i} + k_{c,i}\dot{e}_{c,i} + \dot{\varepsilon}_i\right)\\ &= \omega_n^2/k_{p,i}(e_{c,i} - e_{p,i}) + k_{c,i}/k_{p,i}(k_{p,i}e_{p,i} - k_{c,i}e_{c,i} - \varepsilon_i) + (1/k_{p,i})\dot{\varepsilon}_i\\ &= \left(\frac{\omega_n^2 - k_{c,i}^2}{k_{p,i}}\right)e_{c,i} - \left(\frac{\omega_n^2 - k_{p,i}k_{c,i}}{k_{p,i}}\right)e_{p,i} + \frac{1}{k_{p,i}}(\dot{\varepsilon}_i - k_{c,i}\varepsilon_i).\end{aligned} \tag{29}$$

Fourth, let us consider the following Lyapunov function:

$$V(e_{c,i}, e_{p,i}) = \frac{1}{2}\left[(k_{c,i}^2 - \omega_n^2)e_{c,i}^2 + k_{p,i}^2 e_{p,i}^2\right], \tag{30}$$

where $V(e_c, e_p)$ is the positive definite function for $k_{p,i} > 0$ and $k_{c,i} > \omega_n$, except $e_{c,i} = 0$ and $e_{p,i} = 0$. Now, let us differentiate the above Lyapunov function

$$\begin{aligned}\dot{V} &= (k_{c,i}^2 - \omega_n^2)e_{c,i}\dot{e}_{c,i} + k_{p,i}^2 e_{p,i}\dot{e}_{p,i}\\ &= -k_{c,i}(k_{c,i}^2 - \omega_n^2)e_{c,i}^2 - k_{p,i}(\omega_n^2 - k_{p,i}k_{c,i})e_{p,i}^2 - (k_{c,i}^2 - \omega_n^2)e_{c,i}\varepsilon_i + k_{p,i}e_{p,i}\dot{\varepsilon}_i - k_{p,i}k_{c,i}e_{p,i}\varepsilon_i\\ &= -k_{c,i}(k_{c,i}^2 - \omega_n^2)e_{c,i}^2 - k_{p,i}(\omega_n^2 - k_{p,i}k_{c,i})e_{p,i}^2 + (k_{c,i}^2 - \omega_n^2)\left(\alpha^2 e_{c,i}^2 - \left|\alpha e_{c,i} + \frac{1}{2\alpha}\varepsilon_i\right|^2 + \frac{1}{4\alpha^2}\varepsilon_i^2\right)\\ &\quad + k_{p,i}\left(\beta^2 e_{p,i}^2 - \left|\beta e_{p,i} - \frac{1}{2\beta}\dot{\varepsilon}_i\right|^2 + \frac{1}{4\beta^2}\dot{\varepsilon}_i^2\right) + k_{p,i}k_{c,i}\left(\gamma^2 e_{p,i}^2 - \left|\gamma e_{p,i} + \frac{1}{2\gamma}\varepsilon_i\right|^2 + \frac{1}{4\gamma^2}\varepsilon_i^2\right)\\ &= -(k_{c,i} - \alpha^2)(k_{c,i}^2 - \omega_n^2)e_{c,i}^2 - k_{p,i}[\omega_n^2 - (k_{p,i} + \gamma^2)k_{c,i} - \beta^2]e_{p,i}^2\\ &\quad - (k_{c,i}^2 - \omega_n^2)\left|\alpha e_{c,i} + \frac{1}{2\alpha}\varepsilon_i\right|^2 - k_{p,i}\left|\beta e_{p,i} - \frac{1}{2\beta}\dot{\varepsilon}_i\right|^2 - k_{p,i}k_{c,i}\left|\gamma e_{p,i} + \frac{1}{2\gamma}\varepsilon_i\right|^2\\ &\quad + \left[\frac{(k_{c,i}^2 - \omega_n^2)}{4\alpha^2} + \frac{k_{p,i}k_{c,i}}{4\gamma^2}\right]\varepsilon_i^2 + \frac{k_{p,i}}{4\beta^2}\dot{\varepsilon}_i^2\end{aligned}$$

Therefore,

$$\begin{aligned}\dot{V} &\le -(k_{c,i} - \alpha^2)(k_{c,i}^2 - \omega_n^2)e_{c,i}^2 - k_{p,i}[\omega_n^2 - (k_{p,i} + \gamma^2)k_{c,i} - \beta^2]e_{p,i}^2\\ &\quad + \left[\frac{(k_{c,i}^2 - \omega_n^2)}{4\alpha^2} + \frac{k_{p,i}k_{c,i}}{4\gamma^2}\right]\varepsilon_i^2 + \frac{k_{p,i}}{4\beta^2}\dot{\varepsilon}_i^2\end{aligned} \tag{31}$$

where $e_{c,i}^2$ term is negative definite with any positive constant satisfying $\alpha < \sqrt{\omega_n}$ and $e_{p,i}^2$ term is negative definite under the given conditions (25). Here, since the inequality (31) follows the ISS property (21), we concludes that the proposed balancing controller gives the disturbance input($\varepsilon_i, \dot{\varepsilon}_i$)-to-state($e_{p,i}, e_{c,i}$) stability (ISS) to the simplified control system model of bipedal robot. □

The balancing control input u_i suggested in above Theorem 1 is applied to the term $\dot{c}_d$ in Eq. (17) or Eq. (20). In other words, two equations (17) and (20) should be modified to include the balancing controller (24) to compensate the CoM/ZMP errors as follows:

$$\dot{c}_{\mathrm{fsem},d} = (\dot{c}_d - k_p\, e_p + k_c\, e_c) - \sum_{i=2}^{n} J_{c_i} J_i^{+} \dot{x}_{i,d} \tag{32}$$

$$\dot{c}_{\mathrm{psem},d} = (\dot{c}_d - k_p\, e_p + k_c\, e_c) - J_{c_2} J_2^{+} \dot{x}_{2,d} - \sum_{i=3}^{n} J_{c_i} \dot{q}_{i,d} \tag{33}$$

where k_p and k_c imply the diagonal matrices with the corresponding $k_{p,i}$ and $k_{c,i}$ elements, respectively, and e_p and e_c imply the corresponding vectors with $e_{p,i}$ and $e_{c,i}$ elements, respectively. Also, in order to obtain the desired joint driving velocity, the kinematic resolution method of Eq. (16) or Eq. (18) suggested in previous section should be applied to the real bipedal robot.

Remark 2 *Note that the ZMP controller in above theorem has the negative feedback differently from the conventional controller. Also, for practical use, the gain conditions of balancing controller can be simply rewritten without arbitrary positive constants β and γ as follows:*

$$k_{c,i} > \omega_n \quad \text{and} \quad 0 < k_{p,i} < \omega_n \quad \text{for } i = x, y,$$

because the stability proof is very conservative in above theorem.

The suggested balancing control method can be divided as the kinematic resolution and closed-loop kinematic control of CoM Jacobian with embedded motion. First, the proposed kinematic resolution method has main advantage such that it offers the whole body coordination function such as balance control to humanoid robot automatically. Second, the proposed closed-loop kinematic control method offers the stability and robustness to humanoid motion control system against unknown disturbances. For arbitrary given arm motions such as dancing, the partially specified embedded CoM motion term $\dot{c}_{\mathrm{psem},d}$ is automatically changed with the desired arm motions ($\dot{q}_{3,d}$ and $\dot{q}_{4,d}$) in Eq. (33), and then both desired leg motions ($\dot{q}_{1,d}$ and $\dot{q}_{2,d}$) are generated by using the equations (18) and (19). Like this, the suggested kinematic resolution method offers the whole body coordination function such as balance control to humanoid robot automatically.

5. Experimental results

In this section, we show the performance of proposed kinematic resolution method and the robustness of balancing controller through experiments for humanoid robot `Mahru I' developed by KIST. Its Denavit-Hartenberg parameters for kinematics and centroid/mass data for CoM kinematics are in Table 1. These data are utilized to resolve the CoM Jacobian with embedded motions kinematically.

First, in order to show the automatic balancing (or the function of WBC) by the kinematic resolution method developed in section 3, we implemented the dancing motion of humanoid robot. The desired dancing arm motions shown in Fig. 5 are applied to the dual

Left Leg	α (rad)	a (m)	d (m)	θ (rad)	Centroid(σ_x,σ_y,σ_z)(m)	Mass(kg)
LL0	0	0.09	-0.146	$\pi/2$	(0.0,0.031,-0.0338)	2.067
1	$\pi/2$	0	0	0	(0.0295,0.0,-0.0015)	1.8206
2	$\pi/2$	0	0	$-\pi/2$	(0.2249,0.0187,-0.0156)	3.3586
3	0	0.31	0	0	(0.1451,0.0285,-0.0026)	2.2238
4	0	0.31	0	0	(0.0,-0.0257,-0.0143)	2.6922
5	$-\pi/2$	0	0	0	(0.0951,0.0047,0.0083)	1.9091
6	0	0.103	0	0	x	x
Right Leg						
RL0	0	-0.09	-0.146	$\pi/2$	(0.0,0.031,-0.0338)	2.067
1	$\pi/2$	0	0	0	(-0.0295,0.0,-0.0015)	1.8206
2	$\pi/2$	0	0	$-\pi/2$	(0.2249,0.0187,0.0156)	3.3586
3	0	0.31	0	0	(0.1451,0.0285,0.0026)	2.2238
4	0	0.31	0	0	(0.0,-0.0257,0.0143)	2.6922
5	$-\pi/2$	0	0	0	(0.0951,-0.0047,0.0083)	1.9091
6	0	0.103	0	0	x	x
Left Arm						
LA0	0	0	0	0	(0.0278,0.0,-0.051)	0.823
1	$-\pi/2$	0	-0.061	$-\pi/2$	(0.062,0.000125,-0.0085)	1.437
2	$\pi/2$	0	-0.0055	$\pi/2$	(0.00798,-0.00146,0.2)	0.9
3	$-\pi/2$	0	0.224	$\pi/2$	(0.0,-0.0228,0.00739)	0.11
4	$\pi/2$	0	0	0	(-0.0025,0.000035,0.153)	0.781
5	$-\pi/2$	0	0.225	$-\pi/2$	(0.0,-0.0208,0.017)	0.053
6	0	0.041	0	$-\pi/2$	x	x
Right Arm						
RA0	0	0	0	0	(0.0278,0.0,0.051)	0.823
1	$-\pi/2$	0	0.061	$-\pi/2$	(0.062,0.000125,-0.0085)	1.437
2	$\pi/2$	0	-0.0055	$\pi/2$	(-0.00798,-0.00146,0.2)	0.9
3	$-\pi/2$	0	0.224	$\pi/2$	(0.0,-0.0228,-0.00739)	0.11
4	$\pi/2$	0	0	0	(-0.0025,0.000035,0.153)	0.781
5	$-\pi/2$	0	0.225	$-\pi/2$	(0.0,-0.0208,0.017)	0.053
6	0	0.041	0	$-\pi/2$	x	x
Head						
H0	0	0.04	0.47	0	(0.0,0.0,0.0)	0.5
1	$-\pi/2$	0	0	0	(0.0,0.0,0.0)	0.5
2	0	0	0	0	x	x
Pelvis	x	x	x	x	(-0.056,0.0,-0.0173)	4.6341
Waist	x	x	x	x	(0.0,0.0,0.17)	25.7

Table 1. The DH parameters, centroid and mass data of KIST humanoid `Mahru I'

arms, and then the motions of support left/right legs are generated as shown in Fig. 6 by the suggested kinematic resolution method of CoM Jacobian with embedded dancing arm motions. Actually, we utilized the kinematic resolution method of CoM Jacobian with the partially specified embedded motion, namely, Eq. (18) and Eq. (19). The initial positions of CoM are $c_x = 0.034[m]$, $c_y = 0.0[m]$, $c_z = 0.687[m]$, respectively. The experimental results show the good performance of proposed method. Though the joint configurations of dual arms are rapidly changed with the dancing motion given as shown in Fig. 5, the position of CoM is not nearly changed at the initial position as shown in Fig. 6. The joint configurations of both legs are automatically generated as shown in Fig. 7 to maintain the CoM position constantly. Also, we can see in Fig. 8 that the ZMP has the small changes within the bounds of $\pm 0.01[m]$ approximately. As a result, we could succeed in implementing the fast dancing motion stably thanks to the developed kinematic resolution method of CoM Jacobian with embedded (dancing) motion.

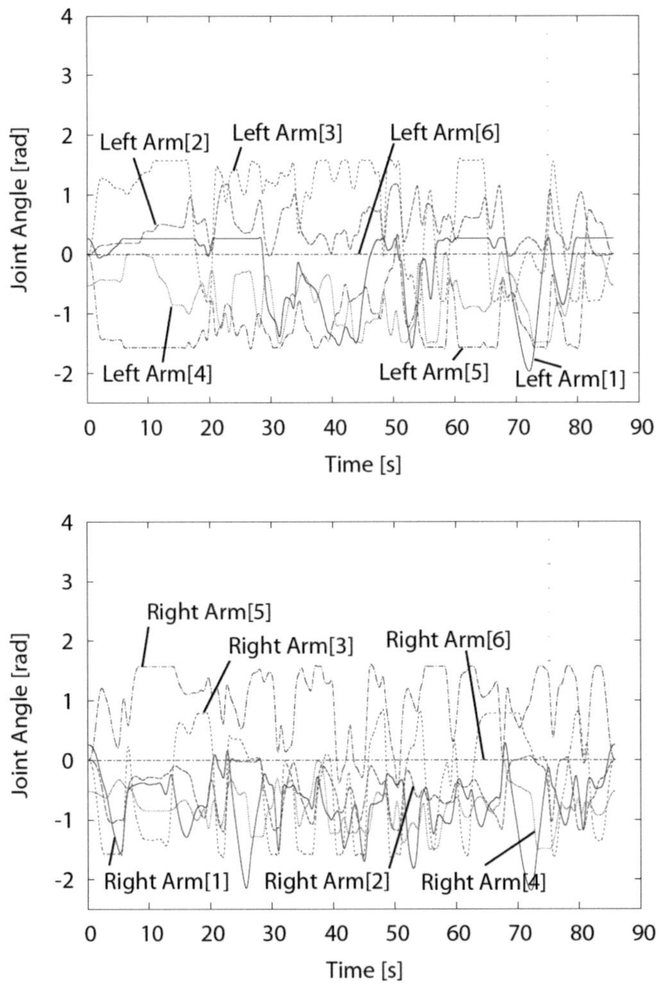

Fig. 5. Desired Joint Trajectories of Left and Right Arms while Dancing

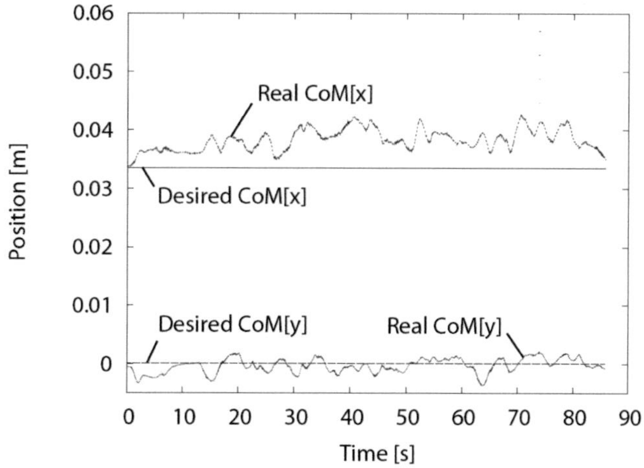

Fig. 6. Experimental Result: Actual CoM Trajectories while Dancing

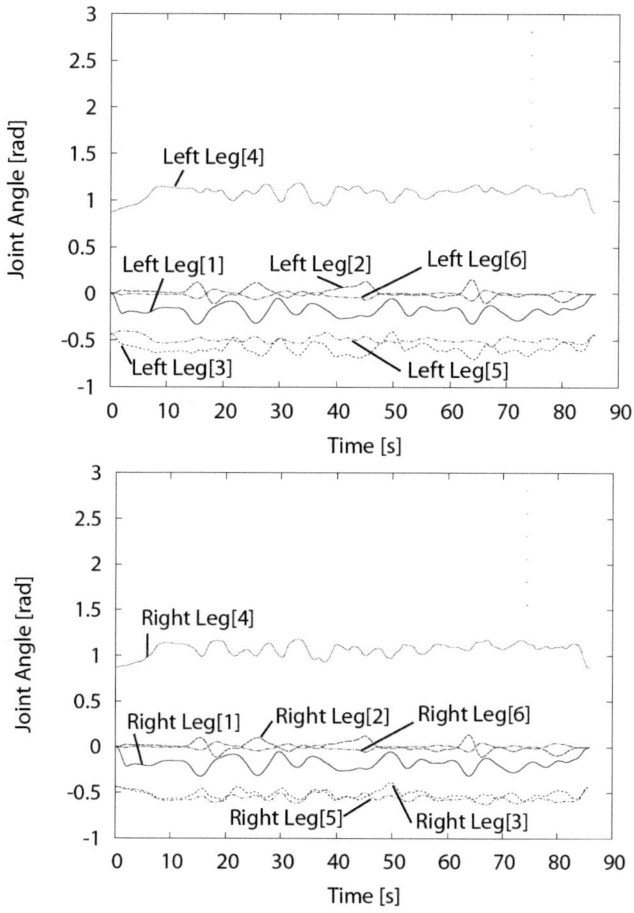

Fig. 7. Experimental Result: Actual Joint Trajectories of Left and Right Legs generated by the developed Kinematic Resolution Method while Dancing

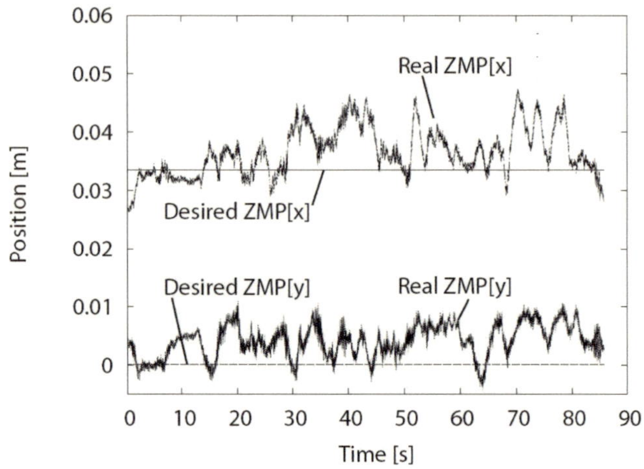

Fig. 8. Experimental Result: Actual ZMP Trajectories while Dancing

Second, in order to show the performance of the suggested balancing controller of Eq. (33) against external disturbances, we realized the corresponding experiments only for both legs as shown in Fig. 9. In experiment, if we push the robot forward or pull it backward, then the ZMP errors are caused and these ZMP errors give rise to the balancing controller input of Eq. (33) and then the balancing control input is resolved using the suggested resolution method of Eq. (18) and Eq. (19). Hence, the robot was able to recover the original posture as shown in Fig. 9. These results demonstrate the robustness and performance of proposed balancing controller.

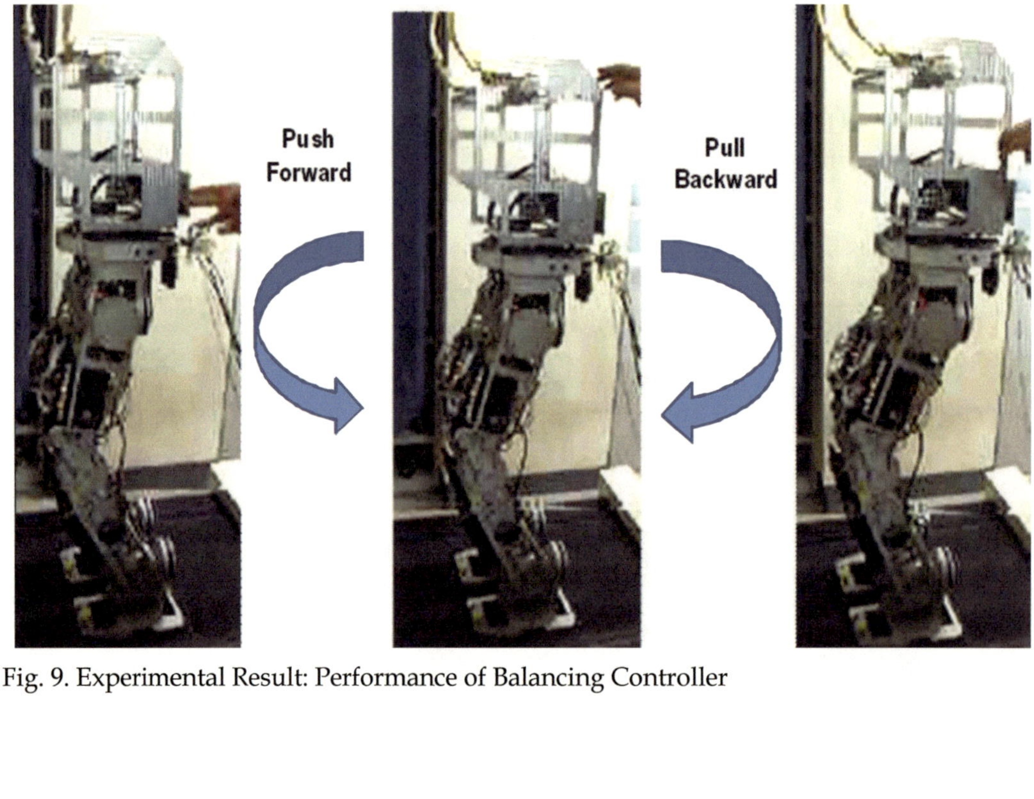

Fig. 9. Experimental Result: Performance of Balancing Controller

6. Concluding remarks

In this chapter, the kinematic resolution method of CoM Jacobian with embedded task motion and the balancing control method were proposed for humanoid robot. The proposed kinematic resolution method with CoM Jacobian offers the whole body coordination function to the humanoid robot automatically. Also, the disturbance input-to-state stability (ISS) of the proposed balancing controller was proved to show the robustness against disturbances. Finally, we showed the effectiveness of the proposed methods through experiments.

7. References

Choi, Y.; Kim, D.; Oh, Y. & You, B-J. (2007). Posture/Walking Control of Humanoid Robot based on the Kinematic Resolution of CoM Jacobian with Embedded Motion, *IEEE Trans. on Robotics*, Vol 23, No. 6, pp. 1285-1293.

Choi, Y.; Kim, D. & You B-J. (2006) On the Walking Control for Humanoid Robot based on the Kinematic Resolution of CoM Jacobian with Embedded Motion, *IEEE Int. Conf. on Robotics and Automation*, pp, 2655-2660.

Choi, Y.; You, B-J. & Oh, S.R. (2004). On the stability of indirect ZMP controller for biped robot systems, *Proc. of IEEE/RSJ Int. Conf. on Intelligent Robots and Systems*, pp. 1966–1971.

Choi, Y. & Chung, W-K.; (2004). *PID Trajectory Tracking Control for Mechanical Systems*, ser. Lecture Notes in Control and Information Sciences (LNCIS). Springer, Vol. 298, New York.

Huang, Q.; Yokoi, K.; Kajita, S.; Kaneko, K.; Arai, H.; Koyachi, N. & Tanie, K. (2001). Planning walking patterns for a biped robot, *IEEE Trans. on Robotics and Automation*, Vol. 17, No. 3, pp. 280–289.

Grizzle, J.W.; Westervelt, E.R. & Wit, C.C. (2003). Event-based PI control of a underactuated biped walker, *Proc., 42nd IEEE Conf. on Decision and Control*, pp. 3091–3096.

Hirai, K.; Hirose, M.; Haikawa, Y. & Takenaka, T. (1998). The development of Honda humanoid robot, *Proc. of IEEE Int. Conf. on Robotics and Automation*, pp. 1321–1326.

Kajita, S.; Saigo, M. & Tanie, K. (2001). Balancing a humanoid robot using backdrive concerned torque control and direct angular momentum feedback, *Proc. of IEEE Int. Conf. on Robotics and Automation*, pp. 3376–3382.

Kajita, S.; Kanehiro, F.; Kaneko, K.; Fujiwara, K.; Yokoi, K. & Hirukawa, H. (2002). A realtime pattern generator for biped walking, *Proc. of IEEE Int. Conf. on Robotics and Automation*, pp. 31–37.

Kajita, S.; Kanehiro, F.; Kaneko, K.; Fujiwara, K.; Harada, K.; Yokoi, K. & Hirukawa, H. (2003). Biped walking pattern generation by using preview control of zero-moment point, *Proc. of IEEE Int. Conf. on Robotics and Automation*, pp. 1620–1626.

Kajita, S.; Kanehiro, F.; Kaneko, K.; Fujiwara, K.; Harada, K.; Yokoi, K. & Hirukawa, H.; (2003). Resolved momentum control : Humanoid motion planning based on the linear and angular momentum, *Proc. of IEEE Int. Conf. on Robotics and Automation*, pp. 1644–1650.

Kim, J-H. & Oh, J-H. (2004). Walking control of the humanoid platform KHR-1 based on torque feedback control, *Proc. of IEEE Int. Conf. on Robotics and Automation*, pp. 623–628.

Lohmeier, S.; Loffler, K.; Gienger, M.; Ulbrich, H. & Pfeiffer, F.; (2004). Computer system and control of biped "Johnnie", *Proc. of IEEE Int. Conf. on Robotics and Automation*, pp. 4222–4227.

Park, J-H. (2001). Impedance control for biped robot locomotion, *IEEE Trans. on Robotics and Automation*, Vol. 17, No. 6, pp. 870–882.

Takanishi, A.; Lim, H.; Tsuda, M. & Kato, I.; (1990). Realization of dynamic biped walking stabilized by trunk motion on a sagittally uneven surface, *Proc. of IEEE/RSJ Int. Conf. on Intelligent Robots and Systems*, pp. 323– 330.

Westervelt, E-R.; Grizzle, J-W. & Koditschek, D-E.; (2003). Hybrid zero dynamics of planar biped walkers, *IEEE Trans. on Automatic Control*, Vol. 48, No. 1, pp. 42–56.

Nagasaka, K.; Kuroki, Y.; Suzuki, S.; Itoh, Y. & Yamaguchi, J.; (2004). Integrated motion control for walking, jumping and running on a small bipedal entertainment robot, *Proc. of IEEE Int. Conf. on Robotics and Automation*, pp. 3189–3194.

Nagasaki, T.; Kajita, S.; Yokoi, K.; Kaneko, K. & Tanie, K.; (2003). Running pattern generation and its evaluation using a realistic humanoid model, *Proc. of IEEE Int. Conf. on Robotics and Automation*, pp. 1336–1342.

Sentis, L. & Khatib, O.; (2005). Synthesis of whole-body behaviors through hierarchical control of behavioral primitives, *International Journal of Humanoid Robotics*, Vol. 2, No. 4, pp. 505–518.

Goswami, A. & Kallem, V.; (2004). Rate of change of angular momentum and balance maintenance of biped robots, *Proc. of IEEE Int. Conf. on Robotics and Automation*, pp. 3785–3790.

Harada, K.; Kajita, S.; Kaneko, K. & Hirukawa, H.; (2003). ZMP analysis for arm/leg coordination, *Proc. of IEEE Int. Conf. on Robotics and Automation*, pp. 75–81.

Sugihara, T. & Nakamura, Y.; (2002). Whole-body cooperative balancing of humanoid robot uisng COG Jacobian, *Proc. of IEEE/RSJ Int. Conf. on Intelligent Robots and Systems*, pp. 2575–2580.

8. Acknowledgement

This work was supported by IT R & D program of MIC & IITA (2006-S-028-01, Development of Cooperative Network-based Humanoids Technology), Republic of Korea.

9. Appendix

9.1 Proof of Eq. (5)

In Fig. 1, the position of *i*-th limb represented on the world coordinate is given by

$$r_i = r_o + R_o{}^o r_i$$

where R_o is the rotation matrix of body center frame with respect to world coordinate frame. Let's differentiate above equation, then

$$\dot{r}_i = \dot{r}_o + \dot{R}_o{}^o r_i + R_o{}^o\dot{r}_i \quad \leftarrow \dot{R}_o = [\omega_o \times]R_o$$

$$\dot{r}_i = \dot{r}_o + [\omega_o \times]R_o{}^o r_i + R_o{}^o\dot{r}_i \quad \leftarrow [a\times]b = -b[\times a]$$

$$\dot{r}_i = \dot{r}_o - [R_o{}^o r_i \times]\omega_o + R_o{}^o\dot{r}_i$$

in which,

$$[a\times] = \begin{bmatrix} 0 & -a_z & a_y \\ a_z & 0 & -a_x \\ -a_y & a_x & 0 \end{bmatrix}$$

Now, if we include the angular velocity, then the total velocity of i-th limb motion represented on the world coordinate can be obtained as follows:

$$\dot{r}_i = \dot{r}_o - [R_o{}^o r_i \times]\omega_o + R_o{}^o\dot{r}_i$$

$$\omega_i = \omega_o + R_o{}^o\omega_i$$

Therefore,

$$\begin{bmatrix} \dot{r}_i \\ \omega_i \end{bmatrix} = \begin{bmatrix} I_3 & -[R_o{}^o r_i \times] \\ 0_3 & I_3 \end{bmatrix} \begin{bmatrix} \dot{r}_o \\ \omega_o \end{bmatrix} + \begin{bmatrix} R_o & 0_3 \\ 0_3 & R_o \end{bmatrix} \begin{bmatrix} {}^o\dot{r}_i \\ {}^o\omega_i \end{bmatrix}$$

in short,

$$\therefore \quad \dot{x}_i = X_i^{-1}\dot{x}_o + X_o{}^o J_i \dot{q}_i$$

9.2 Proof of Eq. (8)

In Fig. 1, the position of CoM represented on the world coordinate is given by

$$c = r_o + \sum_{i=1}^{n} R_o{}^o c_i$$

Let us differentiate above equation, then

$$\dot{c} = \dot{r}_o + \sum_{i=1}^{n} (\dot{R}_o{}^o c_i + R_o{}^o \dot{c}_i)$$

$$\dot{c} = \dot{r}_o + \omega_o \times \sum_{i=1}^{n} R_o{}^o c_i + \sum_{i=1}^{n} R_o{}^o \dot{c}_i$$

$$\dot{c} = \dot{r}_o + \omega_o \times (c - r_o) + \sum_{i=1}^{n} R_o{}^o \dot{c}_i \quad \leftarrow {}^o\dot{c}_i = {}^o J_{ci} \dot{q}_i$$

$$\therefore \quad \dot{c} = \dot{r}_o + \omega_o \times (c - r_o) + \sum_{i=1}^{n} R_o{}^o J_{ci} \dot{q}_i$$

9.3 Derivation of CoM jacobian of *i*-h limb: ${}^o J_{c_i}$

The CoM position of k-th link in i-th limb represented on the body center frame is given by

$$ {}^{o}c_{i,k} = {}^{o}r_{k-1} + {}^{o}R_{k-1} {}^{k-1}c_k $$

where ${}^{o}r_{k-1}$ and ${}^{o}R_{k-1}$ mean the position and rotation matrix of (*k-1*)-th link frame in *i*-th limb represented on the body center frame, respectively, and

$$ {}^{k-1}c_k = R_z(q_k)\sigma_k = \begin{bmatrix} \cos(q_k) & -\sin(q_k) & 0 \\ \sin(q_k) & \cos(q_k) & 0 \\ 0 & 0 & 1 \end{bmatrix} \begin{bmatrix} \sigma_{k,x} \\ \sigma_{k,y} \\ \sigma_{k,z} \end{bmatrix} $$

in which $R_z(q_k)$ is the *z*-directional rotation matrix about *k*-th link driving axis and σ_k means the constant centroid position of *k*-th link with respect to (*k-1*)-th frame, this should be obtained from the design procedures of robot similarly to the Denavit-Hartenberg parameters. Let us differentiate and rearrange above equation like this:

$$ \begin{aligned} {}^{o}\dot{c}_{i,k} &= {}^{o}\dot{r}_{k-1} + {}^{o}\dot{R}_{k-1} {}^{k-1}c_k + {}^{o}R_{k-1} {}^{k-1}\dot{c}_k \\ &= {}^{o}\dot{r}_{k-1} + {}^{o}\omega_{k-1} \times {}^{o}R_{k-1} {}^{k-1}c_k + {}^{o}R_{k-1}({}^{k-1}\omega_k \times {}^{k-1}c_k) \\ &= {}^{o}\dot{r}_{k-1} + {}^{o}\omega_k \times {}^{o}R_{k-1} {}^{k-1}c_k \qquad \textit{since } {}^{o}\omega_k = {}^{o}\omega_{k-1} + {}^{o}R_{k-1} {}^{k-1}\omega_k \\ &= {}^{o}\dot{r}_{k-1} + {}^{o}\omega_k \times {}^{o}R_{k-1} R_z(q_k)\sigma_k, \end{aligned} $$

in which

$$ {}^{o}\omega_k = {}^{o}\omega_{k-1} + \dot{q}_k({}^{o}z_{k-1}) $$

$$ {}^{o}\dot{r}_{k-1} = {}^{o}\dot{r}_{k-2} + {}^{o}\omega_{k-1} \times ({}^{o}r_{k-1} - {}^{o}r_{k-2}). $$

where ${}^{o}z_{k-1}$ means the *z*-direction (or driving axis) vector of (*k-1*)-th link frame represented on body center frame. For instance, for *k=1* (first link of *i*-th limb),

$$ \begin{aligned} {}^{o}\dot{c}_{i,1} &= {}^{o}\dot{r}_0 + {}^{o}\omega_1 \times {}^{o}R_0 R_z(q_1)\sigma_1 \\ &= \dot{q}_1[{}^{o}z_0 \times {}^{o}R_0 R_z(q_1)\sigma_1], \end{aligned} $$

for *k*=2 (second link of *i*-th limb),

$$ \begin{aligned} {}^{o}\dot{c}_{i,2} &= {}^{o}\dot{r}_1 + {}^{o}\omega_2 \times {}^{o}R_1 R_z(q_2)\sigma_2 \\ &= \dot{q}_1[{}^{o}z_0 \times ({}^{o}r_1 - {}^{o}r_0)] + [\dot{q}_1({}^{o}z_0) + \dot{q}_2({}^{o}z_1)] \times {}^{o}R_1 R_z(q_2)\sigma_2 \\ &= \dot{q}_1[{}^{o}z_0 \times ({}^{o}r_1 - {}^{o}r_0 + {}^{o}R_1 R_z(q_2)\sigma_2)] + \dot{q}_2[{}^{o}z_1 \times {}^{o}R_1 R_z(q_2)\sigma_2] \end{aligned} $$

for *k*=3 (third link of *i*-th limb),

$$ \begin{aligned} {}^{o}\dot{c}_{i,3} &= {}^{o}\dot{r}_2 + {}^{o}\omega_3 \times {}^{o}R_2 R_z(q_3)\sigma_3 \\ &= \dot{q}_1[{}^{o}z_0 \times ({}^{o}r_2 - {}^{o}r_0)] + \dot{q}_2[{}^{o}z_1 \times ({}^{o}r_2 - {}^{o}r_1)] + [\dot{q}_1({}^{o}z_0) + \dot{q}_2({}^{o}z_1) + \dot{q}_3({}^{o}z_2)] \times {}^{o}R_2 R_z(q_3)\sigma_3 \\ &= \dot{q}_1[{}^{o}z_0 \times ({}^{o}r_2 - {}^{o}r_0 + {}^{o}R_2 R_z(q_3)\sigma_3)] + \dot{q}_2[{}^{o}z_1 \times ({}^{o}r_2 - {}^{o}r_1 + {}^{o}R_2 R_z(q_3)\sigma_3)] + \dot{q}_3[{}^{o}z_2 \times {}^{o}R_2 R_z(q_3)\sigma_3] \end{aligned} $$

for $k = n_i$ (last link of i-th limb),

$$\begin{aligned}{}^{o}\dot{c}_{i,n_i} &= {}^{o}\dot{r}_{n_i-1} + {}^{o}\omega_{n_i} \times {}^{o}R_{n_i-1}R_z(q_{n_i})\sigma_{n_i} \\ &= \dot{q}_1[{}^{o}z_0 \times ({}^{o}r_{n_i-1} - {}^{o}r_0 + {}^{o}R_{n_i-1}R_z(q_{n_i})\sigma_{n_i})] + \dot{q}_2[{}^{o}z_1 \times ({}^{o}r_{n_i-1} - {}^{o}r_1 + {}^{o}R_{n_i-1}R_z(q_{n_i})\sigma_{n_i})] + \cdots \\ &\quad + \dot{q}_{n_i}[{}^{o}z_{n_i-1} \times {}^{o}R_{n_i-1}R_z(q_{n_i})\sigma_{n_i}]\end{aligned}$$

The CoM position of i-th limb represented on the body center frame is obtained as follow:

$${}^{o}c_i = \sum_{k=1}^{n_i} \mu_{i,k}\, {}^{o}c_{i,k}$$

where $\mu_{i,k}$ means the mass influence coefficient of k-th link in i-th limb defined as Eq. (10). Also, its derivative has the following form:

$${}^{o}\dot{c}_i = \sum_{k=1}^{n_i} \mu_{i,k}\, {}^{o}\dot{c}_{i,k}$$

Therefore,

$${}^{o}\dot{c}_i = \left[j_1, j_2, j_3, \cdots, j_{n_i} \right] \begin{bmatrix} \dot{q}_1 \\ \dot{q}_2 \\ \dot{q}_3 \\ \vdots \\ \dot{q}_{n_i} \end{bmatrix}$$

where

$$j_1 = \sum_{k=1}^{n_i} \mu_{i,k} \{ {}^{o}z_0 \times ({}^{o}r_{k-1} - {}^{o}r_0 + {}^{o}R_{k-1}R_z(q_k)\sigma_k) \}$$

$$j_2 = \sum_{k=2}^{n_i} \mu_{i,k} \{ {}^{o}z_1 \times ({}^{o}r_{k-1} - {}^{o}r_1 + {}^{o}R_{k-1}R_z(q_k)\sigma_k) \}$$

$$j_3 = \sum_{k=3}^{n_i} \mu_{i,k} \{ {}^{o}z_2 \times ({}^{o}r_{k-1} - {}^{o}r_2 + {}^{o}R_{k-1}R_z(q_k)\sigma_k) \}$$

$$j_{n_i} = \mu_{i,n_i} \{ {}^{o}z_{n_i-1} \times ({}^{o}R_{n_i-1}R_z(q_{n_i})\sigma_{n_i}) \}$$

In short,

$$\therefore \quad {}^{o}\dot{c}_i = {}^{o}J_{c_i}\dot{q}_i$$

In addition, the total CoM position represented on the body center frame is also obtained as follow:

$$\therefore \quad {}^{o}c = \sum_{i=1}^{n} {}^{o}c_{i} = \sum_{i=1}^{n}\sum_{k=1}^{n_i} \mu_{i,k}\, {}^{o}c_{i,k}$$

19

Multiresolutional Filter Application for Spatial Information Fusion in Robot Navigation

Özer Ciftcioglu
Delft University of Technology
The Netherlands

1. Introduction

Robot navigation is one of the major fields of study in autonomous robotics (Oriolio, G., Ulivi, G. et al., 1998; Beetz, M., Arbuckle, T. et al., 2001; Wang, M. & Liu, J.N.K., 2004). In the present work, application of a multiresolutional filter for spatial information fusion in vision robot navigation is described. The novelty of the research is the enhanced estimation of the spatial sensory information in autonomous robotics by means of wavelet decomposition into multiresolutional levels and the fusion of the processed information while the processing takes place at the respective levels. Although wavelet-based information fusion is used in different applications (Hong, L., 1994; Hsin, H.C. & Li, A.C., 2006), its application in robotics is not common in literature. A wavelet-based filtering application in perceptual robotics is articulated earlier where human perception was central to the research. In the present work, optimal vision signal estimation for an autonomous robot is presented, where vision information is obtained by appropriate shaping of the visual sensory information and optimal estimation is obtained by a multiresolutional filter. One of the peculiarities of the research is essentially the wavelet-based dynamic filtering rather than static wavelet decomposition for image analysis or off-line decomposition for signal analysis. A multiresolutional dynamic filtering is a rather complex system. In this system vector form of wavelet decomposition is required as briefly mentioned by Hong (Hong, L., 1993) where the description of a multi-sensor fusion rather than wavelet decomposition was central to the study. In contrast to (Hong, L., 1993), in this work the multiresolutional dynamic filtering is central to the study together with the emphasis on application peculiarities in autonomous robotics; namely, several options for vector-wise operations are pointed out and one of the options is described in detail. Also, in autonomous robotics, the estimation of angular velocity is not a measurable quantity and it has to be estimated from the measurable state variables so that obstacle avoidance problem is taken care of. Therefore, the angular velocity estimation in real-time is a critical task in autonomous robotics and from this viewpoint the multiresolutional spatial information fusion process is desirable for enhanced robot performance. For the fusion process extended Kalman filtering can be used and is actually used in the present study. Perceptual vision employed in this study is one of the vision alternatives in robotics providing several essential possibilities characterizing the robot as to its performance. Especially perceptual robotics is gaining importance in the last decade as to its expected connection to human-like behaviour. The organization of the work is as follows.

After brief description of wavelet transform, a detailed description of vector wavelet decomposition and inverse wavelet transform is presented with a vector decomposition and reconstruction example. The optimal processing of decomposed information is briefly given as proposed by Hong (Hong, L., 1993). This is followed by fusion of information from different multiresolutional levels that it is based on minimum fusion error covariance. Finally, autonomous robot implementation is described with the experimental results illustrating the sensory information and estimated trajectory thereby demonstrating the effective navigation of a moving robot. The work is accomplished in a virtual environment with the intention of realization in real-life environment in due course as the work constitutes a common place for diverse robotics areas like autonomous, perceptual, vision, cognitive etc.

2. Perceptual vision and multiresolutional decomposition

In this work a robot with perceptual vision is considered. The robot is provided with vision via sensors which can measure the distance between the robot and the forward direction via rays interacting with the environment. The rays can be conveniently shaped within a vision cone with a narrow solid cone angle. This is illustrated in figure 1.

Fig. 1. Perception measurement in the virtual reality. Perception is determined as probability by means of the rays impinging on an object.

Vision rays interact with the environment which includes side walls and hindrance along the forward direction. For an environment a permanent fictive vertical planes at certain threshold distances are considered. If a distance is more than the threshold value it is taken to be the threshold value. This means if the real environment is in front of that fictive plane the actual distance measured by the sensor is considered as distance. If the measured distances are beyond the predetermined threshold distances the robot moves forward without any angular velocity. The position estimation of robot is accomplished by means of Kalman filtering via the states of the robot navigation dynamics as presented earlier (Ciftcioglu et al., 2007). In this work, the enhanced estimation of the states is investigated via wavelet transform (Mallat, S.G., 1989; Ogden, T., 1997; Mallat, S., 1999; Percival, D.B. & Walden, A.T., 2000).

2.1 Wavelets

The novel feature of wavelets is that they are localized in time and frequency as to signals. This behaviour makes them convenient for the analysis of non-stationary signals. It is an elementary introduction of wavelets by introducing a scaling function, such that

$$\varphi(t) = \sqrt{2}\sum_k h_k \varphi(2t-k) \tag{1}$$

A counterpart of this function is called mother wavelet function obtained from

$$\psi(t) = \sqrt{2}\sum_k g_k \varphi(2t-k) \tag{2}$$

where l_k and h_k are related via the equation

$$g_k = (-1)^k h_{1-k} \tag{3}$$

In signal processing literature (3) is known as the *quadrature mirror relation* and the filter ***h*** and *g* as *quadrature mirror filters*. The coefficients g_k and h_k that is, the quadrature mirror filters are used to compute the wavelet transform. $\phi(t)$ and $\psi(t)$ form orthogonal functional low-pass and high-pass filters respectively which are spaces in $L_2(\mathfrak{R})$ where inner product of functions with finite energy is defined. The orthogonal spaces satisfy the property

$$\varphi_{m,k}(t) \cup \psi(t)_{m,k} = \varphi_{m+1,k}(t) \tag{4}$$

where

$$\varphi_{m,k}(t) = 2^{m/2}\varphi(2^m t - k) \tag{5}$$

$$\psi_{m,k}(t) = 2^{m/2}\psi(2^m t - k) \tag{6}$$

m=0 constitutes the coarsest scale. The simplest filter coefficients are known as Haar filter and given by

$$\begin{aligned} h_h &= [h_1 \;\; h_2] = \frac{1}{\sqrt{2}}[1 \quad 1] \\ g_h &= [g_1 \;\; g_2] = \frac{1}{\sqrt{2}}[1 \;\; -1] \end{aligned} \tag{7}$$

Haar wavelets have generally practical value for several reasons. The building blocks in decomposition are discontinuous functions that are not effective approximating smooth functions. However, because of their very simple form, especially in real-time measurement applications their value is eminent. The present application constitutes such an application and demonstrates their effectiveness. More about wavelets can be found in the literature (Mallat, S.G., 1989; Ogden, T., 1997; Mallat, S., 1999; Vidakovic, B., 1999; Percival, D.B. & Walden, A.T., 2000).

2.2 The Kalman filter

Kalman filter is an estimator of states of a dynamic system with a minimal error variance and in this sense it is optimal. The dynamic system is given in a form which is terminologically referred to as state-space:

$$x(k+1) = A(k)x(k) + B(k)w(k) \tag{8}$$

$$z(k) = C(k)x(k) + v(k) \tag{9}$$

where A is the system matrix, B process noise matrix, C is the measurement matrix. Further, w(k) and v(k) are Gaussian process noise and measurement noise respectively with the properties

$$\begin{aligned} &E\{w(k)\} = 0 \\ &E\{w(k)w(l)^T\} = Q(k) \, for \;\; k = l \\ &\qquad\qquad\qquad\;\; = 0 \; otherwise \end{aligned} \tag{10}$$

In this work, Kalman filter is used to combine the information from the measurements at different resolutional levels and enhance the state estimation rather than to employ single measurement at each time-step. The estimation by Kalman filter is accomplished recursively. As to the conventional notations used in the literature (Anderson, B.D.O. & Moore, J.B., 1979; Maybeck, P.S., 1979; Brown, R.G. & Hwang, Y.C., 1997; Zarchan, P. & Musoff, H., 2005; Shalom, Y.B., Li, X.R. et al., 2006; Simon, D., 2006), the recursion is carried out by standard matrix equations given below.

$$x(k+1 \mid k) = A(k)x(k \mid k) \tag{12}$$

$$P(k+1 \mid k) = A(k)P(k \mid k)A(k)^T + B(k)Q(k)B(k)^T \tag{13}$$

$$K(k+1) = \frac{P(k+1 \mid k)C(k+1)^T}{C(k+1)P(k+1 \mid k)C(k+1)^T + R(k+1)} \tag{14}$$

So that the updated state variables and covariance matrix are

$$\begin{aligned} x(k+1 \mid k+1) = {} & x(k+1 \mid k) + \\ & ([z(k+1) - C(k+1)x(k+1 \mid k)]K(k+1) \end{aligned} \tag{15}$$

$$P(k+1 \mid k+1) = [I - K(k+1)C(k+1)]P(k+1 \mid k) \tag{16}$$

2.3 The Multiresolutional filter

N-level multiresolutional dynamic system can be described by

$$\begin{aligned} & x^{[N]}(k_N + 1) = A^{[N]}(k_N)x^{[N]}(k_N), \\ & z^{[i]}(k_i) = C^{[i]}(k_i)x^{[i]}(k_i) + v^{[i]}(k_i) \\ & i = 1, \ldots, N \end{aligned} \tag{17}$$

where i=N is the highest resolution level, so that

$$\begin{aligned} & E\left[w^{[N]}(k_N)\right] = 0, \\ & E\left[w^{[N]}(k_N)w^{[N]}(l_N)^T\right] = Q^{[N]}(k_N), \quad k = l \\ & \qquad\qquad\qquad\qquad\qquad\qquad = 0 \quad k \neq l \end{aligned} \tag{18}$$

Referring to the measurements $z^{(\text{Shalom, Y.B., Li, X.R. et al.})}(k_i)$ at different resolution levels, we write

$$\begin{aligned} &E\left[w^{[i]}(k_i)\right] = 0, \\ &E\left[w^{[i]}(k_i)w^{[i]}(l_i)^T\right] = Q^{[i]}(k_i), \quad k = l \\ &\qquad\qquad\qquad\qquad\quad = 0 \quad k \neq l \end{aligned} \tag{19}$$

and

$$\begin{aligned} &E\left[v^{[i]}(k_i)\right] = 0, \\ &E\left[v^{[i]}(k_i)v^{[i]}(l_i)^T\right] = R^{[i]}(k_i), \quad k = l \\ &\qquad\qquad\qquad\qquad\quad = 0 \quad k \neq l \end{aligned} \tag{20}$$

Measurements at different resolution levels can be carried out in different ways. One way of doing this is shown with the following scheme.

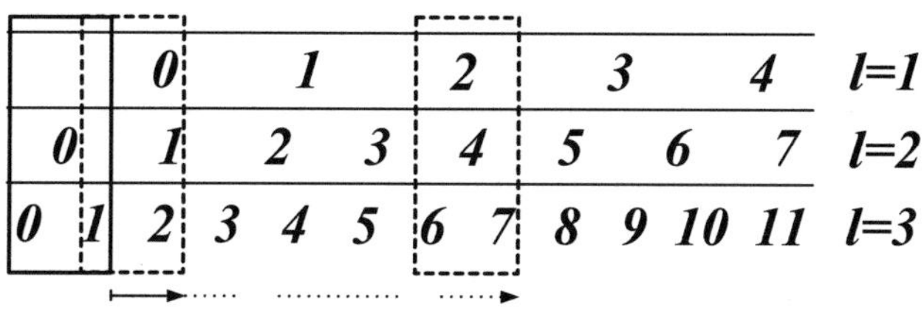

where different resolution levels are indicated by l=1,2,3 while the highest resolution level is for l=3. In this implementation a moving window is used. When samples 0 and 1 are available in the window at l=3, sample 0 is calculated at the resolution level l=1 by wavelet decomposition; when the samples 1 and 2 are available at l=3, the sample 1 is calculated at the resolution level l=2 which is followed by the calculated sample 0 at the level l=1 using the samples 0 and 1 at the level l=2. In this way the moving window multiresolutional dynamic filtering can be executed.

Another alternative for the multiresolutional measurements can be accomplished by the following scheme.

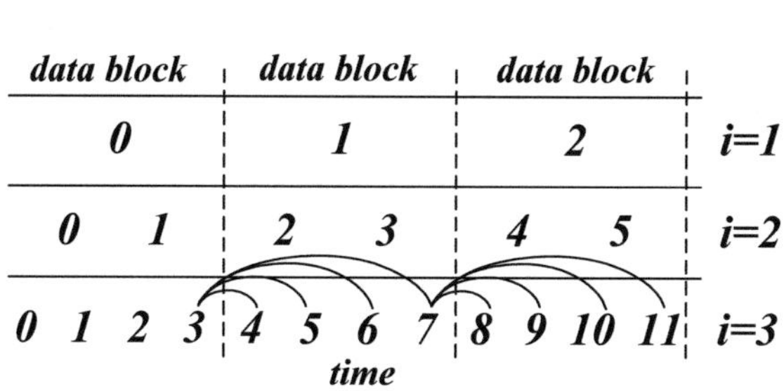

In this scheme each data block at the highest resolution level (i=3) contains 4 samples and at this very level, the Kalman filtering estimations within the data block are carried out using the last data sample of the preceding block, together with the respective data samples within the block. Namely, for instance using the data sample 3 at the level i=3, 1-step ahead, 2-step ahead, 3-step ahead and 4-step ahead Kalman predictions are made and as the data samples 4, 5,6, and 7 are available, the respective estimations are made in sequence at the time points where the measurements are available; that is at the time points 4, 5, 6, and 7.

Another scheme of measurements at different resolution levels are shown in figure 2 and the wavelet decomposition of state variables in a data block is shown in figure 3. This is the actual scheme implemented in this work and therefore explained in details as follows. It is to note that propagation as to state estimation occurs block-wise at the highest resolution level. Decomposition of the states by wavelet transform and reconstruction by inverse wavelet transform is effective in this scheme due to better estimation of the error covariance as to the state estimation as explained below.

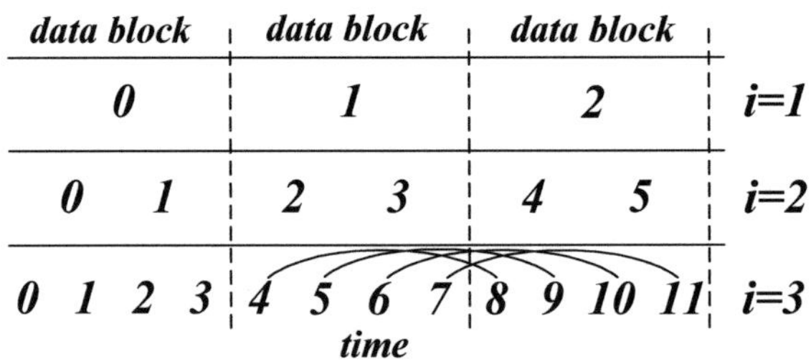

Fig. 2. Measurements at different resolution levels

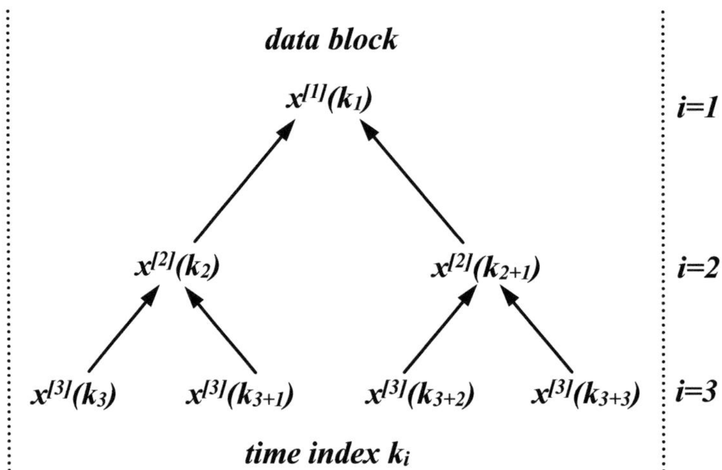

Fig. 3. Wavelet decomposition of state variables in a data block

Within a data block, the state variables at resolution level i are designated as

$$X_m^{[i]} = \begin{bmatrix} x^{[i]}k(i) \\ x^{[i]}k(i+1) \\ \ldots\ldots \\ x^{[i]}k(i+2^{i-1}) \end{bmatrix} \tag{21}$$

$$x^{[i]}k(i) = \begin{bmatrix} x_{k,1}^{[i]} \\ x_{k,2}^{[i]} \\ \cdots \\ x_{k,s}^{[i]} \end{bmatrix} \quad (22)$$

where m is data block index and s is the number of state variables. In a data block, there are 2^{i-1} state variables. Each state variable has s state components. A lower resolution state variable is computed from

$$x^{[i]}(k_i) = h_1 x^{[i+1]}(k_{i+1}) + h_2 x^{[i+1]}(k_{i+1}+1) \quad (23)$$

where h_1 and h_2 are the Haar low-pass filter coefficients. The details component i.e., high frequency part after the decomposition is computed via

$$y^{[i]}(k_i) = g_1 x^{[i+1]}(k_{i+1}) + g_2 x^{[i+1]}(k_{i+1}+1) \quad (24)$$

where g_1 and g_2 are the Haar high-pass filter coefficients. The reconstruction of the states is carried out by combining (22) and (23) in a matrix equation form as given below.

$$\begin{bmatrix} x^{[i+1]}(k_{i+1}) \\ x^{[i+1]}(k_{i+1}+1) \end{bmatrix} = \boldsymbol{h}^T x^{[i]}(k_i) + \boldsymbol{g}^T y^{[i]}(k_i) \quad (25)$$

Wavelet decomposition and reconstruction is carried out according to the scheme shown in figures 4 and 5.

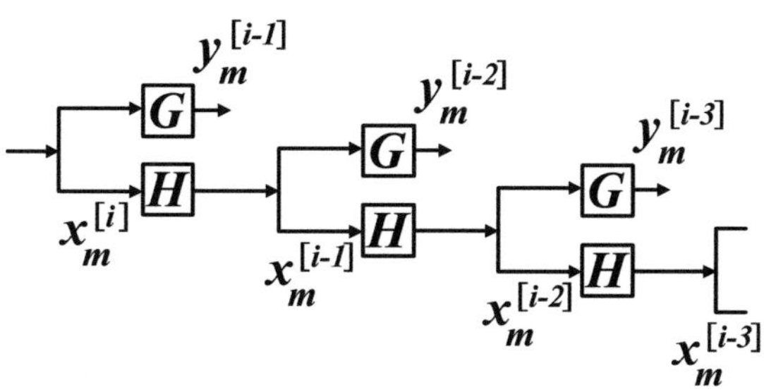

Fig. 4. Wavelet decomposition of state variables in a data block

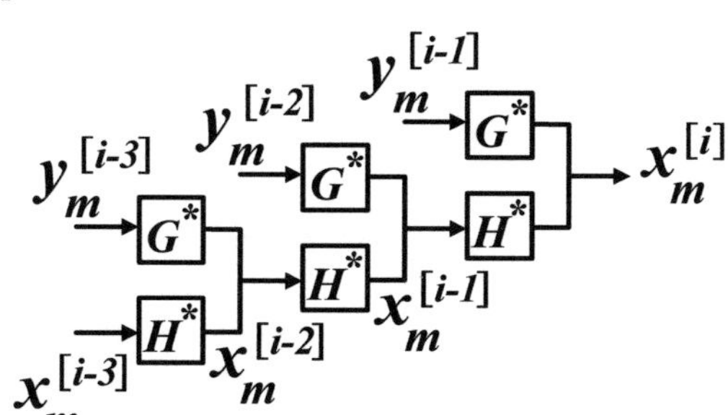

Fig. 5. Wavelet reconstruction of state variables in a data block

The wavelet matrix operator G and the scaling matrix operator H in the decomposition and their counterparts G* and H* in the reconstruction contain two-tap Haar filters and they are related by

$$G^* = G^T \qquad H^* = H^T \tag{26}$$

It is to note that the state variable estimations are carried out at the respective resolution levels, as follows.

$$H^{i+1\to i} = diag\{(1/\sqrt{2})h_h^{(1)},(1/\sqrt{2})h_h^{(2)},....,(1/\sqrt{2})h_h^{(K)}\} \tag{27}$$

where K is the number of filters involved in a decomposition from the resolution level i+1 to i. For instance from 2 to 1, then i=1, and K is given by

$$K = 2^{i-1} = 1 \tag{28}$$

as this is seen in figure 2, as the block length is 2. For the *s* state variable of the system, the *H* matrix is composed of *s* Haar filters at the diagonal as

$$H = diag\{H_{[1]}{}^{i+1\to i}, H_{[2]}{}^{i+1\to i},......., H_{[K]}{}^{i+1\to i}\} \tag{29}$$

Similarly, for the reconstruction filter, we write

$$G^{i+1\to i} = diag\{(\sqrt{2})g_h^{(1)},(\sqrt{2})g_h^{(2)},....,(\sqrt{2})g_h^{(K)}\} \tag{30}$$

$$G = diag\{G_{[1]}{}^{i+1\to i}, G_{[2]}{}^{i+1\to i},......., G_{[K]}{}^{i+1\to i}\} \tag{31}$$

As K is given by (27). For the inverse transform scheme given by figure 5, we write

$$H^* = diag\{H_{[1]}{}^{i\to i+1}, H_{[2]}{}^{i\to i+1},......., H_{[K]}{}^{i\to i+1}\} \tag{32}$$

$$G^* = diag\{G_{[1]}{}^{i\to i+1}, G_{[2]}{}^{i\to i+1},......., G_{[K]}{}^{i\to i+1}\} \tag{33}$$

where

$$H^{i\to i+1} = diag\{(\sqrt{2})h_h^{T(1)},(\sqrt{2})h_h^{T(2)},....,(\sqrt{2})h_h^{T(K)}\} \tag{34}$$

$$G^{i\to i+1} = diag\{(1/\sqrt{2})g_h^{T(1)},(1/\sqrt{2})g_h^{T(2)},....,(1/\sqrt{2})g_h^{T(K)}\} \tag{35}$$

Above T indicates transpose and K is given by (28).

2.3.1 Example block-wise 2-way multiresolutional computation

An example block wise wavelet computation is presented in this subsection. For this we consider figure 2 and the initial data block designated as 0 with the data samples as x_0 and x_1 at the resolution level *i*=2. Also for the sake of simplicity we consider *s*=2; namely only two state variables of the system. Therefore the data samples are denoted as x_{01}, x_{02} and x_{11},

x_{12}; first index is for the data sample, second index for the state variable. We define an auxiliary matrix L as follows

$$L^{[2]} = \begin{bmatrix} 1 & 0 & 0 & 0 \\ 0 & 0 & 1 & 0 \\ 0 & 1 & 0 & 0 \\ 0 & 0 & 0 & 1 \end{bmatrix} \tag{36}$$

which yields

$$L^{[2]} X_m^{[2]} = \begin{bmatrix} 1 & 0 & 0 & 0 \\ 0 & 0 & 1 & 0 \\ 0 & 1 & 0 & 0 \\ 0 & 0 & 0 & 1 \end{bmatrix} \begin{bmatrix} x_{01}^{[2]} \\ x_{02}^{[2]} \\ x_{11}^{[2]} \\ x_{12}^{[2]} \end{bmatrix} = \begin{bmatrix} x_{01}^{[2]} \\ x_{11}^{[2]} \\ x_{02}^{[2]} \\ x_{12}^{[2]} \end{bmatrix} \tag{37}$$

where m is index of the data block which is m=0, above. It is to note that, the sate variables are grouped slightly different for computational convenience; namely the decomposition is carried out by

$$X_m^{[i-1]} = L^{T\,[i-1]} H\, L^{[i]} X_m^{[i]} \tag{38}$$

$$Y_m^{[i-1]} = L^{T\,[i-1]} G\, L^{[i]} X_m^{[i]} \tag{39}$$

where Y_m are the wavelet coefficients. From the definition given by (27) and (30)

$$H^{i+1\to i} = diag\{(1/\sqrt{2})h_h^{(1)}\} = \begin{bmatrix} 1/2 & 1/2 \end{bmatrix} \tag{40}$$

$$G^{i+1\to i} = diag\{(\sqrt{2})g_h^{(1)}\} = \begin{bmatrix} 1 & -1 \end{bmatrix} \tag{41}$$

Also from (29) and (31)

$$\begin{aligned} G &= diag\{G_{[0]}^{\;i+1\to i}, G_{[1]}^{\;i+1\to i}\} \\ H &= diag\{H_{[0]}^{\;i+1\to i}, H_{[1]}^{\;i+1\to i}\} \end{aligned} \tag{42}$$

In (34) substituting i=2, the decomposition at the level i=1, becomes

$$X_0^{[1]} = L^{T\,[1]} H\, L^{[2]} X_m^{[2]} = \begin{bmatrix} 1 & 0 \\ 0 & 1 \end{bmatrix} \begin{bmatrix} .5 & .5 & 0 & 0 \\ 0 & 0 & .5 & .5 \end{bmatrix} \times \begin{bmatrix} 1 & 0 & 0 & 0 \\ 0 & 0 & 1 & 0 \\ 0 & 1 & 0 & 0 \\ 0 & 0 & 0 & 1 \end{bmatrix} \times \begin{bmatrix} x_{01}^{[2]} \\ x_{02}^{[2]} \\ x_{11}^{[2]} \\ x_{12}^{[2]} \end{bmatrix} = \begin{bmatrix} .5\,x_{01}^{[2]} + .5\,x_{11}^{[2]} \\ .5\,x_{02}^{[2]} + .5\,x_{12}^{[2]} \end{bmatrix} \tag{43}$$

which is a low-pass filtered data block by averaging. The wavelet coefficients which are the high-pass filtered data samples are obtained in a similar way as

$$Y_0^{[1]} = L^{T\,[1]} G\, L^{[2]} X_m^{[2]} = \begin{bmatrix} 1 & 0 \\ 0 & 1 \end{bmatrix} \begin{bmatrix} 1 & -1 & 0 & 0 \\ 0 & 0 & 1 & -1 \end{bmatrix} \times \begin{bmatrix} 1 & 0 & 0 & 0 \\ 0 & 0 & 1 & 0 \\ 0 & 1 & 0 & 0 \\ 0 & 0 & 0 & 1 \end{bmatrix} \times \begin{bmatrix} x_{01}^{[2]} \\ x_{02}^{[2]} \\ x_{11}^{[2]} \\ x_{12}^{[2]} \end{bmatrix} = \begin{bmatrix} x_{01}^{[2]} - x_{11}^{[2]} \\ x_{02}^{[2]} - x_{12}^{[2]} \end{bmatrix} \tag{44}$$

(43) and (44) are the multiresolutional decomposition of the data block 0, from level i=2 to the level i=1.

The inverse wavelet transform is carried out according to the general scheme given by figure 5.

$$X_m^{[i]} = L^{T\,[i]} H^*\, L^{[i]} X_m^{[i-1]} + L^{T\,[i]} G^*\, L^{[i-1]} Y_m^{[i-1]} \tag{45}$$

From (32) and (33), we write

$$H^* = diag\{H_{[1]}^{\;1\to 2}, H_{[2]}^{\;1\to 2}\} = \begin{bmatrix} \begin{bmatrix} 1 \\ 1 \end{bmatrix} & \begin{bmatrix} 0 \\ 0 \end{bmatrix} \\ \begin{bmatrix} 0 \\ 0 \end{bmatrix} & \begin{bmatrix} 1 \\ 1 \end{bmatrix} \end{bmatrix} \tag{46}$$

$$G^* = diag\{G_{[1]}^{\;1\to 2}, G_{[2]}^{\;1\to 2}\} = \begin{bmatrix} \begin{bmatrix} .5 \\ -.5 \end{bmatrix} & \begin{bmatrix} 0 \\ 0 \end{bmatrix} \\ \begin{bmatrix} 0 \\ 0 \end{bmatrix} & \begin{bmatrix} .5 \\ -.5 \end{bmatrix} \end{bmatrix} \tag{47}$$

so that, (45) becomes

$$X_0^{[2]} = L^{T\,[2]} H^*\, L^{[1]} X_0^{[1]} + L^{T\,[2]} G^*\, L^{[1]} Y_0^{[i]} \tag{48}$$

The substitution of the respective matrices into (48) yields

$$\begin{bmatrix} 1 & 0 & 0 & 0 \\ 0 & 0 & 1 & 0 \\ 0 & 1 & 0 & 0 \\ 0 & 0 & 0 & 1 \end{bmatrix} \begin{bmatrix} \begin{bmatrix} 1 \\ 1 \end{bmatrix} & \begin{bmatrix} 0 \\ 0 \end{bmatrix} \\ \begin{bmatrix} 0 \\ 0 \end{bmatrix} & \begin{bmatrix} 1 \\ 1 \end{bmatrix} \end{bmatrix} \begin{bmatrix} 1 & 0 \\ 0 & 1 \end{bmatrix} \begin{bmatrix} .5x_{01}^{2} + .5x_{11}^{[2]} \\ .5x_{02}^{[2]} + .5\,x_{12}^{[2]} \end{bmatrix} +$$

$$\begin{bmatrix} 1 & 0 & 0 & 0 \\ 0 & 0 & 1 & 0 \\ 0 & 1 & 0 & 0 \\ 0 & 0 & 0 & 1 \end{bmatrix} \begin{bmatrix} \begin{bmatrix} .5 \\ -.5 \end{bmatrix} & \begin{bmatrix} 0 \\ 0 \end{bmatrix} \\ \begin{bmatrix} 0 \\ 0 \end{bmatrix} & \begin{bmatrix} .5 \\ -.5 \end{bmatrix} \end{bmatrix} \begin{bmatrix} 1 & 0 \\ 0 & 1 \end{bmatrix} \begin{bmatrix} x_{01}^{2} - x_{11}^{[2]} \\ x_{02}^{[2]} - x_{12}^{[2]} \end{bmatrix} = \tag{49}$$

$$\begin{bmatrix} 1 & 0 \\ 0 & 1 \\ 1 & 0 \\ 0 & 1 \end{bmatrix} \begin{bmatrix} .5\,x_{01}^{[2]} + .5\,x_{11}^{[2]} \\ .5\,x_{02}^{[2]} + .5\,x_{12}^{[2]} \end{bmatrix} + \begin{bmatrix} .5 & 0 \\ 0 & .5 \\ -.5 & 0 \\ 0 & -.5 \end{bmatrix} \begin{bmatrix} x_{01}^{[2]} - x_{11}^{[2]} \\ x_{02}^{[2]} - x_{12}^{[2]} \end{bmatrix} = \begin{bmatrix} .5x_{01}^{[2]} + .5x_{11}^{[2]} \\ .5x_{02}^{[2]} + .5x_{12}^{[2]} \\ .5x_{01}^{[2]} + .5x_{11}^{[2]} \\ .5x_{01}^{[2]} + .5x_{12}^{[2]} \end{bmatrix} + \begin{bmatrix} .5x_{01}^{[2]} - .5x_{11}^{[2]} \\ .5x_{02}^{[2]} - .5x_{12}^{[2]} \\ -.5x_{01}^{[2]} + .5x_{11}^{[2]} \\ -.5x_{01}^{[2]} + .5x_{12}^{[2]} \end{bmatrix} = \begin{bmatrix} x_{01}^{[2]} \\ x_{02}^{[2]} \\ x_{11}^{[2]} \\ x_{12}^{[2]} \end{bmatrix}$$

which is $X_m^{[2]}$ as given in (37).
For the same decomposition from i=3 to i=2, the auxiliary matrix L is given by

$$L^{[3]} = \begin{bmatrix} 1 & 0 & 0 & 0 & 0 & 0 & 0 & 0 \\ 0 & 0 & 1 & 0 & 0 & 0 & 0 & 0 \\ 0 & 1 & 0 & 0 & 0 & 0 & 0 & 0 \\ 0 & 0 & 0 & 1 & 0 & 0 & 0 & 0 \\ 0 & 0 & 0 & 0 & 1 & 0 & 0 & 0 \\ 0 & 0 & 0 & 0 & 0 & 0 & 1 & 0 \\ 0 & 0 & 0 & 0 & 0 & 1 & 0 & 0 \\ 0 & 0 & 0 & 0 & 0 & 0 & 0 & 1 \end{bmatrix} \quad (50)$$

For this case i=2, and K is given by

$$K = 2^{i-1} = 2 \quad (51)$$

as this is seen in figure 2, where the block length is 4.

2.3.2 Estimate updating

The basic update scheme for dynamic multiresolutional filtering is shown in Fig. 6 where at each resolutional level, when the measurement is available, the state variables are updated and when the block is complete the inverse wavelet transform and fusion is performed. During the inverse transformation wavelet coefficients $Y_{m+1|m+1}^{[i]}$ saved aside are used.

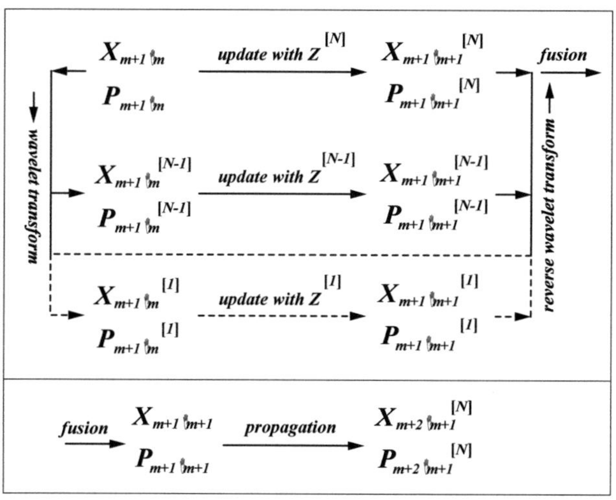

Fig. 6. Wavelet decomposition of state variables in a data block

Explicitly

$$P_{m+1|m}^{[i]} = \begin{bmatrix} P_{XXm+1|m}^{[i]} & P_{XYm+1|m}^{[i]} \\ P_{YXm+1|m}^{[i]} & P_{YYm+1|m}^{[i]} \end{bmatrix} \quad (52)$$

$$X_{m+1|m+1}{}^{[i]} = X_{m+1|m}{}^{[i]} + K_{m+1}{}^{[i]}(Z_{m+1}{}^{[i]} - C_{m+1}{}^{[i]} X_{m+1|m}{}^{[i]}) \tag{53}$$

and

$$P_{XXm+1|m+1}{}^{[i]} = (I - K_{m+1}{}^{[i]} C_{m+1}{}^{[i]}) P_{XXm+1|m}{}^{[i]} \tag{54}$$

As $X_{m+1|m}{}^{[i]}$ and $Y_{m+1|m}{}^{[i]}$, $Y_{m+1|m}{}^{[i+1]}$ are correlated, the covariance matrices $P_{XYm+1|m}{}^{[i]}$ and $P_{YXm+1|m}{}^{[i]}$ are also updated. However, the wavelet coefficients $Y_{m+1|m}{}^{[i]}$ and their covariance matrices $P_{YYm+1|m}{}^{[i]}$ are not updated. The minimum variance Kalman gain matrix $K_{m+1}{}^{[i]}$ at each level, is determined by

$$K_{m+1}{}^{[i]} = P_{XXm+1|m}{}^{[i]} C_{m+1}{}^{[i]T} \left(C_{m+1}{}^{[i]} P_{XXm+1|m}{}^{[i]} C_{m+1}{}^{[i]T} + R_{m+1}{}^{[i]} \right)^{-1} \tag{55}$$

where the measurent matrix $C_{m=1}{}^{[i]}$ and $R_{m+1}{}^{[i]}$ are given by

$$C_{m+1}^{[i]} = diag \begin{bmatrix} C^{[i]}[(m+1)2^{i-1}], C^{[i]}[(m+1)2^{i-1} - 2^{i-1} + 1], \ldots, \\ \ldots\ldots \qquad , C^{[i]}[(m+1) + 2^{i-1} + 1] \end{bmatrix} \tag{56}$$

$$R_{m+1}^{[i]} = diag \begin{bmatrix} R^{[i]}[(m+1)2^{i-1}], R^{[i]}[(m+1)2^{i-1} - 2^{i-1} + 1], \ldots, \\ \ldots\ldots \qquad , R^{[i]}[(m+1) + 2^{i-1} + 1] \end{bmatrix} \tag{57}$$

Once, within the moving window, the sequences of updated state variables and error covariances $X_{m+1|m+1}{}^{[N,i]}$ and $P_{m+1|m+1}{}^{[N,i]}$ for $i=1,2,..,N$ are determined, they must be fused to generate an optimal $X_{m+1|m+1}{}^{[NF]}$ and $X_{m+1|m+1}{}^{[NF]}$. $X_{m+1|m+1}{}^{[NF]}$ and $X_{m+1|m+1}{}^{[NF]}$. For the minimum fusion error covariance $X_{m+1|m+1}{}^{[NF]}$, the fused estimate $X_{m+1|m+1}{}^{[NF]}$ is calculated as

$$X_{m+1|m+1}{}^{[NF]} = P_{m+1|m+1}{}^{[NF\}} \left[\sum_{i=1}^{N} \left(P_{m+1|m+1}{}^{[N,i]} \right) X_{m+1|m}{}^{[N,i]} - \right] (N-1) \left(P_{m+1|m}{}^{[N]} \right)^{-1} X_{m+1|m}{}^{[N]} \tag{58}$$

where the minimum fusion error covariance $P_{m+1|m+1}{}^{[NF]}$ becomes

$$\left(P_{m+1|m+1}{}^{[NF]} \right)^{-1} = \sum_{i=1}^{N} \left(P_{m+1|m+1}{}^{[N,i]} \right)^{-1} - (N-1) \left(P_{m+1|m}{}^{[N]} \right)^{-1}. \tag{59}$$

The fused estimate $X_{m+1|m+1}{}^{[NF]}$ is a weighted summation of both predicted $X_{m+1|m}{}^{[N]}$ and updated $X_{m+1|m+1}{}^{[N,i]}$, for i=1,2,..,N. The sum of the weight factors equal to the identity I. This can be seen by substitution of $P_{m+1|m+1}{}^{[NF]}$ given above into the expression of $X_{m+1|m+1}{}^{[NF]}$ in (58).

3. Autonomous robot navigation

The computer experiments have been carried out with the simulated robot navigation. The state variables vector is given by (22) while taking i=N.

$$x^{[N]}k(N) = \begin{bmatrix} x_{k,1}^{[N]} \\ x_{k,2}^{[N]} \\ \cdots \\ x_{k,s}^{[N]} \end{bmatrix} \tag{60}$$

Explicitly,

$$x^{[N]}k(N) = [x, \dot{x}, y, \dot{y}, \omega]$$

where ω is the angular rate and it is estimated during the move. When the robot moves in a straight line, the angular rate becomes zero. The other state variables are x and y coordinates and the respective velocities.

The experiments are carried out as follows. For a pre-defined trajectory, perception measurements along the trajectory are obtained in a virtual reality environment. This trajectory is designated as reference. The measurements as data samples are used in sequence for a real-time robot navigation by multiresolutional dynamic filter estimation. Designing the experiment in this way allows one to assess the effectiveness of the robot navigation by comparison of the actual navigation path with the estimated one. The overall robot trajectory is shown in figure 7 where there are three lines plotted but only two of them are visible. The line marked by * sign represents the measurement data set. The line marked by • is the multiresolutional dynamic filter estimation. The line indicated by o sign is the reference trajectory. These lines are not explicitly seen in the figure. For explicit illustration of the experimental outcomes the same figure with a different zooming ranges and the zooming powers are given in figures 8-11.

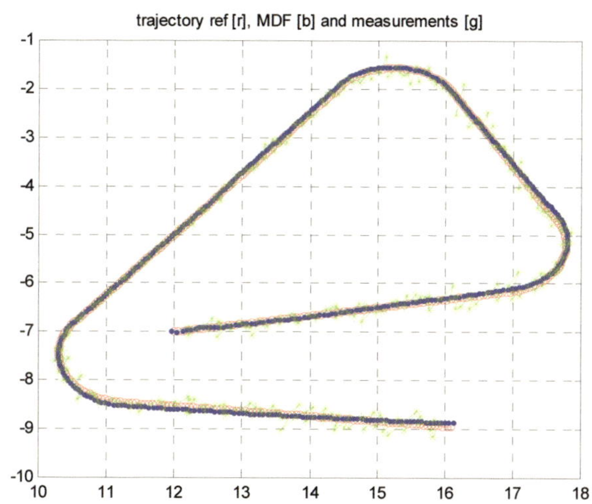

Fig. 7. The overall Robot trajectory with measurement (* signs), and multiresolutinal dynamic filter estimation (• signs).

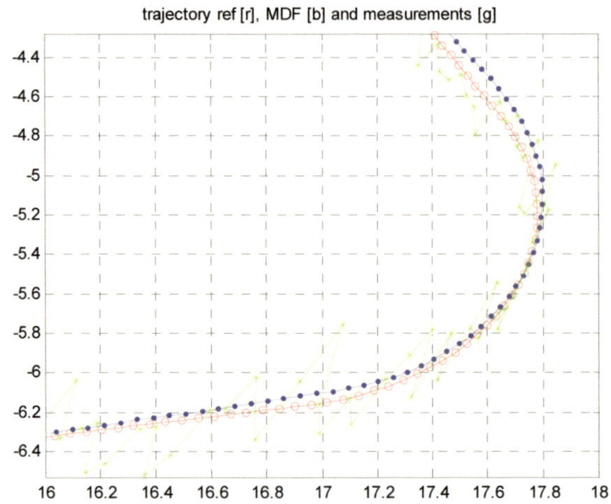

Fig. 8. Enlarged robot trajectory, measurement, and multiresolutinal dynamic filter estimation in angular velocity mode. The * sign is for measurement, o sign for the reference and • sign is the estimated trajectory.

From the experiments it is seen that, the multiresolutional filtering is effective for estimation of the trajectory from perception measurement. Estimations are more accurate in the straight-ahead mode seen in figure 11, compared to the cases where angular velocity does not vanish. This can be explained by noting that in the straight-ahead mode the angular velocity is zero and the system matrix is fixed. However in the non-linear mode the system matrix is approximate due to the estimation error of the angular velocity. In this case the slight difference between the reference trajectory and the estimated trajectory is markedly seen due to the approximation error caused by the Taylor's series expansion and ensuing linearization in (extended) Kalman filtering.

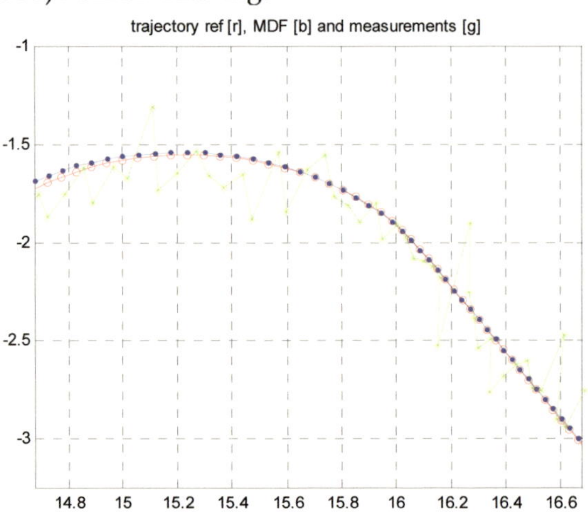

Fig. 9. Enlarged Robot trajectory, measurement multiresolutinal dynamic filter in angular velocity mode. The * sign is for measurement, o sign for the reference and • sign is the estimated trajectory.

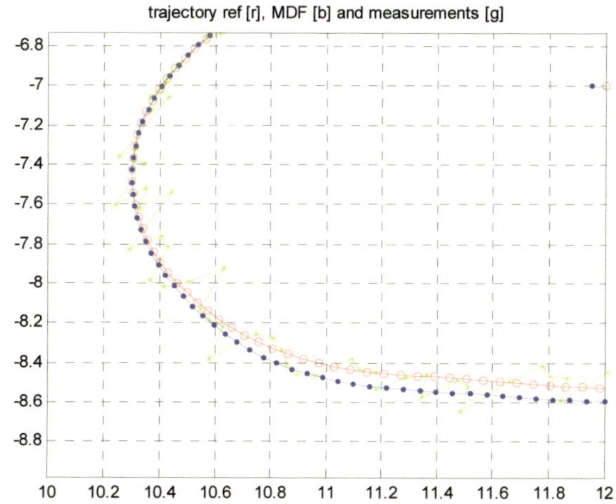

Fig. 10. Enlarged Robot trajectory, measurement multiresolutinal dynamic filter in angular velocity mode. The * sign is for measurement, o sign for the reference and • sign is the estimated trajectory.

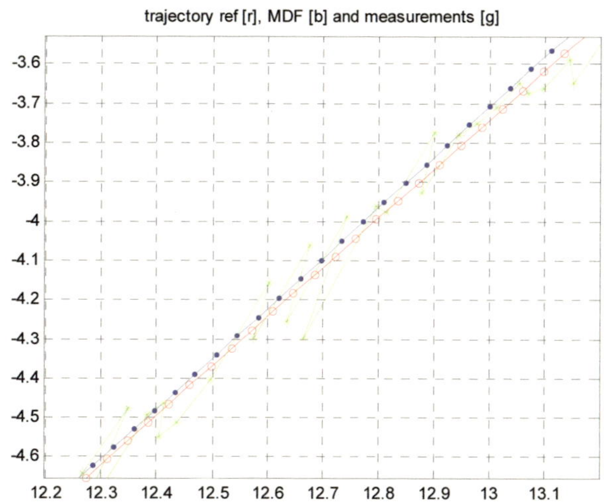

Fig. 11. Enlarged robot trajectory, measurement multiresolutinal dynamic filter straight-ahead mode. The * sign is for measurement, o sign for the reference and • sign is the estimated trajectory.

4. Discussion and conclusion

The detailed description of processing of spatial information for a perceptual robot is described. Especially, the perception is considered to be a probabilistic concept and defined as such (Ciftcioglu, Ö., Bittermann, M.S. et al., 2006) so that the probabilistic consideration in perceptual robot is desirable for human-like robot navigation. This facilitates the robotic movement coupling the cognition of a robot with the environment. In this context, Bayesian

perception and cognition studies (Knill, D.C. & Richards, W., 1996) can be more substantiated with more accurate priors. This is especially important when one deals with a certain scene but from a different vantage point.
In vision robotics autonomous navigation is a challenging issue. Using the perceptual spatial information vision robot can navigate autonomously by perception-based sensory information so that it can react better with respect to the environmental and circumstantial situations. Better is in the sense of obstacle avoidance and way finding without crash. In the present work, the processing the spatial information is accomplished by means of wavelet technology that is the main component in the information fusion process. The simplest form of wavelets is used to cope with the real-time requirements in the navigation and it is demonstrated that, this choice is justified. Kalman filtering is the machinery for the fusion process, as well as it plays the role of optimal observer (Friedland, B., 1986) of the measurements referring to distance (r), angle (θ) and the scalar velocity (v). These may partly be computed from the estimated state variables, in the case of unavailable actual measurement counterparts. For that matter, it is the angular velocity for which the measurements are unavailable, in this work. The work is a clear demonstration of the working state observers and therefore an important hint for the physical robotics exercises. It is reminded that, the present work is an autonomous robotics simulation implemented in a virtual reality and the results reported are the computer experiments.
Applications of vision robot are countless and they are well documented in the literature. Autonomous robotics with perception is a new dimension in a vision robot as such its vision can be shaped for application dependent vision properties. The investigations presented in this work are motivated by measuring the perception aspects of an architectural design from different vantage points and altogether to obtain a compromise among them, as to a final decision-making for that design. Such a task in one hand is a routine work for a robot in virtual reality and the actual (real) scene having been represented in the virtual reality, the scene can be perceived in many different ways as desired. The same task for a human is in the first place extremely tedious but more importantly subjective in terms of perception measurement and therefore can be deemed to be inconclusive due to probable inconsistencies among different observers and even for a single viewer. This is because of the complexity of the task due to the number of objects involved in the scene. Here the objects may be a number of dwelling units subject to marketing by a building project developer. Since a number of locations and architectural requirements impose condition on perception aspects, the task is a multi-objective optimization subject to these impositions as constraints. Also, in this situation one is interested in a solution on a Pareto front. Such complexities are today effectively dealt with thanks to evolutionary algorithms (Deb, K., 2001; Coello, C.A.C., Veldhuizen, D.A. et al., 2003; Sato, H., Aguirre, H.E. et al., 2007). Some endeavours in the context of architecture are reported in (Bittermann, M. et al., 2008; Ciftcioglu, O. & Bittermann, M.S., 2008). The present research is another endeavour along this line revealing the exhaustive details of spatial information fusion as to accurate robot navigation using multiresolutional dynamic filter approach.

8. References

Anderson, B. D. O. & J. B. Moore (1979). Optimal Filtering. Prenctice-Hall, 0-13-638122-7, Englewood Cliffs, New Jersey

Beetz, M., T. Arbuckle, et al. (2001). Integrated, plan-based control of autonomous robots in human environments. IEEE Intelligent Systems Vol. 16, No. 5, 56-65,

Bittermann, M., I. S. Sariyildiz, et al. (2008). Performance-Based Pareto Optimal Design. Proceedings of Tools and Methods of Competitive Engineering (TMCE), Izmir

Brown, R. G. & Y. C. Hwang (1997). Introduction to Random Signals and Applied Kalman Filtering: With MATLAB Exercises and Solutions, 3rd ed. Wiley, 0-471-12839-2, New York

Ciftcioglu, O. & M. S. Bittermann (2008). Solution Diversity in Multi-Objective Optimization: A study in Virtual Reality. Proceedings of IEEE World Congress on Computational Intelligence, Hong Kong, IEEE

Ciftcioglu, Ö., M. S. Bittermann, et al. (2006). Towards computer-based perception by modeling visual perception: a probabilistic theory. Proceedings of 2006 IEEE Int. Conf. on Systems, Man, and Cybernetics, pp. 5152-5159, Taipei, Taiwan

Ciftcioglu, Ö., M. S. Bittermann, et al. (2007). Visual perception theory underlying perceptual navigation.In: Emerging Technologies, Robotics and Control Systems S. Pennacchio, 139-153 Palermo, International Society for Advanced Research. 1, Palermo

Coello, C. A. C., D. A. Veldhuizen, et al. (2003). Evolutionary Algorithms for Solving Multiobjective Problems. Kluwer Academic Publishers, Boston

Deb, K. (2001). Multiobjective Optimization using Evolutionary Algorithms. John Wiley & Sons,

Friedland, B. (1986). Control System Design. McGraw-Hill, London

Hong, L. (1993). Multiresolutional filtering using wavelet transform. IEEE Trans. on Aerospace and Electronic Systems Vol. 29, No. 4, 1244-1251,

Hong, L. (1994). Multiresolutional multiple model target tracking. IEEE Tranasctions on Aerospace and Electronic Systems Vol. 30, No. 2, 518-524,

Hsin, H. C. & A. C. Li (2006). Wavelet-Based Kalman Filtering in Scale Space for Image Fusion.In: Pattern Recognition and Computer Vision. C. H. Chen and P. S. P. WangSingapore, World Scientific, Singapore

Knill, D. C. & W. Richards (1996). Perception as Bayesian Inference. Cambridege University Press, Cambridge

Mallat, S. (1999). A Wavelet Tour of Signal Processing. Associated Press, New York

Mallat, S. G. (1989). A theory for multiresolution signal decomposition:the wavelet representation. IEEE Trans. on Pattern Analysis and Machine Intelligence Vol. 11, No. 7, 674-693,

Maybeck, P. S. (1979). Stochastic Models, Estimation and Control, Vol I. Academic Press, New York

Ogden, T. (1997). Essential Wavelets for Statistical Applications and Data Analysis. Birkhauser, Boston

Oriolio, G., G. Ulivi, et al. (1998). Real-time map building and navigation for autonomous robots in unknown environments. IEEE Trans. on Systems, Man and Cybernetics - Part B: Cybernetics Vol. 28, No. 3, 316-333,

Percival, D. B. & A. T. Walden (2000). Wavelet Methods for Time Series Analysis. Cambridge University Press,

Sato, H., H. E. Aguirre, et al. (2007). Controlling dominance area of solutions and its impact on the performance of MOEAs.In: Lecture Notes on Computer Science, Evolutionary Multi-Criterion OptimizationHeidelberg, Springer, Heidelberg

Shalom, Y. B., X. R. Li, et al. (2006). Estimation and Application to Tracking and Navigation. Wiley, 0-471-41655-X, Montreal

Simon, D. (2006). Optimal State Estimation. Wiley Interscience, 13 978-0-471-70858-2, New Jersey

Vidakovic, B. (1999). Statistical Modeling by Wavelets. John Wiley & Sons, Inc, 0-471-29365-2, New York

Wang, M. & J. N. K. Liu (2004). Online path searching for autonomous robot navigation. Proceedings of IEEE Conf. on Robotics, Automation and Mechatronics, pp. 746-751, Singapore

Zarchan, P. & H. Musoff (2005). Fundamentals of Kalman Filtering: A Practical Approach. Amarican Institute of Aeronautics and Astranautics, Inc. (AIAAA), 1-56347-694-0, Virginia 20191-4344

20

Interception and Rendezvous Between Autonomous Vehicles

Yechiel J. Crispin
Aerospace Engineering Department
Embry-Riddle Aeronautical University
USA

1. Introduction

The problem of pursuit and evasion is a classic problem that has intrigued mathematicians for many generations. Suppose a target or evader is moving along a given curve in the plane. A pursuer or chaser is moving such that its line of sight is always pointing towards the target. The classic problem is to determine the trajectory of the pursuer such that it eventually captures the target. There is extensive literature on this problem, see for example the recent book by Nahin (Nahin, 2007). A special case of the pursuit problem is the case where the curve of the target is a circle. This is a classic problem which was first treated by Hathaway (Hathaway, 1921), see also the book by Davis (Davis, 1962). In the classic problems, the line of sight is always pointing towards the target. This method of pursuit is also known as 'pure pursuit' or 'dog pursuit' or 'courbe de chien' in French. Although there have been some modern works on extensions of the classic problem of pursuit, see for example (Marshall, 2005), here we completely abandon the 'pure pursuit' restriction and we treat the problem where the pursuer is required to capture or intercept the target in a given prescribed time or in minimum time. Also, in the classic problem, the velocities of the target and pursuer are assumed constant and there is no consideration of the physics of the motion, such as thrust forces, hydrodynamic or aerodynamic drag and other forces that might be acting on the target and pursuer.

In an active-passive rendezvous problem between two vehicles, the passive or target vehicle moves passively along its trajectory. The active or chaser vehicle is controlled or guided such as to meet the passive vehicle at a later time, matching both the location and the velocity of the target vehicle. An interception problem is similar to the rendezvous problem, except that there is no need to match the final velocities of the two vehicles. On the other hand, in a cooperative rendezvous problem, the two vehicles are active and maneuver such as to meet at a later time, at the same location with the same velocity. The two vehicles start the motion from different initial locations and might have different initial velocities.

An optimal control problem consists of finding the control histories, i.e., the controls as a function of time, and the state variables of the dynamical system such as to minimize a performance index. The differential equations of motion of the vehicles are treated as dynamical constraints. One possible approach to the solution of the rendezvous problem is to formulate it as an optimal control problem in which one is seeking the controls such as to

minimize the differences between the final locations and final velocities of the vehicles in some mathematical sense, for example in the least squares sense. Here we use a minimax formulation in order to define the objective function that contains the terminal constraints.
The methods of approach for solving optimal control problems include the classical indirect methods and the more recent direct methods. The indirect methods are based on the calculus of variations and its extension to the maximum principle of Pontryagin, which is based on a Hamiltonian formulation. These methods use necessary first order conditions for an optimum, they introduce adjoint variables and require the solution of a two-point boundary value problem (TPBVP) for the state and adjoint variables. Usually, the state variables are subjected to initial conditions and the adjoint variables to terminal or final conditions. **TPBVPs** are much more difficult to solve than initial value problems (IVP). For this reason, direct methods of solution have been developed which avoid completely the Hamiltonian formulation. For example, a possible approach is to reformulate the optimal control problem as a nonlinear programming (NLP) problem by direct transcription of the dynamical equations at prescribed discrete points or collocation points.
Direct Collocation Nonlinear Programming (DCNLP) is a numerical method that has been used to solve optimal control problems. This method uses a transcription of the continuous equations of motion into a finite number of nonlinear equality constraints, which are satisfied at fixed collocation points. This method was originally developed by Dickmanns and Well (Dickmanns, 1975) and used by Hargraves and Paris (Hargraves, 1987) to solve several atmospheric trajectory optimization problems. Another class of direct methods is based on biologically inspired methods of optimization. These include evolutionary methods such as genetic algorithms (Goldberg, 1989), particle swarm optimization methods and ant colony optimization algorithms.
Genetic algorithms (GAs) provide a powerful alternative method for solving optimal control problems. They have been used to solve control problems (Crispin, 2006, 2007), orbital transfer and rendezvous problems (Crispin and Ricour, 2007). GAs use a stochastic search method and are robust when compared to gradient methods. They are based on a directed random search which can explore a large region of the design space without conducting an exhaustive search. This increases the probability of finding a global optimum solution to the problem. They can handle continuous or discontinuous variables since they use binary coding. They require only values of the objective function but no values of the derivatives. However, GAs do not guarantee convergence to the global optimum. If the algorithm converges too fast, the probability of exploring some regions of the design space will decrease. Methods have been developed for preventing the algorithm from converging to a local optimum. These include fitness scaling, increased probability of mutation, redefinition of the fitness function and other methods that can help maintain the diversity of the population during the genetic search.

2. Interception and rendezvous as optimal control problems

We study interception and rendezvous problems for vehicles moving in an incompressible viscous fluid such as water. The vehicle has a propulsion system that delivers a thrust of constant magnitude T and is controlled by varying the thrust direction, i.e., thrust vectoring. Since the fluid is viscous, a drag force acts on the vehicle, in the opposite direction of the velocity. The motion takes place in a horizontal plane, the (x, z) plane, either at the surface of

the water or at a constant depth. The target vehicle is moving along a circle of radius *R*. We describe the motion in a cartesian frame of reference *(x, z)* with its origin at the center of the circle, with *x* positive to the right and *z* positive upwards.
Let the angle β denote the orientation of the thrust vector ***T***, which is also the angle of the velocity vector ***V***. The angle β(t), which depends on the time *t*, is measured positive counter-clockwise from a reference horizontal line, the positive direction of the *x* axis. In the examples we are going to present, the motion of the chaser will start from a point located outside the circle and will be moving from right to left, see Figure 2. In this case, it is convenient to use the complementary angle $\gamma(t)=\pi-\beta(t)$ as the control.
The interception problem is an optimal control problem, in which it is required to determine the control function or control history γ(t) of the chaser vehicle, such that it will meet the target vehicle at a prescribed location at the terminal time t_f. Since GAs deal with discrete variables, we discretize the values of γ(t). The motion of the vehicle is governed by Newton's second law of motion and the kinematic relations between velocity and distance:

$$md\boldsymbol{V}/dt = \boldsymbol{T} + \boldsymbol{D} \tag{1}$$

$$dx/dt = V\cos\beta(t) \tag{2}$$

$$dz/dt = V\sin\beta(t) \tag{3}$$

where $\beta(t)=\pi-\gamma(t)$ is the angle of velocity vector *V*, ***T*** is the thrust and ***D*** is the drag force acting on the body.

$$d\boldsymbol{V}/dt = \boldsymbol{T}/m + \boldsymbol{D}/m \tag{4}$$

Writing this equation for the components of the forces along the tangent to the vehicle's path, we get:

$$dV/dt = T/m - D/m \tag{5}$$

Here *V*, *T* and *D* are the magnitudes of the velocity, thrust and drag vectors, respectively. We introduce the drag coefficient C_D:

$$D = \tfrac{1}{2}\rho V^2 S C_D \tag{6}$$

where ρ is the fluid density, *S* is a typical cross-section area of the vehicle. The coefficient of drag depends on the Reynolds number $Re = \rho\, Vd/\mu$ where *d* is a typical length dimension of the vehicle, e.g., $d = \sqrt{S}$.
Substituting the drag from (6) and writing $T = amg$, where *a* is the thrust to weight ratio *T/mg*, equation (5) becomes:

$$dV/dt = ag - \rho V^2 S C_D/2m \tag{7}$$

Introducing a characteristic length L_c, time t_c and speed v_c as

$$L_c = 2m/\rho S C_D, \quad t_c = \sqrt{L_c/g}, \quad v_c = \sqrt{gL_c} \tag{8}$$

the following non-dimensional variables, denoted by a bar, can be defined:

$$x = L_c \bar{x}, \quad z = L_c \bar{z},$$

$$t = (L_c/g)^{1/2}\,\bar{t}, \quad V = (gL_c)^{1/2}\,\overline{V} \tag{9}$$

Substituting in (7), we have:

$$d\overline{V}/d\bar{t} = a - \overline{V}^2 \tag{10}$$

Similarly, the other equations of motion can be written in non-dimensional form as

$$d\bar{x}/d\bar{t} = -\overline{V}\cos\gamma(t) \tag{11}$$

$$d\bar{z}/d\bar{t} = \overline{V}\sin\gamma(t) \tag{12}$$

For each vehicle the initial conditions are:

$$V(0) = V_0, \quad x(0) = x_0, \quad z(0) = z_0 \tag{13}$$

We now define a rendezvous problem between two vehicles. We denote the variables of the first vehicle by a subscript 1 and those of the second vehicle by a subscript 2. We will now drop the bar notation indicating non-dimensional variables. The two vehicles might have different thrust to weight ratios, which we denote by a_1 and a_2, respectively. The equations of motion for the system of two vehicles are:

$$\begin{aligned}
dV_1/dt &= a_1 - V_1^2 \\
dx_1/dt &= -V_1\cos\gamma_1(t) \\
dz_1/dt &= V_1\sin\gamma_1(t) \\
dV_2/dt &= a_2 - V_2^2 \\
dx_2/dt &= -V_2\cos\gamma_2(t) \\
dz_2/dt &= V_2\sin\gamma_2(t)
\end{aligned} \tag{14}$$

We consider the case where the second vehicle, which is the target vehicle, is constrained to move along a circular trajectory. In this case, the magnitude of the velocity vector V_2 is constant. The angular speed ω, the azimuth angle $\theta(t)$ and the required control angle $\gamma_2(t)$ are given by:

$$\omega = V_2/R, \qquad \theta(t) = \omega t, \qquad \gamma_2(t) = \pi/2 - \theta(t) \tag{15}$$

where R is the radius of the circle. $\theta(t)$ is measured at the center of the circle, positive counter-clockwise, from the positive x direction. The vehicles can start the motion from different locations and at different speeds. The initial conditions are given by:

$$V_1(0) = V_{10}, \quad x_1(0) = x_{10}, \quad z_1(0) = z_{10}$$

$$V_2(0) = V_{20}, \quad x_2(0) = x_{20}, \quad z_2(0) = z_{20} \tag{16}$$

The rendezvous problem consists of finding the control function $\gamma_1(t)$ such that the two vehicles arrive at a same terminal location on the circle and at the same speed in the given time. The terminal constraints are given by:

$$x_1(t_f) = x_2(t_f), \quad z_1(t_f) = z_2(t_f), \quad V_1(t_f) = V_2(t_f) \tag{17}$$

In order to fulfill the terminal constraints using a genetic algorithm, we define the following objective function:

$$f[x_1(t_f), x_2(t_f), z_1(t_f), z_2(t_f), V_1(t_f), V_2(t_f)] =$$

$$= \min\max\{(x_1(t_f) - x_2(t_f))^2, (z_1(t_f) - z_2(t_f))^2, (V_{1x}(t_f) - V_{2x}(t_f))^2,$$

$$(V_{1z}(t_f) - V_{2z}(t_f))^2\} \tag{18}$$

where $V_{1x}(t_f)$, $V_{2x}(t_f)$, $V_{1z}(t_f)$ and $V_{2z}(t_f)$ are the x and z components of the velocity vectors of the two vehicles at the terminal time $t = t_f$. We can also define an interception problem, of the target-chaser type, in which the target vehicle is passive and the chaser vehicle maneuvers such as to match the location of the target vehicle, but not its velocity. Consistent with the above terminal constraints, we define the following objective function for the interception problem:

$$f[x_1(t_f), x_2(t_f), z_1(t_f), z_2(t_f)] = \min\max\{(x_1(t_f) - x_2(t_f))^2, (z_1(t_f) - z_2(t_f))^2\} \tag{19}$$

We have also tried an objective function defined by the distance between the two terminal points of the two vehicles, but it was found that the mini-max objective function works better in satisfying the terminal constraints. We use standard numerical methods for integrating the differential equations. The time interval is divided into N time steps of duration $\Delta t = t_f/N$. The discrete time is $t_i = i\Delta t$. We used a second-order Runge-Kutta method with fixed time step. We also tried a fourth-order Runge-Kutta method and a variable time step and found that the results were not sensitive to the method of integration.

The control function $\gamma(t)$ is discretized to $\gamma(i) = \gamma(t_i)$ according to the number of time steps N used for the numerical integration. Depending on the accuracy of the desired solution, we can choose the number of bits n^i for encoding the value of the control $\gamma(i)$ at each time step *i*. The number of bits used for encoding $\gamma(i)$ and the number of time steps N will have an influence on the computational time. Therefore n^i and N must be chosen carefully in order to obtain an accurate enough solution in a reasonable time. We use smoothing of the control function by fitting a third-order polynomial to the discrete values of $\gamma(i)$. The values of the polynomial at the *N* discrete time points are used as the current values of $\gamma(i)$ and are used in the integration of the differential equations.
An appropriate range for $\gamma(i)$ is $\gamma \in [-\pi, \pi]$. We choose *N=30* or *40* as a reasonable number of time steps. We now need to choose the parameters associated with the Genetic Algorithm. First, we select the lengths of the "genes" for encoding the discrete values of $\gamma(i)$. A choice of $n^i = 8$ bits for $\forall i \in [0, N-1]$ was made. A reasonable size for the population of solutions is typically in the range $n_{\text{pop}} \in [50, 200]$. For this problem, there is no need for a particularly large population, so we select $n_{\text{pop}} = 50$. The probability of mutation is set to a value of 5 percent $p_{\text{mut}} = 0.05$.

3. Chaser-target interception

We present an example of a chaser-target interception problem between two vehicles. The motion takes place in the horizontal plane *(x, z)*. The first vehicle is active (vehicle 1) and the second vehicle is the passive or target vehicle (vehicle 2) and moves at a constant speed V_2 along a circular trajectory. The target vehicle starts from a point on the circle and keeps moving along its circular trajectory. The chaser vehicle starts at a point outside the circle and the interception occurs at a point on the circle at the final time. The interception point is not known a priori. The parameters and initial conditions for this example are given below.

$$t_0 = 0, \quad t_f = 35, \quad a_1 = 4a/9, \quad a_2 = a = 0.05, \quad \gamma_1 \in [-\pi, \pi] \tag{20}$$

The initial conditions are:

$$x_1(0) = 4, \quad z_1(0) = 0, \quad x_2(0) = 1, \quad z_2(0) = 0$$

$$V_1(0) = \sqrt{a_1} = 0.2236, \quad V_2(0) = \sqrt{a_2} = 0.1491 \tag{21}$$

Since this is an interception problem, we do not require matching between the final velocities. The objective function is defined by equation (19). The parameters of the genetic algorithm are summarized in Table 1.

N_v	n_{pop}	n^i	N	p_{mut}	N_{gen}	$\gamma_{1\min}$	$\gamma_{1\max}$
2	50	8	40	0.05	100	$-\pi$	π
a	t_0	t_f	(x_{01}, z_{01})	(x_{02}, z_{02})	(V_{01}, V_{02})	x_f	z_f
0.05	0	35	$(4, 0)$	$(1, 0)$	$(\sqrt{4a/9}, \sqrt{a})$	free	free

Table 1. Parameters for the interception problem with $t_f = 35$

The results of this example are displayed in Figures (1-4). Figure 1 displays the control function of the chaser vehicle as a function of time for a terminal time of 35 units.

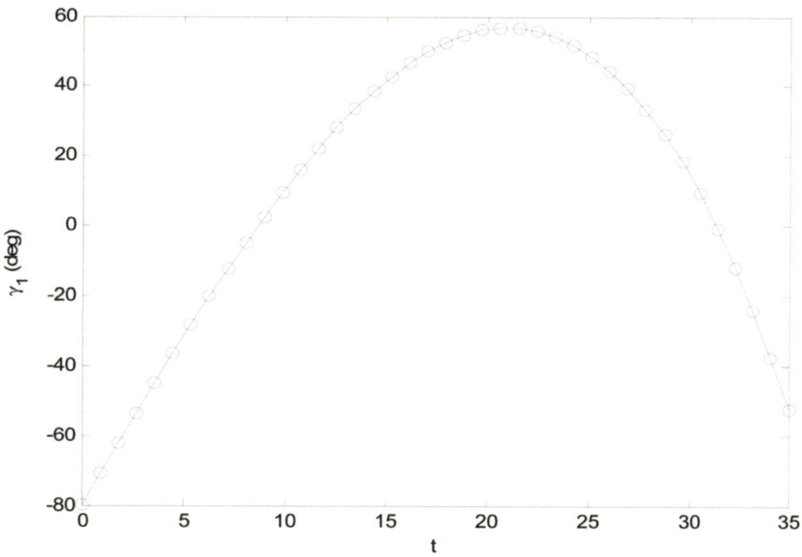

Fig. 1. Control function for the chaser vehicle in a chaser-target interception on a circle with a terminal time of 35 units.

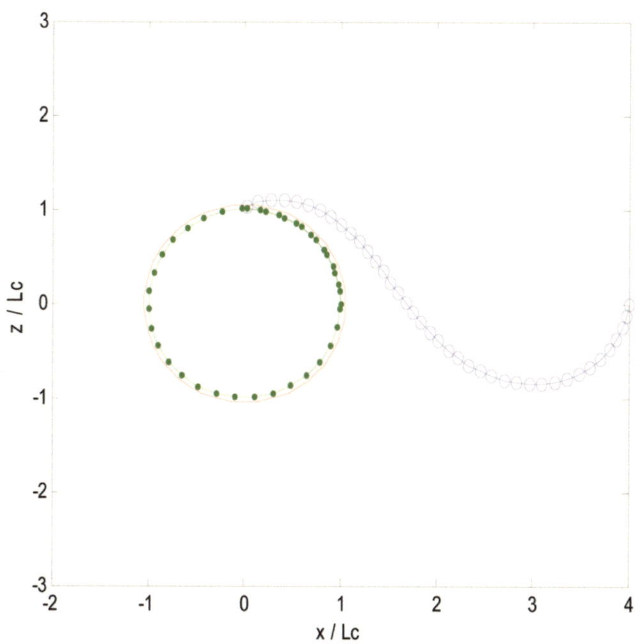

Fig. 2. Trajectories for a chaser-target interception on a circle with a terminal time of 35 units.

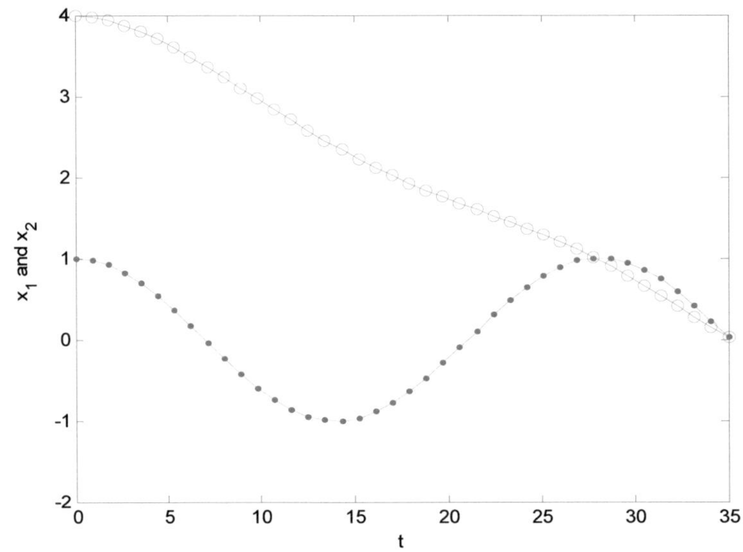

Fig. 3. Trajectories *x(t)* for a chaser-target interception on a circle with a terminal time of 35 units.

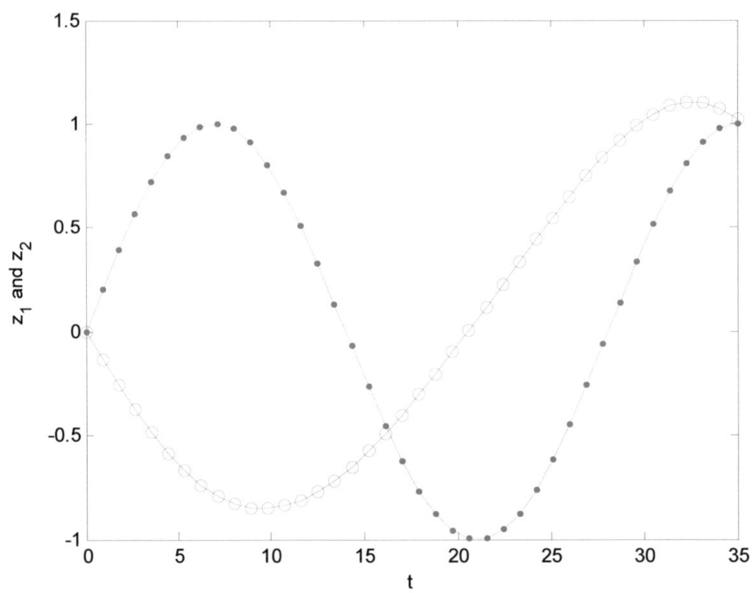

Fig. 4. Trajectories *z(t)* for a chaser-target interception on a circle with a terminal time of 35 units.

The trajectories of the target and chaser vehicles are shown in Figure 2. The target starts the motion at the point (1,0) on the circle and the chaser starts at the point (4,0) outside the circle. The line with dots is the trajectory of the target. The line with circles is the trajectory of the chaser. The continuous curve without dots marks the circle. Figures 3 and 4 show the trajectories of the target and the chaser vehicles in parametric form, with the time as a parameter. Figure 3 displays the *x(t)* coordinates of the two vehicles as a function of time and Figure 4 displays the *z(t)* coordinates for the two vehicles. The lines with dots are the target trajectories and the lines with circles are the chaser trajectories. It can be seen that the final boundary conditions are fulfilled, that is, the coordinates of the two vehicles are equal at the final time.

In the above example, a non-dimensional terminal time of 35 units was used. We now look at the problem of finding the minimum time for interception for the same set of parameters and initial conditions as in the previous example. We decreased the final time to 30 units and then to 25 units and were able to obtain a solution, i.e., the terminal constraints are fulfilled for the shorter times. We then tried to decrease the final time to a value of 23 units but we could not obtain a solution that fulfills the terminal constraints. We then tried a slightly higher value of 24 units and were able to obtain a solution. We therefore conclude that the minimum time for interception in this case is very close to 24 units . Since we have shorter times, we decreased the number of discrete time points from $N=40$ to $N=30$. The results for this example are given in Figures (5-8). Figure 5 shows the control function of the chaser vehicle. Comparing with Figure 1, we can see that the control looks much different because of the shorter final time. Figure 6 shows that the interception occurs as a head-on collision, a phenomenon that occurs also in the classic problem of pursuit. From Figure 7, it can be seen that the horizontal distance between the chaser and the interception point is reduced linearly at maximum speed, because of the minimum final time. In this case, it can also be seen from Figures 7 and 8, that the final boundary conditions are fulfilled.

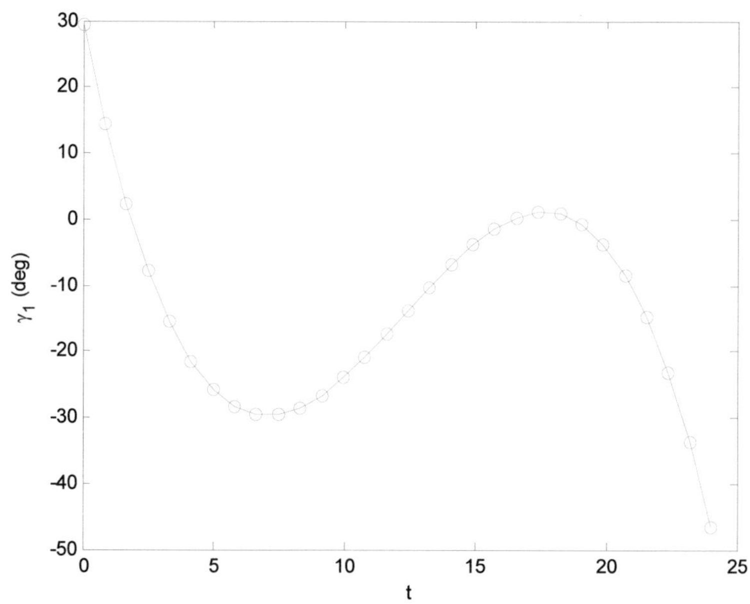

Fig. 5. Control function for the chaser vehicle in a chaser-target interception on a circle with a minimum terminal time of 24 units.

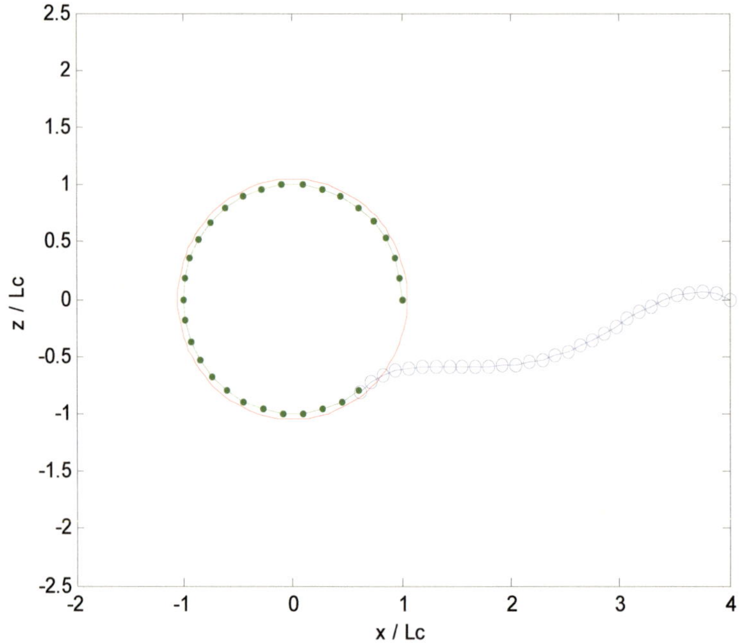

Fig. 6. Trajectories for a chaser-target interception on a circle with a minimum terminal time of 24 units.

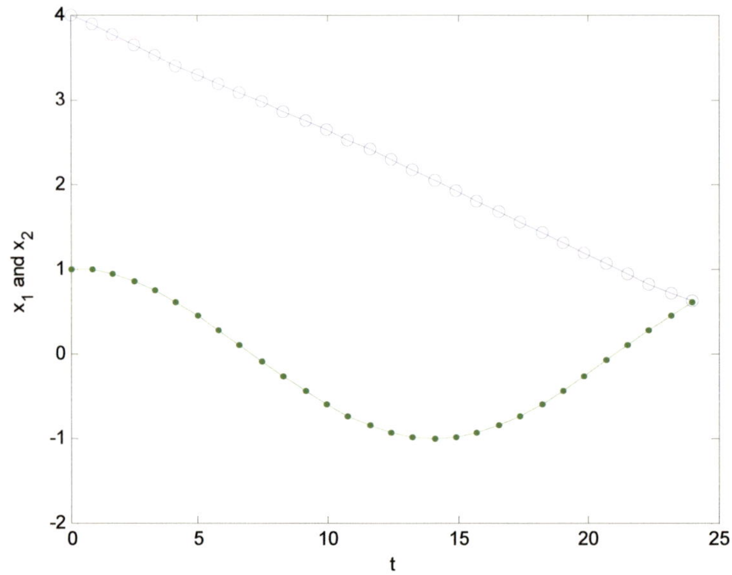

Fig. 7. Trajectories *x(t)* for a chaser-target interception on a circle with a minimum terminal time of 24 units.

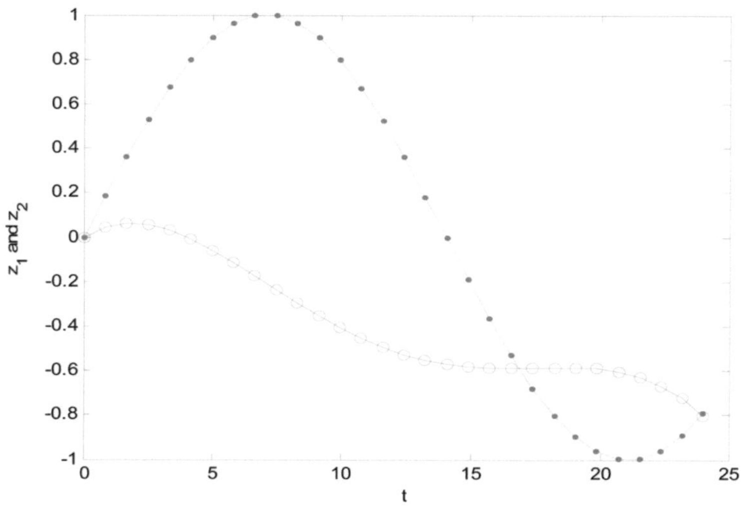

Fig. 8. Trajectories *z(t)* for a chaser-target interception on a circle with a minimum terminal time of 24 units.

4. Rendezvous between two vehicles

We treat a rendezvous problem between two vehicles where one vehicle is moving along a circular trajectory. The first vehicle starts from a point outside the circle. The second vehicle starts at a point on the circle. The final time is given and the rendezvous point can occur at any point on the circle. The vehicles have the same thrust to weight ratio *a*. We present results with a final time of 25 units, which is close to the minimum time.

$$t_0 = 0,\ t_f = 25,\ a_1 = a = 0.05,\quad a_2 = a = 0.05,\quad \gamma_1 \in [-\pi,\ \pi] \tag{22}$$

The initial conditions are:

$$x_1(0) = 4,\ \ z_1(0) = 0,\ \ x_2(0) = 1,\ \ z_2(0) = 0,\ \ V_1(0) = \sqrt{a_1},\ V_2(0) = \sqrt{a_2} \tag{23}$$

The parameters for this test case are summarized in Table 2. The results are given in Figure 9. The curves with dots show the trajectories of the target and the curves with circles show the trajectories of the chaser vehicle. The upper left part of the figure shows the trajectories in the plane *(x, z)*. The chaser vehicle is maneuvering such as to approach the rendezvous point from behind, in order to match the velocity vector of the target vehicle. A similar effect can be seen in Figure 10, which displays results similar to Figure 9, but with the chaser vehicle starting from rest. It is interesting to note that in this case, the *z* coordinate of the chaser, follows the *z* coordinate of the target, as can be seen in the bottom right of Figure 10.

N_v	n_{pop}	n^i	N	p_{mut}	N_{gen}	$\gamma_{\min}$	$\gamma_{\max}$
2	50	8	30	0.05	100	$-\pi$	π
a	t_0	t_f	(x_{01}, z_{01})	(x_{02}, z_{02})	(V_{01}, V_{02})	x_f	z_f
0.05	0	25	$(4,0)$	$(1,0)$	$(\sqrt{a_1}, \sqrt{a_2})$	free	free

Table 2. Parameters for the rendezvous problem

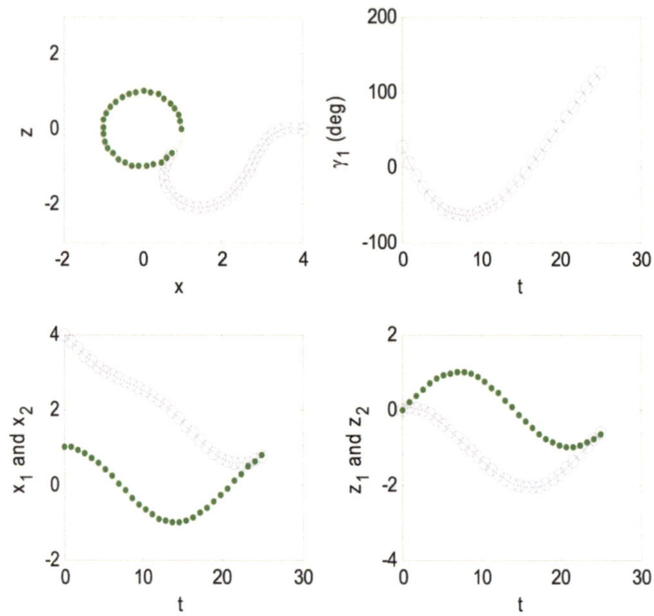

Fig. 9. Rendezvous between two vehicles on the circle when the outer vehicle starts at an initial speed. Top left: Trajectories of the two vehicles. Top right: Control thrust angle of the vehicle starting outside the circle. Bottom left: Horizontal coordinates as a function of time. Bottom right: vertical coordinates as a function of time.

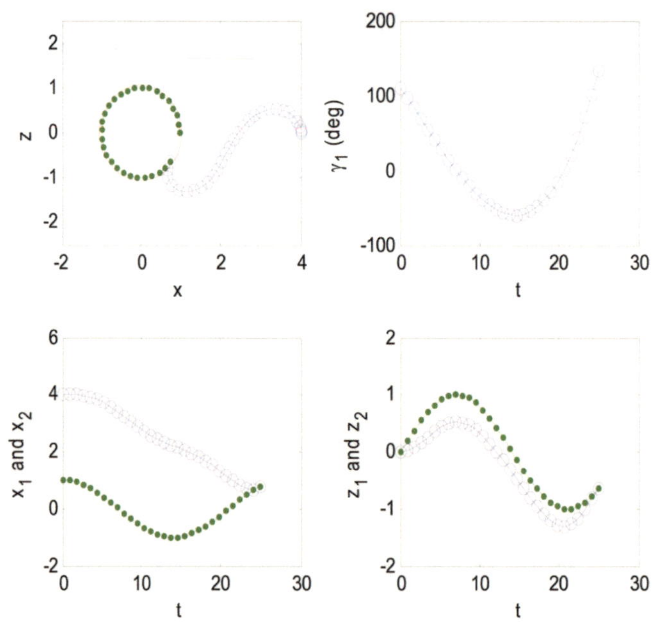

Fig. 10. Rendezvous between two vehicles on the circle when the outer vehicle starts from rest. Top left: Trajectories of the two vehicles. Top right: Control thrust angle of the vehicle starting outside the circle. Bottom left: Horizontal coordinates as a function of time. Bottom right: vertical coordinates as a function of time.

5. Conclusion

The interception and rendezvous problems between two autonomous vehicles moving in an underwater environment has been treated using an optimal control formulation with terminal constraints. The vehicles have a constant thrust propulsion system and use the direction of the thrust vector for steering and control. We use a genetic algorithm to determine directly the control history of the vehicle by evolving populations of possible solutions of initial value problems. In order to fulfill the final boundary conditions as terminal constraints, a mini-max objective function has been defined. An interception problem, where one vehicle moves along a circular trajectory at constant speed and the second vehicle acts as a chaser, maneuvering such as to capture the target in a prescribed time has been solved. The problem of minimum time to interception has also been treated. The rendezvous problem where the target vehicle moves along a circle and the chaser vehicle starts from a point outside the circle either from rest or with an initial velocity has also been solved for a terminal time close to the minimum time for rendezvous. This method can be extended to include multiple autonomous vehicles. Another direction for future research is to include additional realistic effects besides thrust and drag. For example, additional effects can be included to better describe the dynamics of the vehicles, such as finite size, rigid body dynamics, inertia and added mass effects.

6. References

Bryson, A. E., (1999), "Dynamic Optimization", Addison Wesley Longman.

Crispin, Y.J. and Ricour, M.E., (2007), "Cooperative Rendezvous Between Active Low-Thrust Spacecraft", Paper AIAA-2007-6859, AIAA Guidance, Navigation and Control Conference, Hilton Head, SC.

Crispin, Y.J., (2006), "An Evolutionary Approach to Nonlinear Discrete-Time Optimal Control with Terminal Constraints", in Informatics in Control, Automation and Robotics I, pp.89-97, Springer, Dordrecht, Netherlands.

Crispin, Y.J., (2007), "Evolutionary Computation for Discrete and Continuous Time Optimal Control Problems", in Informatics in Control, Automation and Robotics II, Springer, Dordrecht, Netherlands.

Davis, H.T., (1962), Introduction to Nonlinear Differential and Integral Equations, Dover Publications, New York.

Dickmanns, E. D. and Well, H., (1975), "Approximate Solution of Optimal Control Problems Using Third-Order Hermite Polynomial Functions," Proceedings of the 6th Technical Conference on Optimization Techniques, Springer-Verlag, New York

Goldberg, D.E., (1989), "Genetic Algorithms in Search, Optimization and Machine Learning", Addison-Wesley Publishing Company

Hargraves, C. R. and Paris, S. W., (1987), "Direct Trajectory Optimization Using Nonlinear Programming and Collocation", AIAA Journal of Guidance, Control and Dynamics, Vol. 10, pp. 338-342

Hathaway, A.S., (1921), "Pursuit in a Circle", American Mathematical Monthly, Vol. 28, pp.93-97.

Marshall, J.A., Lin, Z., Broucke, M.E. and Francis, B.A., (2005), "Pursuit Strategies for Autonomous Agents", in Cooperative Control, A Post-Workshop Volume, Lecture

Notes in Control and Information Sciences 309, pp.137-151, Springer-Verlag, Berlin, Heidelberg, New York.

Nahin, P.J., (2007), Chases and Escapes, The Mathematics of Pursuit and Evasion, Princeton University Press, Princeton, NJ.

21

Efficient Data Collection with an Automated Robotic Helicopter Using Bayesian Adaptive Sampling Algorithms for Control

Steven M. Crunk [1] and Marian Farah [2]
[1]*San Jose State University*
[2]*University of California, Santa Cruz*
USA

1. Introduction

Traditional methods of data collection are often expensive and time consuming. We propose a novel data collection technique, called Bayesian Adaptive Sampling (BAS), which enables us to capture maximum information from minimal sample size. In this technique, the information available at any given point is used to direct future data collection from locations that are likely to provide the most useful observations in terms of gaining the most accuracy in the estimation of quantities of interest. We apply this approach to the problem of estimating the amount of carbon sequestered by trees. Data may be collected by an autonomous helicopter with onboard instrumentation and computing capability, which after taking measurements, would then analyze currently available data and determine the next best informative location at which a measurement should be taken. We quantify the errors in estimation and work towards achieving maximal information from minimal sample sizes. We conclude by presenting experimental results that suggest our approach towards biomass estimation is more accurate and efficient as compared to random sampling.

Bayesian Adaptive Sampling (BAS) is a methodology that allows a system to examine currently available data in order to determine new locations at which to take new readings. This procedure leads to the identification of locations where new observations are likely to yield the most information about a process, thus minimizing the required data that must be collected. As an example of the application of this methodology, we examine the question of standing woods in the United States. In order to estimate the amount of carbon sequestered by trees in the United States, the amount of standing woods must be estimated with quantifiable uncertainty (Wheeler, 2006). Such estimates come from either satellite images or near ground measurements. The amounts of error in the estimates from these two approaches are currently unknown. To this end, an autonomous helicopter with differential GPS (Global Positioning System), LIDAR (Light Detection and Ranging), stereo imagers, and spectrometers has been developed as a testing platform for conducting further studies (Wheeler, 2006). These instruments are capable of measuring the reflectance data and the location of the Sun and helicopter in terms of the zenith and the azimuth angles (Figure 1). The objective is to develop a controlling software system for this robotic helicopter, which optimizes the required ground sampling. There are a number of methods by which data maybe collected. The first simplistic data collection method is to conduct an exhaustive ground sampling, that is, to send the helicopter to every possible location. The second approach is to perform random sampling until the estimates have acceptable standard errors. Although random sampling presents a possibility that the helicopter will take samples from the locations that offer the greatest amount of information, and therefore reduce the needed sample size, there is no guarantee that such a sample set will be chosen every time. The third and more efficient method is to take only a few samples from "key" locations that are expected to offer the greatest amount of information. The focus of this paper is to develop a methodology that will identify such key locations from which the helicopter should gather data.

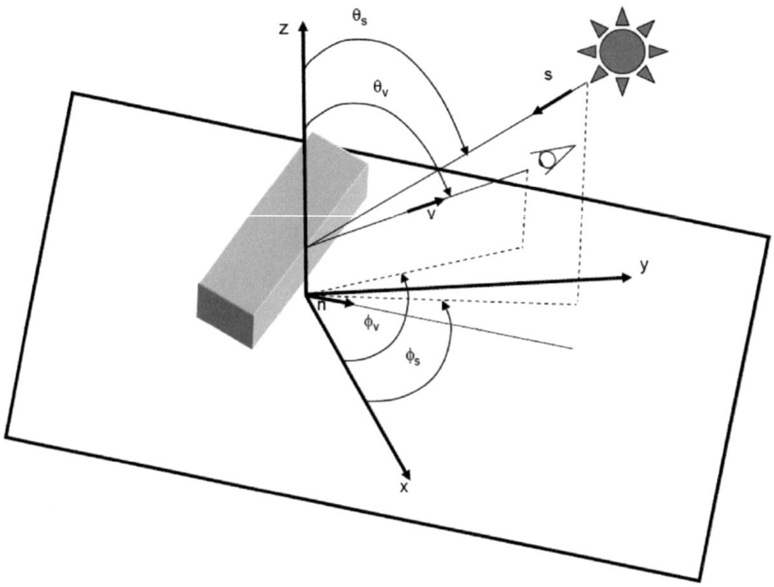

Fig. 1. Viewpoint and Position of the Sun. θ_s, ϕ_s are the zenith and the azimuth angles of the Sun, and θ_v , ϕ_v are the zenith and the azimuth angles of the view, respectively (Wheeler, 2006).

In the work described here, the key locations are identified using current and previously collected data. The software works in tandem with the sampling hardware to control the helicopter's position. Once a sample has been obtained, the data are fed into the system, which then calculates the next best location to gather further data. Initially, the system assumes an empirical model for the ground being examined. With each addition of data from the instruments, the parameter estimates of the model are updated, and the BAS methodology is used to calculate the helicopter's next position. This process is repeated until the estimated uncertainties of the parameters are within a satisfactory range. This method allows the system to be adaptive during the sampling process and ensures adequate ground coverage.

The application employs a bi-directional reflectance distribution function (BRDF), in which the calculation of the amount of reflection is based on the observed reflectance values of the object, and the positions of the Sun and the viewer (Nicodemus, 1970). The advantage of using this function is that it enables the system to compensate for different positions of the

Sun during sampling. Once the reflectance parameters are estimated, BAS uses the principle of maximum entropy to identify the next location where new observations are likely to yield the most information.

In summary, the BAS methodology allows the system to examine currently available data in order to determine new locations at which to take new reflectance readings. This procedure leads to the identification of locations where new observations are likely to yield the most information.

2. Background and related work

Computing view points based on maximum entropy using prior information has been demonstrated by Arbel & Ferrie (1970), where this technique was used to create entropy maps for object recognition. Vazquez et al. (2001) also demonstrated a technique for computing good viewpoints based on Information Theory. Whaite & Ferrie (1994) developed an autonomous explorer that seeks out those locations that give maximum information without using a priori knowledge of the environment. Makay (1992) used Shannon's entropy to obtain optimal sample points that would yield maximum information. The sample points are taken from the locations that have largest error bars on the interpolation function. In our work, the optimal locations that offer the maximum amount of information are identified using the principle of maximum entropy, where the maximization is performed using techniques suggested by Sebastiani and Wynn (2000).

A simplified but elucidating version of an example given in Sebastiani and Wynn (2000) is as follows: Suppose that the relationship between a dependent variable y and an independent variable x is known to be linear over the domain of interest $x \in [1,4]$, and is known to be one of two equations with additive noise, $y_i = 5 + 7x_i + \varepsilon_i$ or $y_i = 17 + 3x_i + \varepsilon_i$, each with a Bayesian prior belief to update of 1/2. Figure 2 shows the two possibilities over

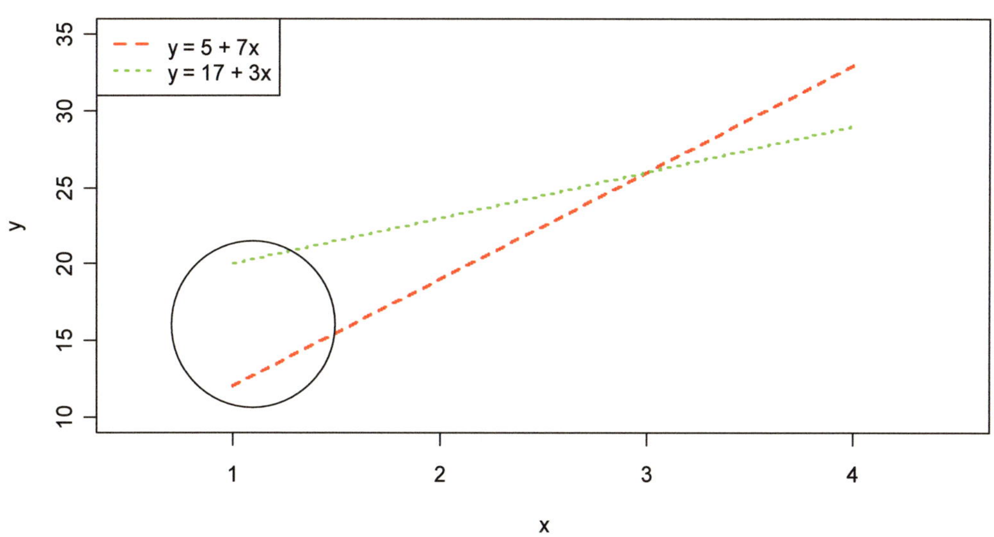

Fig. 2. Two lines, with circle indicating the region where the two models have the greatest diference, and thus where the maximum entropy sampling procedure indicates a new observation should be taken in order to best determine which model is more likely correct.

the domain of interest. Our intuition tells us that an observation taken at $x = 3$ will not help us to discriminate between the two possible models, as the expected resonse, E($y\,|\,x$=3) is the same for each of the two models. Our intuition further tells us that an observation taken at $x = 1$ will yield a response, $y\,|\,x$=1, providing the most "information" as to the correct model, since the two models are most distinct at $x = 1$. An application of the theorems of Sebastiani and Wynn show that the optimal location of the independent variable at which to take an observation that maximizes the entropy (expected information as defined by Shannon (1948)) is at $x^* = \arg_x\min\{(5 + 7x) - (17 + 3x)\}^2 = \arg_x\min\{144 - 96x + 16x^2\} = 1$, which agrees with our intuition.

3. Application and model

Surface BRDF is often used to measure vegitation or other attributes of a surface. By it's very nature, BRDF requires measurements taken from a variety of viewing angles and sun positions (and hence different times of day), as solar radiation reflected by the surface is not uniform in all directions (Zhang et al., 1997), nor is it uniform from the same direction when the sun is shining on the location from different directions, as can been seen in Figure 3. This shows the importance of including not only the viewpoint position but also the position of the sun when taking reflectance measurements in order to accurately measure ground cover. The model for the data used in our framework is based on the semi-empirical MISR (multi-angle imaging spectrometer) BRDF Rahman model (Rahman et al., 1993):

$$
\begin{aligned}
r(\theta_s,\theta_v,\phi_s,\phi_v) &= \rho\left[\cos(\theta_s)\cos(\theta_v)\{\cos(\theta_s)+\cos(\theta_v)\}\right]^{k-1} \\
&\times \exp\left(-b\cdot p(\Omega)\right)h(\theta_s,\theta_v,\phi_s,\phi_v)
\end{aligned} \tag{1}
$$

where

$$
h(\theta_s,\theta_v,\phi_s,\phi_v) = 1 + \frac{1-\rho}{1+G(\theta_s,\theta_v,\phi_s,\phi_v)} \tag{2}
$$

$$
G(\theta_s,\theta_v,\phi_s,\phi_v) = \sqrt{\tan^2(\theta_s)+\tan^2(\theta_v)-2\tan(\theta_s)\tan(\theta_v)\cos(\phi_s-\phi_v)} \tag{3}
$$

$$
p(\Omega) = \cos(\theta_s)\cos(\theta_v)+\sin(\theta_s)\sin(\theta_v)\cos(\phi_s-\phi_v) \tag{4}
$$

where $r = r(\theta_s,\theta_v,\phi_s,\phi_v)$ is the measured reflectance, ρ is the surface reflectance at zenith, k is the surface slope of reflectance, b is a constant associated with the hotspot, or "antisolar point" (the point of maximum reflectivity, which is the position where the sensor is in direct alignment between the Sun and the ground target), θ_s,ϕ_s are the zenith and the azimuth angles of the Sun, respectively (Fig. 1), and θ_v,ϕ_v are the zenith and the azimuth angles of the view, respectively (Fig. 1), where in each case the zenith angle θ ranges from 0 to $\pi/2$ (horizontal to vertical) and the azimuth angle ranges from 0 to 2π (all the way around a circle), thus the viewpoint (and the sun) are assumed on a half-hemisphere centered over the object or in this case region of land to be studied.

With this application, the goal of this study is to estimate the parameters ρ, k and b to a predefined level of accuracy, or as some in the engineering field say, to "minimize the length of the error bars of the paramater estimates", with as few observations as is reasonably possible.

Fig. 3. The effect of forward scattering of light (on the left) and backward scaterring of light (on the right) in two photographs of the same location taken from the same position at different times of day (Lucht & Schaaf, 2006).

4. Methodology

Our framework consists of the following two steps:

1. Parameter Estimation: In this step, we estimate the values of the parameters (ρ, k and b), and their covariance matrix and standard errors, given currently available data of the amount of observed reflected light, and the zenith and azimuth angles of the Sun and the observer.
2. Bayesian Adaptive Sampling (Optimal Location Identification): In this step, we use the principle of maximum entropy to identify the key locations from which to collect the data.

Once the key location is identified, the helicopter goes to that location, and the instruments on the helicopter measure the reflectance information. These data are then fed into the Parameter Estimation stage and the new values of the parameters (ρ, k and b) are calculated. This process is repeated (Fig. 4) until the standard errors of the parameter estimates achieve some predefined small value, ensuring adequacy of the estimated parameters. This process minimizes the required number of observations to be taken by ensuring that each new observation is taken from a location which maximizes the expected information obtained from the forthcoming observation, and using the then observed data in order to determine the location from which to take the subsequent observation.

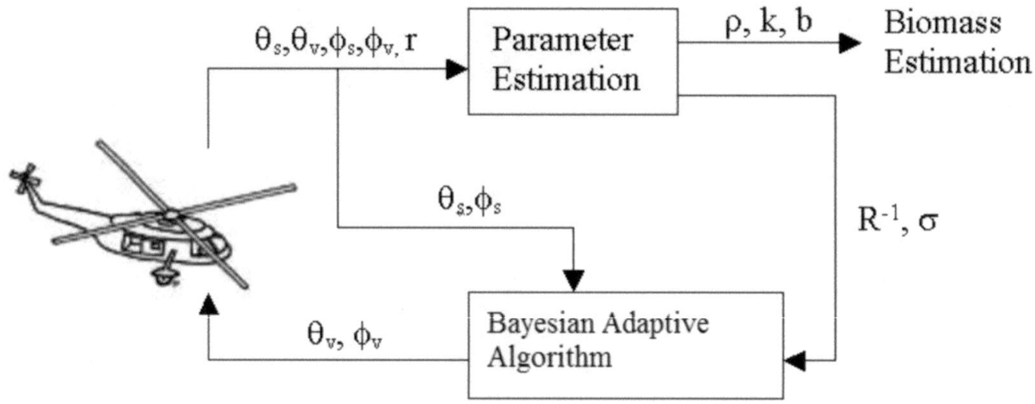

Fig. 4. Overview of Bayesian Adaptive Sampling.

5. Implementation

5.1 Parameter estimation

The input to this step is the observed reflectance value (r), zenith and azimuth angles of the Sun (θ_s, ϕ_s), and zenith and azimuth angles of the observer/helicopter (θ_v, ϕ_v). The parameters (ρ, k, and b) are estimated using the following iterated linear regression algorithm: Taking the natural logarithm of $r = r(\theta_s, \theta_v, \phi_s, \phi_v)$ results in the following near linear version of this model:

$$\ln r = \ln \rho + (k-1)\ln\left[\cos(\theta_s)\cos(\theta_v)\{\cos(\theta_s)+\cos(\theta_v)\}\right] - b\,p(\Omega) + \ln h(\theta_s,\theta_v,\phi_s,\phi_v) \tag{5}$$

Note that aside from the term $\ln h(\theta_s,\theta_v,\phi_s,\phi_v)$, which contains a nonlinear ρ, the function $\ln(r)$ is linear in all three parameters, $\ln(\rho)$, k, and b. "Linearization" of $\ln(h)$ is accomplished by using the estimate of ρ from the previous iteration, where at iteration n in the linear least-squares fit the value of $h(\theta_s,\theta_v,\phi_s,\phi_v)$ from (2) is modified and is taken to be the constant in (6) where $\rho^{(0)}$ is set equal to zero.

$$h^{(n)}(\theta_s,\theta_v,\phi_s,\phi_v) = 1 + \frac{1-\rho^{(n-1)}}{1+G(\theta_s,\theta_v,\phi_s,\phi_v)} \tag{6}$$

We thus obtain as out model, for observation i, the linear model

$$y_i = \beta_0 + k\,x_{i1} + b\,x_{i2} + \varepsilon_i \qquad (7)$$

where we assume that the errors in the model, ε_i, are independent and identically distributed (*iid*) random variables from a Normal distribution with mean zero and variance σ^2; $\beta_0 = \ln\rho$ is the regression constant/intercept to be estimated; $x_{i1,}$ x_{i2} are the independent/ explanatory variables given in (8) and (9), which are functions of the viewpoint and the position of the sun as described above and are measured at observation time i; and y_i is a function of the response and functions which are known/fixed constants at the given time i as shown in (10).

$$x_{i1} = \ln\left[\cos(\theta_{si})\cos(\theta_{vi})\{\cos(\theta_{si}) + \cos(\theta_{vi})\}\right] \qquad (8)$$

$$x_{i2} = -\,p(\Omega) = -\left[\cos(\theta_{si})\cos(\theta_{vi}) + \sin(\theta_{si})\sin(\theta_{vi})\cos(\phi_{si} - \phi_{vi})\right] \qquad (9)$$

$$y_i = \ln(r_i) + \ln\left[\cos(\theta_{si})\cos(\theta_{vi})\{\cos(\theta_{si}) + \cos(\theta_{vi})\}\right] - \ln h^{(n)}\left(\theta_{si}, \theta_{vi}, \phi_{si}, \phi_{vi}\right) \qquad (10)$$

Next, linear regression is performed on the model in (7) (with iteration until the convergence of the parameters due to the nonlinear term $h(\theta_s, \theta_v, \phi_s, \phi_v)$ being treated as the constant $h^{(n)}(\theta_s, \theta_v, \phi_s, \phi_v)$). From the regression, we find estimates of the following quantities: Estimates of the parameters $\beta_0, k,$ and b, an estimate of σ^2 (the variance of the errors, as discussed in and around (7)), and an estimate of the covariance matrix, R^{-1}, of the parameter estimates, i.e., the estimate of the matrix given in (11). Recall that $\beta_0 = \ln\rho$ so that $\rho = \exp(\beta_0)$, and through use of the delta method (Casella and Berger, 2002) an estimate of ρ is $\hat{\rho} = \exp(\hat{\beta}_0)$ and $\hat{\mathrm{V}}(\hat{\rho}) \approx \left(\exp(\hat{\beta}_0)\right)^2 \hat{\mathrm{V}}(\hat{\beta}_0) = \hat{\rho}^2\,\hat{\mathrm{V}}(\hat{\beta}_0)$.

$$R^{-1} = \begin{bmatrix} \mathrm{V}(\hat{\beta}_0) & \mathrm{Cov}(\hat{\beta}_0, \hat{k}) & \mathrm{Cov}(\hat{\beta}_0, \hat{b}) \\ \mathrm{Cov}(\hat{k}, \hat{\beta}_0) & \mathrm{V}(\hat{k}) & \mathrm{Cov}(\hat{k}, \hat{b}) \\ \mathrm{Cov}(\hat{b}, \hat{\beta}_0) & \mathrm{Cov}(\hat{b}, \hat{k}) & \mathrm{V}(\hat{b}) \end{bmatrix} \qquad (11)$$

5.2 Bayesian adaptive sampling

This step identifies the most informative location (θ_v, ϕ_v) to which to send the helicopter to take a new reflectance reading. We employ the principle of maximum entropy, in which the available information is analyzed in order to determine a unique epistemic probability distribution. The maximization is performed as per techniques suggested by Sebastiani and Wynn (2000), where in order to maximize the amount of information about the posterior parameters (ρ, k and b), we should maximize the entropy of the distribution function. Mathematically, maximizing the entropy is achieved by maximizing $\ln\left|X'\Sigma^{-1}X + R\right|$ where $|\mathrm{A}|$ is the determinent of the matrix A, A′ indicates the transpose of the matrix A, Σ is the covariance matrix of the error terms, and X is the regression design matrix where the ith row is associated with the ith observation (either a future or past observation, depending on the context)

$$X = \begin{bmatrix} 1 & x_{11} & x_{12} \\ 1 & x_{21} & x_{22} \\ \vdots & \vdots & \vdots \\ 1 & x_{k1} & x_{k2} \end{bmatrix}.$$

As the natural logorithm is a monotone increasing function, and we have assumed that the errors are *iid* N(0, σ^2) and so $\Sigma = \frac{1}{\sigma^2} I$ where I is the identity matrix , our criteria reduces to maximizing the quantity given in (12).

$$\left| \frac{1}{\sigma^2} X'X + R \right| \tag{12}$$

Note that σ^2 and R are estimated in the Parameter Estimation step and are thus at this stage assumed to be known quantities. As described above, the matrix X contains functions of the zenith and azimuth angles of the Sun, (θ_s, ϕ_s), and the viewpoint/helicopter, (θ_v, ϕ_v) at which future observations are to be taken (i.e., for each time i at which future observations are to be taken). Since the times at which future observations are to be taken are known, and thus the positions of the sun at these times are known, the only remaining unknown quantities in (12) are the values of (θ_v, ϕ_v). Thus, the new location(s) to which the helicopter will be sent are the values of (θ_v, ϕ_v) in the rows of X associated with new observations that maximize (12).

5.3 Initialization

Before data have been collected, one of course cannot calculate parameter estimates as described in section 5.1. Before parameter estimates are collected, one does not have an estimate of R or σ^2 to use in (12) from section 5.2. In order to initialize the procedure, one can use a Bayesian prior for these quantities, or equivalently, any estimate based upon prior knowledge of the quantities. In the absence of any prior knowledge, one may take an uninformative prior for R to be the identity matrix I, and any arbitrary estimate of σ^2. See the appendix for a more thourough discussion of the quantity in (12) to be maximized.

6. Simulation

We conduct two simulated experiments in which the estimates of the model parameters are calculated. In the first experiment, "Estimation Using Random Observations", the data are collected by sending the helicopter to random locations. In the second experiment, "Estimation using BAS", the data are collected using BAS.

We note that the Sun moves through 2π radians in a 24-hour period, *i.e.*, at the rate of $2\pi/(24*60) \approx 0.004363323$ or slightly less than 0.005 radians per minute. We will assume it takes about 2 minutes for the helicopter to move to a new location. Thus, the position of the Sun changes by approximately 0.01 radians between measurements.

In our simulation, the true values of the parameters ρ, k and b are 0.1, 0.9, and -0.1, respectively. For the purpose of this paper, the observed values were simulated with added noise from the process with known parameters. This allows us to measure the efficacy of the algorithm in estimating the parameters and minimizing the standard errors of our estimates. In actual practice, the parameters would be unknown, and we would have no way of knowing how close our estimates are to the truth, that is, if the estimates are as accurate as implied by the error bars.

6.1 Estimation using observations taken at random locations

In this experiment, we send the helicopter to 20 randomly chosen locations to collect data. Starting with the fifth observation, we use the regression-fitting algorithm on the collected input data set (the observed reflectance information, and the positions of the Sun and the helicopter), to estimate the values of the parameters ρ , k , b as well as their standard errors. Table 1 shows the results of this experiment.

Obs #	θ_v	ϕ_v	r	Estimate (se) of ρ	Estimate (se) of k	Estimate (se) of b
1	0.114	1.673	0.157552			
2	0.882	6.013	0.156616			
3	0.761	0.917	0.192889			
4	0.678	1.308	0.180404			
5	0.260	0.114	0.152558	0.0683 (0.1172)	0.8497 (0.0607)	-0.5958 (0.1413)
6	1.195	2.367	0.146659	0.0767 (0.0932)	0.7906 (0.0476)	-0.4506 (0.1040)
7	0.237	2.805	0.149475	0.0830 (0.0746)	0.8268 (0.0404)	-0.3745 (0.0893)
8	0.166	1.700	0.155497	0.0832 (0.0641)	0.8286 (0.0345)	-0.3722 (0.0788)
9	0.320	2.012	0.154191	0.0831 (0.0572)	0.8277 (0.0307)	-0.3735 (0.0713)
10	1.224	4.085	0.129133	0.0917 (0.0465)	0.8369 (0.0381)	-0.2483 (0.0539)
11	1.409	3.442	0.135005	0.0917 (0.0431)	0.8380 (0.0309)	-0.2481 (0.0503)
12	0.092	1.559	0.154096	0.0920 (0.0394)	0.8398 (0.0285)	-0.2462 (0.0471)
13	0.806	0.891	0.200401	0.0888 (0.0402)	0.8129 (0.0284)	-0.2952 (0.0453)
14	1.256	5.467	0.147654	0.0891 (0.0385)	0.8181 (0.0259)	-0.2914 (0.0433)
15	0.227	1.284	0.155373	0.0889 (0.0368)	0.8169 (0.0248)	-0.2919 (0.0418)
16	1.129	5.522	0.148721	0.0889 (0.0354)	0.8174 (0.0236)	-0.2918 (0.0402)
17	0.507	5.696	0.150381	0.0891 (0.0333)	0.8183 (0.0225)	-0.2904 (0.0380)
18	0.119	4.363	0.142232	0.0890 (0.0302)	0.8181 (0.0207)	-0.2908 (0.0357)
19	0.245	0.524	0.151915	0.0889 (0.0299)	0.8172 (0.0205)	-0.2901 (0.0355)
20	0.446	2.408	0.144471	0.0884 (0.0297)	0.8149 (0.0204)	-0.2930 (0.0354)

Table 1. Observations and Estimates Using Random Sampling

6.2 Estimation using BAS

In this experiment, the first five locations of the helicopter are chosen simultaneously using an uninformative prior distribution (*i.e.*, as no estimate of R has yet been formed; it is taken to be σ^2 I, with σ^2 taken to be 10^{-6} based on information regarding the accuracy of the instrumentation) and an X matrix with five rows in which the positions of the Sun (θ_{si}, ϕ_{si}) are known and (12) is maximized over five pairs of helicopter viewpoints (θ_{vi}, ϕ_{vi}). Subsequently, we use BAS to calculate the next single best informative location for the helicopter to move to in order to take a new reflectance observation, in which case (12) is maximized with an X matrix containing a single row in which the position of the Sun (θ_s, ϕ_i) is known, σ^2 and R are taken to be their estimated values from the regression performed on the data collected to this point, and the only unknowns, to be calculated, are a single pair of helicopter viewpoint values, (θ_v, ϕ_v). After the helicopter is sent to this location and takes a reflectance reading, a new regression is performed on all data accumulated to this point, and the procedure is repeated. In practice, a computer onboard the helicopter would perform the regression, calculate the new next-best location, direct the helicopter to that location, take the reflectance reading, and repeat until some predetermined stopping point, which might include some fixed number of observations, or until the error bars have reached some predefined level indicating an acceptable accuracy in the parameter estimates, or perhaps until the helicopter has nearly exhausted it's fuel supply. In this simulation, we stop at the fixed number of 20 observations.

Table 2 shows the results from this experiment. In both experiments estimates of the parameters, along with their standard errors, cannot be formed until at least five observations have been taken.

Obs #	θ_v	ϕ_v	r	Estimate (se) of ρ	Estimate (se) of k	Estimate (se) of b
1	0.460	0.795	0.172364			
2	0.470	0.805	0.177412			
3	1.561	3.957	0.161359			
4	1.561	0.825	0.183571			
5	1.265	3.977	0.129712	0.1041 (0.0325)	0.90904 (0.00879)	-0.1249 (0.0290)
6	0.514	0.845	0.173072	0.1042 (0.0252)	0.90927 (0.00700)	-0.1255 (0.0233)
7	1.561	3.400	0.160130	0.1045 (0.0223)	0.90857 (0.00615)	-0.1220 (0.0199)
8	1.172	4.007	0.130101	0.1029 (0.0192)	0.90547 (0.00577)	-0.1329 (0.0180)
9	0.723	0.875	0.189697	0.1039 (0.0244)	0.90663 (0.00748)	-0.1428 (0.0228)
10	1.561	0.885	0.192543	0.1042 (0.0213)	0.90801 (0.00569)	-0.1394 (0.0185)
11	0.527	0.895	0.172811	0.1042 (0.0193)	0.90796 (0.00523)	-0.1392 (0.0172)
12	1.561	4.047	0.164530	0.1044 (0.0193)	0.90696 (0.00519)	-0.1343 (0.0167)
13	1.561	4.057	0.164822	0.1046 (0.0190)	0.90636 (0.00505)	-0.1314 (0.0161)
14	1.137	4.067	0.131443	0.1038 (0.0169)	0.90483 (0.00471)	-0.1365 (0.0148)
15	0.713	0.935	0.183894	0.1042 (0.0169)	0.90538 (0.00480)	-0.1397 (0.0149)
16	1.561	0.945	0.192280	0.1048 (0.0163)	0.90777 (0.00427)	-0.1333 (0.0136)
17	1.187	4.097	0.134701	0.1047 (0.0146)	0.90757 (0.00399)	-0.1340 (0.0125)
18	0.655	0.965	0.176841	0.1048 (0.0140)	0.90779 (0.00385)	-0.1349 (0.0120)
19	1.561	4.117	0.168819	0.1049 (0.0142)	0.90694 (0.00388)	-0.1321 (0.0120)
20	1.148	4.127	0.132199	0.1045 (0.0132)	0.90617 (0.00373)	-0.1349 (0.0114)

Table 2. Observations and Estimates Using BAS

6.3 Results

In this section, we compare and analyze the results of our two experiments. The comparison results shown graphically in Figures 5-7 show that the estimates using the data from the "well chosen" locations using BAS are closer to the true values, $\rho = .1$, $k = 0.9$ and $b = -0.1$, than the estimates based on data from the randomly chosen locations. Also, the error bars using BAS are much shorter, indicating higher confidence in the estimates of the parameters based on the "well chosen locations", *e.g.*, the length of the error bar for the estimate calculated using data/observations from five well chosen locations is as short as the error bar based on data collected from 20 random locations.

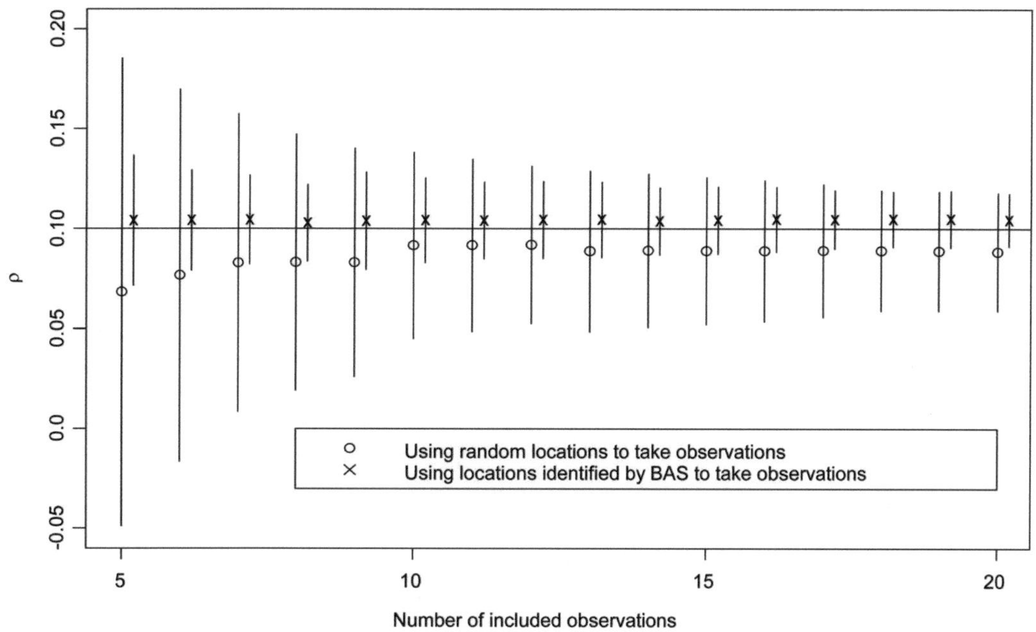

Fig. 5. Estimates and Error Bars for ρ .

Within each figure, the horizontal axis indicates the number of observations between five and twenty that were used in forming the estimates. The vertical axis is on the scale of the parameter being estimated. Above each observation number, an "o" represents the point estimate (using the data from the first observation through the observation number under consideration) of the parameter using the randomly chosen locations and the observations from those locations. The "x" represents the point estimate of the parameter using observations taken at locations chosen through BAS. The bars are "error bars" and extend one standard error above and below the estimated parameter, based on the data collected at locations from the appropriate experiment. The horizontal line represents the true value of the parameter in our simulation.

Note that in Figure 6 and Figure 7, the error bars rarely overlap the true value of the parameter. This can be attributed to two factors. In large part, this is due to the fact that they are engineering "error bars" with a length of one standard error beyond the point estimate. Traditional 95% statistical confidence intervals based on two standard errors would in virtually every case overlap the true values. Additionally, these are cumulative plots, in which the same data are used, adding observations to form the parameter estimates as one moves to the right in each figure. Thus the point estimates and error bars are dependent upon one another within a figure, and not independent point estimates and confidence intervals.

Finally, we see that the estimates using BAS (to select the points from which to take observation) are generally closer to the truth than when we use random points to take observations, and more importantly the standard errors associated with any given number of observations are much smaller.

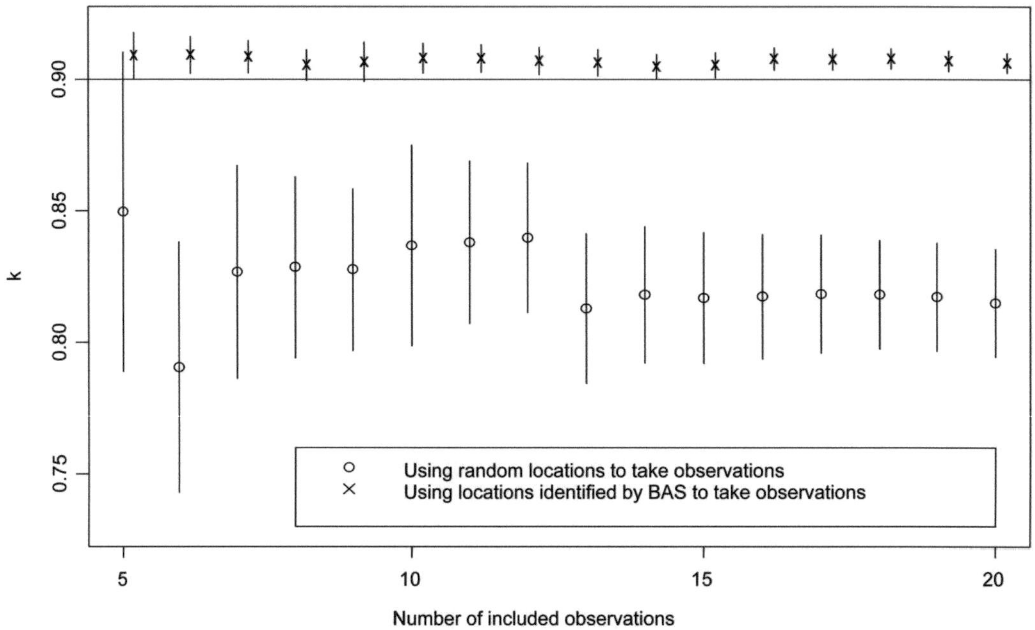

Fig. 6. Estimates and Error Bars for k .

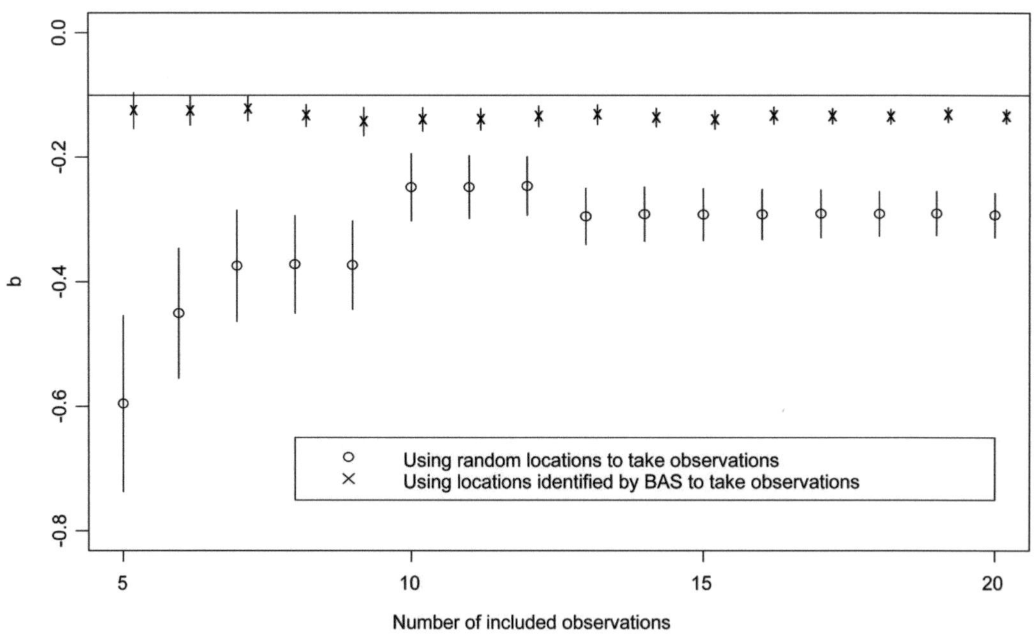

Fig. 7. Estimates and Error Bars for b .

7. Conclusions and future work

Our initial results have shown that BAS is highly efficient compared to random sampling. The rate at which the standard errors, or the error bars, are reduced is much quicker, and hence the significant amount of information is found more quickly compared to other traditional methods. We have also shown that this methodology performs well even in the absence of any preliminary observations. Further simulation has shown evidence that BAS can be three times as efficient as random sampling. This efficiency amounts to savings of time and money during actual data collection and analysis.

In addition to the application discussed in this paper, the theoretical framework presented here is generic and can be applied directly to other applications, such as, military, medical, computer vision, and robotics.

Our proposed framework is based on the multivariate normal distribution. The immediate extensions of this framework will be to accommodate non-normal parameter estimate distributions. As part of our future study, we intend to employ sampling methodologies using Bayesian Estimation Methods for non-normal parameter estimate distributions. We also intend to use cost effectiveness as an additional variable. In this initial work, the focus was to identify the viewpoints that would give us the most information. However, it is not always feasible or efficient to send the helicopter to this next "best" location. As part of our future work, we intend to identify the next "best efficient" location for the helicopter from which it should collect data.

8. Acknowledgements

Much of this work was performed as part of the CAMCOS (Center for Applied Mathematics, Computation, and Statistics) student industrial research program in the Department of Mathematics at San Jose State University, San Jose, California, 95192-0103, USA.

This work was supported in part by the NASA Ames Research Center under Grant NNA05CV42A. We are appreciative of the help of Kevin Wheeler, Kevin Knuth and Pat Castle, who while at Ames suggested these research ideas, provided background materials, and assisted in securing funding.[1]

9. Appendix

In section 5, we noted that maximizing the entropy of the experiment is equivalent to maximizing the quantity $\left| \frac{1}{\sigma^2} X'X + R \right|$ originally given in (12), where R and σ come from a Bayesian prior belief which in this case could be interpreted as the estimates based upon a regression of the data collected up to the given point in time. The matrix X contains a row(s) with functions of known position(s) of the sun (θ_s, ϕ_s) and unknown positions of the

[1] Before this research could be implemented in practice, two of the above mentioned liaisons left NASA and the then current funding expired.

viewpoint/helicopter (θ_v, ϕ_v) which are then the parameters over which the objective function is maximized. In this setting, X is based on future observations. If the X matrix is filled with the appropriate functions (as defined in (8) and (9)) from all of the observations already taken, each observation being associated with one row in the X matrix, where we might call this matrix X_{obs} where the subsript "obs" reminds us that this matrix is based on previous observations rather than future observations, then the quantity

$$\frac{1}{\hat{\sigma}^2} X'_{obs} X_{obs} \tag{13}$$

is known as the observed Fisher Information. In this setting, the observed Fisher Information matrix is the inverse of the estimated variance matrix of the parameters (Cassella and Berger, 2002; Weisberg, 2005; Montgomery et al., 2006), and thus (13) **IS** the matrix R that is used above after the intial set of observations are taken and estimates can be formed.

Any linear algebra textbook will show that stacking two conformable matricies, say

$$\aleph = \begin{bmatrix} X_{obs} \\ \cdots \\ X \end{bmatrix},$$

will give the result $\aleph'\aleph = X'_{obs} X_{obs} + X'X$ and thus, after obtaining intial estimates of the variances from the regression and using them for our prior R, and also estimating the error variance σ^2, our objective function given in (12) can be written as $\left| \frac{1}{\hat{\sigma}^2} X'X + R \right| =$

$$\left| \frac{1}{\hat{\sigma}^2} X'X + \frac{1}{\hat{\sigma}^2} X'_{obs} X_{obs} \right| = \left| \frac{1}{\hat{\sigma}^2} (X'X + X'_{obs} X_{obs}) \right| = \left(\frac{1}{\hat{\sigma}^2} \right)^3 \left| (X'X + X'_{obs} X_{obs}) \right| = \frac{1}{\hat{\sigma}^6} \left| \aleph'\aleph \right|$$

where the third equality follows from the fact that for scalar a and a matrix A of dimnension nxn, $|aA| = a^n |A|$, and that $X'X + X'_{obs} X_{obs}$ is a 3x3 matrix since we are estimating three parameters. Noting that we have no control over $\hat{\sigma}^2$, the estimated variance of the error terms, we see that maximizing (12) is equivalent to maximizing $|\aleph'\aleph|$, where the first, say, n rows of $\aleph$ are based on the functions of $\theta_{si}, \theta_{vi}, \phi_{si}, \phi_{vi}$ in (8) and (9) for observations which have already been taken, and one final row in which the position of the sun, θ_s, ϕ_s, is known and the maximization occurs over the two parameters θ_v, ϕ_v which will indicate the new position to which the helicopter should move in order to take the next observation.[2]

---

[2] Of course, this assumes that we are using prior data to define our Bayesian prior for R. If we have a strong belief for some other prior distribution of the parameters (and thus a different variance matrix R^{-1}), then the above will not hold.

In summary, we find that maximizing the expected Shannon Information, i.e. the entropy, of an experiment, in which the experiment is associated with a regression fit where the errors are assumed to be Normally distributed, is equivalent to maximizing the determinent of the Fisher Information associated with that experiment. Further study is waranted in cases which are not regression based and/or where errors are not neccessarily Normally distributed.

10. References

Arbel, T. & Ferrie, F. (1999). Viewpoint selection by navigation through entropy maps, *Proceedings of 7th IEEE International Conference on Computer Vision*, pp. 248–254, 0-7695-0164-8, Kerkyra Greece, September 1999, IEEE Computer Society, Los Alamitos, CA, USA

Cassela, G. & Berger, R. (2002). *Statistical Inference*, Duxbury, 0-534-24312-6, Pacific Grove, California

Lucht, W. & Schaaf, C. (2006). *BRDF Explained*, retrieved May 2008, from http://www-modis.bu.edu/brdf/brdfexpl.html

MacKay, D. (1992). A clustering technique for digital communications channel equalization using radial basis function networks. *Neural Computation*, vol. 4, no. 4, pp. 590–604

Montgomery, D., Peck, E. & Vinning, G. (2006). *Introduction to Linear Regression Analysis*, Wiley, 0-471-75495-1, Hoboken, New Jersey

Nicodemus, F, (1970). Reflectance Nomenclature and Directional Reflectance and Emissivity. *Applied Optics*, vol. 9, no. 6, June 1970, pp. 1474-1475

Rahman, H., Pinty, B. & Verstaete, M. (1993). A coupled surface-atmosphere reflectance (CSAR) model. Part 1: Model description and inversion on synthetic data, *Journal of Geophysical Research*, vol. 98, no. 4, pp. 20779–20789

Sebastiani P. & Wynn, H. (2000). Maximum entropy sampling and optimal Bayesian experimental design, *Journal of Geophysical Research*, Vol. 62, Part 1, pp. 145-157

Sebastiani, P. & Wynn, H. (2000). Experimental Design to Maximize Information, Bayesian Inference and Maximum Entropy Methods, *Science and Engineering: 20th Int. Workshop AIP Conf. Proc.*, pp. 192-203, , Gif sur Yvette, France, vol. 568

Shannon, C. (1948). A Mathematical Theory of Communication, *Bell System Technical Journal*, vol. 27, pp. 379-423, 623-656, July, October, 1948

Vazquez, P. Feixas, P., Sbert, M. & Heidrich, W. (2001). Viewpoint selection using viewpoint entropy, *Proc. of Vision, Modeling, and Visualization*, pp. 273–280, Germany, 2001

Weisberg, S. (2005). *Applied Linear Regression*, Wiley, 0-471-66379-4, Hoboken, New Jersey

Whaite, P. & Ferrie, F. (1994). Autonomous exploration: Driven by uncertainty, *Proc. of the Conf. on Computer Vision and Pattern Recognition*, pp. 339–346, California, 1994

Wheeler, K. *Parameterization of the Bidirectional Surface Reflectance*, NASA Ames Research Center, MS 259-1, unpublished

Zhang, Z., Kalluri, S. & JaJa, J. (1997). *High Performance Algorithms for Global BRDF Retrieval*, Technical Report, Institute for Advanced Computer Studies, University of Maryland, College Park

MO-Miner: A Data Mining Tool Based on Multi-Objective Genetic Algorithms

Gina M. B. de Oliveira[1], Luiz G. A. Martins[1] and Maria C. S. Takiguti[2]
[1]*Universidade Federal de Uberlândia (UFU)*
[2]*Universidade Presbiteriana Mackenzie*
Brazil

1. Introduction

Data mining (DM) is the process to extract previously unknown and implicit information from large databases. Several techniques have been used to discover such kind of knowledge; most of them derived from machine learning and statistics. The majority of these approaches focus on the discovery of accurate knowledge. However, this knowledge should be useless if it does not offer some kind of surprisingness to the final user. Based on this idea, some investigations were started recently with the aim of extracting accurate and interesting information from datasets. In this sense, data mining can be faced as a multi-objective problem.

Different machine learning approaches have been employed to perform data mining tasks. Most of them are based on evolutionary methods, like genetic algorithms (Goldberg, 1989). Another advantage of genetic algorithms is that they can be adapted to treat multi-objective problems in a Pareto sense. Various multi-objective genetic algorithms have been proposed in the literature, like the method known as *Non-dominated Sorting Genetic Algorithms* (NSGA) (Srinivas & Deb, 1994).

The task performed in a data mining process depending on what kind of knowledge someone needs to extract. The main types of tasks performed by DM algorithms are classification, association, clustering, regression, sequences analysis, summarizing and dependence modelling. Classification task searches for the knowledge able to predict the value of a previously defined goal attribute based on other attributes. This knowledge is often represented by IF-THEN rules and it is the most investigated data mining task. Dependence modelling task can be seen as a generalization of classification. It also aims to discover rules able to predict the goal attribute value, from values of the prediction attributes. However, in dependence modelling, there are more than one goal attribute. Initially, a small set of goal attributes is specified, whose prediction is considered interesting. These attributes may occur in the consequent or in the antecedent parts of the rule. The others attributes occur only in the antecedent.

A multi-objective evolutionary data mining environment named *MO-miner* was implemented based on the family of algorithms called non-dominated sorting genetic algorithms (NSGA). The two desirable properties of the rules being mined - accuracy and interestingness - are simultaneously manipulated. *MO-miner* keeps the metrics related to

these properties separated during the evolution, as different objectives used in the fitness calculus in a Pareto-based approach. This chapter describes the investigation performed using *MO-miner* in the mining of accurate and interesting rules for the dependence modelling task. *MO-miner* was tested in two public domain databases named *Zoo* and *Nursery*. The results obtained by *MO-miner* had been compared with those generated by a standard GA in order to identify the benefits related to the multi-objective approach.
Section 2 presents a brief review on multi-objective genetic algorithms with emphasis on the method used here, the NSGA proposed by Srinivas and Deb (1994). The major data mining tasks are presented in Section 3. This section also presents a review of recent works which have also investigated the application of multi-objective approaches to data mining tasks. Section 4 describes the multi-objective environment named *MO-miner*, which was implemented to perform the dependence modelling task. The two data sets used to evaluate the environment are characterized in Section 5. Section 6 presents the results and their analysis regarding the application of *MO-miner* and a comparative single-objective environment to the data sets. Finally, Section 7 presents the major conclusions of this work.

2. Evolutionary algorithms and multi-objective problems

Evolutionary algorithms (EAs) are computational search methods gleaned from biological evolution. In these methods a population of candidate solutions is generated by random means, and then evaluated in terms of how close they are to the desired solution of the problem at issue. According to the score derived from that evaluation, a selection procedure then determines which subpopulation will be used as the basis for the creation of another population of candidate solutions. The new population is obtained through the application of genetic operators over the selected subpopulation, the expectation being that the new population be, in average, better than its predecessor. This process is repeated over generations, until a population emerges with a satisfactory candidate solution, or until some predefined computational constraint (such as a certain number of generations) is achieved. Evolutionary methods enable us to obtain solutions to problems for which there are no known exact algorithms or, for those where an algorithm exists, but the solution requires a long time of processing to be found (Mitchell, 1997). The genetic algorithm (GA) proposed by Holland is the most well-known EA (Goldberg, 1989). The success of GA search is due to the symbiosis of three operators: selection, crossover and mutation.

2.1 Pareto optimum and multi-objective genetic algorithms

Many real world problems involve simultaneous optimization of multiple objectives, so that it is not always possible to achieve an optimum solution in respect to all of them individually considered. In this kind of problem, there is a set of solutions better than all the other solutions in the search space (Srinivas & Deb, 1994), which is named the Pareto optimum or set of non-dominated solutions.
Suppose that there are *N* objectives $f_1, f_2, \dots f_N$ to be simultaneously optimized. A solution *A* is said to be dominated by another solution *B*, or *B* dominates *A*, if *B* is better than *A* in relation to at least one of the objectives f_i, and is better than or equal to *A* in relation to all other objectives $f_1, \dots f_{i-1}, f_{i+1} \dots f_N$. Two solutions A and B are non-dominated in relation to each other if A does not dominate B and B does not dominate A. For example, suppose that functions f1 and f2 in Figure 1 must be simultaneously maximized. One can affirm that solution A is better than solutions C and D; that is, C and D are dominated by A. However,

in the case of solutions A and B, it is not possible to affirm which one is the best. Therefore, one can say that solutions *A* and *B* are non-dominated and they dominate solutions *C* and *D*.

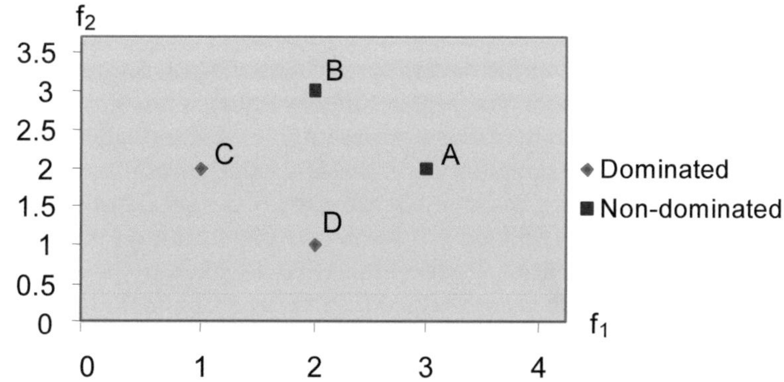

Fig. 1. Example of dominated and non-dominated solutions when maximizing f1 and f2.

The Pareto optimum is the set of non-dominated solutions considering the entire search space, that is, any candidate solution that is better than one of the solutions from the Pareto optimum in respect to one of the objectives, is guaranteed to be worse with respect to another objective.

Multi-objective evolutionary methods try to find this solution set by using each objective separately, without aggregating them as an unique objective (Coello, 1999). Some of the most traditional multi-objective evolutionary methods are VEGA (Vector Evaluated Genetic Algorithm; Shaffer, 1985), NSGA (Non-dominated Sorting Genetic Algorithms; Srinivas & Deb, 1994), MOGA (Multiple Objective Genetic Algorithm; Fonseca & Fleming, 1993), and NPGA (Niched Pareto Genetic Algorithm; Horn & Nafpliotis, 1993); the NSGA is briefly described below, insofar as it provided the procedures used in the present work.

More recently, new evolutionary methods have been developed, characterised by the fact that - unlike the methods mentioned above - an elitist strategy was incorporated in them, leading to methods such as NSGAII (Non-dominated Sorting Genetic Algorithm II; Deb *et al.*, 2000), SPEA (Strength Pareto Evolutionary Algorithm; Zitzler & Thiele, 1999), SPEA2 (Strength Pareto Evolutionary Algorithm; Zitzler *et al.*, 2001), PAES (Pareto Archieved Evolution Strategy; Knowles & Cornes, 1999) and PESA (Pareto Enveloped-based Selection Algorithm; Knowles *et al.*, 2000).

2.2 Non-dominated sorting genetic algorithm

The Non-dominated Sorting Genetic Algorithm (NSGA) was proposed by Srinivas and Deb (1994) and is based on the concept of non-domination in a Pareto sense as explained previously.

The basic difference of NSGA in relation to a standard GA is the way in which the individuals are evaluated and selected. Basically, in order to obtain the fitness value of an individual, instead of using the fitness components associated with each objective involved in the problem, these components are used to rank the individuals according to their degree of domination over the others in the population; this measure of domination is used to define the fitness value that guides the action of the genetic operators and the selection process.

In order to work out that ranking, the population is organised in several layers of non-domination, the outermost layer representing the non-dominated individuals, and the innermost containing the most dominated ones. Initially, all individuals in the population that are non-dominated are separated as the first, outermost layer, and a (dummy) fitness value is assigned to them, whose role is simply to characterise their degree of domination over the others in the population. The highest fitness value is assigned to the outermost layer, as the individuals in it exhibit the highest degree of domination (the actual value assigned is proportional to the population size; this and other details are being omitted here for brevity). Then, the remaining individuals are classified again, also based on the non-domination criterion, and the second layer is formed with a (dummy) fitness lower than the first one. This process continues until all individuals are classified in their respective layers.
Once the non-domination layers were obtained, in order to maintain the diversity of the individuals inside each layer, a sharing function method is used to assign fitness to each individual (Goldberg, 1989). The basic idea of the sharing function is to reduce the fitness of the individuals that have many neighbours in the same layer. The fitness reduction is by a factor that depends on the number and proximity of neighbouring individuals, so that, the larger the number of individuals lying close to each other, the higher their fitness reduction; analogously, the more isolated an individual, the smallest the influence of the fitness reduction. As a consequence, the sharing function induces the formation of niches, thus increasing diversity in the population. The sharing function value between two individuals in the same border is given by:

$$Sh(d_{ij}) = \begin{cases} 1-\left(\dfrac{d_{ij}}{\sigma_{share}}\right)^{\alpha}, & if \ d_{ij} < \sigma_{share} \\ 0, & otherwise \end{cases} \tag{1}$$

where, d_{ij} is the distance between the phenotypes of individuals i and j of the same front; σ_{share} is the maximum distance allowed to two individuals being considered members of the same niche; and α is the power that defines as the sharing degree decreases with the proximity.
A *niche counter* parameter is calculated through the addition of the value of the sharing function mentioned above for all individuals of the current front. The shared fitness values of each individual are calculated by the division of its dummy fitness by its *niche counter*.
After the non-domination classification and fitness assignment are performed, a proportional selection scheme is used so that the outer the layer an individual is in, the likelier its chance to reproduce. Stochastic remainder selection was the actual scheme used, preserved here because it was present in the original proposition of the NSGA. In this method, the next generation is formed by taking as many copies of every individual in the current population as the result of the integer division of its fitness value by the average fitness of the population. Subsequently, if the population needs be filled in, the fractional part of the latter division is used (Goldberg, 1989). The other steps of the NSGA are very similar to the simple GA (Goldberg, 1989).
Figure 2 shows a schematic flow chart of the NSGA.
Therefore, no elitist strategy is used in NSGA, in the sense of an external archive that would retain the best non-dominated solutions found along the evolutionary process, that would directly pass on to the next generation, regardless of how well fitted those individuals might

already be. This is in contrast with various more recent, elitist, multi-objective evolutionary methods (Coello *et al.*, 2002), such as NSGA-II (Deb *et al.*, 2000), SPEA (Zitzler & Thiele, 1999), SPEA2 (Zitzler *et al.*, 2001), PAES (Knowles & Cornes, 1999) and PESA (Knowles *et al.*, 2000).

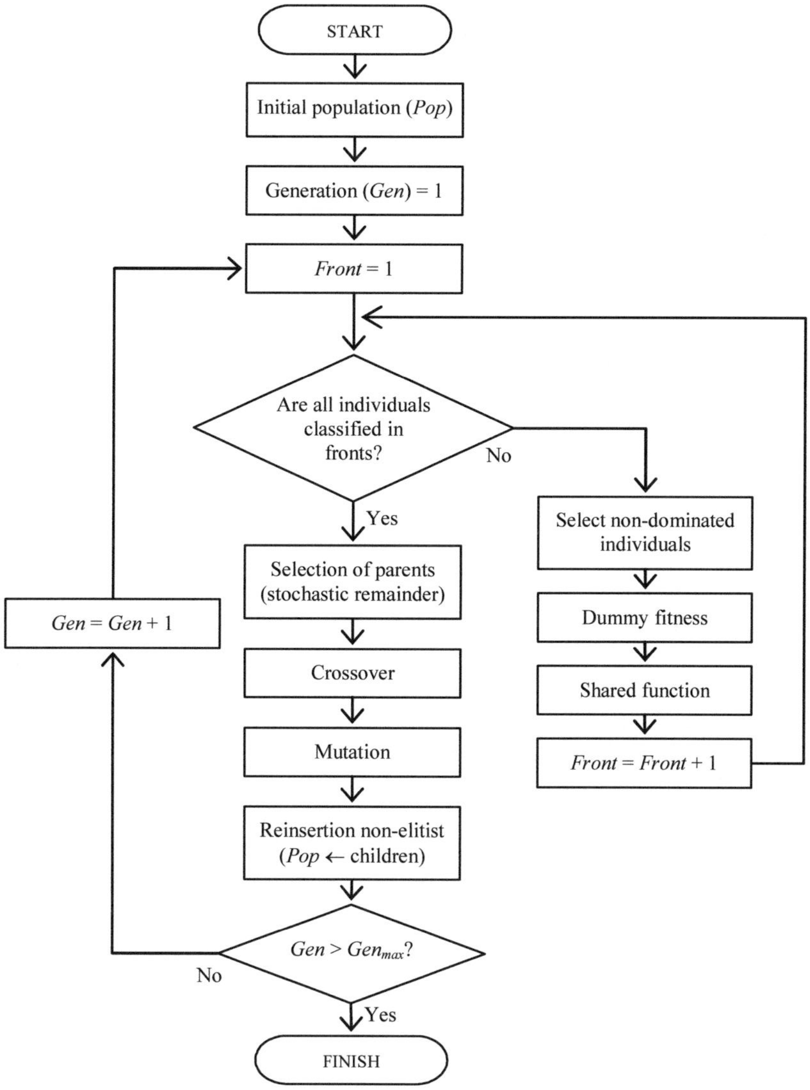

Fig. 2. NSGA's flow chart

3. Data mining

3.1 Knowledge discovery in databases

Data mining (DM) is the process of identification and extraction of useful information in typically large databases. DM aims to automatically discover the knowledge that is not easily perceivable. Data mining is considered the central step of the high-level process called

Knowledge Discovery in Databases (KDD). KDD includes includes the application of several preprocessing methods aimed at facilitating the application of the data mining algorithm and postprocessing methods aimed at refining and improving the discovered knowledge.

Data preprocessing aims to assure the data quality and to reduce the problem complexity, consolidating the relevant information. It commonly includes the following steps (Freitas, 2002):

i. data integration: involves, for instance, removing inconsistencies in attribute names or attribute value names between data sets of different sources.

ii. data cleaning: may involve detecting and correcting errors in the data, filling in missing values, etc.

iii. discretization: consists of transforming a continuous attribute into a categorical (or nominal) attribute, taking on only a few discrete values.

iv. attribute selection: consists of selecting a subset of attributes relevant for classification, among all original attributes.

Data postprocessing goal is to improve the understanding of the knowledge discovered, validating it through quality measures of the solution and the data analyst perception. It can be applied, for example, to simplify the discovered rule set by removing some rules and/or rule conditions, in order to improve knowledge comprehensibility for the user. Another goal of postprocessing can be to extract a subset of interesting rules, among all discovered ones. Methods for selection of interesting rules can be divided into subjective methods, user-driven and domain-dependent; and objective ones.

3.2 Data mining tasks

There are different types of tasks associated to data mining process. Each task can be thought of as a particular kind of problem to be solved by a data mining algorithm.The main types of tasks performed by DM algorithms are:

i. classification: prediction of the value (the class) of a user-specified goal attribute based on the values of other attributes, called the predicting attributes.

ii. association: discovery of association rules typically extracted from databases in which each record consists of a set of binary attributes called items. An association rule is a relationship of the form IF X THEN Y, where X and Y are sets of items and $X \cap Y = \varnothing$.

iii. clustering: creation of groups (clusters) determined by the data set, based on similarity measures or probabilistic models.

iv. Regression: association of an item to one or more prediction variables of real values.

v. sequences analysis: determination of the sequential characteristics, for example, in data with time dependence, which idea is the extraction and register of deviations and trends in the time.

vi. summarizing: description in compact form of a data subgroup.

vii. dependence modeling: prediction of the values of several user-specified goal attributes based on the values of other predicting attributes, being that a goal atribute X can be a predictive atribute of another goal atribute Y.

Dependence modelling is the task investigated in this work. The most popular DM tasks are classification and association. Classification is the most investigated one. It searches for the knowledge able to predict the value of a previously defined goal attribute based on other attributes (prediction attributes). This knowledge is often represented by IF-THEN rules.

Dependence modeling task can be seen as a generalization of the classification task. Dependence modeling task also aims to discover rules able to predict the goal attribute value, from values of the prediction attributes. However, in dependence modeling, there are more than one goal attribute. Initially, a small set of goal attributes whose prediction is considered interesting is specified. These attributes may occur in the consequent part of the rule (THEN) or in the antecedent part of the rule (IF). The others attributes occur only in the rule antecedent. As a consequence, dependence modeling task has a larger search space than the classification task, when one considers the same data set for both tasks.

Dependence modeling task is also different from the discovery of association rules. In association task, any attribute of the analyzed database can occur in the antecedent or in the consequent part of a rule. By other hand, only few pre-selected attributes - the goal attributes - can occur in the consequent part when the dependence modeling is considered. Also, in dependence modeling task, these same goal attributes can occurs in the antecedent part of a rule.

3.3 Evolutionary algorithms applied to data mining

Depending on the goal of the related mining process, a different class of DM algorithm is more appropriate. Some of the main techniques used in data mining are neural networks and decision trees. Evolutionary methods, especially the genetic algorithms (GA) (Goldberg, 1989), come also being investigated in several DM problems. Accordinly to Freitas (2002), the main motivation for using GAs in the discovery of high-level prediction rules is that they perform a global search and cope better with attribute interaction than the greedy rule induction algorithms often used in data mining.

The most relevant aspects in GA specification are the individual's encoding and the fitness function, which must represent the mined information and the quality of this information, respectively.

Two approaches have been used to represent IF-THEN rules as GA individuals: Michigan and Pittsburgh (Freitas, 2002). In the Michigan approach, each individual represents a single rule and the population as a whole represents a set of rules. In Pittsburgh approach, each individual represents a full set of rules. The kind of rule which is wanted to discover and the kind of data mining task addressed in the application can determine the choice between these two approaches. For example, in a classification task it is usually to evaluate the quality of the rule set as a whole, rather than the quality of a single rule. Therefore, the interaction among the rules is important indicating the Pittsburgh approach as more appropriated. The Michigan approach might be more adequate in other kinds of data mining problems as in a task where the goal is to find a small set of high-quality prediction rules, and each rule is often evaluated independently of other rules. Another aspect important to this choice is that using the Michigan approach, GA are more suitable for a good convergence as the individual encoding is simpler than in Pittsburgh's. In this work, the Michigan approach was adopted.

The evaluation of each individual is based on the metrics used to measure the quality of the rule codified. Several metrics can be used but usually only those related to the prediction accuracy of the rules are used, such as confidence factor, coverage, sensitivity and specification. For example, in (Fidelis *et al.*, 2000) the quality of the individual (rule) is measured by multiplying specificity and sensitivity. Freitas (2002) has pointed out the importance to include other metrics in the rules evaluation, such as, comprehensibility and interestingness (Carvalho *et al.*, 2003). This raises the question of how to evaluate the trade-off between the accuracy of a model and other quality criteria. A typical example is a scenario where one wants to maximize both the accuracy and the simplicity of a classification model. This scenario configurates the data mining task as a multiobjective problem. The "conventional" approach to deal with this scenario is transforming the multi-objective problem into a single-objective one using a weighted formula. For example, a standard genetic algorithm was used by Noda and collaborators (1999) to search for rules evaluated by two metrics, one related to interestingness and another related to accuracy. Genetic algorithm aggregates theses two metrics into a single fitness, through of a pondered average.

Freitas (2004) presented a critical analysis about the three main multi-objective approaches to cope with data mining tasks:

i. Transforming the original multi-objective problem into a single-objective one by using a weighted formula, as the single-objective GA previouly cited and described in (Noda *et al.*, 1999).

ii. Manipulating simultaneous and independent objectives with the use of Pareto-based GA, as the approach investigated in this work and described in the next section.

iii. The lexicographical approach, where the objectives are ranked in order of priority, as the work described in (Gonçalves *et al.*, 2005) and (Pila *et al.*, 2006).

Although the weighted formula approach is the most popular approach in the data mining literature, Freitas pointed out some drawbacks related to this approach. One of these drawbacks is that it mixes different non-commensurable model-quality criteria into the same formula. In particular, this has the disadvantage of producing model-quality measures that are not very meaningful to the user, going against the principle that in data mining discovered knowledge should be not only accurate but also comprehensible to the user. This and other drawbacks of the weighted-formula approach are avoided by the Pareto approach.

Pareto-based multi-objective GAs have been previously investigated in DM tasks, as the algorithms used in the search for nonlinear models of direct marketing in (Bhattacharyya, 2000) and to select characteristics for non-supervised learning in (Kim *et al.*, 2000). In (Iglesia *et al.*, 2003), an approach based on the NSGAII is employed in the classification task (only one goal attribute), using metrics related to the rules precision: accuracy and covering. A Pareto-based GA was implemented to mine rules from market-basket type databases (association task), using metrics for accuracy, comprehensibility and degree of interestingness (Ghosh & Nath, 2004). An approach based on the NSGAII was also applied to the association task, for the simultaneous maximization of covering and confidence metrics (Ishibuchi *et al.*, 2006a). Other related works that investigate multi-objective data mining are references (Francisci & Collard, 2003; Pappa *et al.*, 2004; Lane & Gobet, 2005; Ishibuchi *et al.*, 2006b; Berlanga *et al.*, 2006).

4. *Mo-miner*: a multi-objective data mining environment

A data mining environment was implemented to obtain interesting and accurate rules. It was named *MO-miner* (Multi-Objective-based miner). It employs a multi-objective GA inspired by NSGA and NSGAII methods that manipulates the metrics accuracy and degree of interestingness of the rules, which are kept as independent objectives. This environment is the major investigation of this work and it was implemented based on the model

described in (Noda *et al.*, 1999), called GA-nuggets. The major modification refers to the multi-objective approach, because GA-nuggets uses a single-objective GA and evaluates the metrics accuracy and degree of interestingness through a pondered average.
The individual's encoding represents one candidate rule (Michigan approach). The set of goal attributes must be predefined. The antecedent of the rule is formed by a conjunction of conditions of the form $A_i = V_{ij}$, where A_i is the *i-th* attribute and V_{ij} is the *j-th* value of this attribute's domain. The consequent consists on a simple condition of the form $G_k = V_{kl}$, where G_k is the *k-th* goal attribute and V_{kl} is the *l-th* value of its domain. Each individual represents the antecedent of a rule. The individual is encoded by a fixed size string, with *n* genes that represent the values of each one of the attributes, as illustrated in Figure 3. This encoding must represent a variable length rule. If an attribute does not occur in the rule antecedent, the value of its gene is "-1". For each individual, the GA automatically chooses the best goal attribute to put in the consequent ($G_k = V_{kl}$), that is, the values *k* and *l* in order to maximize rule's evaluation.

A_1	A_2	A_3	...	A_n
V_1	V_2	V_3	...	V_n

Fig. 3. Individual representation

The fitness function must reflect the goal of mining accurate and interesting rules. To do this, two metrics were used: *AntInt* and *PredAcc. AntInt* is the degree of interestingness of the antecedent part and *PredAcc* is the predictive accuracy of the rule. They are defined following.
The rule's predictive accuracy (*PredAcc*) is an adaptation of the sensitivity indicator, usually employed to measure the predictive ability of any rule:

$$PredAcc = \frac{\left(|A \,\&\, C| - 0.5\right)}{|A|} \tag{1}$$

where $|A\&C|$ is the number of registers in the database that satisfy both antecedent and consequent of the rule (true positives); and $|A|$ is the number of cases that satisfy only the rule antecedent (true positives plus false positives). The subtraction of 0.5 in the numerator was included to penalize rules covering few training examples. This factor is not used when the *PredAcc* is calculated in the test set, because it contains data unseen during training.
The degree of interestingness of the rule antecedent (*AntInt*) is calculated by an information-theoretical measure (Freitas, 1998) given by:

$$AntInt = 1 - \frac{\left(\sum_{i=1}^{n} InfoGain(A_i) \Big/ n \right)}{\log_2\left(|dom(G_k)|\right)} \tag{2}$$

where *n* is the number of attributes in the antecedent; $dom(G_k)$ is the number of possible values of the attribute G_k occurring in the consequent and $InfoGain(A_i)$ is the information gain associated to the attribute A_i, given by:

$$InfoGain(A_i) = Info(G_k) - Info(G_k|A_i) \quad (3)$$

$$Info(G_k) = -\sum_{l=1}^{m_k}(\Pr(V_{kl})\log_2(\Pr(V_{kl}))) \quad (4)$$

$$Info(G_k|A_i) = \sum_{j=1}^{n_i}(\Pr(V_{ij})(-\sum_{l=1}^{m_k}(\Pr(V_{kl}|V_{ij})\log_2(\Pr(V_{kl}|V_{ij})))) \quad (5)$$

where m_k is the number of possible values of the attribute G_k; n_i is the number of possible values of the prediction attribute A_i; $Pr(X)$ denotes the probability of X and $Pr(X|Y)$ denotes the conditional probability of X given Y.
The fitness calculation starts with two objectives, *AntInt* and *PredAcc*, which are kept separated, instead of working with an aggregation of these metrics. Subsequently, these objectives are used to perform the classification of non-dominated fronts, in the Pareto sense, in a similar way to that performed in NSGA. After the initial population of rules was generated, each individual is evaluated with respect to the two metrics by equations (1) and (2). These values are used to classify the individuals in hierarchical non-dominated fronts. Dummy fitness value is attributed to the individuals from each front. Later, these values are adjusted for each rule in a same front through the calculation of the sharing function and by the niche counter (Srinivas & Deb, 1994). After this, each individual finally receives a scalar fitness and the children generation is processed by four steps:

i. Selection of crossover pairs,
ii. Uniform crossover with eventual repair of invalids chromosomes,
iii. Random mutation,
iv. Insertion and removal of genes.

These steps are detailed as follows.
A simple tournament selection is used, where a group (Tour) of individuals is randomly sorted and the one with the highest fitness is chosen for reproduction. It is important to note that a stochastic remainder selection was used in the original proposition of the NSGA. We adopted Tour = 3 in the tournament selection, in which each pair of individuals are selected from two groups of 3. The criterion used to define the tournament winner is: a pair <goal attribute, value> is drafted, the rules participants of the tournament are compared with respect to this pair and the one with the highest fitness is chosen.
After the selection, a uniform crossover operator was used with a probability of 70% for applying crossover to a pair of individuals and another probability of 50% for swapping each attribute's value in the rule antecedents of two individuals. After crossover is done, GA verifies if any invalid individual was created. If so, it uses a repair operator to produce valid individuals.
The mutation operator randomly transforms the value of an attribute into another value belonging to the domain of that attribute. The mutation rate used was 5%.
Besides crossover and mutation, there are the insertion and removal operators. These operators randomly insert or remove a condition in the antecedent. Their probabilities depend on the number of attributes in the antecedent. The insertion operator has a null probability of application when the rule antecedent has the maximum number of attributes (as specified by the user). The removal operator works in opposite way. The used insertion and removal rates are 50% and 70%, respectively.

The GA uses a kind of elitism (Goldberg, 1989) in the population reinsertion after each generation. Elitism strategy keeps the *M* best prediction rules for each pair <goal attribute, value> to the next generation, among *K* possible pairs, corresponding to an elitism factor of $K \times M$. In the experiments described in section 6, the number of rules kept by elitism adopted was 3 rules for each pair.

It is important to note that the original NSGA is a non-elitist method. The fact that we centred the multi-objective approach herein around a non-elitist method, instead of using its more recent, elitist counterpart, NSGA-II (Deb *et al.*, 2000), derives from the problem being tackled. The point is that, the single-objective genetic algorithm presented in (Noda *et al.*, 1999) (GA-nuggets) - that has been used as the basis of our multi-objective GA - relies on a elitist strategy for reinsertion n of individuals in the population; hence, to maintain all the basic structure of that reference GA was a natural constraint, at the same time that using a non-elitist multi-objective algorithm seemed a sensible decision, so as not to overload the resulting algorithm as far as elitism is concerned. But now that the simpler, and to some extent naive multi-objective approach we adopted has shown to be successful, changing the actual multi-objective algorithm to a more sophisticated and well-established one, is clearly a natural step ahead.

A population size of 100 individuals and a number of 50 generations were used. Although *MO-miner* goal is to find a front with several non-dominated solutions, a criterion to select the best rule, in end of GA execution, must be established. We applied the same function adopted in (Noda *et al.*, 1999) as the single-objective fitness function and it consists of two parts. The first one measures the degree of interestingness of the rule as a whole (antecedent and consequent), while the second measures its predictive accuracy. Equation (6) was applied to decide which is the best rule in the highest non-dominated front:

$$Fitness = w_1 \times \frac{AntInt + ConsInt}{2} + w_2 \times PredAcc \tag{6}$$

where *AntInt* is the degree of interestingness of the antecedent, *PredAcc* is the predictive accuracy of the rules; and w_1 and w_2 are user-defined weights and we used the same values adopted by Noda and collaborators (1999), $w_1 = 1/3$ and $w_2 = 2/3$, that gives a greater weight to the rule accuracy.

ConsInt is the degree of interestingness of the consequent part (Noda *et al.*, 1999) and it is calculated by:

$$ConsInt = \sqrt{1 - \Pr(V_{kl})} \tag{7}$$

where $Pr(V_{kl})$ is the relative frequency of the goal attribute value V_{kl}.

Accordingly to Freitas (2004), one of the arguments against Pareto approaches to data mining algorithms is the difficulty of choosing a single "best" solution to be used in practice. A possible criticism of such kind of approach is that the data mining algorithm returns a set of non-dominated solutions, whereas in practice the user will often use a single solution. Therefore, a difficult problem associated with the Pareto approach is to decide how can one choose the "best" non-dominated solution, out of all non-dominated solutions in the final generation. Choosing the best solution based on Equation (6), we define GA's final output in an objective way and we avoid the problem cited above. Besides, the rules are selected with the same final criterion used in the single-objective environment implemented in (Noda *et al.*, 1999) what turns the results obtained by the two approaches suitable for a comparative analysis.

The metric *ConsInt*, given by equation (7), was not employed as an independent objective of our multi-objective approach; because it returns a constant value that depends only on the consequent pair <*goal attribute, value*> at issue and on the number of occurrences of this pair in the database. However, it is used in the final evaluation of the best individual to weight the degree of interestingness of the rule in the same way as done in (Noda *et al.*, 1999). Figure 4 presents a schematic flow chart of *MO-miner* environment.

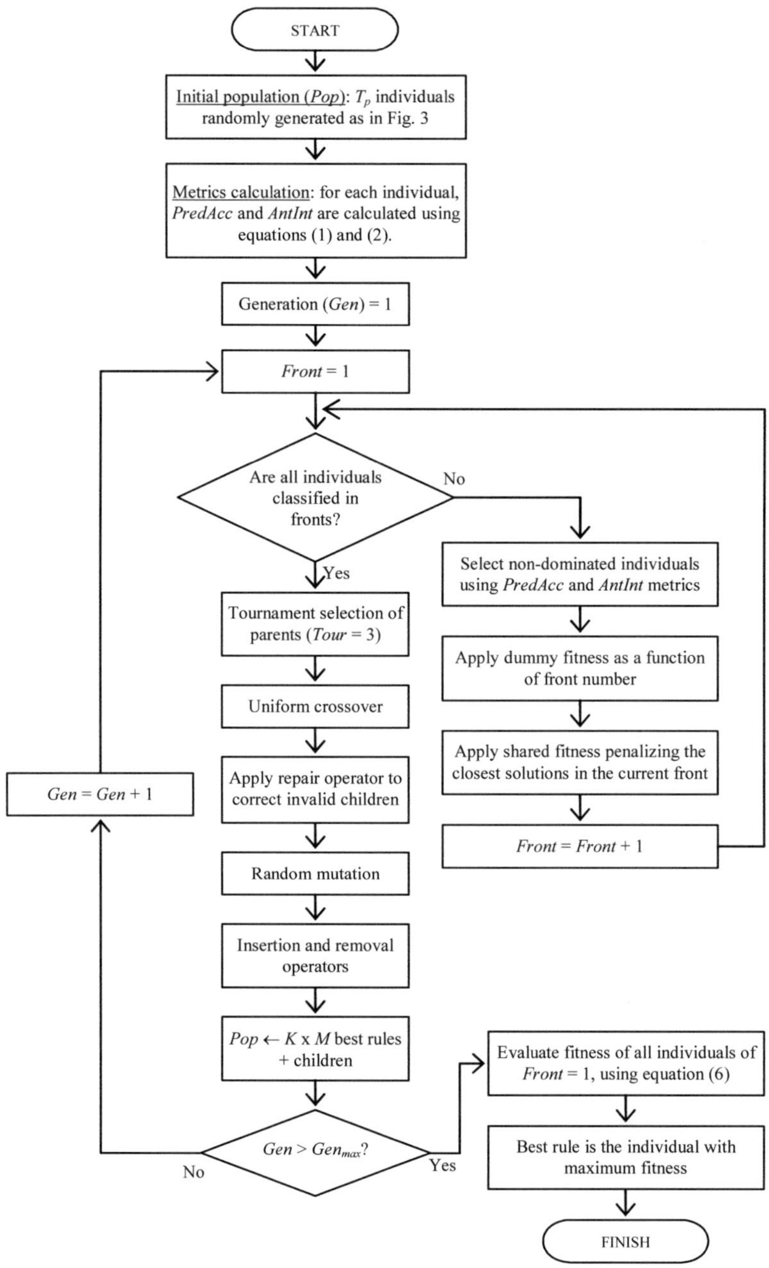

Fig. 4. *MO-miner*'s flow chart

5. Data sets

We used two data sets to evaluate *MO-miner*: *Nursery* and *Zoo*. These data sets were also explored in (Noda *et al.*, 1999). Both databases are public data sets and they were obtained from the UCI repository (www.ics.uci.edu/~mlearn/MLRepository.html) of machine learning databases.

Nursery database was obtained in http://archive.ics.uci.edu/ml/datasets/Nursery, available in this repository since 1997. It was derived from a hierarchical decision model originally developed to rank applications for nursery schools. It was used during several years in 1980's when there was excessive enrollment to these schools in Slovenia, and the rejected applications frequently needed an objective explanation. The final decision depended on three subproblems: occupation of parents and child's nursery, family structure and financial standing, and social and health picture of the family. The database contains 12960 registers and nine categorical attributes: parents, health, form, children, has_nurs, housing, finance, social and recommendation. Three attributes was used as goal: finance, social and recommendation. Finance has two possible values (convenient and inconv), social has three possible values (non-prob, slightly-prob and problematic) and recommendation has five values in its domain (not-recom, recommended, very-recom, priority, spec-prior). It results in rules with ten different consequents:

- Consequent 1: finance = convenient
- Consequent 2: finance = inconv
- Consequent 3: social=non-prob
- Consequent 4: social = slightly-prob
- Consequent 5: social =problematic
- Consequent 6: recommendation = not-recom
- Consequent 7: recommendation = recommended
- Consequent 8: re-commendation = very-recom
- Consequent 9: recommendation = priority
- Consequent 10: recommendation = spec-prior

Zoo is the second public database used and it was also obtained from the UCI repository (http://archive.ics.uci.edu/ml/datasets/Zoo), and it is available in this repository since 1990. Each instance of this database corresponds to an animal. It contains only 101 registers and a total of eighteen categorical attributes: hair, feathers, eggs, milk, backbone, fins, legs, tail, catsize, airbone, aquatic, breathes, venomous, toothed, predator, domestic and type. Three goal attributes was chosen: predator, domestic and type. Predator and domestic are binary attributes and type has seven possible numerical values (from 1 to 7). Therefore, there are eleven different consequents:

- Consequent 1: predator=0
- Consequent 2: predator=1
- Consequent 3: domestic=0
- Consequent 4: domestic=1
- Consequent 5: type=1
- Consequent 6: type=2
- Consequent 7: type=3
- Consequent 8: type=4
- Consequent 9: type=5

- Consequent 10: type=6
- Consequent 11: type=7

6. Experiments

6.1 Single-objective environment

Besides *MO-miner* detailed in Section 4, a second environment was implemented to supply comparative results to evaluate the effectiveness of the multi-objective approach. It was implemented strongly based on the model described in (Noda *et al.*, 1999), called GA-nuggets. It uses a single-objective standard genetic algorithm with the same goal of *MO-miner*: to discover accurate and interesting rules.

Each individual is evaluated through the pondered average described by Equation (6). The components *PredAcc, AntInt* and *ConsInt* of Equation (6) was calculated in the same way as described to *MO-miner* and they are given by equations (1), (2) and (7), respectively. The user-defined weights w_1 and w_2 in Equation (6) was also adopted as in (Noda *et al.*, 1999): $w_1 = 1/3$ and $w_2 = 2/3$.

The single-objective GA also uses a population size of 100 individuals and a number of 50 generations per execution. After the initial population of rules was generated, each individual is evaluated with respect to Equation (6). The children generation is processed as in the multi-objective environment:

i. Simple tournament selection to crossover (*Tour* = 3).
ii. Uniform crossover with eventual repair of invalids chromosomes.
iii. Random mutation.
iv. Insertion and removal of the genes.

Some small modifications had been made in the single-objective GA with respect to the one described in (Noda *et al.*, 1999):

i. We adopted *Tour* = 3 in the tournament selection.
ii. The criterion used to define the tournament winner is: a pair <*goal attribute, value*> is drafted, the rules participants of the tournament are compared with respect to this pair and the one with the highest fitness is chosen.
iii. The number of rules kept by elitism is 3 rules for each pair.

6.2 Experiments description

Two kinds of experiments had been executed with the single-objective GA and with *MO-miner*. The first one used the full data set during GA evolution, getting the better rules to predict each one of the possible consequents $G_k = V_{kl}$. In the second experiment, a cross-validation was used to evaluate the quality of the discovered rules. Hence, the data set is divided into five mutually exclusive and exhaustive partitions. This second kind of experiment is composed by five types of GA evolution, each one using a different partition as test set. All partitions, except the test set, had been used during the rules evolution as training set, corresponding to the data effectively used by the genetic algorithm during rule search. Later, the rules are evaluated concerning their prediction accuracy in the test partition. The final result is defined by the average results considering the five test partitions.

6.3 Results using *Nursery* database

The results of this section was previously discussed in (Oliveira *et al.*, 2008).

6.3.1 Full set experiment

Both single-objective GA and *MO-miner* had been applied to 100% of the registers from *Nursery* database. In each experiment, 10 different GA executions had been performed. Table 1 presents *AntInt* and *PredAcc* values for the best rule obtained in these executions, for each consequent. Figure 5 shows radial charts of these metrics to aid the comparative analysis: (a) *AntInt* and (b) *PredAcc*. In general, both environments had mined rules with a high degree of interestingness and a good accuracy for the majority of the consequents. The single-objective GA found rules with degree of interestingness below of 0.95 in consequents 6, 8, 9 and 10 and *MO-miner* only in consequents 8 and 10. In all the others consequents, *AntInt* exceeds 0.97. Both the single-objective and *MO-miner* had found bad rules (≤ 0.5) for consequents 3, 4 and 7 with respect to accuracy. These results are compatible with those reported in (Noda *et al.*, 1999).

In a comparative analysis, one can observe that the most significant change occurs with respect to the *AntInt*, while the metric *PredAcc* remains practically unchanged for both environments. Concerning degree of interestingness, *MO-miner* returned rules better than single-objective GA in seven consequents, with an increase above to 10% in two of them (consequents 6 and 9). With respect to accuracy, the multi-objective returns rules a little bit higher than the single-objective environment in three consequents (consequents 6, 9 and 10) and equivalent in the others.

We conclude that our multi-objective approach found rules with higher degree of interestingness, without damage on the prediction accuracy.

Table 2 presents the best rules discovered by *MO-miner*. Their quality of prediction is better estimated by the average of *PredAcc* obtained in cross-validation experiments, described as follows.

Consequent	Single-Objective		MO-miner	
	AntInt	*PredAcc*	*AntInt*	*PredAcc*
1	0.999	0.987	0.999	0.987
2	0.999	0.992	0.999	0.992
3	0.993	0.498	0.999	0.498
4	0.993	0.498	0.999	0.498
5	0.996	0.994	0.999	0.994
6	0.891	0.998	0.998	0.999
7	0.978	0.500	0.998	0.500
8	0.921	0.969	0.921	0.969
9	0.907	0.996	0.998	0.998
10	0.908	0.997	0.909	0.999

Table 1. *Nursery* Database: *AntInt* and *PredAcc* obtained with single-objective GA and *MO-miner* in full set experiment.

6.3.2 Cross-validation experiment

The cross-validation method was used to verify the generalization capability of the mined rules. *Nursery* database was divided into five partitions and five different GA evolutions had been performed. In each kind of evolution, a different partition was used as the test set

and the others four partitions are merged and they were used in the evolutionary search for the rules (training). Ten different GA executions had been performed for each test partition. After the best rules for each consequent were obtained, they had been applied to the test partition to evaluate its prediction over examples not seem during the evolution.
Table 3 presents *PredAcc* average for the best rules found for each environment, both in training and in test. Figure 6 shows radial charts of the metric *PredAcc* (a) in training and (b) in test.

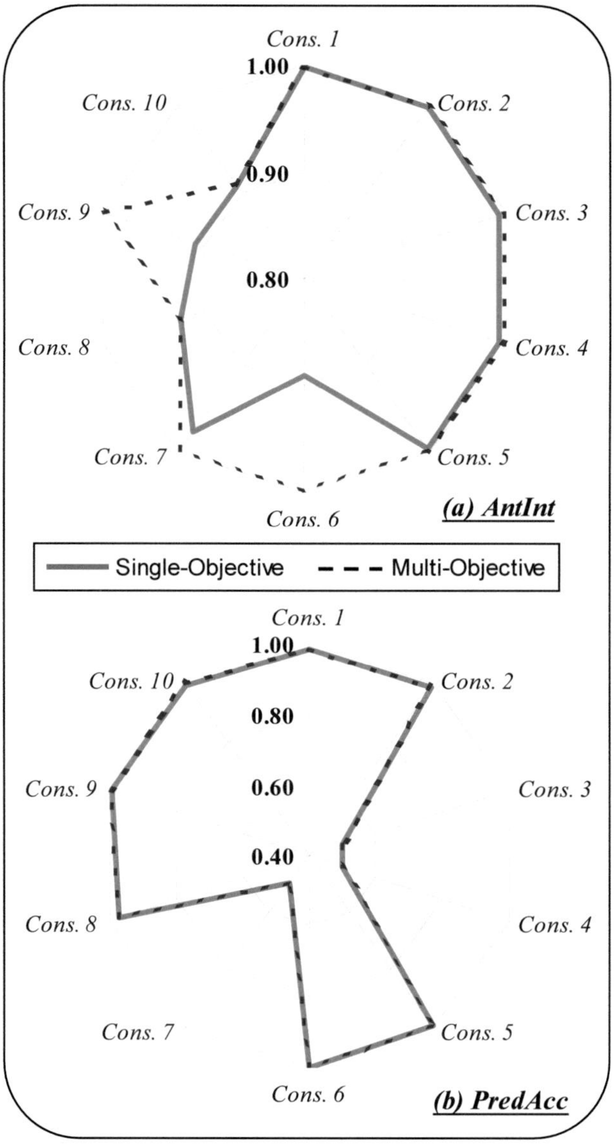

Fig. 5. Single-objective GA and MO-miner results in the full *Nursery* database: (a) *AntInt* (b) *PredAcc.*

IF (*health* = *recommended*) AND (*form* = *foster*) AND (*housing* = *convenient*) AND (*recommendation* = *very_recom*) THEN (*finance* = *convenient*)
IF (*housing* = *convenient*) AND (*social* = *nonprob*) AND (*health* = *recommended*) AND (*recommendation* = *spec_prior*) THEN (*finance* = *inconv*)
IF (*health* = *recommended*) AND (*recommendation* = *very_recom*) THEN (*social* = *non_prob*)
IF (*health* = *recommended*) AND (*recommendation* = *very_recom*) THEN (*social* = *slightly_prob*)
IF (*housing* = *convenient*) AND (*finance* = *convenient*) AND (*health*= *recommended*) AND (*recommendation* = *spec_prior*) THEN (*social* = *problematic*)
IF (*has_nurs* = *proper*) AND (*finance* = *convenient*) AND (*social* = *nonprob*) AND (*health* = *not_recom*) THEN (*recommendation* = *not_recom*)
IF (*parents*=*usual*) AND (*has_nurs*=*proper*) AND (*finance* = *convenient*) AND (*children*=1) AND (*form* = *complete*) AND (*housing* = *convenient*) AND (*social* = *nonprob*) AND (*health* = *recommended*) THEN (*recommendation* = *recommended*)
IF (*parents* = *usual*) AND (*has_nurs* = *less_proper*) AND (*housing* = *convenient*) AND (*finance* = *convenient*) AND (*social* = *slightly_prob*) AND (*health* = *recommended*) THEN (*recommendation* = *very_recom*)
IF (*has_nurs* = *proper*) AND (*housing* = *convenient*) AND (*finance* = *convenient*) AND (*health* = *priority*) THEN (*recommendation* = *priority*)
IF (*has_nurs* = *very_crit*) AND (*finance* = inconvenient) AND (*social* = *problematic*) AND (*health* = *recommended*) THEN (*recommendation* = *spec_prior*)

Table 2. Rules extracted by *MO-miner* from *Nursery* database.

Consequent	Single-Objective		MO-miner	
	Training	Test	Training	Test
1	0.984	0.400	0,985	0,600
2	0.992	0.700	0,991	0,700
3	0.499	0.263	0,499	0,333
4	0.498	0.200	0,498	0,367
5	0.994	0.686	0,993	0,733
6	0.997	0.800	0,999	1,000
7	0.500	0.007	0,500	0,007
8	0.968	0.000	0,972	0.000
9	0.993	0.644	0,999	0,450
10	0.995	0.590	0,998	0,590

Table 3. *Nursery Database: PredAcc* of training and test obtained with single-objective GA and *MO-miner* in cross-validation experiment.

In both environments, the average of *PredAcc* in the training set is close to the values obtained with the full set. However, the average of *PredAcc* in the test set falls substantially, indicating that the rules mined in training do not exhibits good generalization ability. When comparing results of the two environments one can notice that even so the performance has been very similar in the training sets, *MO-miner* returned higher *PredAcc* rules in the test sets. *PredAcc* average measured in test sets for *MO-miner* had been higher in five consequents (1, 3, 4, 5 and 6) and lower in only one (consequent 9). Thus, it is possible to conclude that the rules mined by *MO-miner* present better generalization ability than those obtained by simple-objective GA.

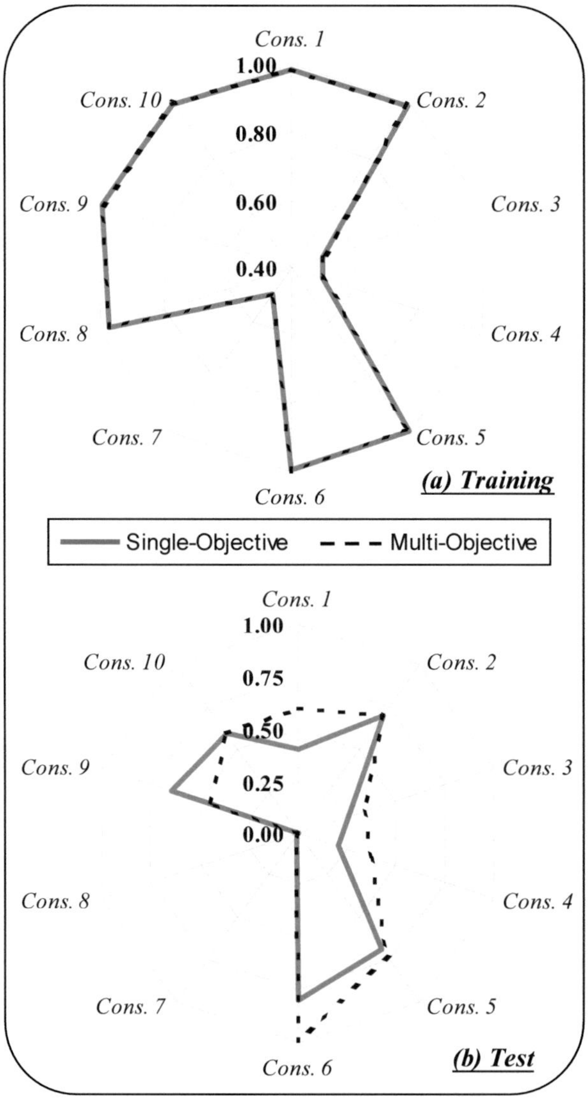

Fig. 6. *PredAcc* average values of the single-objective GA and MO-miner environments in cross-validation experiment using *Nursery* database: (a) training (b) test.

6.4 Results using *Zoo* database

6.4.1 Full set experiment

Table 4 shows the values of *AntInt* and *PredAcc* for the best rules found using the full Zoo database. Both environments (single-objective and *MO-miner*) were able to mine rules with high degree of interestingness and a good accuracy for almost all the consequents; the unique exception was the consequent 4. The single-objective GA found rules with *AntInt* below of 0.9 in consequents 5 to 11 while *MO-miner* only in consequents 7, 8 and 9 (but in all these case *AntInt* was at least 0.84). In the other cases, *AntInt* is above 0.95 for both environments. In respect to accuracy, single-objective and *MO-miner* had found bad rules (< 0.8) only for consequent 4. Again, these results are compatible with (Noda *et al.*, 1999). As observed in Nursery database, the most significant difference occurs with respect to *AntInt*, while *PredAcc* remains practically equivalent for both environments.

Figure 7 shows the radial charts to aid the comparative analysis: (a) *AntInt* and (b) *PredAcc*. Concerning *AntInt*, *MO-miner* returned rules better than single-objective GA in seven consequents, with an increase above to 10% in four of them (5, 6, 10 and 11). With respect to *PredAcc*, *MO-miner* returns rules a little bit higher in four consequents (3, 5, 6 and 8), a lit bit lower in one (consequent 7), and equivalent in the others. As observed in Nursery experiments, the major conclusion is that the multi-objective approach found rules with higher degree of interestingness, without damage on the prediction accuracy.

Table 5 presents the best rules discovered by *MO-miner* using the full Zoo database. Cross-validation experiments described in the next section can better estimate their quality of prediction.

Consequent	Single-Objective		MO-miner	
	AntInt	PredAcc	AntInt	PredAcc
1	0.956	0.929	0.999	0.929
2	0.983	0.958	0.999	0.958
3	0.990	0.980	0.950	0.986
4	0.996	0.500	0.996	0.500
5	0.880	0.984	0.982	0.988
6	0.893	0.971	0.982	0.975
7	0.849	0.875	0.881	0.833
8	0.887	0.958	0.840	0.962
9	0.883	0.875	0.863	0.875
10	0.848	0.937	0.982	0.937
11	0.887	0.937	0.982	0.937

Table 4. *Zoo* Database: *AntInt* and *PredAcc* obtained with single-objective GA and *MO-miner* in full set experiment.

6.4.2 Cross-validation experiment

As performed for the *Nursery* database, the *Zoo* database was divided into five partitions and five different GA evolutions had been performed. For each test partition, ten different GA executions had been performed. The best rules obtained for each consequent had been applied to the test partition to evaluate its prediction.

Table 6 presents *PredAcc* average for the best rules found (over the 5 partitions), both in training and in test. Figure 8 shows radial charts of the metric *PredAcc* (a) in training and (b) in test. The average of *PredAcc* in the training set is close to the values obtained with the full set, as for the single-objective GA as for *MO-miner*. But as observed in *Nursery* experiments, the average of *PredAcc* in the test set falls substantially, indicating that the rules mined in training do not exhibits good generalization ability, for both environments. Once more *MO-miner* returned higher *PredAcc* rules in the test sets even so the performance has been very similar in the training sets. *PredAcc* average measured in test sets for *MO-miner* had been higher in six consequents being this performance significatively in three of them. It is possible to conclude that, as concluded in *Nursery* experiments, the rules mined by *MO-miner* present better generalization ability.

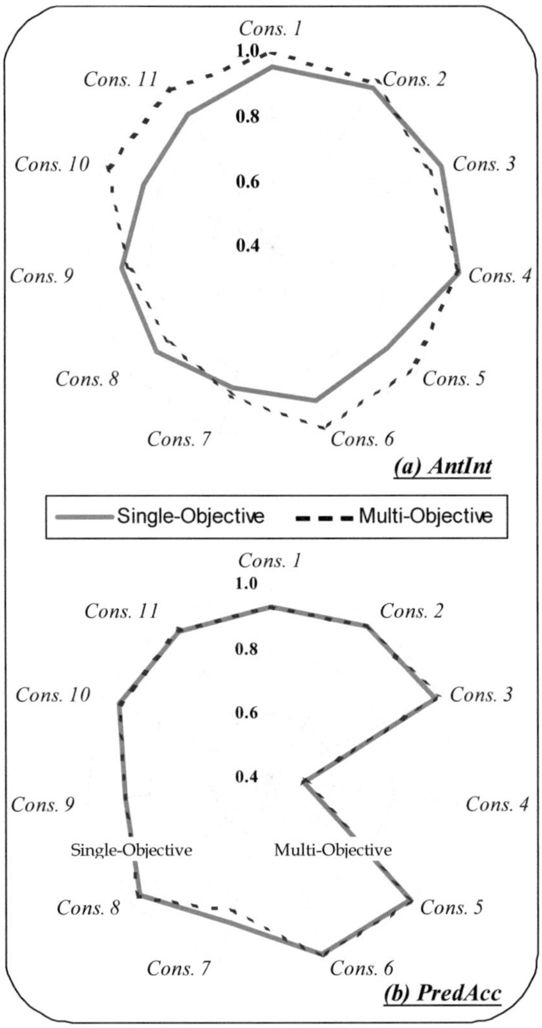

Fig. 7. Single-objective GA and MO-miner results in the full *Zoo* database: (a) *AntInt* (b) *PredAcc*.

IF (*domestic*=1) AND (*catsize*=0) THEN (*predator*=0)
IF (*airbone*=0) AND (*aquatic*=1) AND (*catsize*=1) THEN (*predator*=1)
IF (*hair*=0) AND (*predator*=1) THEN (*domestic*=0)
IF (*eggs*=0) AND (*venomous*=0) AND (*tail*=0) AND (*catsize*=0) THEN (*domestic*=1)
IF (*milk*=1) AND (*venomous*=0) THEN (*type*=1)
IF (*feathers*=1) AND (*venomous*=0) THEN (*type*=2)
IF (*hair*=0) AND (*aquatic*=0) AND (*predator*=1) AND (*toothed*=1) AND (*domestic*=0) THEN (*type*=3)
IF (*eggs*=1) AND (*fins*=1) AND (*domestic*=0) THEN (*type*=4)
IF (*feathers*=0) AND (*airbone*=0) AND (*aquatic*=1) AND (*breathes*=1) AND (*catsize*=0) THEN (*type*=5)
IF(*aquatic*=0) AND (*fins*=0) AND (*legs*=6) AND (*catsize*=0) THEN(*type*=6)
IF (*airbone*=0) AND (*predator*=1) AND (*backbone*=0) AND (*domestic*=0) THEN (*type*=7)

Table 5. Rules extracted by *MO-miner* from *Zoo* database.

Consequent	Single-Objective		MO-miner	
	Training	Test	Training	Test
1	0.919	0.200	0.922	0.433
2	0.947	0.900	0.948	0.933
3	0.979	0.975	0.983	0.978
4	0.471	0.050	0.500	0.080
5	0.980	1.000	0.985	1.000
6	0.964	1.000	0.970	1.000
7	0.842	0.400	0.842	0.430
8	0.948	0.800	0.950	0.800
9	0.833	0.600	0.858	0.600
10	0.915	0.500	0.922	0.600
11	0.917	1.000	0.920	1.000

Table 6. *Zoo Database: PredAcc* of training and test obtained with single-objective GA and *MO-miner* in cross-validation experiment.

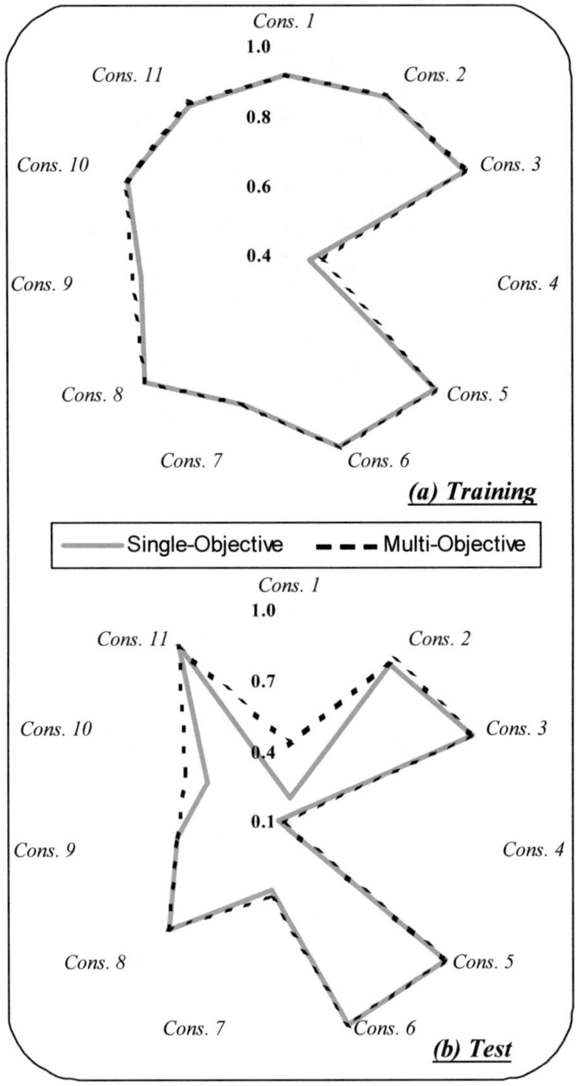

Fig. 8. *PredAcc* average values of the single-objective GA and MO-miner environments in cross-validation experiment using *Zoo* database: (a) training (b) test.

7. Conclusion

The application of a multi-objective GA-based environment named *MO-miner,* inspired by the family of algorithms NSGA and designed to discover accurate and interesting dependence modeling rules, was investigated in this work. Our investigation was performed by applying *MO-miner* to two distinct public data sets named *Nursery* and *Zoo,* composed by 12960 and 101 registers, respectively.

The accuracy and the interestingness of the rules mined by *MO-miner* were compared with those found by a single-objective standard genetic algorithm, based in its turn on the environment implemented by Noda and collaborators (1999) named *GA-nuggets*. The results obtained with our single-objective GA, both for full database and for crossvalidation experiments, are very similar to those reported in (Noda *et al.*, 1999), in which the same data sets were mined.

Comparing the performance of the both environments, it was possible to observe that, when using the full database, they had found rules very similar with respect to accuracy. However, *MO-miner* returned more interesting rules. We believe that this result is due to the kind of rule evaluation performed by *MO-miner*, which kept metrics independents, and to *MO-miner*'s dynamics that drives the search toward the region of the highest values in both metrics. Conversely, in single-objective environment, the balance between the metrics was pre-defined, giving a higher weight to the accuracy with respect to the degree of interestingness.

In respect to the cross-validation experiments, it was evidenced that the results found by *MO-miner* were better, resulting in a lesser accuracy decay between training and test. This indicates that the rules obtained by *MO-miner* have greater generalization than those extracted using the single-objective environment.

Despite the fact that *Zoo* and *Nursery* data sets are very different among each other, specially because the first is a small database and the second is a large one, in the comparative experiments applying the two environments, we arrived at similar conclusions. Using both data sets, *MO-miner* found more interesting rules in the full set experiment and rules with better generalization ability in the cross-validation experiment.

Different paths related to multi-objective genetic algorithms and data mining can be explored as continuity of this study, such as: inclusion of other metrics as comprehensibility (Freitas, 2002); use of others multi-objective techniques (Coello *et al.*, 2002); use of other fitness functions, among others.

8. Acknowledgements

Gina Maira Barbosa de Oliveira thanks Brazilian agencies CNPq and FAPEMIG for financial support.

9. References

Berlanga, F.; del Jesus, M. J.; Gonzalez, P.; Herrera, F. & Mesonero, M. (2006) Multiobjective evolutionary induction of subgroup discovery fuzzy rules: A case study in marketing, *Lecture Notes in Artificial Intelligence (Advances in Data Mining)*, Vol. 4065, pp. 337-349, 2006, Springer-Verlag.

Bhattacharyya, S. (2000). Evolutionary algorithms in data mining: multi-objective performance modelling for direct marketing, *Proceedings of ACM SIGKDD International Conference KDD-2000*, pp. 465-473, 2000.

Carvalho, D. R.; Freitas, A. A. & Ebecken, N.F.F. (2003). A critical review of rule surprisingness measures, *Proceedings of Data Mining IV - Int. Conf. on Data Mining*, pp. 545-556, 2003.

Coello, C. A. C. (1999). A comprehensive survey of evolutionary-based multiobjective optimization techniques, *Knowledge and Information Systems*, Vol. 1, No. 3, 1999, pp. 269-308.

Coello,C. A. C.; Van Veldhuizen, D. A. & Lamont, G. B. (2002). Evolutionary algorithms for solving multi-objective problems, 2002, Kluwer Academic Publishers, New York.

Deb, K.; Agrawal,S.; Pratab, A. & Meyarivan, T. (2000). A fast elitist non-dominated sorting genetic algorithm for multi-objective optimization: NSGA-II, *Proceedings of the Parallel Problem Solving from Nature VI Conference*, pp. 849-858, 2000, Springer, Berlin.

Fidelis, M. V.; Lopes, H. S. & Freitas, A. A. (2000). Discovering comprehensible classification rules with a genetic algorithm, *Proceedings of Congress on Evolutionary Computation (CEC-2000)*, pp. 805-810, 2000, La Jolla, CA, USA.

Fonseca, C. M. & Fleming, P. J. (1993). Genetic algorithms for multiobjective optimization: formulation, discussion and generalization. *Proceedings of the Fifth International Conference on Genetic Algorithms*, Forrest, S. (ed.), pp. 416-423, 1993, San Mateo, California.

Francisci, D. & Collard, M. (2003). Multi-Criteria Evaluation of Interesting Dependencies according to a Data Mining Approach, *Proceedings of Congress on Evolutionary Computation (CEC'2003)*, Vol. 3, pp. 1568--1574, Dec - 2003, IEEE Press, Canberra, Australia.

Freitas, A. A. (1998). On objective measures of rule surprisingness. *Principles of Data Mining & Knowledge Discovery (Proc. 2nd European Symp., PKDD'98). LNAI 1510*, pp. 1-9, 1998, Springer-Verlag.

Freitas, A. A. (2002). A survey of evolutionary algorithms for data mining and knowledge discovery, In: *Advances in Evolutionary Computation*, Ghosh and Tsutsui (ed.), 2002, Springer-Verlag.

Freitas, A. A. (2004). A critical review of multi-objective optimization in data mining: a position paper, *ACM SIGKDD Explorations Newsletter*, Vol. 6, 2004, pp. 2.

Ghosh, A. & Nath, B. (2004). Multi-objective rule mining using genetic algorithms. *Information Sciences*, Vol. 163, No. 1-3, pp. 123-133, 2004.

Goldberg, D. (1989). Genetic Algorithms in Search, *Optimization and Machine Learning*, Addison-Wesley.

Gonçalves, A. S.; Freitas, A. A.; Kato, R. E. & de Oliveira, R. (2005). Using genetic algorithms to mine interesting dependence modeling rules, *Proceedings of 23rd IASTED Int. Multi-Conference on Databases and Applications (DBA-2005)*, 2005.

Horn, J. and Nafpliotis, N. (1993). Multiobjective optimization using the Niched Pareto Genetic Algorithm. *IlliGA1 Technical Report 93005*, University of Illinois at Urbana-Champaign, 1993, Urbana, Illinois.

Iglesia, B.; Philpott, M.; Bagnall, A. & Smith, V. (2003). Data mining rules using multi-objective evolutionary algorithms, *Proceedings of Congress on Evolutionary Computation*, pp. 1552-1559, 2003.

Ishibuchi, H.; Kuwajima, I. & Nojima, Y. (2006a). Multiobjective association rule mining. *Proceedings of PPSN Workshop on Multiobjective Problem Solving from Nature*, 12 pages (Reykjavik, Iceland), 2006.

Ishibuchi, H.; Nojima, Y. & Kuwajima, I. (2006b). Multiobjective Genetic Rule Selection as a Data Mining Postprocessing Procedure, *Proceedings of Genetic and Evolutionary Computation Conference (GECCO'2006)*, Maarten Keijzer *et al.* (ed.), pp. 1591-1592, Vol. 2, 2006, ISBN 1-59593-186-4, ACM Press, Seattle, Washington, USA.

Kim, Y.; Street, W. N. & Menczer, F. (2000). Feature selection in unsupervised learning via evolutionary search, *Proceedings of ACM SIGKDD International Conference KDD-2000*, pp. 365-369, 2000.

Knowles, J. and Corne, D. (1999). The Pareto archived evolution strategy: a new baseline algorithm for multiobjective optimisation, *Proceedings of Congress on Evolutionary Computation*, Angeline, P. J. *et al.* (ed.), pages 98-105, 1999, IEEE Press, Washington, D.C.

Knowles, J.; Corne, D. & Oates, M. (2000). The Pareto-envelope based selection algorithm for multiobjective optimization, *Proceedings of the Sixth International Conference on Parallel Problem Solving from Nature (PPSN VI)*, Schoenauer, M., Deb, K., Rudolph, G., Yao, X., Lutton, E., Merelo, J. J. and Schwefel, H.-P. (ed.), pp. 839-848, 2000, Springer, Berlin.

Lane, P. C. R. & Gobet, F. (2005). Discovering predictive variables when evolving cognitive models, Pattern Recognition and Data Mining, *Lecture Notes in Computer Science*, Vol. 3686, pp. 108--117, 2005, Springer.

Mitchell, M. (1997). An Introduction to Genetic Algorithms, *MIT Press*, Massachusetts.

Noda, E.; Freitas, A. A. & Lopes, H. S. (1999). Discovering Interesting Prediction Rules with a Genetic Algorithm, *Proceedings of Congress on Evolutionary Computation*, pp. 1322-1329, 1999.

Oliveira, G. M. B.; Takiguti, M. C. S. & Martins, L. G. A. (2008). Dependence Modeling Rule Mining using Multi-Objective Generic Algorithms. *Proceedings of 26th IASTED International Conference ARTIFICIAL INTELLIGENCE AND APPLICATIONS (AIA08)*, pp. 278-283, Feb - 2008, Innsbruck.

Pappa, G. L.; Freitas, A. A. & Kaestner, C. A. A. (2004). Multi-Objective Algorithms for Attribute Selection in Data Mining, in *Applications of Multi-Objective Evolutionary Algorithms*, Carlos A. C. Coello and Gary B. Lamont (ed.), pp. 603-626, 2004, World Scientific, Singapore.

Pila, A.; Giusti, R.; Prati, R. & Monard, M. (2006). Multi-Objective Evolutionary Algorithm to Build Knowledge Classification Rules with Specific Properties, *Proceedings of 6th International Conference on Hybrid Inteliggent Systems*, 2006, IEEE Computer Society, NY.

Shaffer, J. D. (1985). Multiple objective optimization with vector evaluated genetic algorithms. In *Genetic Algorithms and their Applications: Proceedings of the First International Conference on Genetics Algorithms*, pp. 93-100, 1985.

Srinivas, N. & Deb, K. (1994). Multiobjective Optimization using nondominated sorting in genetic algorithms, *Evolutionary Computation*, Vol. 2, No. 3, 1994, pp. 221.

Zitzler, E. & Thiele, L. (1999). Multiobjective evolutionary algorithms: a comparative case study and the strength Pareto approach. *IEEE Transactions on Evolutionary Computation*, Vol. 3, No. 4, 1999, pp. 257-271.

Zitzler, E.; Laumanns, M. & Thiele, L. (2001). SPEA2: Improving the {S}trength {P}areto {E}volutionary {A}lgorithm. *EUROGEN 2001: Evolutionary Methods for Design, Optimization and Control with Applications to Industrial Problems*, K. Giannakoglou, D. Tsahalis, J. Periaux, P. Papailou and T. Fogarty (ed.), pp. 12-21, Sep - 2001, Greece.

23

Practical Computer Vision Techniques for Interactive Robotic Heads

Oscar Deniz[2], Javier Lorenzo[1], Modesto Castrillon[1], Luis Anton[1],
Mario Hernandez[1] and Gloria Bueno[2]
[1]*Universidad de Las Palmas de Gran Canaria,*
[2]*Universidad de Castilla-La Mancha*
Spain

1. Introduction

In the last years there has been a surge in interest in a topic called social robotics. As used here, social robotics does not relate to groups of robots that try to complete tasks together. For a group of robots, communication is simple, they can use whatever complex binary protocol to "socialize" with their partners. For us, the adjective social refers to humans. In principle, the implications of this are much wider than the case of groups of robots. Socializing with humans is definitely much harder, not least because robots and humans do not share a common language nor perceive the world (and hence each other) in the same way. Many researchers working on this topic use other names like human-robot interaction or perceptual user interfaces. However, as pointed out in (Fong et al., 2003) we have to distinguish between conventional human-robot interaction (such as that used in teleoperation scenarios or in friendly user interfaces) and socially interactive robots. In these, the common underlying assumption is that humans prefer to interact with robots in the same way that they interact with other people.

Human-robot interaction crucially depends on the perceptual abilities of the robot. Ideal interaction sessions would make use of non-invasive perception techniques, like hands-free voice recognition or computer vision. Computer vision is no doubt the most useful modality. Its non-invasiveness is the most important advantage.

In this paper, a number of computer vision techniques for human-robot interaction are described. All of them have been used in a prototype social robot called CASIMIRO (Figure 1), an animal-like head that stands on a table and has the goal of interacting with people, see (Deniz, 2006) for details.

2. Omnidirectional vision

Most social robots built use two types of cameras: a wide field of view camera (around 70 deg), and a foveal camera. Recently, interest in omnidirectional vision has increased in robotics. Omnidirectional vision allows to capture images that span 360°. Four main techniques are being used to achieve this:

- Cameras with fish-eye lenses, which have a very short focal length (called dioptric systems.

- Cameras with curved (convex) mirrors mounted in front of a standard lens (called catadioptric systems). This is the most common variant.
- Sets of cameras mounted in a ring or sphere configuration. The information flow is time consuming.
- An ordinary camera that rotates around an axis and takes a sequence of images that span 360°. Mechanical looseness can appear.

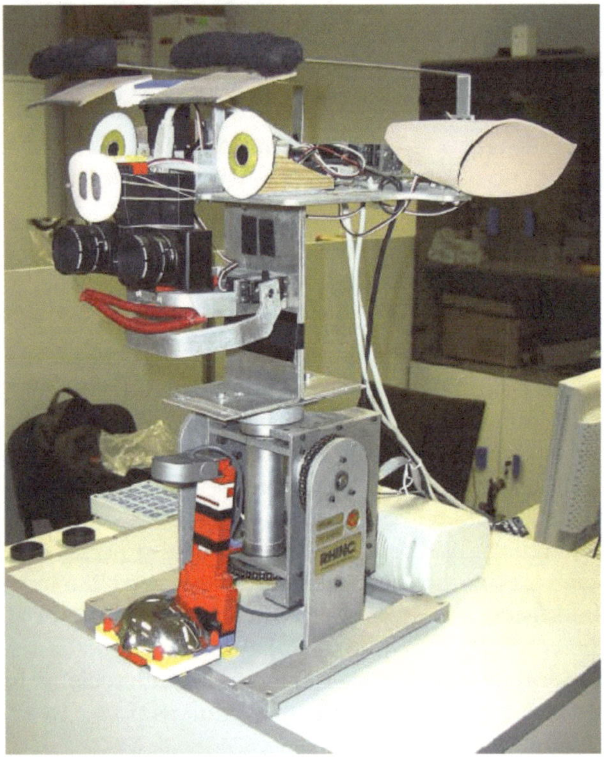

Fig. 1. CASIMIRO.

The omnidirectional camera shown in Figure 2 is a catadioptric setup that gives the robot a 180 deg field of view, which is similar to that of humans. The camera is to be placed in front of the robot. The device is made up of a low-cost USB webcam, construction parts and a curved metallic surface looking upwards, in this case a kitchen ladle. As for the software, the first step is to discard part of the image, as we want to watch only the frontal zone, covering 180 degrees from side to side. Thus, the input image is masked in order to use only the upper half of an ellipse, which is the shape of the mirror as seen from the position of the camera.

A background model is obtained as the mean value of a number of frames taken when no person is present in the room. After that, the subtracted input images are thresholded and the close operator is applied. From the obtained image, connected components are localized and their area is estimated. Also, for each connected component, the Euclidean distance from the nearest point of the component to the centre of the ellipse is estimated, as well as the angle of the centre of mass of the component with respect to the centre of the ellipse and its largest axis. Note that, as we are using an ellipse instead of a circle, the nearness measure obtained (the Euclidean distance) is not constant for a fixed real range to the camera, though it works well as an approximation. The robot uses this estimate to keep an appropriate interaction distance. The background model M is updated with each input frame:

$$M(k+1) = M(k) + U(k) \cdot [I(k) - M(k)] \quad (1)$$

, where I is the input frame and U is the updating function:

$$U(k) = \exp(-b \cdot D(k)) \quad (2)$$

$$D(k) = a \cdot D(k-1) + (1-a) \cdot |I(k) - I(k-1)| \quad (3)$$

a (between 0 and 1) and b control the adaptation rate. Note that M, U and D are images, the x and y variables have been omitted for simplicity. For large values of a and b the model adaptation is slow. In that case, new background objects take longer to enter the model. For small values of a and b, adaptation is faster, which can make animated objects enter the model.

Fig. 2. Omnidirectional camera.

The method described up to this point still has a drawback. Inanimate objects should be considered background as soon as possible. However, as we are working at a pixel level, if we set the a and b parameters too low we run the risk of considering static parts of animate objects as background too. This problem can be alleviated by processing the image D. For each foreground blob, its values in D are examined. The maximum value is found, and all the blob values in D are set to that level. Let the foreground blobs at time step k be represented as:

$$B_i = \{x_{ij} , y_{ij}\} ; i = 1, ..,NB ; j = 1, ..,N_i \quad (4)$$

There are NB blobs, each one with N_i pixels. Then, after (3) the following is applied:

$$m_i = \max_{j=1,..,Ni} D(x_{ij} , y_{ij} , k) ; i = 1, ..,NB \quad (5)$$

$$D(x_{ij} , y_{ij} , k) = m_i ; i = 1, ..,NB ; j = 1, ..,N_i \quad (6)$$

With this procedure the blob only enters the background model when all its pixels remain static. The blob does not enter the background model if at least one of its pixels has been changing.

3. Face detection

Omnidirectional vision allows the robot to detect people in the scene, just to make the neck turn towards them (or somehow focus its attention). When the neck turns, there is no guarantee that omnidirectional vision has detected a person, it can be a coat stand, a wheelchair, etc. A face detection module should be used to detect people (and possibly facial features). Facial detection commonly uses skin-color as the most important feature. Color can be used to detect skin zones, though there is always the problem that some objects like furniture appear as skin, producing many false positives. Figure 3 shows how this problem affects detection in the ENCARA facial detector (M. Castrillon-Santana et al., 2005), which (besides other additional cues) uses normalized red and green color components for skin detection.

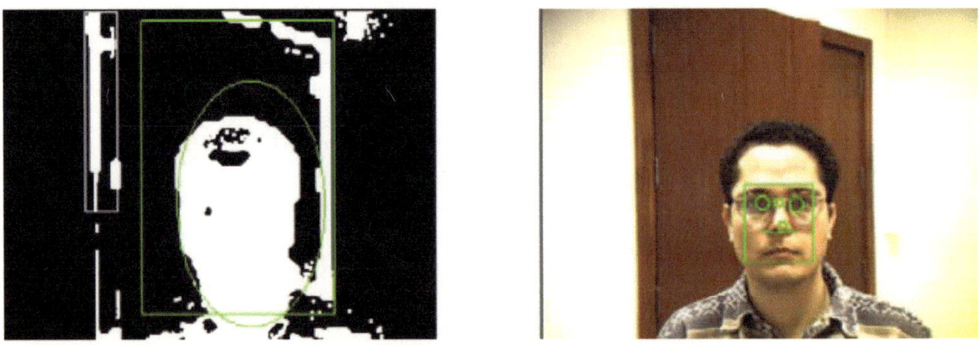

Fig. 3. Skin color detection. Note that wooden furniture is a distractor for facial detection. Both the bounding box and the best-fit ellipse are rather inaccurate (left).

In order to alleviate this problem, stereo information is very useful to discard objects that are far from the robot, i.e. in the background. Stereo cameras are nowadays becoming cheaper and faster. A depth map is computed from the pair of images taken by a stereo camera situated under the nose of the robot. The depth map is efficiently computed with an included optimized algorithm and library. The map is thresholded and an AND operation is performed between this map and the image that the facial detector uses. Fusion of color and depth was also used in (Darrell et al., 1998; Moreno et al., 2001; Grange et al., 2002). The results are shown in Figure 4. Note that most of the undesired wood colored zones are filtered out.

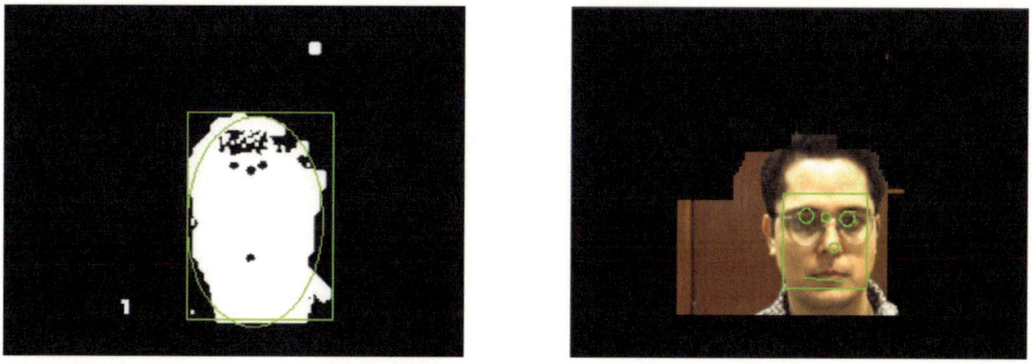

Fig. 4. Skin color detection using depth information.

4. Person recognition

In (Schulte et al., 1999) three characteristics are suggested as critical to the success of robots that must exhibit spontaneous interaction in public settings. One of them is the fact that the robot should have the capability to adapt its human interaction parameters based on the outcome of past interactions so that it can continue to demonstrate open-ended behaviour. CASIMIRO is intended to interact with people. Humans will be the most important "object" in its environment. Data associated to humans (gathered throughout the interaction) should be stored in memory, so that the robot could take advantage of previous experiences when interacting with them. Breazeal (Breazeal, 2002) argues that to establish and maintain relationships with people, a sociable robot must be able to identify the people it already knows as well as add new people to its growing set of known acquaintances. In turn, this capacity will be part of the robot's autobiographical memory.

In order to make this person memory possible, gathered data should be unambiguously associated to the correct person. Facial recognition would be the perfect approach. However, the experience of the author with face recognition is somewhat negative: face recognition still does not work well in unrestricted scenarios. Recognition rates fall as more time passes since the training samples were taken. Illumination, pose and expression variations normally reduce recognition rates dramatically.

Colour histograms of (part of) the person's body could also be used as a recognition technique. Colour histograms are simple to calculate and manage and they are relatively robust. The price to pay is the limitation that data in memory will make sense for only one day (at the most). Colour histograms of a person's body were used for short-term identification people in (Maxwell, 2003; Kahn, 1996; Maxwell et al., 1999) and also for people tracking (Krumm et al., 2000; Collins & Dennis, 2000).

CASIMIRO achieves person identity maintenance by using colour histograms in conjunction with a simple person tracking algorithm. Tracking is done in 1D, for the interesting position is the angle of the person with respect to the robot.

The implemented tracking algorithm is very simple. Each person is represented as a single point in two sets of horizontal positions (positions range from 0 to 180) at times t - 1 and t. The association of points between the two sets is obtained as that which minimizes the total sum of distances between points of the two sets. This minimization involves a factorial search, though it is practical for the number of people that will be expected to interact with the robot.

Ties can appear, for example in the case of crossings, see the example of Figure 5. These ties are broken by selecting the association with lowest variance of distances, 1 with A and 2 with B in the case of the example. This always selects non-crossings.

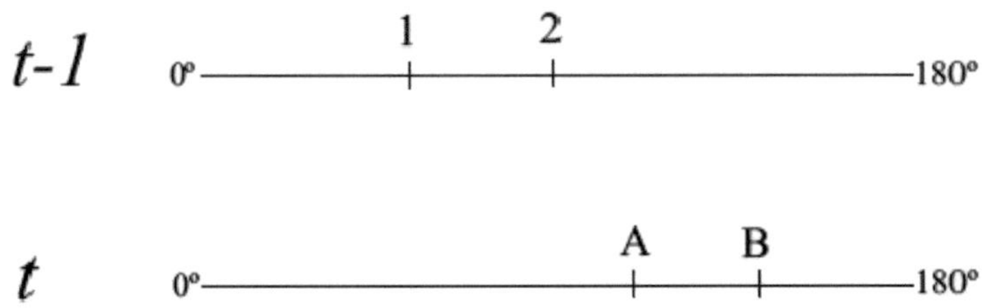

Fig. 5. Tie in sum of distances. The sum of distances |1 - A| + |2 - B| is equal to |1 - B| + |2 - A|. Without further information, we cannot know if the two individuals have crossed or not.

Crossings are detected by considering that, in a crossing, there is always a fusion and a separation of person blobs. Person blobs are detected by the omnidirectional vision system (see above). Fusions and separations are detected as follows:

- A blob fusion is detected when the number of blobs in the whole omnidirectional image decreases by one at the same time that one of the blobs increases its area significantly.
- A blob separation is detected when the number of blobs in the image increases by one at the same time that a fusioned blob decreases its area significantly.

The only way to know if a there is a crossing is by maintaining some sort of description of the blobs before and after the fusion. Histograms of U and V colour components are maintained for each blob. The Y component accounts for luminance and therefore it was not used. Whenever a separation is detected, the histograms of the left and right separated blobs are compared with those of the left and right blobs that were fusioned previously. Intersection (Swain and Ballard, 1991) was used to compare histograms (which must be normalized for blob size). This procedure allows to detect if there is a crossing, see Figure 6.

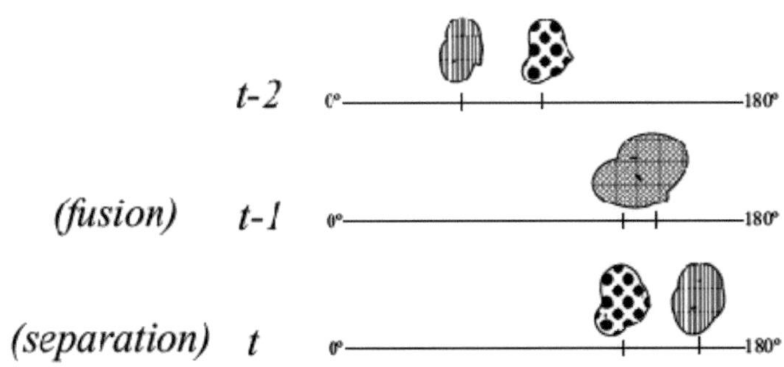

Fig. 6. Crossings can be detected by comparing blob histograms at fusion and separation events.

The histogram similarities calculated are shown in Figure 7. A crossing is detected if and only if (b + c) > (a + d). Note that in the comparison no threshold is needed, making crossing detection relatively robust.

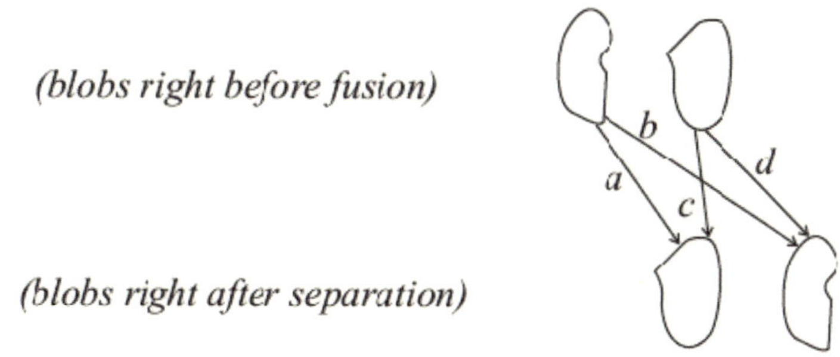

Fig. 7. Blob similarities calculated.

In order to achieve person identification, a set of Y-U histograms are stored for each person detected. The zone from which these histograms are calculated is a rectangle in the lower part of the image taken from the stereo camera placed under the nose of the robot. The rectangle is horizontally aligned with the centre of the face rectangle detected, and extends to the lower limit of the image (chest and abdomen of standing people will always occupy that lower part of the image), see Figure 8. The upper edge of the rectangle is always under the lower edge of the face rectangle detected. The width of the rectangle is proportional to the width of the face rectangle detected.

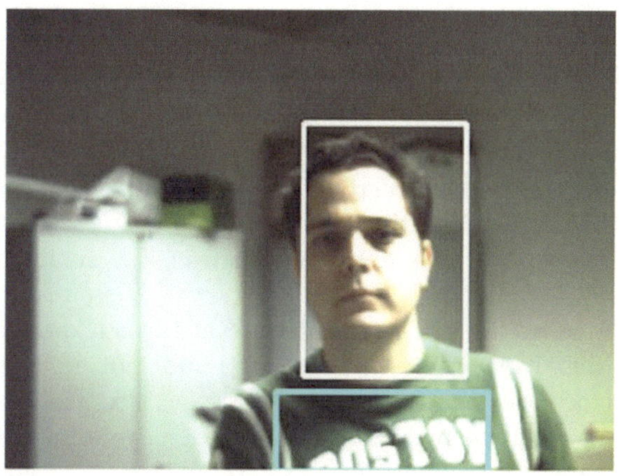

Fig. 8. Region used for person identification.

When the robot fixates on a person that the tracking system has labelled as new (the tracking system detects a new person in the scene when the number of foreground blobs increases and no blob separation is detected), it compares the histograms of the fixated individual with those of previously met individuals. This search either gives the identity of a previously seen individual or states that a new individual is in the scene. In any case the set of stored histograms for the individual is created/updated.

5. Head Nod/Shake detection

Due to the fact that practical (hands-free) voice recognition is very difficult to achieve for a robot, we decided to turn our attention to simpler (though useful) input techniques such as head gestures. Head nods and shakes are very simple in the sense that they only provide yes/no, understanding/disbelief, approval/disapproval meanings. However, their

importance must not be underestimated because of the following reasons: the meaning of head nods and shakes is almost universal, they can be detected in a relatively simple and robust way and they can be used as the minimum feedback for learning new capabilities.
The system for nod/shake detection described in (Kapoor & Picard, 2001) achieves a recognition accuracy of 78.46%, in real-time. However, the system uses complex hardware and software. An infrared sensitive camera synchronized with infrared LEDs is used to track pupils, and a HMM based pattern analyzer is used to the detect nods and shakes. The system had problems with people wearing glasses, and could have problems with earrings too. The same pupil-detection technique was used in (Davis and Vaks, 2001). That work emphasized the importance of the timing and periodicity of head nods and shakes. However, in our view that information is not robust enough to be used. In natural humanhuman interaction, head nods and shakes are sometimes very subtle. We have no problem in recognizing them because the question has been clear, and only the YES/NO answers are possible. In many cases, there is no periodicity at all, only a slight head motion. Of course, the motion could be simply a 'Look up'/'Look down'/'Look left'/'Look right', though it is not likely after the question has been made.
For our purposes, the nod/shake detector should be as fast as possible. On the other hand, we assume that the nod/shake input will be used only after the robot has asked something. Thus, the detector can produce nod/shake detections at other times, as long as it outputs right decisions when they are needed. The major problem of observing the evolution of simple characteristics like intereye position or the rectangle that fits the skin-color blob is noise. Due to the unavoidable noise, a horizontal motion (the NO) does not produce a pure horizontal displacement of the observed characteristic, because it is not being tracked. Even if it was tracked, it could drift due to lighting changes or other reasons. In practice, a horizontal motion produces a certain vertical displacement in the observed characteristic. This, given the fact that decision thresholds are set very low, can lead the system to error. The performance can be even worse if there is egomotion, like in our case (camera placed on a head with pan-tilt).
The proposed algorithm uses the pyramidal Lucas-Kanade tracking algorithm described in (Bouguet, 1999). In this case, there is tracking, and not of just one, but multiple characteristics, which increases the robustness of the system. The tracker looks first for a number of good points to track over the whole image, automatically. Those points are accentuated corners. From those points chosen by the tracker we attend only to those falling inside the rectangle that fits the skin-color blob, observing their evolution. Note that even with the LK tracker there is noise in many of the tracking points. Even in an apparently static scene there is a small motion in them.
The method is shown working in Figure 9. The LK tracker allows to indirectly control the number of tracking points. The larger the number of tracking points, the more robust (and slow) the system. The method was tested giving a recognition rate of 100% (73 out of 73, questions with alternate YES/NO responses, using the first response given by the system).
What happens if there are small camera displacements? In order to see the effect of this, linear camera displacements were simulated in the tests. In each frame, an error is added to the position of all the tracking points. If (D_x,D_y) is the average displacement of the points inside the skin-color rectangle, then the new displacement is D_x+e_x and D_y+e_y. The error, which is random and different for each frame, is bounded by $-e_{max} < e_x < e_{max}$ and $-e_{max} < e_y < e_{max}$. Note that in principle it is not possible to use a fixed threshold because the error is unknown. The error also affects to the tracking points that fall outside the rectangle.

Assuming that the objects that fall outside the rectangle are static we can eliminate the error and keep on using a fixed threshold, for $(D_x + e_x) - (F_x + e_x) \approx D_x$ and $(D_y + e_y) - (F_y + e_y) \approx D_y$. For the system to work well it is needed that the face occupies a large part of the image. A zoom lens should be used. When a simulated error of $e_{max} = 10$ pixels was introduced, the recognition rate was 95.9% (70 out of 73). In this case there is a slight error due to the fact that the components F_x and F_y are not exactly zero even if the scene outside the rectangle is static.

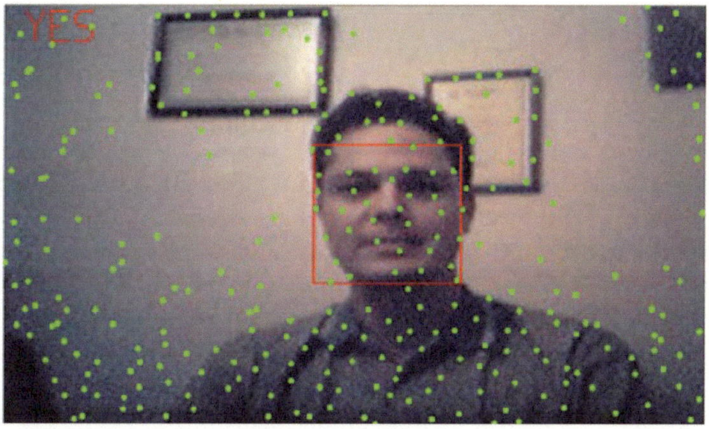

Fig. 9. Head nod/shake detector.

Another type of error that can appear when the camera is mounted on a mobile device like a pan-tilt unit is the horizontal axis inclination. In practice, this situation is common, especially with small inclinations. Inclinations can be a problem for deciding between a YES and a NO. In order to test this effect, an inclination error was simulated in the tests (with the correction of egomotion active). The error is a rotation of the displacement vectors **D** a certain angle α clockwise. Recognition rates were measured for different values of α, producing useful rates for small inclinations: 90% (60 out of 66) for $\alpha = 20$, 83.8% (57 out of 68) for $\alpha = 40$ and 9.5% (6 out of 63) for $\alpha = 50$.

6. Habituation

Habituation is a filtering mechanism that has received a lot of attention in physiology and psychology. In particular, some researchers have investigated the mechanisms of habituation in animals, being one of the most known works the study of the Aplysia's gillwithdrawal reflex (Castellucci et al., 1970). When the animal's siphon is touched, its gill contracts for a few seconds. If the siphon is stimulated repeatedly, the gill-withdrawal effect tends to disappear. Crook and Hayes (Crook & Hayes, 2001) comment on a study carried out on two monkeys by Xiang and Brown who identified neurons that exhibit a habituation mechanism since their activity decreases as the stimulus is shown repeatedly. Stanley's model (Stanley, 1976) of habituation, proposed to simulate habituation data obtained from the cat spinal cord, has been widely used in the literature. This model describes the decrease efficacy y of a synapsis by the first-order differential equation:

$$\tau \frac{dy(t)}{dt} = \alpha\big(y_0 - y(t)\big) - S(t) \quad (7)$$

where y_0 is the normal, initial value of y, S(t) represents the external stimulation, τ is a time constant that governs the rate of habituation and α regulates the rate of recovery. Equation (7) ensures that the synaptic efficacy decreases when the input signal S(t) increases and returns to its maximum y0 in the absence of an input signal.
The model given by (7) can only explain short-term habituation, so Wang introduced a model to incorporate both short-term and long-term habituation using an inverse S-shaped curve (Wang, 1995):

$$\tau \frac{dy(t)}{dt} = \alpha z(t)\big(y_0 - y(t)\big) - \beta y(t)S(t) \tag{8}$$

$$\frac{dz(t)}{dt} = \gamma z(t)(z(t) - l)S(t) \tag{9}$$

where α, y_0 and γ have the same meaning than in (7), & regulates the habituation and z(t) decreases monotonically with each activation of the external stimulation S(t), and models the long-term habituation. Due to this effect of z(t) after a large number of activations, the recovery rate is slower.
Note that novelty detection is a concept related to habituation. Novelty detection is the discovery of stimuli not perceived before and so habituation serves as a novelty filter (Stiles & Ghosh, 1995). From an engineering viewpoint, perceptual user interfaces, like humanlike robots, should be endowed with a habituation mechanism. The interest is twofold. First, it would be a filtering mechanism, discarding (or minimizing the importance of) repetitive information while paying attention to new experiences. This is in part motivated by the desire to distinguish between artificial and human signals. Artificial signals are often static or repeat with a fixed frequency. We do not want our robot to pay much attention to the hands of a wall-mounted clock. Instead, it would be more interesting to detect nonrepetitive stimuli, such as a conversation or a sudden loud noise. Note that we generally consider monotonous signals as those having a fixed frequency or frequencies (which can be zero, that is, the signal does not change) but signals whose frequency changes in a periodic pattern could also be considered monotonous. Higher scales are also possible but we do not consider them in this work because they are very hard to visualize and real examples of them are not so common. Second, habituation would lead to a more human-like behaviour, as perceived by users of the interface. As an example of this, consider Kismet. Someone can catch the eye of the system while waving a hand in its visual field of view, but if the stimulus is repetitive for a long time the system can show a lack of interest in it. Many aspects of Kismet's mental architecture are directly or indirectly influenced by the detection of monotonous sensory signals: stimulation and fatigue drives and the arousal dimension of its affect space (and in turn some emotional states, like surprise, boredom or interest). If we use the model of Equation (7) we can obtain undesired effects with certain stimuli. A periodic input signal (with frequency greater than zero) can produce a response that does not exhibit habituation. This is due to the fact that the model does not account for changing stimuli, but for continuous ones. In order to include this fact in the model, we propose to use an auxiliary signal which will be zero when the stimulus is stationary or with a fixed frequency, and one otherwise, and use this signal as an input to the habituation model (7). The auxiliary signal, which basically detects monotonous stimuli, is obtained from the

spectrogram of the stimulus itself. The spectrogram is a time-frequency distribution of a signal, and it is based on the Fourier Transform with a sliding window, see Figure 10 (Holland et al., 2000).

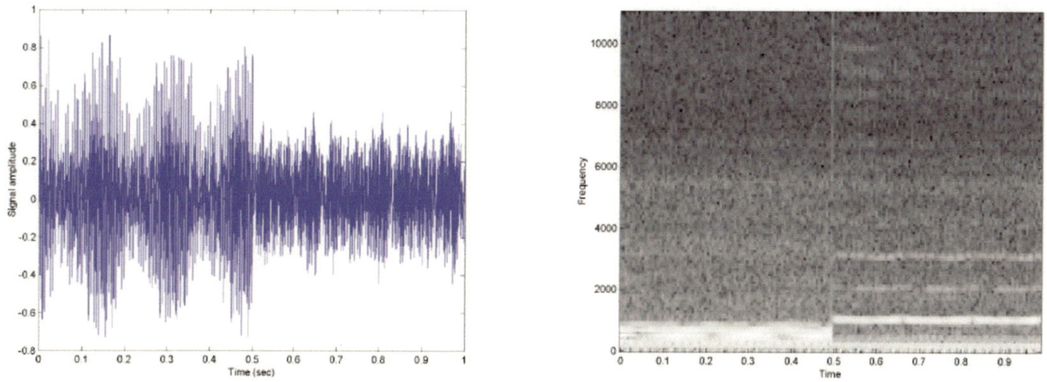

Fig. 10. Audio signal (left) and its corresponding spectrogram (right).

Spectrograms are computed from windows of the input signal. These windows, of length l, overlap by l - 1 samples. Let each spectrogram be represented as a matrix M, in which rows represent frequencies and columns represent time. We calculate the variance of each row of M, which produces a column vector v. The norm of this vector v is a measure of how monotonous the input signal is. The norm will be high when the signal is changing, and low otherwise. Thus, the auxiliary signal needed is simply the thresholded norm of v. The amplitude of the input signal affects the power content of the spectrograms, and in turn the norm of v. Thus, prior to calculating the FFT the input signal must be normalized dividing each input window by the sum of its absolute values. A value of 1 for the auxiliary signal will mean that there are changes in the input signal, while a value of 0 indicates that the input signal is monotonous. Once the auxiliary signal is available, the model (7) is used to get the desired habituation behaviour, as controlled by parameters τ and α. The auxiliary signal is then, for a given threshold T:

$$A = \begin{cases} 1 \ \ if \ \ |v| > T \\ 0 \ \ if \ \ |v| \leq T \end{cases} \tag{10}$$

With this method both static and fixed frequency stimuli can be detected. The algorithm described above was implemented to test it with different input signals. The first experiments that we present use only the first level mentioned above.

In order to gather signals from the visual domain, we recorded video containing a yellow bright stimulus (a yellow card) that was moved in a repetitive fashion, see Figure 11-a). Using simple segmentation techniques we extracted the centroid of the card on each frame (384x288) and summed the x and y pixel coordinates to form the one-dimensional signal of Figure 5.29-b). The sequence of card movements throughout the recording was: horizontal movement, random (aperiodic) movement, vertical movement and vertical movement at a different frequency than the previous one. The results appear in Figure 12. Windows of 128 samples were used, and the variance threshold was set to 1000.

The habituation mechanism described here was implemented in CASIMIRO, for signals in the visual domain only, i.e. images taken by the stereo camera (see Section 3). The difference between the current and previous frame is calculated. Then it is thresholded and filtered with Open and Close operators. Also, blobs smaller than a threshold are removed. Then the centre of mass of the resultant image is calculated. The signal that feeds the habituation algorithm is the sum of the x and y components of the centre of mass. This way when the image does not show significant changes or repetitive movements are present for a while the habituation signal grows. When it grows larger than a threshold, an inhibition signal is sent to the Attention module, which then changes its focus of attention. Neck movements produce changes in the images, though it was observed that they are not periodic, and so habituation does not grow.

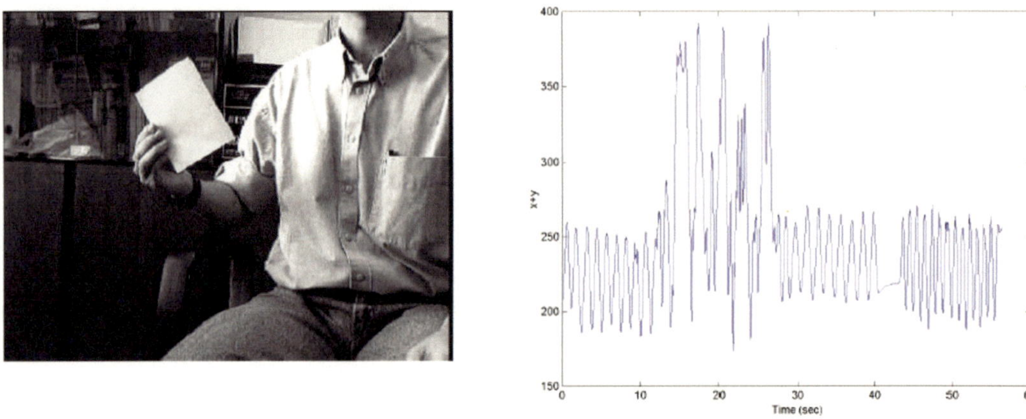

a) b)

Fig. 11. a) Video recording used for the visual habituation experiment, b) unidimensional signal extracted from it.

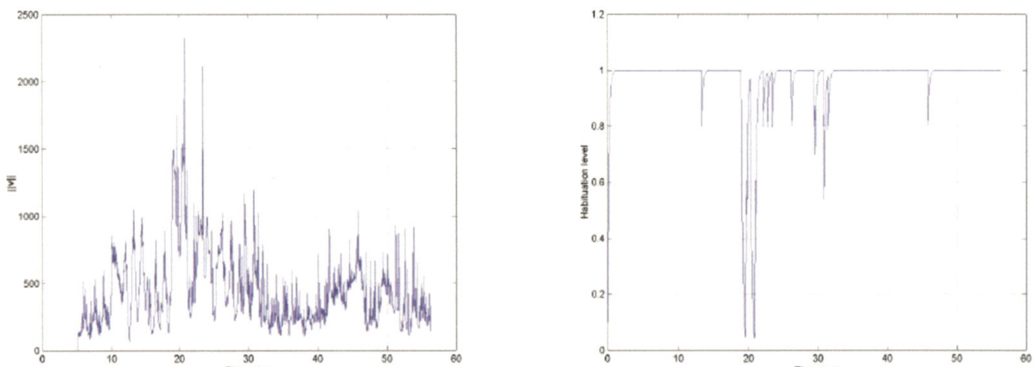

a) b)

Fig. 12. a) Evolution of the (l2) norm of the variance vector v, b) habituation level, using τ = 5, α = 1.

7. Other abilities

In this section two minor abilities of the robot are described: an owner recognition method that is simple yet powerful enough and a "hand waving" detector.
Owner recognition:
The recognition of the robot's owner or caregiver is a potentially important feature. The recognition may not affect the creator at all, but it could impress other people present in the room. Person recognition is difficult to achieve, the identification of the robot's owner may be more feasible, for the problem reduces to two classes (owner and not-owner). Still, using face or voice for owner recognition would lead to essentially the same lack of robustness of face and speech recognition for multiple individuals. Who is the owner of the robot? The correct answer is: the person who buys it. However, we are more interested in the role of the person who cares the robot and uses it more frequently (this person is generally the owner too). This person is the one who switches the robot on, which is usually done from a certain part of the robot or from a certain computer. That cue may be exploited to recognize the robot owner or caregiver. *Amazing Amanda,* a doll released in autumn of 2005, is able to recognize the girl that takes the mother role. Once the doll is activated, it starts asking questions. That way, the doll can "learn" the girl's voice patterns. From that moment on, the doll is able to recognize the utterances of its "mommy". Other voices can lead Amanda to say "You don't sound like Mommy".
Such technique may seem rather ad hoc. However, the approach finds striking examples in nature. Lorenz, one of the founders of ethology, found that, upon coming out of their eggs, geese follow and become attached to the first moving object that they encounter. He showed this by rearing the geese from hatching. From that moment on the geese would follow him. Such phenomenon, which also appears in mammals, is known as imprinting (Lorenz, 1981).

Fig. 13. The computer from where CASIMIRO is started. The interaction space is on the right.

In the case of CASIMIRO, the main computer (from which the robot is switched on) is situated behind the robot, on the same table, see Figure 13. A camera was placed on top of that computer. The owner detection module uses that camera to search for a skin coloured blob in the image. When the robot is switched on this module will detect a skin coloured blob. The camera has a wide-angle lens, and a relatively low resolution of 160x120 is used. When no blob is encountered in the image the module notifies the Attention module of that event. At that moment the owner detection module exits in order to free CPU resources. Once it has been notified by the owner detection module, the Attention module considers the owner as the first blob that "enters" the omnidirectional camera image from the left. The "owner" property is stored along with the individual in the tracking process. This simple procedure is a form of imprinting. In a sense, the robot finds its owner-caregiver in the first human it sees. It does not stores any biometric features to recognize the owner after being switched on, only its position.

Hand-waving detector:

In initial tests with the robot we saw that people tended to wave his/her hand to the robot. We decided to detect this and make the robot respond with a funny phrase. The contour of each foreground blob is obtained and compared with the contour of the same blob in the previous frame. The comparison is made using Hu moments (Hu, 1962). When the comparison yields too much difference between the contours (i.e. a threshold is exceeded) a flag is activated. When the face detection module (Section 3) detects more than one skin colour blob in the image and a certain amount of motion and the flag is activated then the Boolean perception HANDWAVING is set to TRUE.

8. Conclusion

A number of simple but useful computer vision techniques have been described, suitable for human-robot interaction. First, an omnidirectional camera setting is described that can detect people in the surroundings of the robot, giving their angular positions and a rough estimate of the distance. The device can be easily built with inexpensive components. Second, we comment on a color-based face detection technique that can alleviate skin-color false positives. Third, a simple head nod and shake detector is described, suitable for detecting affirmative/negative, approval/disapproval, understanding/disbelief head gestures. Fourth, a habituation-detection scheme has been described, allowing the robot to focus on natural signals in the environment. Finally, two minor owner recognition and hand-waving detection abilities are also described. All the techniques have been satisfactorily implemented and tested on a prototype social robot.

Future work should focus on how computer vision could aid in hands-free speech recognition, an all-important ability for an interactive robot. Progress has been already made by syncing lip motion with audio speech signals, although the processing cost is still too long for useful real-time interaction.

9. References

Bouguet, J. (1999). Pyramidal implementation of the Lucas Kanade feature tracker. *Technical report*, Intel Corporation, Microprocessor Research Labs, OpenCV documents.

Breazeal, C. L. (2002). *Designing social robots.* MIT Press, Cambridge, MA.

M. Castrillon-Santana, H. Kruppa, C. G. & Hernandez, M. (2005). Towards real-time multiresolution face/head detection. *Revista Iberoamericana de Inteligencia Artificial,* 9(27):63–72.

Castellucci, V., Pinsker, H., Kupfermann, I., & Kandel E.R. (1970). Neuronal mechanisms of habituation and dishabituation of the gill-withdrawal reflex in Aplysia. *Science,* 167:1745–1748.

Collins, G. & Dennis, L. A. (2000). System description: Embedding verification into Microsoft Excel. In *Conference on Automated Deduction,* pages 497–501.

Crook, P., & Hayes, G. (2001). A robot implementation of a biologically inspired method of novelty detection. In *Proceedings of TIMR 2001 - Towards Intelligent Mobile Robots,* Manchester, April.

Darrell, T., Gordon, G., Harville, M., & Woodfill, J. (1998). Integrated person tracking using stereo, color, and pattern detection. In *Procs. of IEEE Computer Society Conference on Computer Vision and Pattern Recognition,* pages 601–608, Santa Barbara, CA.

Davis, J. & Vaks, S. (2001). A perceptual user interface for recognizing head gesture acknowledgements. In *Proc. of ACM Workshop on Perceptual User Interfaces,* Orlando, Florida.

Deniz, O. (2006). *An Engineering Approach to Sociable Robots.* PhD Thesis, Department of Computer Science, Universidad de Las Palmas de Gran Canaria.

Fong, T., Nourbakhsh, I. & Dautenhahn, K. (2003). A survey of socially interactive robots. *Robotics and Autonomous Systems,* 42(3-4).

Grange, S., Casanova, E., Fong, T., & Baur, C. (2002). Vision based sensor fusion for humancomputer interaction. In *Procs. of IEEE/RSJ International Conference on Intelligent Robots and Systems.* Lausanne, Switzerland.

Holland, S., Kosel, T., Waver, R., & Sachse, W. (2000). Determination of plate source, detector separation fron one signal. *Ultrasonics,* 38:620–623.

Hu, M. (1962). Visual pattern recognition by moment invariants. *IRE Transactions on Information Theory,* vol. 8(2), 179–187.

Kahn, R. (1996). Perseus: An Extensible Vision System for Human-Machine Interaction. PhD Thesis, University of Chicago.

Kapoor, A. & Picard, R. (2001). A real-time head nod and shake detector. In *Proceedings from the Workshop on Perspective User Interfaces.*

Krumm, J., Harris, S., Meyers, B., Brumitt, B., Hale, M., & Shafer, S. (2000). Multicamera multiperson tracking for easyliving. In *3rd IEEE International Workshop on Visual Surveillance,* Dublin, Ireland.

Lorenz, K. (1981). *The Foundations of Ethology.* Springer Verlag, Heidelberg.

Maxwell, B. (2003). A real-time vision module for interactive perceptual agents. *Machine Vision and Applications,* (14):72–82.

Maxwell, B., Meeden, L., Addo, N., Brown, L., Dickson, P., Ng, J., Olshfski, S., Silk, E., & Wales, J. (1999). Alfred: the robot waiter who remembers you. In *Procs. of AAAI Workshop on Robotics.*

Moreno, F., Andrade-Cetto, J., & Sanfeliu, A. (2001). Localization of human faces fusing color segmentation and depth from stereo. In *Procs. ETFA,* pages 527–534.

Schulte, J., Rosenberg, C., & Thrun, S. (1999). Spontaneous short-term interaction with mobile robots in public places. In *Procs. of the IEEE Int. Conference on Robotics and Automation.*

Stanley, J.C. (1976). Computer simulation of a model of habituation. *Nature,* 261:146–148.

Stiles, B., & Ghosh, J. (1995). A habituation based neural network for spatio-temporal classification. In *Proceedings of the 1995 IEEE Workshop In Neural Networks for Signal Processing,* pages 135–144, Cambridge, MA, September.

Wang, D. (1995). Habituation. In M. A. Arbib, editor, The *Handbook of Brain Theory and Neural Networks,* pages 441–444. MIT Press.

24

Progress in Speech Recognition for Romanian Language

Corneliu-Octavian Dumitru and Inge Gavat
Faculty of Electronics Telecommunications and Information Technology,
University POLITEHNICA Bucharest
Romania

1. Introduction

In this chapter we will present the progress made in automatic speech recognition for Romanian language based on the ASRS_RL (Automatic Speech Recognition System for Romanian Language) research platform.

Speech recognition is a research domain with a long history, but despite this fact, still open for new investigations and answers to the not yet finally solved questions. This situation can be explained by the difficulty of the task, underlying on the fact that speech is a human product, with a high degree of correlation in content, but with a great variability in the formal manifestation as an acoustic signal. Great difficulties cause also the imperfection of the audio chain and the noises in the environment.

The best-known strategies for speech recognition are the statistical and the connectionist ones, but fuzzy sets can also play an important role.

Based on HMM's the statistical strategies have many advantages, among them being recalled: rich mathematical framework, powerful learning and decoding methods, good sequences handling capabilities, flexible topology for statistical phonology and syntax. The disadvantages lie in the poor discrimination between the models and in the unrealistic assumptions that must be made to construct the HMM's theory, namely the independence of the successive feature frames (input vectors) and the first order Markov process.

Based on artificial neural networks (ANNs), the connectionist strategies for speech recognition have the advantages of the massive parallelism, good adaptation, efficient algorithms for solving classification problems and intrinsic discriminative properties. However, the neural nets have difficulties in handling the temporal dependencies inherent in speech data.

The learning capabilities of the statistical and the neural models are very important, classifier built on such bases having the possibility to recognize new, unknown patterns with the experience obtained by training.

The introduction of fuzzy sets allows on one hand the so-called fuzzy decisions, on other hand the "fuzzyfication" of input data, often more suitable for recognition of pattern produced by human beings, by speaking, for example. In a fuzzy decision, the recognizer realizes the classification based on the degree of membership to a given class for the pattern to be classified, a pattern belonging in a certain measure to each of the possible classes. This relaxation in decision leads to significant enhancements in recognition performances, situation that can also be obtained by looking in a fuzzy way to the input data

The learning capabilities offered by the statistical and the connectionist paradigms and also a "nuanced" inside in the reality of the input and output domains of the speech recognizers contribute to a kind of "human likely" behaviour of this automata.

These three main strategies were applied in our speech recognition experiments and the developed algorithms were incorporate in our research platform, the ASRS_RL. The statistical strategies played the most important role in the development of continuous speech recognition based on HMMs. The neural strategies are applied in form of multilayer perceptrons (MLP) and Kohonen maps (KM) for tasks like vowel or digit recognition. Algorithms for hybrid structures like fuzzy HMM, fuzzy MLP or MLP -HMM are also incorporated in our research platform.

The chapter will be structured as: **Section 2** will describe the "state of the art" of the techniques applied for Romanian language. The **Section 3** is dedicated to the capabilities of the ASRS_RL system. The data bases for training and testing, the feature extraction methods and the learning strategies incorporated in the system are briefly introduced. In **Section 4** are shown and commented the experimental results of the ASRS_RL platform in two tasks: first a task of continuous speech recognition, next a telephone dial experiment. In the first task are applied HMMs with context dependent modelling and the performance is expressed in word recognition rate (WRR) for the recognition experiments and in phrase recognition rate (PRR) for speech understanding experiments. For the second task are implemented HMMs without context dependent modelling, for monophones. **Section 5** presents results and comments for experiments done applying alternative strategies to HMM. These strategies are investigated for simple tasks like vowel and digit recognition. The error rate is evaluated for vowel recognized by using MLP, KM, SVM (Support Vector Machine), fuzzy - MLP and fuzzy-HMM strategies and for digits recognized by the HMM - MLP.

The last Section (**Section 6**) end the chapter with conclusions extracted from the experimental results and with perspectives for our future work.

2. Techniques applied for Romanian language

A brief history of the speech processing techniques for Romanian language was given by the president of the Romanian Academy, Prof. Dr. ing. Mihai Draganescu in the introductory speech to the third conference dedicated to man-machine communication by natural languages, SPED 2003, held at Bucharest: *"Almost 20 years ago, the Romanian Academy organized the first session on the analysis and synthesis of the speech signal. It was a moment of recognition of the activity of Romanian scientists in the domain of speech technology, with works beginning in 1963 (Edmon Nicolau, Inge Weber, Stefan Gavat), 1973 (Aurelian Lazaroiu), 1976 (Eugeniu Oancea, with a volume on analysis and synthesis of speech) and with new papers of Corneliu Burileanu, Horia Nicolai Teodorescu, Eugeniu Oancea, Grigore Stolojanu, Virgil Enatescu and others. Since then the domain evolved from Speech technology to spoken language technology. This had to be foreseen however from the very beginning involving the use of artificial intelligence, both for natural language processing and for acoustic-phonetic processes of the spoken language."* (Draganescu, 2003).

Continuing this tradition, our team has experimented a set of Computational Intelligence algorithms for Automatic Speech Recognition (ASR) extended also for speech understanding in the frame of the three main paradigms mentioned in the introduction.

We started caring out experiments with neural recognizers for vowels (Grigore et al., 1996), (Grigore et al., 1998), for isolated words (Valsan et al., 1998-a) and word spotting (Valsan et al., 1998-b). Then we introduced the HMM modelling in our experiments, first for isolated word recognition, further for continuous speech. The refinement of the HMM based recognizers was our permanent goal, and today we can report a system with reasonable performances in continuous speech recognition. for Romanian language (Gavat et al. 2003, Dumitru et al,, 2006, Dumitru et al,2007, Gavat et al, 2008)

But also very attractive in order to improve recognition performance was the development of hybrid systems. One of the firsts was a neuro-statistic hybrid, in different ways proposed by many authors (Bourlard et al., 1990), (Richard et al., 1991), (Lippmann et al., 1993), (Bourlard et al., 1994). We realized such a system for Romanian language as a variant consisting in a HMM assisted by a MLP as *a posteriori* probability estimator (Gavat et al., 1996-a). This combination, realized between components with complementary properties (i.e. the good integrating time successions HMM and the high discriminative MLP), leads to an increase in recognition rates that exceeds 2%. Furthermore, as suggested in (Juang et al., 1992), (Reichl et al., 1994) we refined our hybrid structure by a supplementary discriminative training (Gavat et al., 1998), obtaining a further improvement with around 3.5%. We applied also fuzzy variants for neural and hard classifiers. By applying fuzzy decisions, we have obtained improved performances in classical algorithms like k-NN or ISODATA (Gavat et al., 1996-a) and also in neural recognizers realized with MLP or self-organizing feature maps (Gavat et al., 1997), (Grigore et al., 1999). To apply fuzzy concepts to HMMs was the next natural step to be followed in our studies, experimenting in speech recognition the hybrid called Fuzzy-HMM (FHMM). (Gavat et al., 2001-a), (Gavat et al., 2001-b).This developed hybrid systems were applied until now only in tasks for vowel and isolated word recognition, the implementation for continuous speech recognition remaining as a future desiderate.

3. ASRS_RL system

Speech is a communication modality, so how the communication is accomplished is a question of high interest. Traditionally (Juanhg et al., 2000) it can be admitted that the speech communication chain is organized in four stages: detection of acoustic-phonetic cues to form words, syntactic and grammatical analysis to build sentences, semantic evaluation to determine the possible meanings of a sentence and pragmatic evaluation in order to select the convenient meaning.

An ASR (Automatic Speech Recognition) system is a machine that performs in two stages (the first two) and an ASRU (automatic speech recognition and understanding) system is a machine for which all stage are necessary. But, a lot of work must still be done to acquire the whole knowledge necessary to entirely realize this last task.

3.1 Capabilities of the ASRS_RL

Automatic Speech Recognition System for Romanian Language (ASRS_RL) is a viable structure, offering multiple options for the tasks that can be performed, for the features that can be used to characterize speech frames, for the learning strategies applied, for the speech databases necessary in training and testing stage. It allows also multilingual (Romanian, English, German and can be extended to other languages) and multimodal (speech and image entries) experiments. One window of the MATLAB interface to choose the working options is shown in Fig.1

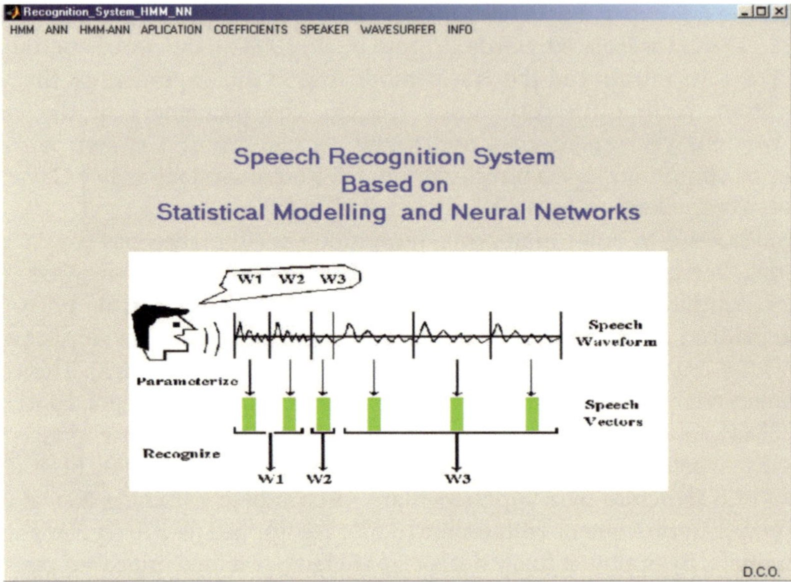

Fig. 1. The recognition system ASRS_RL

a. The possible tasks that can be performed with this system are: continuous speech recognition; digit recognition and vowel recognition.
b. The possible databases for training/testing are: CDRL (Continuous Database for Romanian Language); DDRL (Digit Database for Romanian Language); DDEL (Digit Database for English Language); VDRL (Vowel Database for Romanian Language). The data are sampled by 16 kHz, quantified with 16 bits, and recorded in a laboratory environment (for CDRL, DDRL and VDRL) (Dumitru et al., 2007). For DDEL the database used is the AMP database (www.amp.ece. cmu.edu/_download /Intel/feature_data.html). The DDRL and DDEL are available for three different SNR: 19dB, 25dB and 30dB.
c. The phonetic transcription for each database was created based on the SAMPA standard (Sampa).
d. The features vectors to characterize speech frames (and the number of coefficients for each type) can be constituted (Huang at al., 2001) (Hermansky, 1990) by: formants, obtained from cepstral analysis; mel frequency cepstral coefficients (MFCC) with or without first (D-delta) and second (A-acceleration) order variations, obtained from perceptual cepstral analysis; linear prediction coefficients (LPC) obtained by linear prediction; perceptual linear prediction (PLP) coefficients obtained by perceptual linear prediction. To this features frame energy (E), zero crossing rate (Z_0), and other information can be added.
e. The possible learning strategies (Goronzy, S. (2002) that can be chosen in the system in order to build acoustical models are: hidden Markov models (HMM) for monophones (phonemes without context) and triphones (phonemes with left and right context) (Young, 1992); Support Vector Machines (SVM); artificial neural networks in form of multilayer perceptrons (MLP) and Kohonen maps (KM); hybrid structures like the neuro-statistical hybrid (HMM-MLP), or the fuzzy variants for the MLP or the HMM.

3.2 Databases

In this sub-section we present our databases for Romanian language with the principal characteristics, as well as a "state of the art" on the most important international databases. Based on the investigation of the existing standardized databases for different languages, the most important databases are: for English - TIMIT, WSJO, AMP, etc; for German - Vermobil; for French - NEOLOGOS, for Spanish - Entropic Latino40; for Hungarian - OASIS; but there are also some multilingual databases like CSLU, SpeechDat, Appen , etc.

TIMIT Acoustic-Phonetic Continuous Speech Corpus was created by DARPA-ISTO (Defense Advanced Research Projects Agency - Information Science and Technology Office), and the text of this corpus was realized by MIT (Massachusetts Institute of Technology), SRI (Stanford Research Institute) and TI (Texas Instruments).

The database contains 630 speakers from 8 region of the United States of America, 438 male speakers and 192 female speakers, each of them spoken 10 phrases (Timit).

WSJO (Wall Street JOurnal) is the biggest and variat corpus for English, created by ARPA (Advanced Research Projects Agency). The phrases are from *Wall street Journal*. The initial number of words was 37 million, but finally reduced to 64000 words, the most frequent words in the journal. There are two corpora:J0 (84 speakers, 42 male speakers and 42 female speakers, 7193 phrases) and WSJ1 (200 speakers, 100 male speakers and 100 female speakers, 29320 phrases) (Douglas et al., 1992); WSJCAM0 (for British English created at the Cambridge University in 1995; the corpus was selected from the *Wall street Journal* between 1987-1989; the speakers are: between 18-23 years old, 46, between 24-28 years old, 30, between 29-40 years old, 7, more then 40 years old, 9; each of them uttered 110 phrases) (Robinson et al., 1995).

AMP (Advanced Multimedia Processing) is a database created at the Carnegie Mellon University in Advanced Multimedia Processing Laboratory. The database is sampled by 44,1kHz and quantified with 16 bits, mono system (Amp).

The database contains 10 speakers (7 ale speakers and 4 female speakers) each spoken 78 words for 10 times.

CSLU (Centre of Spoken Language Understanding) is a database crated by Oregon Health & Science University. This database contains many corpora: *22 languages* (a database for 22 language); *Alphadigits* (3025 speakers uttered digits and stream of digits, 78044 audio files); *Kids Speech* (this database is created to help the children with hearing problems; about 100 kids); *Multilanguage Telephone Speech* (for 11 language and contains 2052 speakers, spoken fixed phrases or free continuous speech); etc (Cslu).

SpeechDat is created to develop systems for *voice driven teleservice* and *speech interfaces*.

SpeechDat(M) is a database for 7 languages from West Europe (SpeDatM), 1000 speakers; **SpeechDat(II)** is for 11 languages, official languages of the Europeen Union, each of this databases contains between 500-5000 speakers; **SpeechDat(E)** is for 5 languages from Est Europe, contains between 1000-2500 speakers (SpeDatE).

Appen is a collection of databases for 20 languages for different applications like: telephony, broadcast audio, desktop, pocket PC (Appen). The recorded files are realized in different location: car, studio, office, street or public place. We present the database for one application, telephony. The database is for spoken English in Australia, recorded in office environment. The first part of the database contains 500 speakers and immigrant speakers, each reading 165 phrases; the second part contains 1000 speakers (Australian speakers) reading 75 phrases.

Others databases are for spoken English in Canada (contains 49 speakers each uttered about 99 phrases in office environment) and for spoken French in Canada (contains 48 speakers each spoken about 100 phrases) in office, car, home environment and public place. The phrases are constructed by digits, phonemes, names places jobs, answers yes / no), commands.
Neologos was recorded on the French telephone network (Charlet et al., 2005) and two databases were obtained. The first database is IDIOLOGOS (with two sub-databases: *Bootstrap* - contains 1000 speakers (470 male speakers and 530 female speakers) with ages from 18 to 61 and up to 61, with a very good distribution, each of them uttered 45 phrases; *Eingenspeakers* - contains 200 speakers (97 male speakers and 103 female speakers) for different age, from 18 to 61 and up to 61, each of them spoken 45 phrases/appeal telephonic, 10 appeal are considered).
The second database PAIDIALOGOS is for the children between 7 to 16 years, with one child having more than 16 years. The number of the children is 1010 (510 male children and 500 female children), from different regions of France. The database contains: digits, numbers of credit cards, names of cities, etc. The data are sampled by 8 kHz, quantified with 8 bits.
Vermobil (Elra) was created between 1993-1996 by the research institutions and companies: Institut für Phonetik und digitale Sprachverarbeitung (IPDS) - Kiel, Institut für Kommunikation und Phonetik (IKP) - Bonn, Institut für Phonetik und sprachliche Kommunikation - München and Universität Karlsruhe. The database was divided in two databases: one for training (contains 12000 turns) and one for testing (1800 turns), with different dialects (12 dialects). One turn is equal with around 22.8 words.
Entropic Latino40 (Entro40) was recorded in 1994, in Palo Alto - California, the speakers are Spanish, from Latino America. Database contains 5000 phrases, spoken by 40 speakers (20 male speakers and 20 female speakers) each of them reading 125 phrases. The database is sampled by 16kHz, quantified with 16 bits and recorded in office environment. All the speakers are not specialized speakers and are between 18-59 years old.
OASIS (Kocsor et al., 1995) was created at Hungarian Academy of Sciences (Research Group on Artificial Intelligence). The database contains speech files from 26 speakers (16 male speakers and 10 female speakers) each of this speakers uttered the digits from 0-9, from 10-100, 1000, 1000000 and variants of composed numbers. The database is sampled by 22,05kHz, quantified with 16 bits and recorded in office environment.
Because in Romanian language no standard database is available, we create our database for our recognition tasks. The databases created are: OCDRL (Old Continuous Database for Romanian Language), CDRL (Continuous Database for Romanian Language), DDRL (Digit Database for Romanian Language) and VDRL (Vowel Database for Romanian Language). The data are sampled by 16 kHz, quantified with 16 bits, and recorded in a laboratory environment (Dumitru, 2006).
OCDRL: for continuous speech recognition, is constituted by two databases: the training database contains 500 phrases, spoken by 10 speakers (8 males and 2 females), each speaker reading 50 phrases; the testing database contains 350 phrases spoken by the same speakers. The phrases from the database are for an application of telephone dial.
A statistical investigation of the existing phonemes in the training database, based on phonetic transcription of the SAMPA standard (Sampa), is shown in the Table 1. We can see easy that the preponderant are the vowels (i, i_O, e, a, @, o, u, 1) and semi-vowels (j, e_X, w,

o_X). The distribution of phonemes is very similar to that given by linguists for Romanian language, so that we trusted the data base for the acoustical modelling.

a	**55**	l	21	p	9	b	3	ts	4
sp	102	**j**	**18**	**e_X**	**4**	S	2	g	3
tS	10	f	10	z	4	**1**	**5**	dZ	3
e	**45**	n	29	@	**17**	d	14	**o_X**	**3**
s	24	**o**	**36**	r	35	k_	3	g_	3
t	32	k	8	m	12	**i_O**	**4**	h	4
i	**27**	**u**	**23**	**w**	**6**	v	3	Z	3

Table 1. Number of phonemes existing in the training OCDRL database

CDRL: for continuous speech recognition, the database is constituted for training by 3300 phrases, spoken by 11 speakers (7 males and 4 females), each speaker reading 300 phrases, and for testing by 220 phrases spoken by the same speakers, each of them reading 20 phrases. The phrases are from: information, telecommunication, geography, history, and sports domain. The training database contains over 3200 distinct words; the testing database contains 900 distinct words.
In order to carry out our experiments the database was reorganized as follows: one database for male speakers (MS), one database for female speakers (FS) and one mixed database for male and female speakers (MS and FS). In all cases we have excluded one MS and one FS from the training and used for testing.
Conducting the same statistical investigation of the phonemes distribution, (Table 2), we obtained very similar results with that obtained for the OCDRL data base.

a	**85**	l	48	p	23	b	9	ts	9
sp	152	**j**	**18**	**e_X**	**7**	S	5	g	6
tS	14	f	13	z	10	**1**	**9**	dZ	6
e	**95**	n	41	@	**27**	d	23	**o_X**	**5**
s	38	**o**	**45**	r	64	k_	5	g_	5
t	51	k	18	m	29	**i_O**	**7**	h	7
i	**53**	**u**	**41**	**w**	**10**	v	6	Z	5

Table 2. Number of phonemes existing in the training CDRL database

DDRL: for digit recognition, the database contains speech data from 9 speakers (6 males and 3 females) each speaker reading 9 digits (unu, doi, trei, patru, cinci, şase, şapte, opt, nouă).
VDRL: for vowel recognition, the database contains speech data from 19 speakers (9 males and 10 females) each reading the same 5 vowels (a, e, i, o, u).
The database based on **formants** contains for the training 500 formant vectors, 100 for each vowel and for the testing 250 formant vectors, 50 for each vowel.

3.3 Speech analysis

First step in all recognition tasks is speech analysis, where the speech signal is processed in order to obtain important characteristics, further called features. By features extraction data compression is done, so that the amount of data used for comparisons is greatly reduced and thus, less computation and less time is needed for comparisons (Huang et al., 2001).

Our features extraction is based on perceptual cepstral coding and perceptual linear predictive coding, methods that will be presented further.
The block scheme of the acoustical processor is shown in Fig. 2. Few blocks are common in both linear prediction and cepstral coding (Gavat et al., 2003). The first block in the scheme is the frame blocking, used because speech is fundamentally a non-stationary signal, so we cut short fragments of the speech signal, which are called frames and the speech is approximated as a quasi-stationary random process during a frame. Then we passed each frame through a Hamming window. We can compute at this time the energy of each frame and we can use the energy set of coefficients in the recognition process for more accuracy.

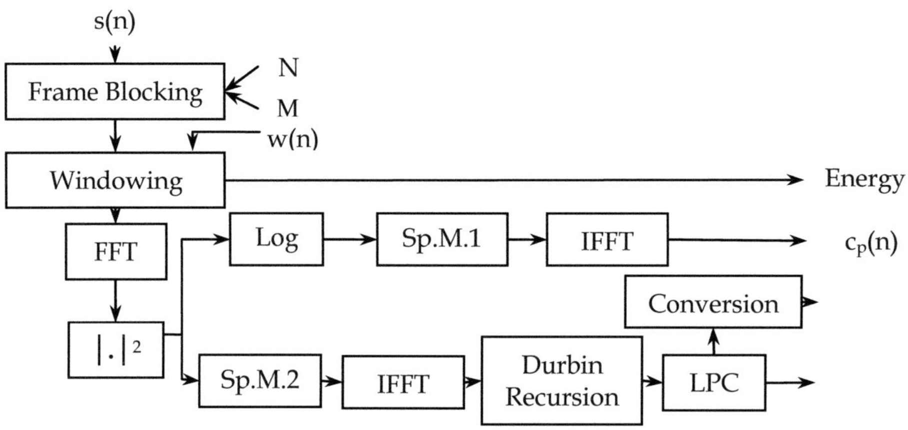

Fig. 2. The block scheme of the acoustical processor

Cepstral analysis is a very reliable method to speech analysis and it can be realized applying FFT (Fast Fourier Transform) to the blocked and windowed time discrete signal *s(n)*. After that, the modulus of the signal is calculated and the logarithm is taken, the result being proportional in fact to the power spectrum of the speech signal. Applying IFFT (Inverse Fast Fourier Transform) leads to the real cepstrum.
Cepstral coefficients, are obtained by re-sampling of the real cepstrum; alone or in addition with the energy *E*, and/or the first and second order differences constitutes a feature vector successfully applied in speech recognition
Mel-frequency cepstral coefficients are calculated from the power spectrum after a spectral manipulation Sp.M.1 in form of a filter bank that is a model for the critical band perception of the human cochlea. Often, in noisy environments the first (D–Delta) and second (A–Acceleration) order variations of the MFCC are applied to enhance the feature vector.
Formants can be calculated from the smoothed spectrum, resulted applying FFT to the windowed cepstrum near the origin.
Linear Predictive coding requires computation of the autocorrelation coefficients possible to be obtained according to the Wiener-Hinchin theorem, by applying IFFT to the power spectrum.
Linear Prediction coefficients are obtained by low order all-pole modelling. The Levinson–Durbin recursive algorithm is used to solve the Yule–Walker equations.

$$\begin{bmatrix} R(1) & R(2) & \dots & R(N) \\ R(2) & R(3) & \dots & R(N-1) \\ \dots & \dots & \dots & \dots \\ R(N) & R(N-1) & \dots & R(1) \end{bmatrix} \times \begin{bmatrix} A(1) \\ A(2) \\ \dots \\ A(N) \end{bmatrix} = \begin{bmatrix} -R(2) \\ -R(3) \\ \dots \\ -R(N+1) \end{bmatrix} \quad (1)$$

where *R(n)* are the autocorrelation coefficients, and *A(n)* are the all-pole model coefficients (the predictor), and *A(1) =1*; through the conversion block the lasts can be converted in other types of coefficients like the reflection coefficients or the LPC cepstral coefficients.

Perceptual Linear Prediction (PLP) coefficients are obtained by spectral manipulation Sp.M.2, displayed in Fig.3. The PLP analysis method (Hermansky, 1990) is more adapted to human hearing, in comparison to the classic Linear Prediction Coding because of the critical band processing of the analysed signal. The LP all-pole model approximates power distribution equally well at all frequencies of the analysis band. This assumption is inconsistent with human hearing, because beyond 800 Hz, the spectral resolution of hearing decreases with frequency; hearing is also more sensitive in the middle frequency range of the audible spectrum (Dumitru et al., 2006).

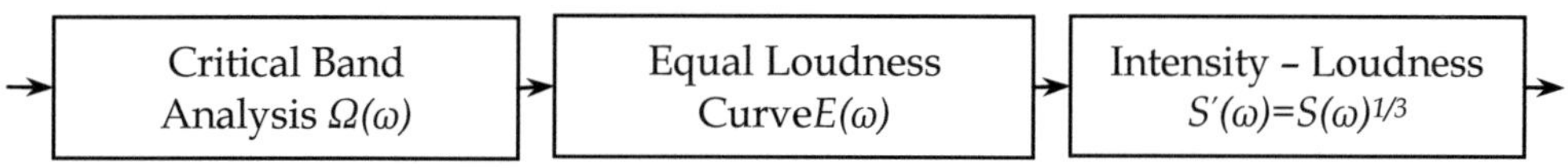

Fig. 3. Block representation for Sp. M. 2

The power spectrum is computed as follows:

$$P(\omega) = \mathrm{Re}(S(\omega))^2 + \mathrm{Im}(S(\omega))^2 \quad (2)$$

The first step is a conversion from frequency to bark, which is a better representation of the human hearing resolution in frequency. The bark frequency corresponding to an audio frequency is:

$$\Omega(\omega) = 6 \ln\left(\frac{\omega}{1200\pi} + \left(\left(\frac{\omega}{1200\pi} \right)^2 + 1 \right)^{0.5} \right) \quad (3)$$

The resulting warped spectrum is convoluted with the power spectrum of the critical band-masking curve, which acts like a bank of filters centred on Ω_i.

The spectrum is pre-emphasized by an equal loudness curve, which is an approximation to the non-equal sensitivity of human hearing at different frequencies, at about 40dB level. A filter having the following transfer function gives the curve:

$$E(\omega) = \frac{(\omega^2 + 56.8 \times 10^6)\omega^4}{(\omega^2 + 6.3 \times 10^6)^2 \times (\omega^2 + 0.38 \times 10^9)} \quad (4)$$

The last operation prior to the all-pole modelling is the cubic-root amplitude compression (Intensity - Loudness Conversion), which simulates the non-linear relation between the intensity of sound and its perceived loudness. Together with the psychophysical equal-loudness pre-emphasis, this operation also reduces the spectral amplitude variation of the critical-band spectrum (Gavat et al., 2008).

3.4 Learning strategies

In the same way into which a human learns to perceive speech from examples by listening, the automatic speech recognizer does apply a learning strategy in order to decode the pronounced word sequence from a sequence of elementary speech units like phonemes, with or without context. By learning are created models for each elementary speech unit, serving in the comparisons to make a decision about the uttered speech unit. We will describe further the learning strategies implemented in our research platform: hidden Markov models (HMM), artificial neural networks (ANN) in form of multilayer perceptrons (MLP) or Kohonen maps (KM). Some hybrid learning strategies are also implemented in form of a HMM-ANN combination and a fuzzy perceptron or fuzzy HMM.

HMMs are finite automata, with a given number of states; passing from one state to another is made instantaneously at equally spaced time moments. At every pass from one state to another, the system generates observations, two processes taking place: the transparent one represented by the observations string (features sequence), and the hidden one, which cannot be observed, represented by the state string (Gavat et al., 2000).

In speech recognition, the left - right model (or the Bakis model) is considered the best choice. For each symbol, such a model is constructed; a word string is obtained by connecting corresponding HMMs together in sequence (Huang et al., 2001).

For limited vocabulary, word models are widely used, since they are accurate and trainable. In the situation of a specific and limited task they become valid if enough training data are available, but they are typically not generalizable. Usually for not very limited tasks are preferred phonetic models based on monophones (which are phonemes without context), because the phonemes are easy generalizable and of course also trainable.

Monophones constitute the foundation of any training method and we also started with them (for any language). But in real speech the words are not simple strings of independent phonemes; these phonemes are affected for the immediately neighboring phonemes by co-articulation. This monophone models are changed now with triphone models (which are phonemes with context) that became actually the state of the art in automatic speech recognition for the large vocabularies (Young, 1992).

A triphone model is a model that takes into consideration the left and the right context of the phonemes. Based on the SAMPA (Speech Assessment Methods Phonetic Alphabet) in Romanian language there are 34 phonemes; the number of necessary models (triphones) to be trained is about 40000, situation which is unacceptable.

In the continuous speech recognition task we modelled only internal - word triphones and we adopted the state tying procedure, conducting to a controllable situation. If triphones are used in place of monophones, the number of needed models increases and it may occur the problem of insufficient training data. To solve this problem, tying of acoustically similar states of the models built for triphones corresponding to each context is an efficient solution (Fig.4).

For example, in Fig. 4b four models are represented for different contexts of the phoneme "a", namely the triphones "$k - a + S$", "$g - a + z$", "$n - a + j$", "$m - a + j$". In Fig. 4c and 4d the clusters formed with acoustically similar states of the corresponding HMMs are represented.

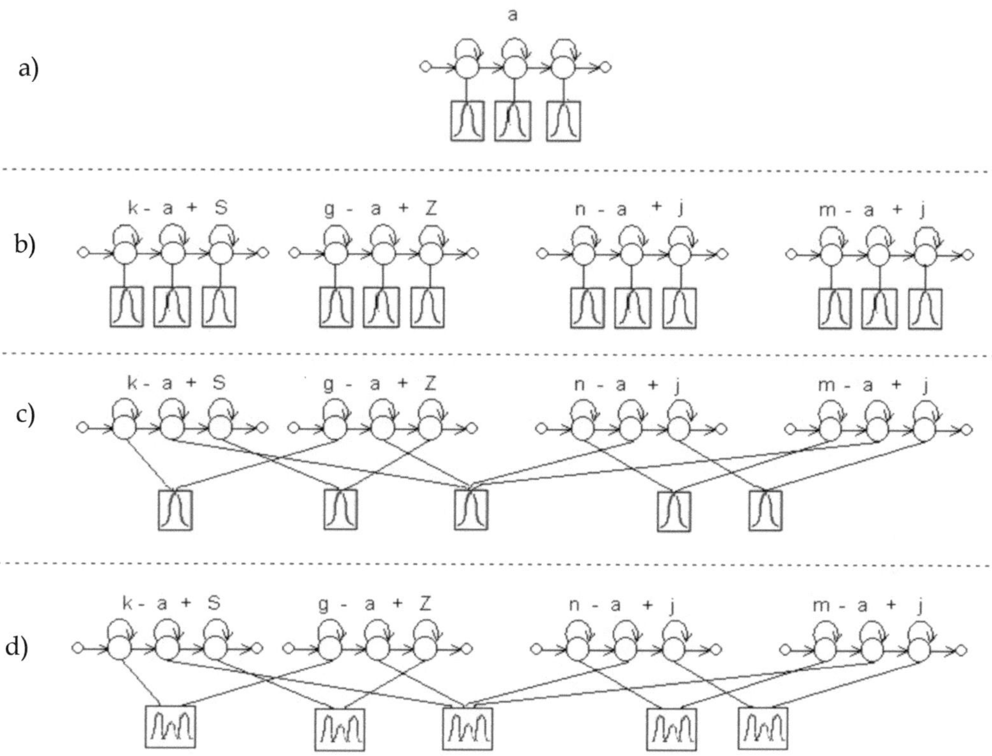

Fig. 4. Different models for triphones around the phoneme "a".

The choice of the states and the clustering in phonetic classes are achieved by mean of phonetic decision trees. A phonetic decision tree is a binary tree (Fig.5), where for each node of the tree questions are associated concerning the contexts of the phoneme (the number of the selected questions is 130 based on the knowledge about phonetic rules for our language).

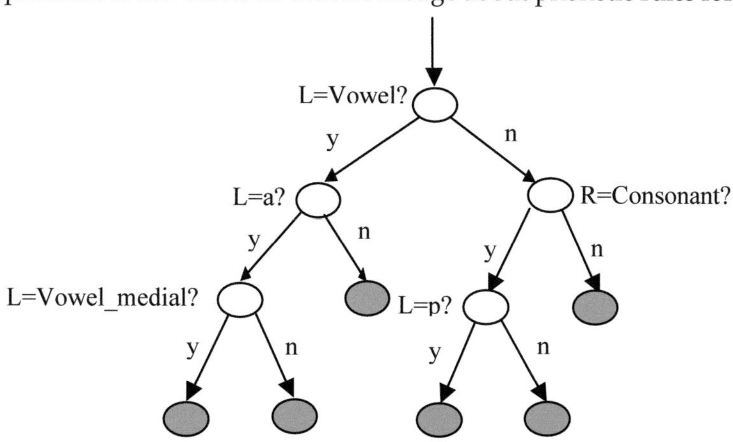

Fig. 5. Phonetic tree for phoneme "m" in state 2.

The questions are chosen in order to increase the log likelihood of the data after splitting. Splitting is stopped when increasing in log likelihood is less than an imposed threshold. In the leaf nodes are concentrated all states having the same answer to the question made along the corresponding path.

Fuzzy - HMM: The generalized model $\overline{\lambda} = (\overline{A}, \overline{B}, \overline{\pi})$ can be characterized by the same parameters (Huang et al., 2001) like the classical, well known model. The major difference in the fuzzy variant, is the interpretation of the probability densities for the classical HMM, as fuzzy densities. On this way the probabilistic similarity measure applied in the classical HMM is replaced by a more suitable fuzzy similarity measure.

The succession of feature vectors, called the observation sequence, *O*, produces the state sequence *S* of the model, and, visiting for example at the moment *t+1* the state $q_{t+1} = S_j$, the symbol b_j is generated. The corresponding symbol fuzzy density $b_j(O_t)$ measures the grade of certainty of the statement that we observed O_t given that we are visiting state S_j .To perform classification tasks, the fuzzy similarity measure must be calculated. Based on the fuzzy forward and backward variables, a fuzzy Viterbi algorithm is proposed in (Mahomed et al., 2000) for the case of the Choquet integral with respect to a fuzzy measure and multiplication as intersection operator.

The fuzzy formulation of the forward variable *a*, bring an important relaxation in the assumption of statistical independence.

The joint measure $\overline{\alpha}_{\Omega_y}\left(\{O_1,...,O_t\}\times\{y_j\}\right)$ can be written as a combination of two measures defined on $O_1, O_2,..., O_t$ and on the states respectively, no assumption about the decomposition of this measure being necessary, where $Y = \{y_1, y_2,..., y_N\}$ represent the states at time *t+1* (*Ω* is the space of observation vectors).

For the standard HMM, the joint measure $P(O_1, O_2,..., O_t, q_{t+1} = S_j)$ can be written as the product $P(O_1, O_2,..., O_t) \cdot P(q_{t+1} = S_j)$, so that two assumptions of statistical independence must be made: the observation at time *t+1*, O_{t+1}, is independent of the previous observations $O_1, O_2,..., O_t$ and the states at time *t+1* are independent of the same observations, $O_1, O_2,..., O_t$.

These conditions find a poor match in case of speech signals and therefore we hope in improvements due to the relaxation permitted by the fuzzy measure.

Training of the generalized model can be performed with the re-estimation formulas also done in (Mahomed et al., 2000) for the Choquet integral. For each model we have trained with the re-estimation formulas the corresponding generalized models, GHMMs, with 3-5 states, analog to the classical case.

After the training, we have calculated the fuzzy measure $\overline{P}(O/\overline{\lambda})$, with the fuzzy Viterby algorithm and made the decisions for recognition in the same manner like for the classical HMM: the correct decision corresponds to the model for which the calculated measure has a maximum.

MLP is the most common ANN architecture used for speech recognition. Typically, the classical MLPs have layered feed-forward architecture, with an input layer, one or more intermediate (hidden) layers, and one output layer. The structure without hidden layer is called Boolean network and is a simple perceptron.

Each layer computes a set of linear discriminative functions, followed by a non-linear function, which is often a sigmoid function.

The numbers of neurons in the hidden layer was experimentally determined, trying to achieve an optimum between the following two opposite requirements: (a) lower computing volume and more rapid process of convergence in the learning period; (b) better performances from the correct classification of input patterns percentage.

In the learning phase are determined the optimum values for weights (Goronzy, 2002), (Valsan et al., 2002) connecting the pairs of neurons from the adjoint layers in the input-output direction using the Back-Propagation algorithm.

Introducing a fuzzy processing of the input layer of the MLP (**fuzzy-MLP**) is a solution to improve the MLP performances.

The structure of the fuzzy neural network (Fig. 6) like the classical one is composed from an input layer, a hidden layer and an output layer.

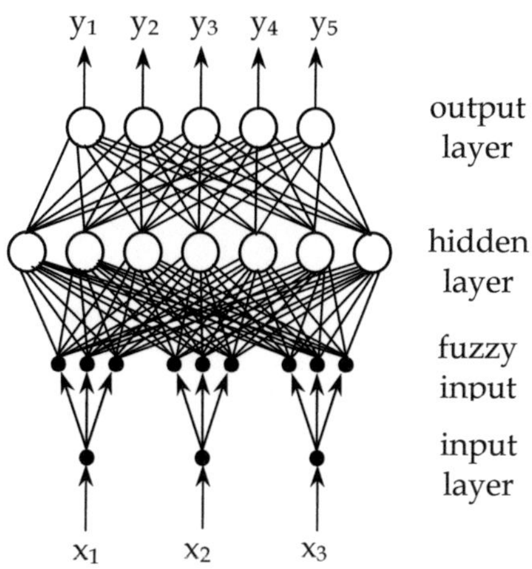

Fig. 6. Schematic diagram of fuzzy MLP

First, the input values are described through a combination of 3 membership values in the linguistic properties set: low, medium and high. For doing this the π membership function is used:

$$\pi(r,c,\lambda) = \begin{cases} 2(1-\|r-c\|/\lambda)^2 & for\ 0 \le \|r-c\| \le \lambda/2 \\ 1-2(\|r-c\|/\lambda)^2 & for\ 0 \le \|r-c\| \le \lambda/2 \\ 0 & otherwise \end{cases} \quad (5)$$

where: $\lambda>0$ is the radius of the π function with c as the central point, || . || denotes the Euclidian norm.

For each component F_{ji} of the input vector F_j the parameters of the π membership function for each linguistic property: low (l), medium (m) and high (h) are computed using the relations:

$$\lambda_{m(Fij)} = (F_{ji\max} - F_{ji\min})/2$$

$$c_{m(Fij)} = F_{ji\min} + \lambda_{m(Fij)}$$

$$\lambda_{1(Fji)} = (c_{m(Fji)} - F_{ji\min})/f_{dn}$$

$$c_{1(Fji)} = c_{m(Fji)} - 0.5\lambda_{1(Fji)} \qquad (6)$$

$$\lambda_{h(Fji)} = (F_{ji\max} - c_{m(Fji)})/f_{dn}$$

$$c_{h(Fji)} = c_{m(Fji)} + 0.5\lambda_{h(Fji)}$$

where $F_{ji\max}$, $F_{ji\min}$ denote the upper and lower bounds of the observed range of feature F_{ji} and f_{dd} is a parameter controlling the extent of overlapping.
After this, the structure of the fuzzy neural network, like the classical one, is composed from a hidden layer and an output layer .
The output vector is defined as the fuzzy class membership values. The membership value of the training $F_i = (F_{i1}\,F_{i2}\,...F_{in})^t$ to class C_k is computed using:

$$\mu_k(F_i) = 1/(1 + z_{ik}/f_d))^{f_c} \qquad (7)$$

where: f_d, f_c are constants controlling the amount of fuzziness in the class-membership set, z_{ik} is the weighted distance between the input vector F_i and the mean $O_k = (O_{k1}\,O_{k1}\,...O_{k1})^t$ of the k-th class, defined as:

$$z_{ik} = \sqrt{\sum_{j=1} [(F_{ji} - O_{kj})/v_{kj}]^2} \qquad (8)$$

where: v_{kj} is the standard deviation of the *j*-th vectors' component from the C_k class.
In the training stage, the Back-Propagation algorithm is used to determine the weights which minimized the mean square error (*mse*) between the real output d_j and the desired one y_j :

$$mse = \sum_{\substack{j=1 \\ F \in train}}^{n} \left(\sum (d_j - y_j)^2\right) \qquad (9)$$

Kohonen maps are competitive neural networks with topological character. The setting up of the winner neurons at output is done with keeping the topological relations between the input vectors. That is the reason for which this neural network is successfully used in pattern recognition (Fig. 7).
It was applied the following structures to recognize vowels: (1) the input layer with 3 neurons, corresponding to the three formant frequencies; (2) three case for the output layer: one-dimensional with 25 neurons, bidimensional with 4x4 neurons, 5x5 neurons, 6x6 neurons, and toroidal with 25 neurons.

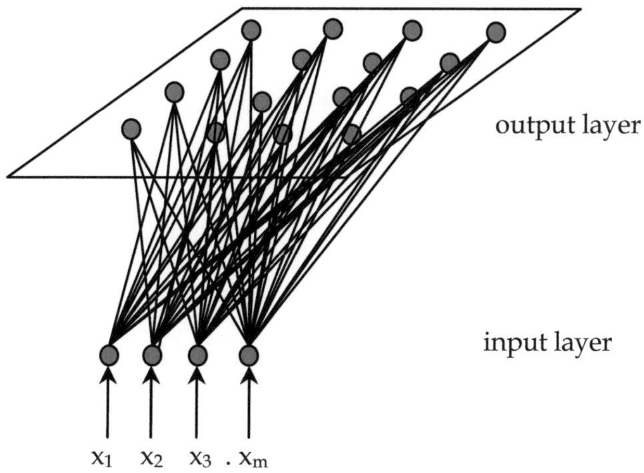

Fig. 7. Schematic diagram of Kohonen maps

In the learning phase (Gold et al., 2002) the structures are trained and the weights of the networks are established in two steps: (a) the determination of the winner neurons; (b) the adaptation of the weights for the winner neurons and for the neurons existing in a certain neighbourhood. In this step important are: the neighbourhood dimension $r(t)$ decreasing during the training, and the learning rate $\eta(t)$ following in our experiments one of the laws:

$$\eta(t) = t^{-1} \text{ or } \eta(t) = t^{-1/2} \tag{10}$$

Hybrid systems: the HMM-based speech recognition methods make use of a probability estimator, in order to approximate emission probabilities $p(x_n/q_k)$, where x_n represents the observed data feature, and q_k is the hypothesized HMM state (Rabiner, 1989). These probabilities are used by the basic HMM equations, and because the HMM is based on a strict formalism, when the HMM is modified, there is a great risk of losing the theoretical foundations or the efficiency of the training and recognition algorithms. Fortunately, a proper use of the MLPs can lead to obtain probabilities that are related with the HMM emission probabilities.

In particular, MLPs can be trained to produce the *a posteriori* probability $p(x_n/q_k)$, that is, the *a posteriori* probability of the HMM state given the acoustic data, when each MLP output is associated with a specific HMM state. Many authors have shown that the outputs of an ANN used as described above can be interpreted as estimates of *a posteriori* probabilities of output classes conditioned by the input, so we will not insist on this matter, but we will mention an important condition, useful for finding an acceptable connectionist probability estimator: the system must contains enough parameters to be trained to a good approximation of the mapping function between the input and the output classes.

Thus, the *a posteriori* probabilities that are estimated by MLPs can be converted in emission probabilities by applying Bayes' rule (11) to the MLP outputs:

$$\frac{p(x_n / q_k)}{p(x_n)} = \frac{p(q_k / x_n)}{p(q_k)} \tag{11}$$

That is, the emission probabilities are obtained by dividing the *a posteriori* estimations from the MLP outputs by estimations of the frequencies of each class, while the scaling factor $p(x_n)$ is considered a constant for all classes, and will not modify the classification.

This was the idea that leads to hybrid neuro-statistical methods, that is, hybrid MLP - HMM methods (Fig. 8), applied for solving the speech recognition problem.

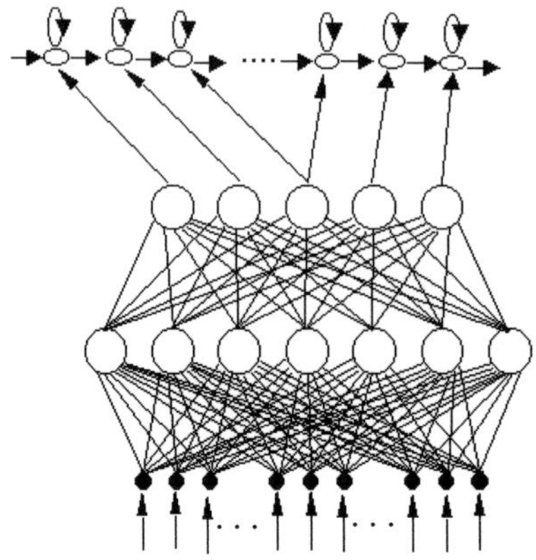

Fig. 8. Schematic diagram of HMM-MLP

4. Speech recognition using ASRS_RL system based on statistical strategies

4.1 Continuous speech recognition

To assess the performance of our ASRS_RL system we initiated comparative tests on the OCDRL and CDRL database for the word recognition rate (WRR) under different conditions for feature extraction, like perceptive cepstral analysis (MFCC_D_A-mel frequency cepstral coefficients with the corresponding first and second order variations -> having 36 coefficients), and linear prediction (LPC-Linear Prediction Coding -> having 12 coefficients). The evaluations are made in the case of acoustical model based on HMM with monophones (phonemes without context) (Dumitru, 2006).

In order to carry out our experiments about speaker independence, the CDRL database was reorganized as follows: one database for male speakers (MS), one database for female speakers (FS) and one database for male and female speakers (MS&FS). In all cases we have excluded one MS and one FS from the training and used for testing.

The words for both databases (OCDRL and CDRL) are chosen taking into account the characteristics of the Romanian language so that the phrase contains enough phonemes to create the models.

The results obtained with the two databases are summarized in Table 3. After that, chosen the second database (CDRL) other evaluations are made.

	OCDRL		CDRL	
	MFCC_D_A	LPC	MFCC_D_A	LPC
Monophone	96.10%	73.80%	57.44%	26.10%

Table 3. WRR for the two cases of databases, OCDRL and CDRL

Trying to create a multi-domain recognition system, we extended the numbers of speakers and of spoken phrases for each speaker (extending OCDRL to CDRL) introducing vocabulary from other domains. In the same time we introduced for WRR evaluation new feature extraction methods, like PLP and new acoustical models based on the HMM with triphones (phonemes with context) (Young et al., 1994).

Table 4 shows the comparative results (average WRR) for three feature extraction methods, MFCC_D_A, LPC and PLP and for two modelling situations, based on monophones and triphones.

Database	Type	MFCC_D_A	LPC	PLP
CDRL	Monophone	57.44%	26.10%	47.00%
	Triphone	78.24%	51.50%	70.11%

Table 4. WRR for monophone and triphone modelling and for three feature extraction methods

The next experiments performed on a Romanian language corpus prove that context-dependent models perform better than context-independent models. The recognition system was trained with 3000 phrases collected from ten speakers (more than 3000 distinct words). Gaussian output probability distribution was assumed for the 36 mel-frequency cepstrum coefficients (12 MFCC and first and second order variation).
Firstly, the 34 context-independent models (monophones) were trained, and the system was tested with an unenrolled speaker. The testing utterances contained distinct words and a loop-grammar was assumed, i.e. any word could occur after any word, anytime.
The core of our experiments is the construction of the decision tree for each state of the triphones derived from the same monophone. The monophones were cloned initially, and the resulted triphones were trained by embedded Baum-Welch procedure. Then, the decision tree was build for different thresholds (TL) in terms of log-likelihood resulting different size systems (example of the evolution of the log-probability values - in the training phase is presented in Fig. 9). The results are presented in Table 5. For a small threshold (TL) of 300, the trees are big and the system is large having 2954 tied states with a huge number of parameters. For a big threshold of 6000, the trees are much smaller, implying a great reduction in the system size, from 7521 triphone states to 416 states, (5.5% remained size) while the performance is degrading with less than 1%. In Table 5, are given also the word recognition rate (WRR), the accuracy.

TL	Initial states / final states	Remained size	WRR
300	7521 / 2954	39.30%	90.14%
900	7521 / 1448	19.30%	89.60%
1200	7521 / 1164	15.50%	90.31%
1800	7521 / 908	12.10%	90.51%
2400	7521 / 747	9.90%	90.02%
3000	7521 / 643	8.50%	89.97%
3600	7521 / 573	7.60%	90.07%
4200	7521 / 522	6.90%	89.79%
4800	7521 / 480	6.40%	89.60%
5400	7521 / 446	5.90%	88.85%
6000	7521 / 416	5.50%	88.75%

Table 5. The results obtained for different thresholds for constructing the phonetic trees

Now, we evaluate the WRR for the CDRL database taking into account the following situations: triphone modelling/monophone modelling; gender based training/mixed training and LPC and PLP/MFCC coefficients.
The speech files from these databases were analysed in order to extract the interesting features. The feature extraction methods used are based on LPC, PLP and MFCC.

The results obtained in the experiments realized under these conditions are summarized in the next three Tables.

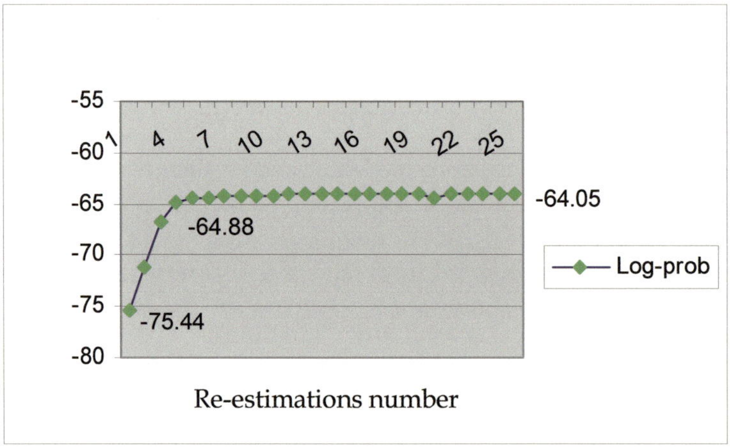

Fig. 9. Log-probability evolutions

Training MS	Type	WRR		
		MFCC_D_A	LPC	PLP
Testing MS	Monophone	56.33%	30.85%	34.02%
	Triphone	81.02%	49.73%	68.10%
Testing FS	Monophone	40.98%	23.23%	25.12%
	Triphone	72.86%	47.68%	59.00%

Table 6. WRR training MS and testing MS and FS

Training FS	Type	WRR		
		MFCC_D_A	LPC	PLP
Testing MS	Monophone	53.56%	26.72%	23.78%
	Triphone	69.23%	49.73%	53.02%
Testing FS	Monophone	56.67%	31.11%	34.22%
	Triphone	78.43%	61.15%	58.55%

Table 7. WRR training FS and testing MS and FS

Training MS and FS	Type	WRR		
		MFCC_D_A	LPC	PLP
Testing MS	Monophone	57.44%	26.10%	47.00%
	Triphone	78.24%	51.50%	70.11%
Testing FS	Monophone	49.89%	24.06%	41.22%
	Triphone	74.95%	50.49%	69.65%

Table 8. WRR training MS&FS and testing MS and FS

The following comments can be made:

a. for 12 LPC coefficients the word recognition rates are low: 30.85% (monophone) training and testing with MS and 49.73% (triphone); 31.11% (monophone) training and testing with FS and 61.15% (triphone); 26.10% (monophone) training MS and FS and testing with MS and 51.5% (triphone);
b. for 5 PLP coefficients the obtained results are very promising, giving word recognition rates about 58.55% (triphone training and testing FS), 68.10% (triphone training and testing MS) and 70.11% (triphone training MS and FS and testing MS);
c. for 36 MFCC_D_A coefficients (mel-cepstral coefficients with first and second order variation) we obtained the best results, as we expected: monophone 56.33% and triphone 81.02%, training and testing with MS; monophone 56.67% and triphone 78.43%, training and testing with FS; monophone 57.44% and triphone 78.24%, training MS and FS and testing with MS.

In the following part (ASRU system) (Juanhg et al., 200) are displayed the results obtained for PRR (Phrase Recognition Rate) in the same condition as well as evaluate the WRR: the CDRL database; the MFCC with the first and second order variation, or the PLP or the LPC feature vector; the acoustical model without and with context.

The experimental results for PRR are shown in the next Tables, using for the word recognition order a loop grammar.

Training MS	Type	PRR		
		MFCC_D_A	LPC	PLP
Testing MS	Monophone	20.00%	5.00%	10.00%
	Triphone	66.25%	11.25%	37.50%
Testing FS	Monophone	21.25%	6.25%	16.25%
	Triphone	48.75%	12.50%	32.50%

Table 9. PRR training MS and testing MS and FS

Training FS	Type	PRR		
		MFCC_D_A	LPC	PLP
Testing MS	Monophone	21.25%	3.75%	6.26%
	Triphone	37.50%	11.25%	21.25%
Testing FS	Monophone	30.00%	12.50%	22.50%
	Triphone	57.50%	33.75%	30.00%

Table 10. PRR training FS and testing MS and FS

Training MS and FS	Type	PRR		
		MFCC_D_A	LPC	PLP
Testing MS	Monophone	23.75%	3.75%	18.75%
	Triphone	60.00%	10.00%	36.25%
Testing FS	Monophone	31.25%	6.25%	20.00%
	Triphone	55.00%	11.25%	46.25%

Table 11. PRR training MS&FS and testing MS and FS

After the experiments made on continuous speech recognition (Dumitru, 2006) we have following observations:

a. the triphone modelling is effective, conducting to increasing in WRR between 15% and 30% versus the monophone modelling. The maximal enhancement exceeds 30% for training MS and testing FS for MFCC_D_A (Fig. 10a).
b. A gender based training is conducting to good result for tests made with speakers from the same gender (training MS / testing MS: 81.02%, testing FS: 72.86%; training FS/testing FS: 78.43%, testing MS: 69.23%); changing gender in testing versus training leads to a decrease in WRR around 10%. For a mixed trained data base changing gender determines only variations around 5% in WRR (Fig. 10b).
c. The PRR variation follows the WRR line, but the results are drastically lower when we compare with WRR.

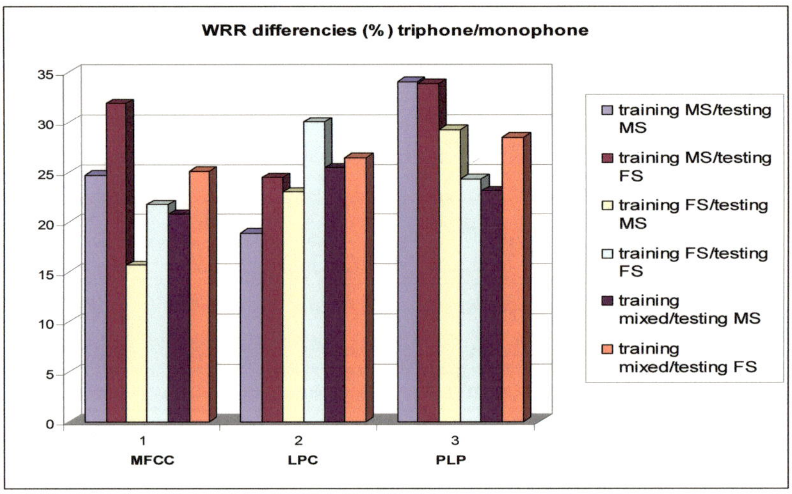

a)

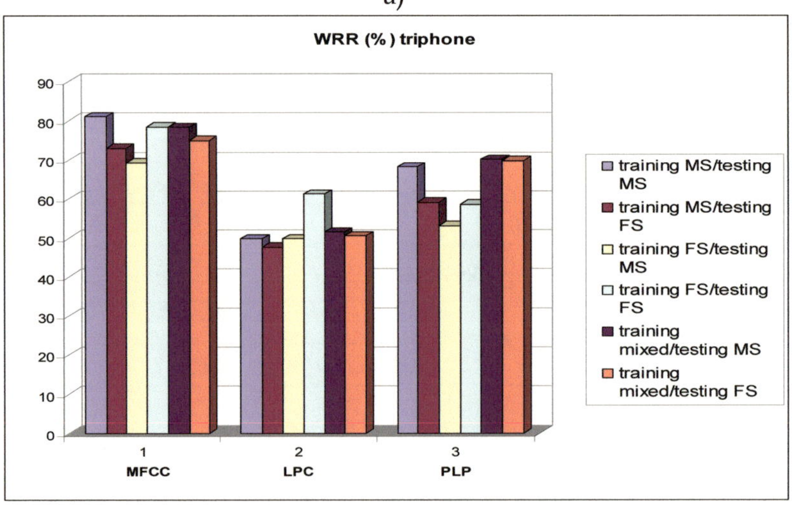

b)

Fig. 10. WRR triphone *vs.* monophone

4.2 Telephone dial application

Other investigations using HMM are done for telephone dial domain. For this application the HMM without context modelling and MFCC coefficients are considered.

Our system was developed for a telephone numbers dialling application (Fig.11). The training data are constituted by the 850 phrases (OCDRL database), spoken by ten speakers, and the testing data are constituted by the 30 phrases, spoken by in training enrolled and also unenrolled speakers. The data are in the typical "wave" format, sampled by 16 kHz, quantified with 16 bits, recorded in a laboratory environment.

Examples of training phrases are:

Formeaza opt sapte sase cinci sase patru doi trei;

Telefoneaza Octavian;

Formeaza opt sapte cinci patru trei doi unu.

Examples of testing phrases are:

Formeaza sapte doi opt cinci unu sase patru cinci unu unu;

Telefoneaza Octavian Dumitru.

In the first experiments the speech signal was parameterized with 39 coefficients (MFCC_D_A): 13 mel cepstral coefficients; 13 delta coefficients; 13 accelerations coefficients. The recognition rates are summarized in Table 12.

Speaker	No. test words	No. recognised words	Recognition rate
Enrolled	103	98	95.1%
Unenrolled	103	93	90.3%

Table 12. Recognition rate

Fig. 11. Telephone dial application

For the enrolled persons the recognition rate was 95.1%, and for the unenrolled speakers the recognition rate was lower as in the first case, but still quite satisfactory (Dumitru, 2006).

Our next experiments were conducted for different choices in speech parameterization. The experiments made in the further described conditions lead to the recognition rates showed for the corresponding parameters in Fig. 12. The chosen parameter sets are the following: MFCC_0 (13) - 13 mel cepstral coefficients; MFCC_0_E (14) - 13 mel cepstral coefficients with log energy; MFCC_0_D (26) -13 mel cepstral coefficients and 13 delta coefficients; MFCC_0_D_A (39) -13 mel cepstral coefficients, 13 delta coefficients and 13 acceleration coefficients; MFCC_D_A_E (39) -12 mel cepstral coefficients, 12 delta coefficients, 12 acceleration coefficients, log energy, delta energy, acceleration energy.
The best rate, 96%, was obtained with 13mel-cepstral coefficients and 13 delta coefficients (MFCC_0_D).
Now, our interest is in medical domain to help the dentists in their job creating a system for completing the examination and treatment dental charts (a project between our laboratory and the University of Medicine and Pharmacy of Bucharest).

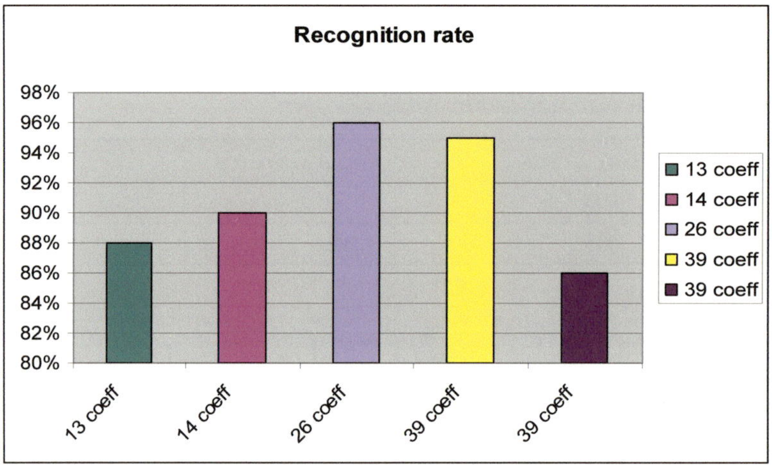

Fig. 12. Recognition rate for different number of coefficients

5. Digit and vowel recognition using ASRS_RL system based on neural and hybrid strategies

5.1 Digit recognition using neuro-statistical strategies

In the first digit recognition experiment, we compare the performance of two kinds of classifiers, namely the SVM (only to compare with other structure) and HMM, the performance being appreciated by their recognition rate and by their generalization capacity. For that, we performed two types of tests on the DDRL database: first, with enrolled speakers, which mean that the speakers were involved both in training and testing. We used five speakers for training and four for testing. For the second type of tests, with unenrolled speakers we used the leave-one-out method: for each word, we trained the classifier with 8 speakers and tested with the 9th repeating the procedure for each speaker.
The conditions for feature extraction are: perceptive cepstral analysis giving a 13 - dimensional vector having as components 13 MFCCs and perceptual linear prediction giving a 5 - dimensional feature vector having as components 5 PLP coefficients.

In the second digit recognition experiment, we evaluate the performance obtained with HMM and with the hybrid neuro-statistical system (HMM - MLP) for unenrolled and enrolled speaker (Dumitru, 2006).
The digit parameters were extracted by cepstral analysis, in form of 12 mel-frequency cepstral coefficients.

a. HMM: for each digit we constructed a left-right hidden Markov model with 3 states;
b. HMM - MLP: the system consists of 9 hybrid models corresponding to 9 digits. Each hybrid model is made of 5 states, each state being associated with one output node of the MLP. The MLP has one hidden layer (100 nodes), and the input layer consisting of 12 nodes (Dumitru et al., 2007).

The results for digit recognition are presented in Table 13.
The performance of the HMM, SVM and hybrid system are validate using other databases, for example Advanced Multimedia Laboratory from the Carnegie Mellon University. The results show the same behaviour, but the WRRs are slightly higher for the English database, recorded in a studio with specialized speakers.

Type	Features extraction	Enrolled	Unenrolled
SVM	MFCC	97.70%	91.70%
HMM	MFCC	98.00%	97.50%
HMM-MLP	MFCC	98.50%	98.30%
SVM	PLP	91.70%	84.70%
HMM	PLP	95.10%	94.20%

Table 13. WRR for digit recognition using SVM, HMM and HMM-MLP

Based on the experimental results obtained, the following conclusions can be extracted:

a. It is to seen that SVM performance is slightly after that of the HMM, but is really promising, taking into account that the HMM has the benefit of a so long refinement time.
b. Trying to reduce these limitation effects of HMMs (the model training is not discriminative), chosen an alternative approaches can be a good solution. This approaches combine the HMM with MLP into a hybrid system. The results for HMM are lower than the results for the hybrid system.

5.2 Vowel recognition using neural and fuzzy strategies

The learning strategies applied in our recognition experiments are: the Kohonen maps, the MLP, the fuzzy-MLP and the fuzzy-HMM.
There are three experiments can be made:
(A) In the first experiment, the vowel recognition rate using the VDRL database and the MFCC coefficients (in form of 12 mel-frequency cepstral coefficients) was determined; the results obtained with MLP and HMM as learning strategies are comparatively presented in Table 14.
The VDRL database was organized as follows: one database for male speakers (MS), one database for female speakers (FS). In booth cases one male speaker (MS) and one female speaker (FS) was excluded from the training database and used their data for the testing.
The experimented MLP is a two-layer perceptron trained with Back-Propagation algorithm, having in the output layer 5 nodes corresponding to the 5 vowels to be classified and 100

nodes in the hidden layer (experimentally chosen). The number of the input nodes is equal to the number of features (12 MFCC).
The HMMs chosen for comparison are Bakis (or left-right) structures with five states and for each vowel one model is created (Gavat et al., 2000).
In Table 14, are displayed only the results in the case of training MS and testing with MS and FS. Similarly results were obtained for the training with FS (Dumitru et al., 2007).

Vowel	HMM		MLP	
	MS	FS	MS	FS
a	100.00%	80.71%	100.00%	82.28%
e	85.81%	43.25%	94.82%	50.67%
i	85.71%	85.71%	95.15%	92.41%
o	90.90%	51.33%	97.00%	52.78%
u	88.88%	71.42%	94.83%	77.85%
Mean	*91.26%*	*66.48%*	*96.36%*	*71.20%*

Table 14. Vowel recognition rate in the case of training with MS and testing with MS and FS.

(B) In the second experiment, the vowels were described by three formant frequencies and the error rates obtained with different learning strategies are given in Table 15.
The database for formants contains 500 formant vectors, 100 for each vowel for the training and 250 formant vectors, 50 for each vowel for the testing.
The learning structures applied for these investigations are: (a) KM with the input layer with 3 neurons, corresponding to the three formant frequencies and three variants for the output layer: unidimensional with 25 neurons, bidimensional with 5x5 neurons, and toroidal with 25 neurons; (b) MLP with 3 layers organized as it follows: (1) the input layer with 3 neurons, corresponding to the three formant frequencies; (2) the hidden layer with 0 (Boolean network) or 4 neurons, (3) the output layer with 5 neurons corresponding each to a processed class (in our case the vowels *a, e, i, o, u*); (c) fuzzy-MLP.

Vowel	KM 1-dim	KM 2-dim	KM toroidal	Boolean	MLP	Fuzzy MLP
a	2.60%	1.20%	2.00%	4.00%	0.00%	1.50%
e	3.20%	2.40%	2.40%	6.00%	2.50%	1.00%
i	2.20%	1.60%	1.20%	4.00%	1.00%	0.50%
o	1.80%	1.20%	1.20%	10.00%	1.00%	1.00%
u	2.20%	2.10%	1.80%	10.00%	1.00%	0.00%
Mean	*2.40%*	*1.70%*	*1.72%*	*6.80%*	*1.10%*	*0.80%*

Table 15. Error rates in for different learning strategies for the case of formant

(C) In the third experiment the parameterization is realized with the mel cepstral coefficients and the first and second order differences of these coefficients deduced from homomorfic filtering. The error rates obtained in the vowel recognition tests are given in Table 16, comparatively for the generalized HMM (fuzzy-HMM) and the classical HMM.

Vowel	Fuzzy - HMM	Classical HMM
a	5.10%	6.90%
e	2.40%	4.80%
i	3.80%	7.30%
o	2.50%	5.90%
u	0.70%	3.90%
Mean	*2.90%*	*5.78%*

Table 16. Error rates for generalized and for classical HMMs

For vowel recognition in Romanian language the following conclusion can be reported:
(A) The recognition rates in the case of MLP are higher than in the case of HMM. A possible explanation can be the fact that the model training is discriminative, while in the case of HMM the training is not discriminative, which represents a disadvantage of HMM utilization.
(B) The KM 2-dimensional structures and the toroidal have the same performance, weaker is the performance of the 1-dimensional structure. The best balanced situation corresponds to the 2-dimensional map 5x5, in which all neurons are associated to a vowel to be recognized. The performance obtained in the case of the Boolean network is unacceptable, but the MLP acts well.
Using fuzzy-MLP structure it is an improvement with a mean value of 0.30% comparative with the non-fuzzy structure.
(C) A mean decreasing of nearly 3% is realized in the error rate by adopting the fuzzy-HMM instead of the probabilistic one.

6. Conclusions

In this chapter we presented the work done until now to implement a tool for research in speech recognition for Romanian language. What we realized can be summarized as follows:

- We have implemented basic algorithms for feature extraction for perceptual and simple cepstral analysis and perceptual and simple linear prediction
- We have implemented learning strategies based on HMMs, ANNs, and hybrid strategies like fuzzy-MLP, fuzzy HMM or HMM-MLP in the framework of the statistical, connectionist and fuzzy paradigms of Computational Intelligence
- We have constructed databases to train the models and to test our recognizers that obey the same statistical rules as given from linguists for Romanian language.
- We have trained the models and tested the recognizers on this data bases

What we achieved and what is further to be done is sketched bellow:

- The obtained results in ASR for WRR are acceptable, what can not be said for ASRU experiments, so that much work must be done to enhance PRR by extending the knowledge resources to the semantic and pragmatic levels. Our resources for the moment cover the acoustic and phonetic level by the trained models and the dictionary and in a very primitive manner the syntactic level, by a loop grammar. A more complex grammar and a language model are our next objectives

- For validation of our algorithms, a professional data base for Romanian language could be very helpful. Some universities in Romania try to do that by cooperation and we hope to succeed in our attempt.
- To extend the possibilities of our platform we are planning to enhance the feature extraction methods and also try to develop neural and hybrid strategies suitable for continuous speech recognition experiments

7. References

AMP Advanced Multimedia Processing Laboratory: http://amp.ece.cmu.edu/_download/Intel/feature_data.html

Appen Appen Database http://www.appen.com.au/services/

Bourlard, H. and Wellekens, C.J. (1990). Links between Markov models and multilayer perceptrons. IEEE Transactions on Pattern Analysis and Machine Intelligence, Vol. 12, pp. 1167-1178.

Bourlard, H. and Morgan, N. (1994). *Connectionist Speech Recognition - A Hybrid Approach.* Kluwer Academic Press.

Charlet, D., Krstulovic, S,. Bimbot, F., Boëffard, O., & others. (2005). Neologos : an optimized database for the development of new speech processing algorithms, *Proceedings Interspeech'05*, pp. 1549-1552, Lisbonne, Portugal.

Cslu Center of Spoken Language Understanding: http://cslu.cse.ogi.edu/

Douglas, B.P., Janet, M. B. (1992). The design for the wall street journal-based CSR corpus, *Proceedings ICSLP*, pp. 899-902.

Draganescu, M., (2003). Spoken language Technology, *Proceedings of Speech Technology and Human-Computer-Dialog (SPED2003)*, pp. 11-12, Bucharest, Romania.

Dumitru, C.O., Gavat, I. (2006). A Comparative Study of Features Extraction Methods Applied for Continuous Speech Recognition in Romanian Language, *Proceedings the 48th International Symposium ELMAR 2006*, pp. 115-118, Zadar, Croatia.

Dumitru, O. (2006). *Modele neurale si statistice pentru recunoasterea vorbirii*, Ph.D. thesis.

Dumitru, C.O., Gavat, I. (2007). Vowel, Digit and Continuous Speech Recognition Based on Statistical, Neural and Hybrid Modelling by Using ASRS_RL, *Proceedings EUROCON 2007*, pp.856-863, Warsaw, Poland.

Elra European Language Ressources Association, http://www.elra.info/

Entro40 Entropic Latino40 Speech Database http://noble.gs.washington.edu/~noble/latino40.html

Gavat, I., Zirra, M. and Enescu, V. (1996-a). A hybrid NN-HMM system for connected digit recognition over telephone in Romanian language, *Proceedings IVTTA '96*, pp. 37-40, Basking Ridge, N.J.

Gavat, I. and Zirra, M. (1996-b). Fuzzy models in vowel recognition for Romanian language, *Proceedings Fuzzy-IEEE '96*, pp. 1318-1326, New Orleans.

Gavat, I., Grigore, O., Zirra, M. and Cula, O. (1997). Fuzzy variants of hard classification rules, *Proceedings NAFIPS'97*, pp. 172-176, New York.

Gavat, I., Zirra, M. and Cula, O. (1998). Hybrid speech recognition system with discriminative training applied for Romanian language, *Proceedings MELECON '98*, pp. 11-15, Tel Aviv, Israel.

Gavat, I., & all. (2000). *Elemente de sinteza si recunoasterea vorbirii*, Ed. Printech, Bucharest.

Gavat, I., Valsan, Z., Sabac, B., Grigore, O. and Militaru, D. (2001-a). Fuzzy similarity measures - alternative to improve discriminative capabilities of HMM speech recognizers, *Proceedings ICA 2001*, pp. 2316-2317, Rome, Italy.

Gavat, I., Valsan, Z. and Grigore, O. (2001-b). Fuzzy-variants of hidden Markov models applied in speech recognition, *Proceedings SCI 2001, Invited Session: Computational Intelligence In Signal And Image Processing*, pp. 126-130, Orlando, Florida.

Gavat, I., Dumitru, C.O., Costache, G., Militaru, D. (2003). Continuous Speech Recognition Based on Statistical Methods, *Proceedings of Speech Technology and Human-Computer-Dialog (SPED2003)*, pp. 115-126, Bucharest.

Gavat,I., Dumitru, C.O. (2008). The ASRS_RL - a Research Platform, for Spoken Language Recognition and Understanding Experiments, *Lecture Notes in Computer Science (LNCS)*, Vol. 5073, Part II, pp.1142-1157.

Gold, B., Morgan, N. (2002). *Speech and audio signal processing*, John Wiley&Sons, N. Y.

Goronzy, S. (2002). *Robust Adaptation to Non-Native Accents in Automatic Speech Recognition*, Springer - Verlag, Berlin.

Grigore, M. and Gavat, I. (1996). Vowel recognition with nonlinear perceptron, *Proceedings CAS '96* , pp. 155-158, Sinaia, Romania.

Grigore, O., Gavat, I. and Zirra, M. (1998). Neural network vowel recognition in Romanian language, *Proceedings CONTI '98*, pp. 165-172, Timisoara, Romania.

Grigore, O. and Gavat, I. (1999). Neuro-fuzzy models for speech pattern recognition in Romanian language, *Proceedings ESIT'99*, pp. 98-103, Rhodos, Greece.

Hermansky, H. (1990). Perceptual Linear Predictive (PLP) Analysis of Speech, *Journal Acoustic Soc. America*, Vol. 87, No. 4, pp. 1738-1752.

Huang, X., Acero, A., Hon, H.W. (2001). *Spoken Language Processing–A Guide to Theory, Algorithm, and System Development*, Prentice Hall, 2001.

Juang, B.H. and Katagiri, S. (1992). Discriminative learning for minimum error classification. *IEEE Transactions on Signal Processing*, Vol. 12, pp. 3043-3054.

Juanhg, B.H., Furui, S. (2000). Automatic Recognition and Understanding of Spoken Language–A First Step Toward Natural Human–Machine Communication, *Proceedings IEEE*, Vol. 88, No. 8, pp. 1142-1165.

Kocsor, A., Kovacs, K., Kuba Jr., A., Toth, L. (1999). An Overviey of the OASIS Speech Recognition Project, *Proceedings 4th International Conference on Applied Informatics*, Eger-Noszvajm Hungary, 30 August - 3 Sepetember.

Lippmann, R. and Singer, E. (1993). Hybrid neural network / HMM approaches to word spotting, *Proceedings ICASSP '93*, pp. 565-568, Minneapolis.

Mahomed, M., Gader, P. (2000). Generalized Hidden Markov Models, *IEEE Transactions on Fuzzy Systems*, Vol. 2, pp. 67-93.

Rabiner, L.R. (1989). A Tutorial on Hidden Markov Models and Selected Applications in Speech Recognition , *Proceedings of the IEEE*, Vol. 77, No. 8, pp. 257-286 .

Richard, M. and Lippmann, R. (1991). Neural network classifiers estimate Bayesian a posteriori probabilities, *Neural Computation*, Vol. 4, pp. 461-483.

Reichl, W., Caspary, P. and Ruske, G. (1994). A new model-discriminant training algorithm for hybrid NN-HMM systems, *Proceedings ICASSP '94*, pp. 677-680, Adelaide, Australia.

Robinson, T., Fransen, J., Pye, D., Foote, J., Renals, S. (1995°. WSJCAM0: A British English Speech Corpus for Large Vocabulary Continuous Speech Recognition, *Proceedings Int. Conf. Acoustics, Speech and Signal Processing*, pp. 81-84, Detroit.

Sampa. http://www.phon.ucl.ac.uk/home/sampa

SpeDatM The SpeechDat projects home page: http://www.speechdat.org/

SpeDatE SpeEastern European Speech Databases for Creation of Voice Driven Teleservices, http://www.fee.vutbr.cz/SPEECHDAT-E/

Timit DARPA TIMIT Acoustic-Phonetic Continuous Speech Corpus http://www.ldc.upenn.edu/Catalog/readme_files/timit.readme.html

Valsan, Z., Sabac, B. and Gavat, I. (1998-a). Combining self organizing feature map and multilayer perceptron in a neural system for fast key-word spotting. Proceedings SPECOM '98, pp. 303-308, St. Petersburg, Russia.

Valsan, Z., Sabac, B., Gavat, I. and Zamfirescu, D. (1998-b). Combining self-organizing map and multilayer perceptron in a neural system for improved isolated word recognition, *Proceedings Communications '98*, pp. 245-251, Bucharest, Romania.

Valsan, Z., Gavat, I., Sabac, B., Cula, O., Grigore, O., Militaru, D., Dumitru, C.O., (2002). Statistical and Hybrid Methods for Speech Recognition in Romanian, *International Journal of Speech Technology*, Vol. 5, No. 3, pp. 259-268.

Young, S.J. (1992). The general use of tying in phoneme-based HMM speech recognizers, *Proceedings ICASSP'92*, Vol. 1, pp. 569-572, San Francisco.

Young, S.J., Odell, J.J., Woodland, P.C. (1994). Tree based state tying for high accuracy modelling, *ARPA Workshop on Human Language Technology*, Princeton.